Statistical Methods for Materials Science

The Data Science of Microstructure Characterization

Statistical Methods for Materials Science

The Data Science of Microstructure Characterization

Jeffrey P. Simmons
Lawrence F. Drummy
Charles A. Bouman
Marc De Graef

CRC Press
Taylor & Francis Group
Boca Raton London New York

CRC Press is an imprint of the
Taylor & Francis Group, an **informa** business

Disclaimer
The views presented by Drs. Simmons and Drummy are theirs alone and do not necessarily represent the views of the Department of Defense, AFRL, or the United States Air Force.

CRC Press
Taylor & Francis Group
6000 Broken Sound Parkway NW, Suite 300
Boca Raton, FL 33487-2742

First issued in paperback 2021

Version Date: 20181220

ISBN 13: 978-0-367-78028-9 (pbk)
ISBN 13: 978-1-4987-3820-0 (hbk)

Library of Congress Cataloging-in-Publication Data

Names: Simmons, Jeffrey P., editor. | Bouman, Charles Addison, editor. | De Graef, Marc, editor. | Drummy, Lawrence F., editor.
Title: Statistical methods for materials science : the data science of microstructure characterization / edited by Jeffrey P. Simmons, Charles A. Bouman, Marc De Graef, Lawrence F. Drummy, Jr.
Description: Boca Raton, Florida : CRC Press, [2019] | Includes bibliographical references.
Identifiers: LCCN 2018029225| ISBN 9781498738200 (hardback) | ISBN 9781498738217 (ebook adobe reader) | ISBN 9781351647380 (ebook epub) | ISBN 9781351637879 (ebook mobipocket)
Subjects: LCSH: Materials science--Mathematical models. | Materials science--Statistical methods.
Classification: LCC TA404.3 .S77 2019 | DDC 620.1/10727--dc23
LC record available at https://lccn.loc.gov/2018029225

**Visit the Taylor & Francis Web site at
http://www.taylorandfrancis.com**

**and the CRC Press Web site at
http://www.crcpress.com**

Invert, always invert.

Carl Gustav Jacob Jacobi

Contents

IV Structure Formation in Materials

Contents

7 Statistical Reconstruction and Heterogeneity Characterization in 3-D Biological Macromolecular Complexes **111**

by Qiu Wang and Peter C. Doerschuk

8 Object Tracking through Image Sequences **127**

by Song Wang, Hongkai Yu, Youjie Zhou, Jeffrey P. Simmons, and Craig Przybyla

IV Structure Formation in Materials

Contents

V Microstructure

12 Estimating Orientation Statistics

by Stephen R. Niezgoda

13 Representation of Stochastic Microstructures

by Stephen R. Niezgoda

Contents

Preface

This book grew out of a collaboration between the editors and others whose home institutions were the Air Force Research Laboratory (AFRL), Purdue University, and Carnegie Mellon University. Although the collaboration was between three different institutions, our specialties represented four different disciplines: (1) physics, (2) microscopy, (3) phase transformations, and (4) signal processing. Our own particular take on data science in materials characterization comes from the interdiffusion of these four disciplines, augmented by ideas of colleagues. Our collaboration began with a series of workshops, where we invited speakers, in roughly equal proportions, from materials science and imaging science. The speakers were instructed to make presentations in their own "native tongue." That is, they were asked not to simplify their presentations so as to be understood by the broad audience, but to use language and format appropriate for their respective technical conferences. The result was a sort of "Tower of Babel" mixture of people all speaking different languages and trying desperately to understand and to be understood. While the material was never fully understood by any one person, enough was understood to make the workshops a success. The general comment from those in attendance was often something like "I learned a *lot* from this workshop."

We prepared this book on the suggestion of Lou Han, of Taylor & Francis. To keep with our tradition with the workshops, we decided to invite authors, again in equal proportions, from the materials science and imaging science fields, with particular depth in the four disciplines represented by our interests. Again, we encouraged the authors to present the material as appropriate to be published in the journals of their respective fields, but we put a greater emphasis on developing a roadmap of the literature in their areas with some illustrative examples of their own work.

Integrating the disparate fields into a single book was the responsibility of the editors. Time will tell how well we accomplished this. The general structure of the book provides chapters that we believe would be easily understood by those in the materials science field as well as chapters that would be easily understood by those in the imaging science field. The materials-centric chapters occur early and the imaging-centric chapters towards the end. It may be that the best order in which to read the book would be in the forward direction for the materials scientist and in the reverse direction for the imaging scientist. Our goal was to allow anyone in the broad audience to have a place where the material was readily understandable, from which they could venture out into the less familiar. Like the workshops, it is not expected that every chapter would be understood at first reading by any one person.

The particular blend of disciplines covered in this book yielded some surprises as our collaborations matured. The most notable of these is that there are two sorts of disciplines: one, which has many difficult problems and associated frustrations, where researchers can only hope that, with sufficient study, these problems will ultimately yield to solution. This is the nature of materials science. Imaging science, on the other hand, consists of researchers who are quite adept at finding algorithmic solutions to problems and have, as a rite of passage in the journals, the necessity to show how their new algorithm beats the previously published ones. Researchers go from one problem class to the next with great rapidity, as compared to materials science, and are ever in search of a new problem to solve. Researchers can only hope that somebody downstream from their efforts can adopt their algorithms and put them to work in practical applications. The result was that the collaboration was between a problem-rich field and a solution-rich field with the resulting combination a sort of "thermite reaction," that rapidly yielded results that were years ahead of what we had expected.

A second surprise experienced, at least by those more steeped in physics, was that the modes of investigation between materials science and imaging science were radically different in a deceptive way. Whereas materials science adopted the great tradition of the scientific revolution of formulating a quantitative law, making future predictions, and then testing these predictions with experiment, imaging science starts with the data and makes inferences as to its origin. It was not immediately obvious that these were opposite directions. We in materials science were in the habit of thinking that, if we find conditions that will result in what we observed, those must have been the actual conditions at the beginning. In imaging science, the response would be, "Yes, those *could have been* the initial conditions, but almost certainly, they were not." Bridging *that* gap was the key element responsible for the unexpected advances we had made.

This particular cross-disciplinary approach has been fruitful and we, the editors, have benefited greatly from the interdiffusion of methods and concepts between the various fields. We put together this book, in the same spirit as our original collaboration, in the hopes that we can further open the field, either to those with problems in search of solutions or those with solutions in search of problems. It was our intention that the book would be equally valuable to the materials scientist and the imaging scientist. That is, as useful to those wanting to learn new ways of getting results from the ever-increasing data stream afforded by the proliferation of sensors and automation as it would be to those interested in adding a new data-rich field to their list of customers. The book was designed to have a distribution of maturities of topics so that anyone in either field could find something that they clearly understood as well as many ideas that neither they nor anybody else had considered as research topics in this combination. The resulting mixture, we hope, will make the book a valuable resource with a long shelf life.

For access to figures in full color format, please visit the book's home page at the publisher's website: https://www.crcpress.com/9781498738200.

Jeffrey P. Simmons
Lawrence F. Drummy
Charles A. Bouman
Marc De Graef

About the Editors

Jeffrey P. Simmons is a scientist with the Materials and Manufacturing Directorate of the Air Force Research Laboratory (AFRL). He received a B.S. degree in metallurgical engineering from the New Mexico Institute of Mining and Technology, Socorro, NM, and M.E. and Ph.D. degrees in metallurgical engineering and materials science and materials science and engineering, respectively, from Carnegie Mellon University, Pittsburgh, PA. After receiving a Ph.D. degree, he began work at AFRL as a post-doctoral research contractor. In 1998, he joined AFRL as a research scientist. His research interests are in computational imaging for microscopy, and he has developed advanced algorithms for analysis of large image datasets. Other research interests have included phase field (physics-based) modeling of microstructure formation, atomistic modeling of defect properties, and computational thermodynamics. He has led teams developing tools for digital data analysis and computer resource integration and security. He has overseen execution of research contracts on computational materials science, particularly in prediction of machining distortion, materials behavior, and thermodynamic modeling. He has published in both the materials science and signal processing fields. He is a member of ACM and a senior member of IEEE.

Lawrence F. Drummy is a senior materials engineer in the Soft Matter Materials Branch, Functional Materials Division, Materials and Manufacturing Directorate, Air Force Research Laboratory in Dayton, OH. Dr. Drummy received his B.S. in physics at Rensselaer Polytechnic Institute, Troy, NY, while researching scanning tunneling microscopy and image processing of silicon growth on surfaces. In 2003 he received his Ph.D. from the Department of Materials Science and Engineering at the University of Michigan, Ann Arbor, while performing research on defect structures in organic molecular semiconductor thin films for flexible electronics.

Dr. Drummy's research interests include three-dimensional morphology characterization of biological, polymeric, and nanostructured materials, the structure of materials at interfaces, and data analytics for materials science applications such as microscopy. His research in tomographic reconstruction methods has made important advances for materials research, and the methods are currently in use in government, academic, and industrial labs. Dr. Drummy has authored over 75 peer-reviewed publications and 1 issued patent, with an h-index of 30, and over 3,000 citations. He has organized symposia, workshops, and conferences on image processing, data science, tomography, electron microscopy, materials chemistry, and nanocomposite materials. He was guest editor for *Polymer Reviews*, *Clays and Clay Minerals*, and *Materials* special issues on topics ranging from electron microscopy to nanocomposites. Dr. Drummy has been the recipient of the Arthur K. Doolittle Award from the American Chemical Society Polymer Materials Science and Engineering Division, Air Force Office of Scientific Research Star Team, Rensselaer Medal Award, and a Padden Award Finalist in the Division of Polymer Physics of the American Physical Society.

Charles A. Bouman received a B.S.E.E. degree from the University of Pennsylvania in 1981 and an M.S. degree from the University of California at Berkeley in 1982. From 1982–1985, he was a full staff member at MIT Lincoln Laboratory and in 1989 he received a Ph.D. in electrical engineering from Princeton University. He joined the faculty of Purdue University in 1989 where he is currently the Showalter Professor of Electrical and Computer Engineering and Biomedical Engineering.

Professor Bouman's research is in statistical signal and image processing in applications ranging from medical to scientific and consumer imaging. His research resulted in the first commercial model-based iterative reconstruction (MBIR) system for medical X-ray computed tomography (CT), and he is co-inventor on over 50 issued patents that have been licensed and used in millions of consumer imaging products. Professor Bouman is a member of the National Academy of Inventors, a fellow of the IEEE, a fellow of the American Institute for Medical and Biological Engineering (AIMBE), a fellow of the Society for Imaging Science and Technology (IS&T), and a fellow of the SPIE professional society. He is the recipient of the 2014 Electronic Imaging Scientist of the Year Award, and the IS&Ts Raymond C. Bowman Award. He has been a Purdue University Faculty Scholar and received the College of Engineering Engagement/Service Award, and Team Award. He was also founding co-director of Purdue's Magnetic Resonance Imaging Facility from 2007–2016; and chair of Purdue's Integrated Imaging Cluster from 2012–2016. He was previously the Editor-in-Chief for the *IEEE Transactions on Image Processing*; a Distinguished Lecturer for the IEEE Signal Processing Society; and a vice president of technical activities for the IEEE Signal Processing Society, during which time he led the creation of the *IEEE Transactions on Computational Imaging*. He has been an associate editor for the *IEEE Transactions on Image Processing*, the *IEEE Transactions on Pattern Analysis and Machine Intelligence*, and the *SIAM Journal on Mathematical Imaging*. He has also been a vice president of publications and a member of the board of directors for the IS&T Society, and he is the founder and co-chair of the SPIE/IS&T conference on computational imaging.

Marc De Graef received his B.S. and M.S. degrees in physics from the University of Antwerp, Belgium, in 1983, and his Ph.D. in physics from the Catholic University of Leuven, Belgium, in 1989, with a thesis on copper-based shape memory alloys. He then spent three and a half years as a post-doctoral researcher in the materials department at the University of California at Santa Barbara before joining Carnegie Mellon in 1993 as an assistant professor. He is currently professor and co-director of the J. Earle and Mary Roberts Materials Characterization Laboratory. His research interests lie in the area of microstructural characterization of structural intermetallics and magnetic materials, and include the development of numerical techniques to model a variety of materials characterization modalities. Prof. De Graef has published 2 textbooks and more than 280 publications.

Contributors

Hyrum Anderson
Endgame, Inc.
Albuquerque, New Mexico

Hossein Beladi
Deakin University
Geelong, Australia

Charles A. Bouman
Purdue University
West Lafayette, Indiana

Stephen Bricker
University of Dayton
Dayton, Ohio

David B. Brough
Georgia Institute of Technology
Atlanta, Georgia

Patrick G. Callahan
University of California Santa Barbara
Santa Barbara, California

Mary Comer
Purdue University
West Lafayette, Indiana

Brian L. DeCost
Carnegie Mellon University
Pittsburgh, Pennsylvania

Marc De Graef
Carnegie Mellon University
Pittsburgh, Pennsylvania

Peter C. Doerschuk
Cornell University
Ithaca, New York

Lawrence F. Drummy
Air Force Research Laboratory
Dayton, Ohio

David Furrer
Pratt & Whitney
East Hartford, Connecticut

Russell Hardie
University of Dayton
Dayton, Ohio

Elizabeth A. Holm
Carnegie Mellon University
Pittsburgh, Pennsylvania

Sushil R. Kanel
Air Force Institute of Technology
Dayton, Ohio

Kurt Larson
Sandia National Laboratories
Albuquerque, New Mexico

Emanuel A. Lazar
University of Pennsylvania
Philadelphia, Pennsylvania

Jian Luo
University of California at San Diego
San Diego, California

Dhriti Nepal
Air Force Research Laboratory
Dayton, Ohio

Stephen Mick
Gatan, Inc.
Austin, Texas

R. Rao Nadakuditi
University of Michigan
Ann Arbor, Michigan

Stephen R. Niezgoda
The Ohio State University
Columbus, Ohio

Ryan Noraas
Pratt & Whitney
East Hartford, Connecticut

Craig Przybyla
Air Force Research Laboratory
Dayton, Ohio

Saiprasad Ravishankar
University of Michigan
Ann Arbor, Michigan

Gregory S. Rohrer
Carnegie Mellon University
Pittsburgh, Pennsylvania

Justin Romberg
Georgia Institute of Technology
Atlanta, Georgia

Jeffrey P. Simmons
Air Force Research Laboratory
Dayton, Ohio

David J. Srolovitz
University of Pennsylvania
Philadelphia, Pennsylvania

Veera Sundararaghavan
University of Michigan
Ann Arbor, Michigan

Ming Tang
Rice University
Houston, Texas

James Theiler
Los Alamos National Laboratory
Los Alamos, New Mexico

S.V. Venkatakrishnan
Oak Ridge National Laboratory
Oak Ridge, Tennessee

Qiu Wang
Cornell University
Ithaca, New York

Song Wang
University of South Carolina
Columbia, South Carolina

Jason Wheeler
Sandia National Laboratories
Albuquerque, New Mexico

Rebecca Willett
University of Chicago
Chicago, Illinois

Hongkai Yu
University of South Carolina
Columbia, South Carolina

Youjie Zhou
University of South Carolina
Columbia, South Carolina

Part I

Introduction

Chapter 1

Materials Science vs. Data Science

Jeffrey P. Simmons
Air Force Research Laboratory

Lawrence F. Drummy
Air Force Research Laboratory

Charles A. Bouman
Purdue University

Marc De Graef
Carnegie Mellon University

Materials science and statistical science are two complementary disciplines that, owing to different histories, approach problems from different directions–in some senses, opposite directions. We titled this book, "Data Science for Microstructure Characterization," using the rather vague term *data science* to describe the fusion of these two fundamental approaches that allows the now emerging large stream of data to be used to fundamentally affect the way we do business in materials science.

Materials science is an amalgam of metallurgy, chemistry, physics, mechanics, and other fields that have a common objective of controlling the behavior of substances and a common *scientific* approach towards achieving that objective. These fields owe their origins to the emergence of the scientific method, which was founded on the principle of, not just asserting theories and hypotheses, but also testing them against empirical evidence. The more "pure" sciences, such as physics, seek to derive the theories from first principles and to use experimentation as a validation (or a rejection) of the theories advanced.

On the other hand, statistical science emerged from the mathematics of random processes and is, essentially, the inverse of probabilistic methods. Knowing the sources of uncertainties and how the probabilities are affected when they are combined, distributions such as the Boltzmann distribution, the Fermi–Dirac distribution, or the normal distribution have been derived. These have been validated by direct observation as well as tested by Monte Carlo methods. This is the realm of probabilistic methods.

Statistical science starts with the probabilistic rules, such as those distributions mentioned above, *and* the results of observations. It then works backwards to estimate more fundamental quantities in these rules. For example, the *central limit theorem*, asserts that, in the asymptotic limit, the average of n *independent and identically distributed* (*iid*) observations will approach a Gaussian distribution. This is used by engineers in virtually all fields to estimate the *mean value* of the process that generated the observations. This is not quite the same as the average of the observations, but the two are closely enough related that averages are used to estimate mean values, more or less unconsciously. Most engineers have worked with this for so long that they no longer recognize one as being separate from the other. But, buried in this standard method is an *inverse problem*.

Taking n observations and computing the average gives an engineering estimate of the mean value. But, taking a different set of n observations from the same sort of experiment yields a value for the average that is slightly different from the first. Performing this many times on different data points of the same distribution leads to a *distribution of average values computed*, which, by the central limit theorem, is Gaussian. Note: this does *not* imply or require that the original distribution be normal. By the rules of random variables, the standard deviation of the average of *iid* random variables will have a standard deviation $\frac{1}{\sqrt{n}}$ times that of the distribution of the observations. The standard deviation of the average is sometimes carelessly called the "standard deviation," but is more properly known as the *standard error* associated with the estimate of the mean value. This is the basis of the error bars that are put on data with the simplest statistical analysis.

But, note what the engineer has done: estimation of the mean value of the observations. That is to say, the (unknown) distribution was inverted to give an estimate of its mean as one of its parameters. From this, we could construct a corresponding *forward problem*, involving the mean just estimated, and *an assumption* as to the distribution of the observations. A Monte Carlo model could be used to generate synthetic data that mimics the original dataset. This would be a normal way in which a physical scientist proceeds.

Statistical inversion is a favorite of statistical scientists. The regression method described above for mean value estimation was developed by statistical science. We could say that physical science typically constructs a *forward model* and statistical science typically constructs an *inverse model*. This is the point to which we alluded above: the two are complementary and proceed in opposite directions. This "closed loop" is made possible by using both physical science for forward modeling and statistical science for inverse modeling. This book includes both of these elements and has introductory chapters by Bouman on statistical inversion and by De Graef on physics-based forward modeling.

This example, though, is an over simplification that obscures a very serious complication. Physical science, having been developed from an understanding of the phenomena, has identified the fundamental quantities necessary to produce the "correct answer." This is a bias generated by the fundamental scientific principle that one researcher's results must be repeatable by another if enough information is reported. The consequence of this is that theories tend to predict one, and only one, end result. That is, they are *well-posed* in the sense of having a unique solution. In statistical science, this would be known as a *consistency* constraint in the sense that the answer is consistent among different researchers. Theories of physical science are *biased* in favor of consistency, which leads to the erroneous assumption that every problem is well-posed, having a unique solution. Inverse problems, very commonly, are ill-posed, possessing even an infinite number of perfectly legitimate solutions from the standpoint of solving the forward model. But, the vast majority of them can be immediately recognized as "wrong."

A very simple example of an ill-posed inverse comes from high school algebra. Consider the equation $x^2 = 1$. From the methods of quadratic equations, it is easily shown that there are two roots: $x = +1$ and $x = -1$. But, if x is a distance, for example, it is clear that not all solutions are correct. If we bias the results to only give positive results, we obtain the unique solution of $x = 1$. This is a simple example of *regularization*. Note that, even though the answer is biased, the values did not change: the bias was used to identify the subset of the solutions that were *reasonable solutions*.

The statistical example above where the mean of the measurements was estimated is well-posed because the researcher had an *abundance of data*. In that case, the number of parameters to be estimated is much smaller than the amount of data to be used in their estimation. In a pure mathematical problem, this would be an over-defined problem, but the assumed presence of noise allows for a unique "best estimate" to be given of the mean value. Consistency is straightforward in the case of an abundance of data.

The more interesting, and more practical in the modern world, case is where there are fewer data points than quantities to be estimated. This is the usual case with microscopy. An observation with modern digital microscopy consists of a million or so of intensity values, corresponding to pixel intensities, and the microscopist wants *them all*. Regression techniques are well developed around the problem of reducing these million points to a homogeneous average such as an average intensity, a volume fraction achieving a given intensity, morphological descriptors, or other homogeneous values. This used to be referred to as "data reduction." In modern times, and in microscope imagery, it is not acceptable to simply give an average descriptor; a reconstruction of the local features of the observation is what is required. This is particularly the case if rare or anomalous events are of interest.

Theoretically, one might expect to be able to faithfully reconstruct the structure, with one (significant) pixel intensity resulting from each detector array pixel observation. Practically, though, noise and other problems reduce the actual information contained within the pixel values so that there is actually less information than needed to estimate all variables. Further, sensors are developed based on the physics that may be exploited to obtain information, leaving the challenging problems of "teasing" more information from a set of measurements than the average researcher might think would be possible. Dosage limitations necessary to reduce beam damage can lead to low signal-to-noise ratios, leading to another complication. There are many conditions that arise in real data collection that conspire to make the imaging problems in microscopy illposed.

In imaging science, the classical approach for solving ill-posed problems was illustrated above with the trivial algebra example. In that case, unbeknownst to the mathematics, the human student knows that distances must always be positive. Applying this regularization constraint allows for consistency in the answers reported by students. And so it is with microscopy: the materials scientist has much more domain knowledge than the microscope or the computer charged with solving the problems. When physically realistic regularization is used, the results of the ill-posed inversion problems that arise in microscopy give way to consistent methods that give unique solutions. They are not guaranteed to be "the correct" answer, but they are reasonable and consistent from researcher to researcher. The chapters in Part III of this book were chosen to highlight reasonable solutions made possible by applying some sort of regularizing constraints to materials problems.

Of course, in science, the regularization is much more complex than a simple "lengths must be positive" constraint. Although not usually known by that name, the analytical process is not unfamiliar. For example, in the CALPHAD method for modeling phase diagrams, the analyst critically assesses thermodynamic data and chooses a model of the free energy. The critical part, comprising the intellectual property, is the estimation of the parameters in the free energy model. A novice would choose an approach that reproduces the input data and consider the work completed. But a true expert will look at the other implications of the model so calibrated. It is the extrapolations outside of the fitted data and their reasonableness that actually make for a good set of parameters. We could say, figuratively, that, all things being equal, the model that does not violate the Second Law of Thermodynamics is the better one. And this approach works fine if the amount of data is relatively small (hundreds, not millions of points to be fit).

For imaging science, the tests for reasonableness to provide the regularization to solve an ill-posed problem become too complex to be applied by hand. But, this is exactly where statistical science flowers with its formal approach. With the formal approach, regularization typically takes the form of a *penalty* applied for violating known physics (Tikhonov regularization) or in the form of a high probability that the solution is valid, based on the likelihood that the solution is physical (Bayesian regularization). When the possible solutions of a problem are formulated in terms of a statistical model, an ill-posed problem will have solutions represented by a *distribution* of acceptable solutions. That is, the ill-posedness will lead to an uncertainty in the final reported answer. Done correctly, regularization reduces this uncertainty by applying principles that are already known and accepted to the distribution of acceptable solutions to reduce this uncertainty about the solutions that actually obey these principles.

In modern image processing, an extremely common bias used is to regularize the solution with the Markov random field (MRF). MRF regularization is equivalent to requiring that unphysical situations be penalized according to the energy that they would have if the pixels interacted according to an equilibrium Ising or Potts model. The reasonableness of this approach is justified because interfaces do have interfacial energies. Where it is expected that interfacial energy will smooth an interface, it is reasonable to apply this regularization. But interface smoothness cannot be the conclusion of the study, since the regularization caused this effect during the analysis.

But, regularization does have the advantage over traditional approaches to analysis in that it forces the analyst to *explicitly* acknowledge the biases being applied, where they can be properly critiqued. It is always possible that the bias applied may not actually be correct and it is always possible that the bias is misunderstood and, therefore, misapplied. This is simply the nature of science and those who recognize these misunderstandings have been rewarded.[183]

But, there is a much more serious source of bias: the unconscious inclusion of things taken to be facts because they were never questioned. Statistical methods run rife with opportunities for these sorts of errors. Mark Twain popularized an expression that, "There are three kinds of lies: lies, damn lies, and statistics,"[1] which illustrates how easily the mind can be misled in this realm. The human mind is not logical, nor rigorous, nor self-consistent. It can be trained to perform disciplined, logical thought. But, in its natural, creative state, it is a mysterious environment given to all sorts of magical interpretations that, though powerful, can often be shown to be completely erroneous in the light of experimental evidence. Something about the nature of uncertainty and probabilities leads the mind to feel the comfort of understanding, even when principles are completely false.

The laws of probability have been developed over the years and, unless they are followed, the intuitive results often will be erroneous. Those experienced with probability generally become very careful about what they say whenever the topic comes up, precisely because it is so counterintuitive. We have included a chapter by Comer et al. (Chapter 3) that describes the rigid framework of probability.

The following set of examples is intended to serve as an illustration of the interaction of the human mind with random processes that can produce surprising results.

- Most engineers are familiar with the dubious "Law of Averages," that would have it that, if a coin were tossed ten times and each time, it landed on heads, that the eleventh toss would be less likely to land on heads than tails. True, a long streak is unlikely, but if the streak has already occurred, it says nothing of future events. This is part of the training of engineers and most would not be swayed by this sort of reasoning.

- Engineers are more susceptible to faulty reasoning as is illustrated in the classic "Monty Hall Problem," patterned after the game show hosted by Monty Hall. On that show, at some point, the contestants were asked to choose what is behind one of three doors. Behind one of them is a prize of great value and, behind the others, a worthless prize. Once a contestant has chosen one of the doors, Hall reveals the fact that one of the unchosen doors had a worthless prize behind it. Hall then asks the contestant if he/she would like to change their original choice to the other whose contents are as yet unknown.

Most engineers trained in probabilities have been conditioned to ignore all the extra verbiage and stay with the original choice. But, in fact, the other door is twice as likely to have the desired prize than the original choice. This can be shown, informally, to be correct, because the contestant was only 1/3 likely to have chosen the correct door in the first place. It remains 2/3 likely that the prize was behind another door. Hall's information allowed the contestant to devise a strategy, whereby the correct door could be chosen, if it was among the other doors. Since the desired prize was behind these two doors with 2/3 likelihood, the contestant could capitalize off of this knowledge and gain the prize with 2/3 likelihood. More rigorous proofs of this result exist. Monte Carlo simulations can also show this to be correct. But, to many, it still does not "sound correct."

- "Correlation" is often used interchangeably with "causation" and people often try to control the uncontrollable by seizing on correlated variables. Richard Feynman often told a story which he described as an example of "cargo cult science." [560] During World War II, the US armed forces set up a base on an island in the South Pacific and built an airstrip by which many goods were delivered to the island, This resulted in prosperity for the inhabitants. When the war was over, the planes stopped delivering goods and the prosperity ended. The natives reacted by recreating the

---

[1]Twain attributed this quote to Benjamin Disraeli.

appearance of the airstrip and the actions of military personnel. In their minds, they could bring back the prosperity by imitating the actions of the people involved in the airstrip. In other words, they were trying to recreate one condition by recreating a condition that was correlated with it. This may seem somewhat silly to educated people, yet we do it all the time unconsciously. We often hear research results quoted in the news such as exposure to a particular substance *leading to a statistically significant* increase in the cancer rate. The inference we are left to make is that the substance "causes" the cancer. It may very well be that one causes the other, but it also may be caused by a root cause of both. Is it that the substance is causing their cancers or are people exposing themselves to the substance as a reaction to some environmental condition which itself causes cancer? With only a correlation to go on, it is impossible to say.

This reasoning could be illustrated in another example, where we are not so susceptible to magical thinking, but it uses the same thought process. Diamonds are often found with *kimberlite*, a particular form of lava. Discovery of kimberlite aids a geologist in locating diamonds because their occurrences are correlated. But, it would be absurd to salt the ground with kimberlite in hopes of "causing" diamonds to appear. This would be "cargo cult science."

Now, it is, in principle, possible to increase the flow of kimberlite to the surface, which may in return bring up more diamonds. But, geologists have identified the root cause of both. Kimberlite forms more than twice as deep in the earth's crust as ordinary magmas, where the temperatures and pressures are considerably higher. At those temperatures and pressures, graphite undergoes a phase change to diamond. Knowing this allowed engineers to develop processes involving subjecting graphite to extremely high temperatures and pressures and produce, on demand, one of the most precious and sought after stones in history.

- Equating conditional probabilities to their Bayesian inverses is also very common. Consider the example of a person reading the newspapers and, over the course of several days finds that there were five instances where a dog attacked a person. In all cases, the vicious dog was of a particular breed. It is natural to assume that that particular breed of dog was vicious and that, therefore, additional controls must be placed on that breed of dog.

The newspaper reports indicate that the likelihood that a dog is of this breed, given that it bit someone. Our minds want to flip this upside down and equate this to the likelihood that the dog would bite someone, given that it was of this breed. But, suppose that breed was the only breed in town? Any dog attack would necessarily have been done by this breed of dog.

In probability, the first would be written as $P(\text{breed}|\text{attack})$ and the second as $P(\text{attack}|\text{breed})$. The rule for relating these two is Bayes' rule, a foundational rule of probability. Comer's chapter (Chapter 3) shows that these two probabilities are actually equal under the exact right conditions, but they are not equal in general. This sort of reasoning appears in many investigations of microstructure and its effects on properties.

The examples given above are illustrative of how the intuition, when not tempered with the rigorous structure of mathematical probability, can lead to erroneous thinking. We use our intuition for our ideas, but it is always a good idea to do a check to see whether or not these can be justified formally before building on them.

Finally, in the area of possible pitfalls in statistical science, there is the whole topic of *noise*. Practitioners of the physical sciences are often blind to the presence of noise. This is not so much due to any human perception problem as it is to the mindset that the noise is added in the forward problem and is generally eliminated by the homogeneous averaging performed in data analysis. But, for inverse problems, noise can produce pronounced and erroneous results. Statistical science recognizes that noise will always be present and that, often it is not possible to collect enough data that would enable homogeneous averaging. In tomography, for example, the common technique of filtered back projection amplifies noise on the reconstruction, producing streaking in the reconstructed volumes. Modern tomographic techniques, such as the one outlined in the chapter by Venkatakrishnan and Drummy (Chapter 6), usually use some sort of denoising in the algorithm. In fact, most of the statistical science chapters in this book deal, in one way or another, with denoising. The chapter by Willett gives an introduction to the topic, but denoising is still an evolving research topic and could easily constitute a book unto itself.

As a final note on statistical science, the topic of *sparsity* is a very active research area and we have dedicated the final part (Part VII) to this topic. It is easy to carelessly assume that the more data collected, the more accurate the answer will be. But, while detectors are continuously becoming capable of measuring higher and higher resolutions, leading to more and more degrees of freedom in the measurements, the underlying physical phenomenon being characterized does not become more complex as a consequence. When the sensors have been developed to the point where they have many more degrees of freedom than the underlying phenomenon, this is referred to as sparsity. That is, the information being collected could actually be concentrated in a much smaller number of carefully chosen measurements. Beyond a certain point, adding more sensor degrees of freedom just leads to better characterization of the noise in the observation.

The word "sparsity" is familiar to many computational modelers in the sense of "sparse matrices," which have many zero entries. This is the same concept as in sparse data, except there are not so many zero entries. Rather, many measurements may be formed from combinations of others. In the proper coordinate system, they would show zeroes as entries, but the computational cost of rotating coordinates this way is prohibitive. For example, when diagonalizing a matrix, i.e., expressing the measurements in terms of linear combinations of *basis functions*, on the order of N^3 operations must be performed, where N is the number of dimensions of the matrix. In imaging, N is of the order of millions of pixel intensities, so sparse methods generally do not seek a diagonal representation, but estimate both the basis functions and their coefficients from the data.

This description of sparsity is somewhat abstract. Let us illustrate the concept with an example that is perhaps more familiar to the physical scientist. If we consider the historical development of the ideal gas law, it started with individual researchers such as Dalton, Boyle, Avogadro, and others, each of whom developed laws governing the behaviors of rarified gases. These were combined by Clapeyron [208] into what we now know as the *ideal gas law*. This whole process took a number of years and formed one of the foundational areas of science.

Were this work to be undertaken in modern times, we could generate large databases of measurements of gas content, volumes, temperatures, pressures, and a host of attributes of mixtures of gases and then do all of the analysis of the data offline. But, we would not know of the ideal gas law from the outset. If we visualized the data in different coordinate systems, somewhere in the data, it would be possible to visualize a pattern as shown in Figure 1.1(a). It would turn out that pressure and volume were not independent, but would be related by some (in this example, unknown) functional relationship. Even though the measurements would have been made for all values of pressure and volume, the only ones that would have actually ever been observed would lie along a single curve in the space. That is, within experimental error, all of the data would lie along a one-dimensional *manifold* embedded in the $p - v$ plane. A manifold is a curve or hypersurface of lower dimensionality than the space in which it is embedded that is infinitely thin in at least one dimension. With real data, there is some scatter about this manifold due to noise; often the manifold in this case is referred to as a "thin manifold," where the noise creates some "thickness."

If the data were represented, say as sums of powers of p or v, the pattern would be discovered immediately and could be idealized as shown in Figure 1.1(b). This is the basic concept of sparsity: even though the data may have millions of dimensions, the actual phenomenon being characterized would have considerably fewer dimensions.

A corollary to the existence of sparsity is that experiments may be optimized to produce the same result with a drastic reduction of required measurements. This is the basic strategy behind *model based reconstruction* and the special case known as *compressed sensing*. The hypothesis is that, just as in the ideal gas case above, all images of a particular type lie on a manifold embedded in a very high dimensional space. The strategy is to make fewer measurements than are needed to unambiguously determine the intensities at every pixel in the image, and use a model of the image manifold to produce a unique answer. This is similar to the idea of calibrating a physics-

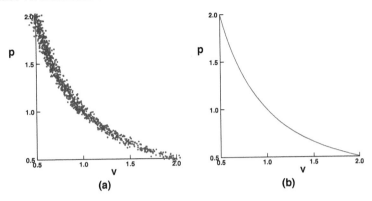

Figure 1.1: Hypothetical observation of the ideal gas law in data. (a) Data collected. (b) Idealized manifold that would contain the data if there were no noise.

based model, where significantly fewer actual measurements need be made because the physics is accurate enough to interpolate within or extrapolate outside of the data points collected.

The basic strategy behind model based reconstruction is illustrated in Figure 1.2 and is described presently. If we have an image with a large number N of pixels, we can describe it in terms of an N-dimensional space. Typically, N is on the order of a million pixels. Each pixel has an intensity somewhere on the interval $[0, 1]$, which is represented as a point in the N-dimensional space, as in Figure 1.2(a). Rather than attempt to visualize a million dimensional space, we can visualize a two-dimensional subspace with $\{x_1, x_2\}$ as coordinates. The image appears as a pixel in this subspace, as shown.

For clarity, we can recognize that there is the *set of measurements*, which would have values $\{y_1, y_2, y_3, \ldots y_n\}$ and the *true image*. The true image is actually an idealization of what the actual structure is and is more properly represented as a continuum. In order to compute a unique answer, let us imagine this true image discretized into n samples, giving values $\{x_1, x_2, x_3, \ldots, x_n\}$. Figure 1.2(a) is referring to the discrete samples of the true image.

With conventional methods, all pixels of a two-dimensional image would be measured at once with a CCD camera, where there is a one-to-one correspondence between a location in the observed image and pixels in the CCD array. But, it would work equally well to measure the *weighted sum* of pixel intensities, so long as we measured N of them and they were independent. Simple matrix inversion would recover the image of the CCD array. In the two-dimensional example discussed, there are only two intensities in the reconstructed image, x_1 and x_2, so there are two independent ways of measuring the intensities of the pixels, corresponding to two different sets of weightings for the sum. That is, we make two measurements, y_1 and y_2 according to the following formulae:

$$y_1 = a_{1,1}x_1 + a_{1,2}x_2 + w_1$$
$$y_2 = a_{2,1}x_1 + a_{2,2}x_2 + w_2 \tag{1.1}$$

where $a_{1,j}$ are the weights for the first measurement, $a_{2,j}$ are those for the second, and w_1 and w_2 are noise added to the signals for measurements 1 and 2, respectively. The weights can be viewed as vectors in the reconstructed image space, as shown in Figure 1.2(b), which we will call the *measurement directions*.

In N dimensions, this may be represented as a matrix equation:

$$Y = AX + W \tag{1.2}$$

where Y, X, and W are the measurement vector, reconstructed image vector, and the noise vector, respectively and A is an $N \times N$ (non-singular) matrix of weights. This could be inverted directly to

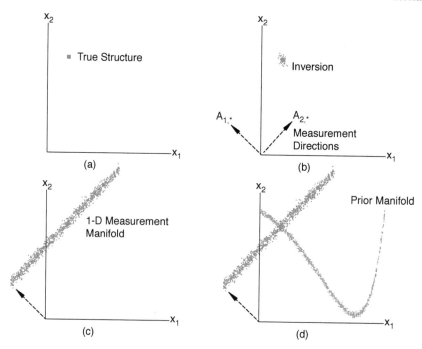

Figure 1.2: Schematic of model based reconstruction of an N-pixel image. Image is represented as a single point in an N-dimensional space, two dimensions of which are illustrated here. (a) The image of the true structure, measured at the precision of the voxels in the measurement (red). (b) Measurements of two linear combinations of pixel intensities to produce two measurements taken in directions, $A_{1,*}$, and $A_{2,*}$ (arrows), which are rows of the measurement matrix (see text). The inversion of these measurements (blue) is noisy due to propagation of measurement noise. (c) Inversion if only one measurement were taken. It is impossible to disambiguate the second dimension, leading to a manifold of acceptable solutions. (d) Unique solution obtainable by invoking a prior model of characteristics of the true structure (brown).

give X for a given measurement Y, but X would have some noise, making it a "zero-dimensional thin manifold," i.e., a point that is blurred out by the presence of noise; see Figure 1.2(b). We will refer to this manifold as the *measurement manifold*.

If we, say, only measured $M-1$ weighted sums, then A would be singular, with a row rank of $M-1$ and a one-dimensional null space. It is impossible to unambiguously invert Equation 1.1: the solution would be something in the row space added to any multiple of something in the null space. That is, the measurement manifold is a one-dimensional thin manifold. This is shown in Figure 1.2(c). We can only measure the projection of the image hyper-voxel on the measurement direction, leading to any position in the manifold yielding the same measurement.

We could reconstruct the image by taking M measurements directly, but, since the reconstructed image is believed to be sparse, it is more expedient to model the sparsity and use that to construct a *prior manifold*, i.e., a manifold within the space on which we believe all of the reconstructed images will lie. This adds the additional constraint to provide a unique solution. This is illustrated in Figure 1.2(d). For model based reconstruction as well as compressed sensing, the strategy is to make $N \ll M$ independent measurements and use a sparse prior to reconstruct the full set of M voxels.

Why would we believe we could model the manifold of reconstructed images? We can use the example of a polycrystalline material to illustrate this. If M voxels were measured of a polycrystalline material that only had hundreds of grains, the *vast majority* of the measured voxels would have the same value. This same value could be inferred by the neighboring voxels. For example, if a voxel to be reconstructed were surrounded by voxels, all having the same intensity but for noise, it is reasonable to assume that the central voxel had this same value. In fact, it is probably more reasonable to assign that same intensity to the center voxel than to accept the result from a measurement. If there were a significant difference, it would be more likely due to an imaging problem than it would be that this isolated piece of material that happened to occupy a single voxel had actually been observed. This is the assumption made when making the elementary *median filter* image processing step on electron backscatter diffraction (EBSD) data. If the sparsity is formulated as a prior (for example, the L_1-norm having a penalty), all of the assumptions behind the median filter would hold, but, because compressed sensing chooses an optimal value, a much more accurate reconstruction would be made.

The chapter by Venkatakrishnan and Drummy (Chapter 6) gives a specific example of model based reconstruction, as applied to TEM tomography and that by Romberg (Chapter 22) gives an overview of compressed sensing with sparsity priors.

With this introduction to "data science," as it is used in this book, it is appropriate to digress and discuss "microstructure characterization," the target of the data science in this book. Microstructure characterization can be subdivided into two broad subfields: (1) microstructure observation and (2) microstructure formation. This is somewhat of a subtlety in the field of materials science. Whereas the microscopist works with the observation of the microstructure, the microstructure modeler, phase transformations specialist, or process developer works with microstructure formation. When the microscopist is interested in modeling, the most relevant type of modeling is forward modeling. That is, the modeling of photons, electrons, ions, or other probing beams as they interrogate the material. The modeling appropriate for microstructure formation could be termed *evolution modeling* to distinguish the evolution of structure from the interaction of probing beams with the existing structure. Both types of modeling are so ubiquitous in the field that they are usually referred to by the same term, "modeling."

There is also a subtlety in the dual meaning of the word "microstructure," that varies according to intended use. Often, a microscopist will use this term to describe the structure actually observed in the microscope and a process modeler, for example, will use it to describe a class of *typical structures* that would be observed in the microscope. The first is very literally the structure observed and the second has a more abstract meaning. In statistical science, the term for the second "microstructure" is *texture*: all of the features that characterize the random process that results in the observed set of structures. Materials science has borrowed this term "texture" to describe the same thing, though for historical reasons, it is generally restricted to the orientation distribution functions of grains in a polycrystalline material. This book covers both meanings of "microstructure" along the lines described above: those working directly with microscopes usually will refer to an actual structure and those working with representations and quantifications usually will refer to the textures, though not always by that name.

This book is put together as a fusion of both the materials science and imaging science fields and we have attempted to maintain the connections between the two fields as best we could. After a brief introduction to the field (Part II, with chapters by Mick, Comer, De Graef, and Bouman) and some examples of statistical science as they have been applied to microscopy (Part III, with chapters by Venkatakrishnan, Wang, and Wang), we have included two chapters on investigations of microstructure in the materials science field. These cover both microscope observations and texture characterizations and descriptions (Part IV, with chapters by Tang, Nepal, and Beladi and Part V, with chapters by Niezgoda, Callahan, Sundararaghavan, Furrer, Lazar, and Holm). We then proceed to two areas of statistical science that are not well known in materials science yet, but that represent future growth areas: anomalies (Part VI, with chapters by Theiler and Przybyla) and sparse methods (Part VII, with chapters by Nadakuditi, Willett, Romberg, and Larson).

The chapters in each part of the book, all by invitation, were written independently of one another, so that one does not need to be read in advance of another for a clear understanding. But, the chapters within the parts are interrelated and, to the degree practicable, we have indicated in each chapter where the topic overlaps with others, at least insofar as the immaturity of the field will allow.

But, to be perfectly honest, this book only presents a skeleton of an emerging field. As new applications are found, additional interrelations yet to be discovered will come to light. In this book, some chapters were chosen to give background that is wellknown in materials science for those in the statistical science and imaging fields, for whom these principles are virtually unknown. Similarly, some chapters cover state-of-the-art algorithms in signal processing and other data science fields that have yet to be exploited in the materials characterization field. But, we have also included a core of efforts that reflect a true blending of the two fields that really can be viewed as a seed for what is to come. The mixture of data science with materials characterization is still very much in its infancy, with the best yet to come. If you can dream it, it probably has not been done!

Part II

Emerging Data Science in Microstructure Characterization

Microstructure characterization is one of the areas in materials science where Big Data is emerging, partly because what we call "images" are really arrays of individual measurement results displayed so to appear to be image data. Add to this the inevitable scripting of the machines to collect many "images" of the material in one run and the proliferation of sensors that give "images" observed in different modalities and the data size starts adding up. How to take advantage of those opportunities is the subject of this book. To this end, we put together this part of the book to describe the scope of developments, at least in our part of the field.

The chapter by Mick describes emerging and anticipated advances in the field of microscopy, from the viewpoint of an engineer with a major original equipment manufacturer (OEM) of microscopes. We had intended to only augment this with chapters by Bouman and De Graef. Bouman's chapter describes inverse methods as a way of solving the inevitable problems of having many more variables to fit than values. This is a problem, even in the "Big Data age." De Graef's chapter describes forward methods, i.e., those methods that start with an assumed structure and compute the expected observations. These both address the same question, "From what I observed, what is the structure?" except one proceeds in the forward direction and the other in the backward direction.

It became apparent that we also needed a chapter in the middle that described the differences in thinking between the inverse method heavy imaging science, from which Bouman hails and the forward method heavy physics of De Graef's background. So, we asked Mary Comer to prepare a chapter describing the language of both fields, the key points needed to appreciate when learning the language of the new field, and some of the concrete concepts that must be appreciated by those new to the field.

The chapters in this section should not be viewed as the all-inclusive, "everything you would ever want to know about..." type chapters, but as introductions to those new to one or the other field. It was our hope that this book would be a starting point and that the readers would proceed to complete the picture.

Chapter 2

Emerging Digital Data Capabilities in Electron Microscopy

Stephen Mick
Gatan, Inc.

2.1 Introduction

Historically, the detection of images has been the bottleneck in high-speed, quantitative data acquisition in electron microscopy. Before the advent of digital cameras, images were collected sequentially on photographic plates and then developed in a darkroom after the microscopy session had been completed. While high-resolution images and diffraction patterns could be acquired on film, it was not uncommon to use long exposure times to collect data. For example, exposure times from several seconds to minutes were expected for collecting a single diffraction pattern on film [1146].

The amount of data which could be collected with film was therefore limited by the rate at which data could be physically captured and developed. Moreover, the cost of the negatives presented a practical bottleneck to capturing large amounts of imaging data.

In the 1980s, TV cameras for the electron microscope were developed by Peter Swann [994]. These cameras used a YAG scintillator, an image intensifier tube, and a television camera to provide an analog TV-rate signal with 493×768 or 581×756 elements. The data output from the camera could be recorded with frame-grabber technologies such as a VCR. While these cameras could be used to capture images up to 30 frames per second, the use of frame-grabbing for image capture resulted in the loss of the raw, pixel-by-pixel data.

Digital data capture for electron microscopy that preserved data for every pixel was reported in the mid-1980s when Mochel & Mochel and then Spence & Zuo adapted CCDs for electron microscopy imaging and readout [695, 957]. The Spence & Zuo adaptation used a parallel detection system based on a 576×382 pixel CCD that required liquid-nitrogen temperatures for operation, and it operated at a frame rate of 2.5 frames per second [957]. The CCD camera became widely available in the early 1990s, when Krivanek and Mooney of Gatan introduced the first commercial digital cameras for capturing TEM images [513]. These early cameras used a single data port for information readout which limited data capture to a few frames per second. As a result of their speed and data architecture, this class of camera was known as slow-scan.

Despite the slow data capture rates and limited fields of view of these slow-scan CCDs, digital imaging increased the rate at which data could be captured as well as the quality of the collected data. That is, digital data capture eliminated the time required for film development and allowed study of captured images within seconds of acquisition. This near-real-time feedback of images and experimental results allowed the microscopist to observe and quickly adjust imaging parameters and experimental conditions to maximize the relevance of the captured data. In addition, storing data in a digital format paved the way for image processing during a microscopy session and automation of TEM tuning [513]. Moreover, the introduction of CCD cameras within electron microscopy was one of the transformative advancements that helped justify the TEAM project and development of a practical spherical aberration correctors for electron microscopy [980].

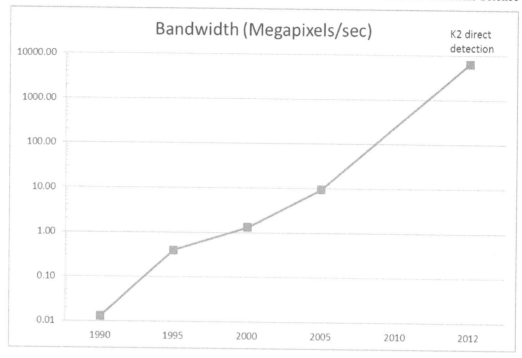

Figure 2.1: Evolution of camera data rate since the first generation of commercial digital cameras in 1990. Current state-of-the-art K2 direct detection camera tops the chart at 6.4 Gigapixels/sec [758].

As digital imaging matured, the data rate grew due both to an increase in the number of pixels on the camera sensors and the frame rates at which data could be captured and recorded to permanent storage. Whereas early digital cameras for electron microscopy had sensor sizes on the order of 600×600 pixels, the most common modern digital cameras for electron microscopy now have sensors that are 4k×4k pixels with some cameras providing approximately 8k×8k pixel readout areas. Regarding frame rates, modern direct detection digital cameras are available that provide up to 1,600 frames per second. Given these advancements, the overall data capture rate has scaled quickly from approximately 1 Megapixel/second with the earliest CCD cameras to over 6,400 Megapixels per second with modern cameras. Figure 2.1 shows how the data rate has scaled since the early 1990s.

The effect of this data rate increase is pronounced for the microscopist who is trying to collect data for the purpose of simply answering a scientific question. At the advent of digital imaging for electron microscopy, a researcher might collect several thousand images over the duration of an entire project. But with modern cameras, it is now quite common to generate more than one hundred thousand, high-quality images and a Terabyte of data from a single microscopy session [346]. In some applications, these cameras collect tens of Terabytes of imaging data in a single microscopy session. In the near future, as the size of the sensors and the speed of readouts continue to increase, the capability to generate dataset sizes will exceed hundreds of Terabytes per microscopy session. While there are many benefits to collecting large amounts of high-speed, high-quality imaging data, there are also challenges to address.

Not only has the quantity of data increased, but also, the quality of the data has greatly increased from the early cameras. The quality of data from a camera for TEM can be characterized by three primary performance metrics: modulation transfer function (MTF), noise transfer function (NTF), and the detective quantum efficiency (DQE) ([703, 702, 566, 518]). The MTF describes the attenuation of image contrast with spatial frequency; the NTF describes the spatial frequency-dependent attenuation of the Poisson noise that is inevitably part of all electron microscopy images; and the

DQE describes the capacity of an imaging device to generate signal (contrast) above the noise inherent in the imaging system ([678, 758]). The DQE is directly related to the square of the MTF and inversely related to the NTS, and better data quality is related to higher values for both MTF and DQE ([314, 875]).

Data quality from TEM cameras has been improved by pursuing camera design choices that maximize both DQE and MTF. The physical design choices that have led to better DQE and MTF (and thereby better data quality) include optimizing the semiconductor processes used for the camera sensors, improving the radiation hardness of the individual pixels, increasing the single-electron sensitivity of pixels, and increasing the readout speed of the sensors ([220, 314, 671]).

The physical attributes that have led to improved data quality have also led to being able to extract quantitative data in every recorded image. In an effort to improve DQE and MTF, for example, modern cameras have leveraged advanced semiconductor processing to provide complementary metal oxide semiconductor (CMOS) sensors where each pixel is a sophisticated circuit capable of high-speed operation. With CMOS sensors, the charge in each pixel is converted to an output voltage at the pixel, and high levels of functionality can be integrated on-chip (or downstream immediately off-chip) including digitization, real-time linearization, and gain correction on full frame images. These gain-corrected, linearized images contribute to improved DQE and MTF and also allow for quantitative analysis on every frame of the collected data.

Data quality has further improved with the advent of direct detection cameras. Whereas most cameras for electron microscopy use a scintillator and lens- or fiber-coupled light guide in front of the sensor, direct detection cameras are directly exposed to the electron beam when acquiring an image. Historically it has been challenging to create a direct detection sensor that can withstand the intense radiation of the direct electron beam in the microscope [220]. However, direct detection cameras have recently become commercially viable because of advanced semiconductor process nodes that enable radiation hard pixels to be reliably manufactured. With robust pixels, direct detection sensors have a much higher signal-to-noise ratio than previous cameras because each primary electron arriving at each pixel generates a large number of electron-hole pairs relative to the read noise of the pixel electronics. Moreover, they offer almost order of magnitude increases in data quality due to the elimination of the noise and image degradation that occurs from the traditional scintillator and fiber optics. Data quality can further be improved with direct detection cameras operating in electron-counting and super-resolution imaging modes.

The quality improvement across successive generations of imaging technology can be seen by plotting the detective quantum efficiency (DQE) for different detector technologies. DQE indicates the capacity of an imaging device to generate signal (contrast) above the noise inherent in the imaging system, and it is recognized as the objective measure for both the ultimate resolution and sensitivity of an imaging device [703, 702, 573]. Figure 2.2 shows the DQE measured with a 300 keV electron beam for SO-163 Film, the Gatan Ultrascan US4000 CCD camera with an ultrahigh-sensitivity scintillator, and the Gatan K2 direct detection camera operating in both intensity integration (Base / *In Situ*) and super-resolution (Summit) modes.

This graph makes it easy to understand why film remained in use for many years after CCD cameras became available for electron microscopy. Although CCD-based imaging provided time efficiency and cost reduction in data collection, the image quality of film remained superior to digital cameras. However, direct detection cameras in the intensity integration mode can match the data quality of film and vastly exceed the quality of film when operating in electron counting and/or super-resolution modes.

2.2   Benefits of Large Data Volumes in Electron Microscopy

The large volume of high-quality data that now can be routinely captured by modern cameras for electron microscopy provides many benefits. These benefits are especially pronounced for direct detection cameras and derive from two key attributes of these cameras: speed and sensitivity.

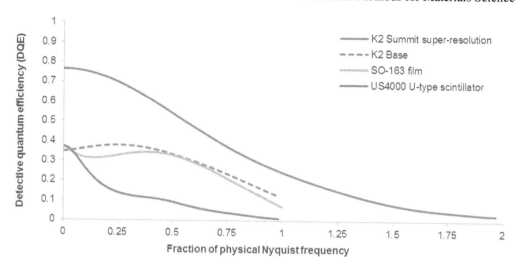

Figure 2.2: DQE measured with a 300 keV electron beam for SO-163 Film, the Gatan Ultrascan US4000 CCD camera with an ultrahigh-sensitivity scintillator, and the Gatan K2 direct detection camera operating in both intensity integration (Base / *In-Situ*) and super-resolution (Summit) modes.

A direct benefit of high-speed modern cameras is the capability to collect data at increasingly smaller time scales. As an example of this benefit, specimen drift has historically been an inherent problem in electron microscopy limiting the resolution of images especially for long exposure times. Drift has many sources but generally is caused by mechanical instabilities of the microscope or the specimen [1125]. When multiple frames from traditional cameras that each have drift were averaged (in an effort to increase the overall signal-to-noise ratio), resolution was further degraded. However, high-speed cameras allow the acquisition of a single image to be captured as a stack of images with short, millisecond time-scale exposures. These image stacks allow for drift to be corrected before accumulation. In the case of drift correction, large amounts of data are acquired, but the data that must be stored is reduced into a final, single image.

An immediate benefit of improved sensitivity is that the microscopist can choose to use less dose while imaging and still achieve significantly better images than with the previous generation of digital cameras. This increased sensitivity (especially from direct detection cameras) has helped lead to resolution records being shattered and new fundamental learning in the study of beam-sensitive samples. As an example, the single particle analysis (SPA) technique within structural biology is currently experiencing a renaissance and as a result was chosen by Nature Methods as the Method of the Year, 2015 [289]. The SPA technique acquires tens of thousands of images of different but similar particles in random orientations. The dataset is then analyzed and a 3-D reconstruction of the underlying macromolecule is computed. The sensitivity and image quality of direct detection cameras have made it possible to develop advanced data processing algorithms for structure reconstruction. As a result, it is now almost routine to resolve biological macromolecules to resolutions much better than 4 Ångstroms — a limit that had rarely been surpassed until direct detection cameras became commercially available [303, 518]. Using data extracted from the EM DataBank underscores this point (Figure 2.3). From this data it can be seen that until direct detection cameras were first introduced in 2012, results exceeding the 4 Ångstrom resolution barrier were achieved between 2–6 times per year. Within three years of the introduction of direct detection cameras, more than 90 structures were resolved to better than 4 Ångstroms — approximately a 20-fold increase — and through July 2016, approximately 130 structures exceeding this resolution barrier had been published. On an annualized rate, the increase in high-resolution structures is on pace to be almost 40× the historical, pre-2012 average.

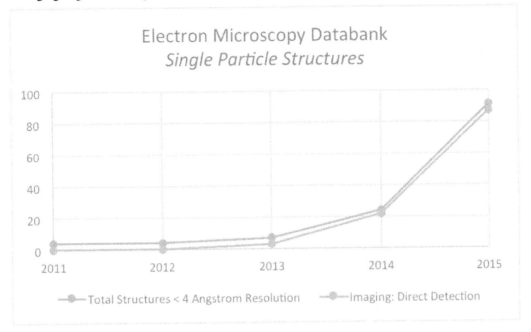

Figure 2.3: Application example: Benefit of large, high-quality datasets in structural biology. Shown in graph: total number of resolved single particle structures submitted to the EM Databank with resolution < 4 Ångstroms and number of these submitted structures where the dataset was collected on a direct detection camera.

The benefits of improved sensitivity and high-speed acquisition from these advanced cameras can also be seen in current material science research including recent work on zeolites [1089, 757], polymers and carbon [961, 481], organic electronics [19], metallic glasses [345], and holography [185].

2.3   Challenges of Large Data Volumes in Electron Microscopy

The speed and sensitivity of these new cameras are also being used to study dynamic, or *in situ*, events with millisecond temporal resolution. The study of material properties through *in situ* electron microscopy is becoming an increasingly important technique [501], but it is with the study of dynamic events that the problem of managing large datasets is most challenging.

Unlike the case discussed above for drift correction where many images are acquired but reduced into a single, final image frame, data acquisition for dynamic reactions often requires that every frame acquired by the camera be saved to disk for subsequent analysis. That is, *in situ* experiments are conducted with the goal of understanding the real-time, structure-property and structure-function relationships of materials while the material is being subjected to near-real-world conditions. Understanding these structure-property-function relationships currently requires that experiments capture images during an entire dynamic process (e.g., grain growth, defect formation/motion, sintering, catalysis, charge/discharge cycle, etc.).

Until recently, digital cameras for electron microscopy had been capable of capturing only tens of frames per second and the amount of data captured could be analyzed manually. With modern cameras, data can be captured at rates exceeding 1,000 frames per second (approaching 10,000 Megapixels per second as shown in Figure 2.1), and the length of an *in situ* experiment can range from a few minutes to the extreme case of several days of continual data collection [687]. For dynamic reactions lasting on the order of a few minutes — the most common *in situ* experimental duration — the 2TB+ dataset generated from fast cameras presents a data processing, storage, handling, and analysis burden for the scientist. At present, multi-day experiments are only rarely attempted as these extreme durations could generate more than a Petabyte of data and exceed the limits of current data capture, analysis, and storage technology available to most labs.

To help mitigate this data burden and to extend the duration of experiments that can be practically performed, modern high-speed cameras are designed with features that allow smart, targeted collection of data. These acquisition features, along with data collection and data handling strategies must be incorporated into the experimental plan to collect, as nearly as possible, only the required data.

Prior to the advent of high frame rate digital cameras, cameras did not integrate any particular data management features. Instead, it was common at the beginning of a dynamic experiment to start recording images and to record continuously during time periods when a researcher was likely to observe a dynamic reaction. This was the common practice for two main reasons: 1) the exact moment when a reaction would take place and need to be captured was not known *a priori*, and 2) data rates were low so saving all the images was not a burden. Thus the standard way to record data during an *in situ* experiment with slow cameras was to "record everything" with the goal of missing no interesting data.

Fast cameras make this "record everything" approach untenable since data capture rates are too high to simply start recording images and capture continuously. Fortunately, the need to record everything is mitigated by some modern, high-speed cameras that provide a "look-back" feature or record buffer. With this approach, the camera continuously streams data to a buffer in computer RAM capturing a rolling window of up to 20 seconds of continuous image data. This allows the researcher to watch the reaction and wait to start saving images to a hard disk until after the reaction of interest begins. Once the reaction has already started, then using this look-back capability, the researcher can initiate a data capture from the beginning of the look-back buffer and record images until the reaction has completed. This simple approach of incorporating a data buffer into the camera architecture can prevent the capture of thousands of unnecessary image frames that would otherwise be collected if the researcher had to follow the former "record everything" approach.

Another camera feature available in only the highest performance direct detection cameras allows electrons to be counted in real-time when images are collected using low electron doses. This electron counting transforms each frame from the camera into a sparse image and provides a $40\times$ reduction in the number of the final images saved to permanent storage.

A third feature set provided by a variety of cameras is software allowing researchers to explore and reduce the collected data. Some of the numerous software tools that are available include:

- cropping image sequences in both time and space
- binning image sequences by $2\times$ up to $16\times$ to reduce overall image size
- summing consecutive frames to reduce the total number of images as well as to improve the signal-to-noise ratio of the final images
- skipping frames to remove images where no dynamic events of interest occur

2.4 Emerging Techniques

While features such as hardware-look-back, electron counting and basic software tools attempt to mitigate the burdens of dealing with large datasets, these techniques will prove insufficient to reduce the amount of data as data rates continue to grow. Simple strategies will not scale to the multi-Terabyte and Petabyte dataset sizes that are anticipated in the future. Additional data management and experimental strategies must be developed to make data capture more efficient and targeted. While much research is ongoing, the following sections describe several key focus areas where developments must occur.

2.4.1   Multi-Instrument Coordination

The emphasis of this chapter so far has been on the development of camera technology and the concomitant growth of imaging data. However, the data that must be captured for most experiments is multidimensional, extending beyond imaging data to include data from and control of the microscope itself, additional detectors, and external components (e.g., power supplies, gas manifolds, light sources, liquid pumps, etc.). Since the goal of an experiment is to capture the data needed to answer a scientific question, there is an opportunity to use data from these other sources to filter imaging data that has been acquired (i.e., to ensure data that is exported for analysis was captured in the desired set of conditions) or even to control and trigger the acquisition of data (i.e., to limit data collection to sequences when the experimental conditions are most likely to produce the required data). The challenge is therefore to integrate the control and operation of multiple instruments/detectors, aggregate and synchronize the resulting data, and create user interfaces for experimental control that enable researchers to efficiently manage the volume and variety of experimental data. This instrument integration, control, and data aggregation would ideally occur in a way that minimizes the amount of isolated or unusable data that is acquired and maximize the amount of information captured at each point in time. Since the equipment used in a typical experiment is often made by multiple vendors, a practical challenge to coordinating instruments is creating well-defined interfaces between all of the instruments and ensuring that the look-back record buffer (present on the camera) is implemented across all data channels.

An example of this multi-instrument coordination can be seen in the development of electron microscopes such as the DTEM and UEM. These systems combine the control of TEM electron optics, a laser excitation system, timing circuitry, and image acquisition to probe ultrafast material science reactions from microsecond to nanosecond timescales [162]. These systems are now being implemented with beam control systems that use a fast electron deflector to raster multiple images onto the face of a camera during a single exposure in a so-called movie mode [530]. In the future, there is an opportunity to implement this movie mode acquisition on faster cameras to stream data continuously. While this integration and coordination would increase both the time resolution and movie length that can be acquired in these experiments, sophisticated timing circuits as well as robust software interfaces, component synchronization, and control are needed.

2.4.2   Upstream Data Analysis

Currently, most image analysis is done downstream of the camera — that is, on data after it has been saved to a hard disk drive. However, as the amount of data grows, there is a need to move data analysis closer to the camera itself (i.e., upstream) and to provide at least some measure of real-time analysis as data is coming off of the camera. By enabling data analysis at or close to the camera rather than after data has been saved to disk, several advantages can potentially be realized. One possible advantage of upstream data processing would be to scan a large field of view while applying rare event or event onset detection algorithms to the acquired images. As soon as a sub-region of the sample is found where the event of interest is imminently expected to occur, then the field of view could be changed to that specific area and the reaction of interest could be captured at higher magnification. This approach would maximize the chance of collecting data of a reaction of interest while minimizing the amount of data that would have to be captured.

Another possibility for upstream data processing is to send data from the camera directly to a high-performance computer cluster for data analysis and experimental feedback. The most basic analysis could include a measure of data quality to give a researcher real-time insight as to whether the dataset being collected is of sufficient quality to warrant subsequent extended analysis. HPC analysis could also be used to augment human interpretation of real-time data to help the researcher adjust *in situ* experimental conditions and parameters to maximize the likelihood of experimental success. Upstream HPC analysis could also include near-real-time comparison to theoretical models and generation of feedback to adjust imaging parameters to maximize data quality and model comparisons. As an example, scientists at Oak Ridge National Laboratory have proposed physics-based analysis of images with a goal of systematically exploring imaging data to understand materials functionality, but the approach necessitates direct data streaming from the camera onto a computer platform for analysis and model comparisons [476].

2.4.3 Data Mining

Regardless of the techniques used to minimize the amount of data that is collected, as data rates and image sizes increase, the challenge will remain to find and extract information from large datasets. Thus data mining approaches and technologies need to be developed. The challenge with electron microscopy data is that each experiment tends to be different so general purpose data mining approaches can be difficult to envision. However, if multi-instrument integration, control, and data aggregation is done, then the data and the metadata about the experiment can be used to select the correct sequences of interest (SOI) and reduce the amount of data that must be post-processed. Analysis of electron beam sensitive materials can provide an example of this approach. That is, materials such as zeolites, polymers and metal organic frameworks quickly degrade under the electron beam and there is a maximum dose to which these samples can be subjected before they start to deteriorate. The most basic data mining process would be to track the cumulative dose to which every region on a specimen is subjected during the entire course of an electron microscopy experiment. If the cumulative dose is tracked, then a filter could be applied to extract (mine for) only data from regions on the sample where a critical dose has not been reached. Data and metadata filters could also be used to select data meeting a variety of conditions such as desired specimen orientations, groups of particles with appropriate statistics, and images acquired only when the correct set of experimental conditions is met.

Another approach to data mining and subsequent data reduction can make use of the fact that the information needed is not always the image but rather a critical measurement or metric from the image. For example, when acquiring a large number of diffraction patterns (or diffraction images), the key information from the entire image might be just a few bytes of information indicating orientation, or the presence or absence of specific diffraction spots, or the relative distance of chosen diffraction spots. In cases like this, data can be mined and extracted based upon one or more critical measurements and the subsequent dataset size can be greatly reduced.

Another development requirement for efficient data mining is an approach to effectively extract information from large, multidimensional datasets. A recent hardware development in electron microscopy has been the advent of fast, hardware-synchronized STEM imaging. In this technique, the electron microscope is configured to scan a focused probe across an x-y area and at each pixel location a complete diffraction pattern is captured. The resulting dataset is 4-dimensional since each pixel in a 2-D grid has associated with it a 2-D diffraction pattern. But because this technique can now be performed in relatively short timespans, it is possible to continuously acquire these 4-D datasets as a function of additional variables such as stress, temperature, electric field, illumination, etc. Thus the final dataset could be of a very high dimension. Many other experiments which result in high-dimensional datasets are also possible [81].

2.4.4 Data Curation

Finally, the problem of curating data needs attention. The large datasets of spectrum images were described as a "data mountain" in the late 1990s [800]. Yet the data sizes from modern, fast cameras and dynamic experiments truly dwarf these previously described mountains of data. It is currently common for researchers to collect large datasets while visiting a collaborator site and then find that the most effective way to transport the data to their home institution is via external hard disk and a courier service. As the size of datasets increases, practical considerations must include "building" the mountain of data in the right place as the data is being acquired, minimizing the number of copies of the data, optimizing the required fidelity of each data copy, and minimizing the number of times the data must be moved.

2.5   Conclusions

Advancements in instrumentation and detector technology have led to a rapid increase in the amount and quality of data captured during electron microscopy experiments. Datasets from a single electron microscopy session are commonly collected that contain thousands of images and consume hundreds of Gigabytes to a few Terabytes of space. Datasets of this size present data processing, storage, handling, and analysis burdens for most scientists. As the number of pixels and data rates continues to increase, a number of challenges must be addressed so that data can be collected in a targeted and efficient manner, analysis has a maximal chance of gleaning information from the large data, and the answers to challenging scientific problems can be discovered.

Chapter 3

Cultural Differences between Materials Science and Image Processing

Mary Comer
Purdue University

Charles A. Bouman
Purdue University

Jeffrey P. Simmons
Air Force Research Laboratory

3.1 What Makes Modern Image Processing So Modern?

While image processing is a diverse field of research that has incorporated contributions from a wide variety of communities, there is no denying that, since its inception, image processing research has played a central role in the signal processing community. Perhaps this is best reflected in the variety of standard image processing texts [344, 820, 707, 427] that represent various snapshots in its early development in fields such as satellite remote sensing where it was a primary driver for development.

Signal processing represents the confluence of the three dynamically interacting fields of communication, radar, and computer science and, not surprisingly, the evolution of the field has been influenced by the various currents within each of these. Traditionally, communication has been concerned with getting a message from the sender to the receiver. Therefore, it has been biased towards finding correct solutions that are deterministic. If the message received is different from that sent, and the correct message cannot be recovered, it is simply *wrong*. Radar, on the other hand, has always worked with noisy signals and it was not acceptable to eliminate the signal simply because it was noisy. For this reason, the radar field has been considerably more biased towards stochastic solutions and the methods of probability, and *inference* of the origin of a signal. Computer science has played the role of the "game changer," by its development of more powerful machines, algorithms, and architectures that allowed for the development of increasingly complex algorithms. Whereas most of the methods of classical image processing use integer arithmetic because of the limitations of computers at the time, modern image processing uses more expensive methods such as floating point computations, Monte Carlo sampling, or optimization methods, on architectures that can vary from desktop microcomputers to supercomputers or graphic processing units (GPUs).

Where the future lies is a matter of debate. The radar field works with the sensor. In a world dominated by very inexpensive sensors collecting large datasets in a small amount of time, it can be argued that this end will dominate in the future. On the other hand, in a world where large datasets are archived and many groups work to compare results of various algorithms to find "the correct" inference, the communication end will dominate. Or, in a world that is awash in disposible data, where, "A really stupid algorithm plus lots of data equals inexplicably good results,"[293] raw horsepower of processing will make the difference and computer science will dominate.

This book lies at the confluence of signal and image processing and materials science and microscopy. We fully expect the interaction between these fields to be equally fruitful. This chapter was included in the book in an attempt to reduce the barriers due to cultural and, frankly, language differences between the fields. After much deliberation, we reached the conclusion that the three main barriers to interdiffusion of knowledge between the fields were (1) the differences in mathematical notation, (2) the concepts associated with probability, specifically, Bayesian probability, and (3) thermodynamics. Within their own native fields, these are part of the *lingua franca*. This chapter is an attempt to ease the difficulty of crossing between the two fields. We did not attempt to be complete, but rather to uncover the structure of these three topics.

3.2 Language of Image Processing: Set Theory and Probability

A key difference in image processing and materials science is that the former is more mathematical. It uses the more modern notation such as that commonly used in set theory. Also, modern image processing often uses Bayesian methods, which contribute to the differences in language. Physics traditionally uses the so-called frequentist statistics, which relies on asymptotic laws like the laws of large numbers or the central limit theorem to estimate means, standard deviations, covariances, and similar quantities. By contrast, Bayesian statistics allows well-established principles to be involved in modeling, at the expense of admitting a bias into the analysis. This section gives introductions to the modern notation as well as Bayesian statistics, with the intent being to illuminate the structure of the methods, rather than to replicate material already in the many texts available on the topics.

3.2.1 Notational Differences

Signal and image processing, like most other mathematical fields, generally uses the modern set theory notation, as opposed to the calculus notation used in most engineering fields. This can give the appearance that publications are mathematical and impenetrable. There are two reasons for this: (1) lack of familiarity and (2) terseness.

Lack of familiarity can be overcome by noticing that, in each field, there really are only a small number of basic constructs used in equations. Most equations represent different reuses among these constructs. Rather than attempting to learn all of the subtle nuances of everything that the notation can express, it is more fruitful to learn the constructs most typically used and to learn to recognize the key indicators of what an equation is actually describing. For example, in image processing, an inverse would be indicated by the use of the argmax_x or argmin_x constructions. These symbols mean "the argument that maximizes (minimizes) the value of the following expression." That is, this is the familiar problem in differential calculus: find the x that maximizes (minimizes) $f(x)$. Another example is the symbol for an estimate of a value, namely a circumflex over it, so that $\hat{x}$ is an estimate of the value of x. In this section, we attempt to cover many of the common constructs likely to be seen in the modern image processing literature.

Terseness in modern notation is intentional. In an attempt to eliminate the possibility of ambiguity, the notation has removed much of the redundancy that exists in algebraic notation. While this has the advantage of eliminating ambiguity, the redundancy is valuable to the reader because it provides a check of understanding. This is the same problem that is encountered in digital representations of data. Plain text has so much redundancy that, in the case of minor corruptions of the data, a human can usually correct the mistakes manually. But an aggressively compressed data file has eliminated the redundancy, so if a single bit is wrong, the whole dataset can be lost. In the calculus notation, on reading $r = \sqrt{x^2 + y^2 + z^?}$, most people would automatically recognize or even correct the exponent for z because it should be symmetrical with the other terms. The modern notation, in pursuit of terseness, has eliminated many of these symmetries, putting the onus on the reader to be careful to notice every term in the equation.

3.2.1.1 Sets

Signal and image processing use sets for describing almost everything. An equation is commonly expressed as a set of points. The calculus expression $y = x^2$, in set theory could be represented as a set $Y = \{(x, y) \in \mathbb{R}^2 : y = x^2\}$. That is, the set Y represents the set of ordered pairs of all real numbers and their squares. In the traditional notation, it would be implicit that this was defined over the reals.

Here, we have used the standard symbol $\mathbb{R}$ to describe the *set* of real numbers. Other very common sets are similarly defined: $\mathbb{C}$ is the set of complex numbers, $\mathbb{I}$ is the set of pure imaginary numbers, $\mathbb{Z}$ is the set of integers, and $\mathbb{N}$ is the set of *natural numbers*, that is, the set of integers that are zero or greater.

Modern set theory notation originated in the 19th century,[1] and has several advantages in signal and image processing. Sets make no implicit assumption of continuity, whereas the traditional notation assumes continuity and explicitly calls out locations of discontinuities. In image processing, a segmentation, for example, is designed to give solutions that have discontinuities whose exact nature is difficult to describe. The modern theory is also very good for problems where there are multiple solutions to the equations. Physics historically is concerned with testing or validating theories, where initial conditions are produced and the system allowed to evolve forward in time. This is compared with the predictions of a model—a *forward model*. In image processing, the end state of observation is recorded as a signal and the challenge is to work backwards to determine what gave rise to the signal, i.e., the challenge is to develop an *inverse model*. These are usually *ill posed* in that the inverse solutions are not usually unique.

A very familiar example of an ill-posed problem from algebra is to find x if $x^2 = 1$. In algebraic notation, this requires two parallel derivations, one for each root: $x = +1$ and $x = -1$. In set theory, a simple specification of a set, such as $A = \{x \in \mathbb{R} : x^2 = 1\}$ will describe all inverse solutions and the derivation can be performed symbolically on all of the elements of the set A. This is a simple example, involving only two entries. But, with an image with a million pixels, for example, a segmentation is a set of a million entries. In that case, treating the two classes of pixel intensity below a threshold and pixel intensity equal to or above the threshold would require two million equations, which would never be done with algebraic notation.

Images as Set Functions. Images are typically represented as functions defined on some index set S. For example, we could describe an image y consisting of $N \times N$ pixels as a real-valued function defined on a 2D lattice S. The value of y at a pixel $(i, j) \in S$ could then be written as $y(i, j)$, or $y_{i,j}$. In this case, the image y can also be viewed as a 2D array, but in the general case that may not be true, depending on the nature of S. To simplify notation, the elements of the index set are sometimes written using a single letter, so the value of the image at site $s \in S$ would be written as y_s.

For an image that can be written as a 2D array, if a particular algorithm were to apply an operation on the first column of the image, then the second, and so on, it may be simpler to define the image y as a vector. It is common practice to represent the image as a vector instead of a 2D array for simplicity, but it destroys the *topological* relationships between the pixels. For example, $y_{1,1}$ is a near neighbor of $y_{1,2}$, but y_1 is not a near neighbor of y_{N+1}, even though these describe exactly the same pixel pair. Sometimes biases can be introduced by being careless with the representations that do not have the same *topology* as the data. But, so long is this is kept in mind, both approaches can be used without creating problems.

The symbol y is commonly used to represent the measured data. This data is used to infer, say, something about the structure being imaged. The symbol typically used for this inferred image is x. A common operation is to recognize that the acquired data has noise. That is, $y = x + noise$ and the challenge is to remove the noise, leaving x as the result. Denoising is a major component of modern image processing. See Chapter 21 for specific examples of denoising.

3.2.1.2 Operations on Sets

Sets do not need to be explicitly described as above, but may be constructed from other sets. For example, $\mathbb{N}$ is just the positive and zero elements of $\mathbb{Z}$. There is a notation developed for this

---

[1] Some of the history can be found at http://plato.stanford.edu/entries/settheory-early.

purpose:

$$\text{derived set} = \{\textit{dummy variable from a given set} :$$
$$\textit{dummy variable } \text{satisfies some condition required for set membership}\} \tag{3.1}$$

so, $\mathbb{N}$ may be specified as:

$$\mathbb{N} = \{i \in \mathbb{Z} : i \geq 0\} \tag{3.2}$$

In English, this would be read "$\mathbb{N}$ is the set of all values of i in $\mathbb{Z}$, *such that $i \geq 0$.*" Sometimes the vertical bar ('|') is used instead of the colon.

This notation is particularly well suited for problems where there is more than one solution. In the example above, we expressed the roots of the quadratic problem $x^2 = 1$ in terms of a set. The solution could also be described more simply in terms of its roots, e.g., $A = \{-1, 1\}$. Similarly, the solution of the nth root of 1 could be expressed as $B_n = \{x \in \mathbb{R} : x^n = 1\}$ or $B_n = \{e^{2\pi i \frac{j}{n}} : j = 1, 2, \ldots, n\}$.

Note that it is not necessary to know how many roots there are to the problem with this notation, or that they even be countable. For example, $P_0 = \{\mathbf{r} \in \mathbb{R}^3 : \mathbf{r} \cdot [1, 1, 1] = 0\}$, describes the set of points in the (111) plane in a cubic crystal that passes through the origin. Here, we have used the symbol $\mathbb{R}^3$, which means that $\mathbf{r}$ is a 3-dimensional vector contained in a 3-dimensional real space. Generalizing this, we could describe each (111) plane by its index (n), describing its distance from the origin by the equation $P_n = \{\mathbf{r} \in \mathbb{R}^3 : \mathbf{r} \cdot [1, 1, 1] = n\}$.

If we were interested in the set of *all the points* in all the (111) *planes*, without losing their identity in terms of the plane to which they belonged, this would be a *collection of sets* or, simply, a *collection*. We could describe this as:

$$\mathscr{P} = \{P_n : n \in \mathbb{N}\}$$
$$P_n = \{\mathbf{r} \in \mathbb{R}^3 : \mathbf{r} \cdot [1, 1, 1] = n\} \tag{3.3}$$

where the symbol $\mathscr{P}$ is used to describe the collection. Similarly, if we were interested in all of the points in the {111} point symmetry-equivalent family of planes that pass through the origin, without losing their identities, these would be a collection, as well. This could be specified, for example, as:

$$\mathscr{V} = \{\{\mathbf{r} \in \mathbb{R}^3 : \mathbf{r} \cdot (g[1, 1, 1] = 0)\} : g \in \mathscr{G}\} \tag{3.4}$$

where $g \in \mathscr{G}$ is an element of the point group of symmetry group of the cubic system, $m3m$, for example. That is, for each element $g \in \mathscr{G}$, we form the set of all points in the gth variant of the {111} plane passing through the origin. $\mathscr{V}$ is the collection of the sets of {111} points over all elements of $\mathscr{G}$.

So far, the derived sets have been subsets of preexisting sets. The roots of equations in the previous example were subsets of $\mathbb{R}$ (set A) or $\mathbb{C}$ (set B). The crystallography examples have all derived the sets as subsets of $\mathbb{R}^3$. But sets can be constructed as combinations of existing sets, as well. For example, the points in two (111) planes can be combined to form another set. For example, if we wanted to merge the points in set P_{-1} with those in P_1, this would be a union:

$$P^* = P_{-1} \cup P_1 \tag{3.5}$$

describes the set of all points in the two (111) planes that sandwich the origin, but that does not include the points in the plain containing the origin. The symbol $\cup$ describes the union of two sets.

The points in common between two sets are formed by the intersection of the two sets, which uses the symbol $\cap$. For example, set B_2 describes the square roots of 1 and the set B_4 describes the fourth root of 1. The intersection of these two sets is:

$$B_2 \cap B_4 = \{-1, 1\} \tag{3.6}$$

This is actually just set B_2, so B_2 is a *subset* of B_4. In symbols, this is $B_2 \subset B_4$. Note that i is an *element* of B_4, because it is not a set. The symbol $i \in B_4$ would be appropriate, as would $\{i\} \subset B_4$ because, here, we have constructed a set that contains i.

We could also separate B_4 into two sets, one containing only the real entries $B_4^{\mathbb{R}} = \{-1, 1\}$ and one containing the imaginary entries $B_4^{\mathbb{I}} = \{-i, i\}$. But, we have done more than this. We have divided B_4 into two *disjoint* sets, i.e., sets that do not overlap: $B_4^{\mathbb{R}} \cap b_4^{\mathbb{I}} = \emptyset$, where the symbol $\emptyset$ denotes the *empty set*. This is a special case of a derived collection called a *partition*, in this case, $\{B_4^{\mathbb{R}}, B_4^{\mathbb{I}}\}$. A partition is a collection of disjoint sets that, when unioned together forms the original set. If we were to recombine the sets, the appropriate way to write this is:

$$B_4 = B_4^{\mathbb{R}} \cup B_4^{\mathbb{I}} \tag{3.7}$$

where the $\cup$ forms the union of the two sets.

Partitions come into image processing in a very natural way. The segmentation is an operation that forms a partition of the index set S of an image y into sets that belong to the different classes of material in the image. In image processing, these regions are typically called *labels*. For example, in an image with sharp contrast, if we estimate that a threshold of 125 will separate the matrix from the precipitate phases, we can partition S into two sets:

$$
\begin{aligned}
B_{mtx} &= \{s \in S : y_s < 125\} \\
B_{ppt} &= \{s \in S : y_s \geq 125\}
\end{aligned}
\tag{3.8}
$$

S has been partitioned into the two label sets B_{mtx} and B_{ppt} such that $S = B_{mtx} \cup B_{ppt}$.

On the other hand, suppose the contrast was not so sharp and, for two different applications, you needed the precipitate phase and the matrix phase. You did not want to underestimate the area fraction of each of these, so you set the thresholds so that each set of pixels is slightly larger than it should have been. In other words, B_{mtx} and B_{ppt} were no longer disjoint and the collection of the sets is not a partition, but still represents every pixel, although some of them are represented twice. This collection is known by the name *covering*, because it "covers" the original system. If the situation were reversed and you were conservative in the segmentation so that $B_{mtx} \cap B_{ppt} = \emptyset$, but $S \neq B_{mtx} \cup B_{ppt}$, the two sets would be disjoint, but not cover S.

3.2.1.3 Computations on Sets

The previous section described constructions of sets, but, of course, the point of sets is to perform computations on their elements. Computations on the set y are generally performed element-by-element. For example, if it were desirable to form the total of all pixel intensities, this could be expressed as:

$$y_{tot} = \sum_{s \in S} y_s \tag{3.9}$$

The number of elements of a set is generally expressed with absolute value signs, e.g., $|S|$, so an estimate of the mean intensity of an image may be computed as $\frac{1}{|S|} \sum_{s \in S} y_s$.

The use of dummy variables to represent elements of a set is typical in computations involving sets. This replaces arbitrary constructions involving counting as well as the order in which certain computations are performed. The details of, say, looping over a set, are generally left to the programmer to use whatever method is convenient.

This separation of the mathematics from the actual implementation leads to a terseness that actually makes for more readability. For example, in a traditional calculus approach, if it were desired to find the value of the minimum of a function $f(x)$, the standard approach would be to set the derivative $f'(x)$ equal to zero. That expression is either solved explicitly or implicitly to find the roots, which are then substituted back into $f(x)$ to give the set of minima. The smallest of these would be the minimum.

In set theory notation, this would be specified as follows. Let f be a mapping from $f : \mathbb{R} \to \mathbb{R}$ and find $y_{min} = \min_x f(x)$. The mapping notation means that f takes a real number as an argument and produces a real number. The implementation details of actually finding the minimum are described separately, e.g., Newton–Raphson, some gradient approach, or whatever the user chooses.

A somewhat more interesting case is when one is actually interested in the x value where f assumes a minimum. The argmin function is used in this case:

$$x^* = \underset{x}{\operatorname{argmin}} f(x) \tag{3.10}$$

That is, x^* is the value of x such that $y_{min} = f(x^*)$. This is a very familiar concept to a student of calculus, but, for some reason, no symbol is introduced to describe it in the elementary courses. The equivalent function for the maximum of $f(x)$ is the argmax.

Note that both argmin and argmax produce values that, when substituted into $f(x)$ have some desired effect. That is, they are *inverse* operations. One thing that is very different between the fields of signal and image processing and physics is that the former makes copious use of inverses, and the argmin (argmax) statements serve as keys to indicate the equations on which inverses are performed.

The argmin is commonly used on expressions involving penalty functions. One of the most common ways to construct a penalty function is to develop a metric that describes the "distance" between a known and estimated quantity. The most familiar to an engineer would be the least squares fit, which produces a *root mean square* (RMS) error.

For example, imagine that we would like to estimate an unknown image x from a measured image y, where y is formed from a linear transformation of x by a matrix A. Then we could compute the RMS error using:

$$||Ax - y||_2 = \sqrt{\sum_s (Ax_s - y_s)^2} \tag{3.11}$$

Since this is a sum of the squares of the difference in Ax and y, this has been given the name, the L_2 norm and is given the symbol $||\cdot||_2$.

The L_2-norm is commonly used, both in physics and signal and image processing and has some well-known drawbacks. The most common of these is that it smooths boundaries between discontinuities. Balancing the "energy" of the sharp jump at the discontinuity with a more gradual increase over several sample points tends to be a characteristic solution returned by the algorithm.

In cases where x is sparse, that is, x assumes some relatively small number of values, separated by jump discontinuities, the L_1-norm is more effective. The L_1-norm is defined as:

$$||f(x)||_1 \triangleq \sum_{s \in S} |f(x_s)| \tag{3.12}$$

Chapter 22 deals with the topic of *sparse reconstruction*, which uses the L_1-norm extensively. The chapter also gives a specific comparison between results using the L_1- and L_2-norms, showing the artifacts created by the latter.

3.2.2 Bayesian Probability and Image Processing

While traditional image processing methods tend to treat images as deterministic quantities, modern image processing methods often use stochastic modeling instead. This is clearly valuable in characterizing systems with random structures and noisy measurements, but actually represents a philosophical departure from the traditional approach. Images are treated as outcomes of random experiments. This follows the very successful approach of Shannon's information theory, [224], which models signals being communicated as random quantities.

Modern probability theory dates back to the first half of the 20th century. With the publication of his book *Foundations of the Theory of Probability* in 1933, Kolmogorov[26] began a revolution in the way that random phenomena are quantified. Kolmogorov's *axiomatic* approach to probability theory provides a mathematical framework that is both rigorous and practical, and forms the approach to probability that is in common use today.

Before Kolmogorov, there were two commonly used methods: (1) the *counting* approach and (2) the *relative frequency* approach. The counting approach simply defines the probability of an event to be the number of outcomes in the event divided by the total number of possible outcomes. For example, the probability of rolling an even number on a six-sided die would be $1/2$, since 2, 4, and 6 are even and there are 6 possibilities. Note that the counting approach assumes that all events that contain the same number of outcomes are equally likely. In the relative frequency approach, the probability of an event is defined as the relative frequency with which the event occurs in repeated independent trials of an experiment. For example, if we are interested in the accuracy of a test on a known standard, we can run n tests and count the number of times that the correct result was obtained (n_c). The relative frequency of success is $\frac{n_c}{n}$. Practically, this number will be dependent upon the specific experiments run, but it can be extended using an *asymptotic relative frequency*, $\lim_{n\to\infty}\frac{n_c}{n}$.

Both these approaches have important limitations. The counting approach assumes that the probability of an event depends only on the number of elements in the event relative to the total number of outcomes, which is often not a reasonable assumption. It also requires that all possible outcomes of an experiment be enumerated, which can severely limit its applicability. The relative frequency approach becomes impractical for complex systems with many random variables. Also, it offers limited modeling capabilities, since all probabilities are defined in terms of relative frequencies. Kolmogorov's axiomatic approach is based on a set of very general rules that, by virtue of their simplicity, provide a much more general and flexible way to treat probability problems.

3.2.2.1 Modern Probability and Sets

In modern probability theory, a random experiment is defined mathematically in terms of sets. The set of *outcomes* is the set of possibilities that could occur from the experiment. A textbook example would be throwing a die, in which case, the set of possible outcomes would be $S = \{1,2,3,4,5,6\}$. Equivalently, an image may be of a microstructure containing multiple phases, say $\alpha, \beta, \gamma, \alpha', \mu$, in which case, you could define $S = \{\alpha, \beta, \gamma, \alpha', \mu\}$. If you did an experiment to determine which phase is under a certain probe position (and did not have any activation volume effects), the outcome of that experiment would be one, and only one, of those phases.

An *event* is a *set of outcomes*. So, in the example above, a particular image may only contain α and β phases. This describes the event that the phases observed were from the set $\{\alpha, \beta\}$. The distinction between an outcome and an event is that an outcome is a single realization and an event is a *class of realizations*. In the above example, if you had a set of images and you divided them into those that showed only α and β phases and those that did not, you would have partitioned S into two events: the event that the image contained only those two phases and the event that it contained something else. In a more pragmatic sense, the microscope gives us outcomes of experiments, and that is all. When we recognize the "microstructure" that is represented by all of the observations, we are working with the event that we have observed the microstructure. The human mind forms the perception of an event from a set of outcomes so effortlessly that we barely even notice that we are doing it. But computers usually get stuck with comparing outcomes and, being unable to form the abstraction, find themselves insisting that two images of the same microstructure are radically different because the pixels are not identical.

In modern probability, the probability is associated with the *events*. The association between events and probability is the *probability mapping*. The axioms describing the properties that *all* probability mappings must satisfy comprise the four axioms of modern probability theory. Combined with the famous Bayes' rule, which follows from the axioms, this forms the basic foundation of the probability methods used in this book.

3.2.2.2 Foundational Rules of Modern Probability

Probability Axioms.   The probability mapping has been generalized in terms of a set of axioms that allow much more freedom in the assignment of probabilities. These axioms are:

1. $P(A) \geq 0$ for any event A,
2. $P(S) = 1$, where S is the set of all possible outcomes, and
3. $P(A \cup B) = P(A) + P(B)$ for any events A and B for which A and B are disjoint sets.
4. The property stated in Axiom (3) for two sets A and B also applies for a countably infinite number of sets that are pairwise disjoint.

Axiom (4) is necessary since, by simple induction, it is possible to show, using Axiom (3), that the probability of the union of any finite number of disjoint sets is the sum of their individual probabilities, but this is not sufficient to guarantee this property if there is an infinite number of sets (e.g., the collection of sets $\{A_n\}$, $n \geq 1$, where A_n is the event that the concentration of a given component of a certain material is $1/n$).

Conditional Probabilities and Bayes' Rule. Bayes' Rule applies to *conditional probability*, that is, the probability of an event occurring *given that* another event also occurred. Note, this is distinctly different from the probability that both events occurred at the same time–that would be the *joint probability*. The joint probability of tossing a (fair) coin twice and having it land heads-up twice is $\frac{1}{2} \cdot \frac{1}{2} = \frac{1}{4}$, but the probability that the coin lands heads-up the second time, given that it did the first time (if the tosses are statistically independent) is still $\frac{1}{2}$. Statistical independence is a common assumption made: many classical *mean field* approximations, such as the *regular solution model* [606], were developed before the advent of modern computing and use this approximation. Today, more accurate results are obtainable by accounting for statistical dependencies.

This is particularly valuable when you have some *prior knowledge* of the system, outside of the experiment just run. For example, if you want to segment particles from a matrix, a simple threshold will provide some sort of answer, but the boundaries of the particles will likely be jagged. We know that is incorrect if we also know that the interfacial energy is relatively high and the sample is well annealed. Bayesian statistics allows the researcher to incorporate this knowledge into the solution, so as to form a balance between what was just measured and what was already known.

Bayes' rule is a straightforward implication of the axiomatic approach and the definition of conditional probability, which is given by:

$$P(B|A) \triangleq \frac{P(A \cap B)}{P(A)} \tag{3.13}$$

for events A and B. Note the complete symmetry between this relation and that for the conditional probability $P(A|B)$:

$$P(A|B) \triangleq \frac{P(A \cap B)}{P(B)} \tag{3.14}$$

Eliminating the unwanted $P(A \cap B)$ gives Bayes' rule:

$$P(B|A) = P(A|B)\frac{P(B)}{P(A)} \tag{3.15}$$

In Equation (3.15), $P(A|B)$ is often referred to as the *likelihood*; $P(B)$ is referred to as the *prior* or the *belief*. The denominator is often referred to as the *evidence*, but in many cases in image processing its value does not change during solution, in which case it can be ignored. The term on the left is referred to as the *posterior probability*.

One final note: the probability estimates given in Bayesian probability can describe different things than those given in the frequentist probabilities most often used in physics. In physics, like quantum mechanics, a phenomenon is measured and it is found that a particular state is occupied with the measured probability. This would be a phenomenological probability because it describes the probability of observing a phenomenon. Bayesian probabilities can be phenomenological, but can also include a measure of the "reasonableness" of an estimate. For example, if the prior knowledge that "the Third Law of Thermodynamics is obeyed," is used as a prior, the Bayesian probability could be interpreted as a "probability that the solution reflects the phenomenon." That is, the estimate would almost never be *exactly* the right answer, but a large deviation from the Third Law would lower the Bayesian probability (through the prior), and thus our confidence in the legitimacy of the estimate.

3.2.2.3 Mathematical Constructs

Before proceeding to the discussion of Bayesian methods for image processing, it will be helpful to describe two mathematical constructs: (1) *random variables* and (2) *probability distributions*. With these, it is possible to develop the methods without explicit reference to sets and to proceed in a more natural calculus-based approach.

Random Variables.   The results above were obtained by explicitly using sets, but a calculus-based approach requires that we use variables. Instead of referring to an event A, we can use a variable X that can take on some value from a set A every time an experiment is run. For example, if we were interested in classifying the pixels in an image containing α, β, and μ phases, and if X represents the random phase at a given pixel, then X can take on the values from the set $\{\alpha, \beta, \mu\}$. If we were interested in estimating the maximum temperature a mixture experienced during an unusual thermal event, X could represent this temperature, $T_{\max}$.

This last example draws the distinction between *discrete random variables* and *continuous random variables*. Discrete random variables take on countable values that we could index with a variable i, say. Continuous random variables take on values that cannot be numbered. Quantities such as phases, crystal structures, crystallographic variants, or magnetic domains, for example, would be modeled as discrete variables. Most of the thermodynamic variables, such as temperature, pressure, chemical potential, entropy, volume, or composition, would be modeled with continuous random variables.

Probability Distributions.   The probability distribution of a random variable can be viewed as a law that governs the stochastic behavior of the variable. Any probability of interest related to a random variable can be computed from its probability distribution. There are several ways to represent the probability distribution of a random variable. Two of them, the *probability mass function* and the *probability density function*, are described in this section.

When working with random variables, the concept of events is still important, but events take a special form. If X is a discrete random variable, then a common type of event of interest is one of the form $A_x = \{X = x\}$ for some value x that the random variable can take. For example, if X is the classification of a pixel example above, $A_\alpha = \{X = \alpha\}$, $A_\beta = \{X = \beta\}$, and $A_\gamma = \{X = \gamma\}$. The event A_x is different for different values of x, so $P(X = x)$ can be viewed as a function of x, and is given a special name: the *probability mass function* (pmf) of X. The pmf of X is often written as $p_X(x)$.

The above paragraph has a number of subtleties that can be missed on first reading. First off, the random variable X is stochastic, meaning that we never actually know what value it would have in a particular experiment. All we know is that the result will be one of the allowed values of X. Conventionally, the random variable is referred to in terms of a capital letter and a particular outcome as a lower case, in this case, x. So the lower case x represents one arbitrary instantiation of the random variable X. The probability mass function $p_X(x)$ may appear redundant, having both the subscript and the argument being the same variable. But, note from the above that $P\{X = x\} \triangleq p_X(x)$. *The subscript simply indicates which random variable the pmf belongs to, whereas the argument is a value that that random variable can take.* In fact, this notation is slightly more compact than the set notation because the $=$ sign is omitted, but all other symbols are present. Omitting the subscript would be the equivalent of asking, "What is the probability that the temperature is 300°?" That would beg the question, "From what random variable? The temperature of a thermocouple in a furnace? the thermometer on the wall?" In cases where it is clear from context which random variable is referred to, the subscript may be dropped from the notation.

If X is a continuous random variable, it does not have a probability mass function. Instead, its distribution is typically written as a probability density function, or pdf. In this case, probabilities are computed by integrating the density function. The value of the density function, which will also be denoted here as $p_X(x)$, is *not* a probability itself for a fixed value of x. The function p_X must instead be integrated over a region to find the probability that X lies within that region.

From this slight detour, we can now proceed to the description of Bayesian image processing, as it is commonly developed.

3.2.2.4   *Bayesian Probability in Image Processing*

Typically, with Bayesian image processing, the likelihood is estimated from the image being processed, and some sort of prior is used to influence the result using prior knowledge. The true solution generally cannot be known, so some sort of *estimation criterion* is applied to define the *most likely* solution(s) to the Bayesian problem. Therefore, three elements of Bayesian image processing are (1) the estimation criterion, (2) the likelihood model, and (3) the prior model. These are discussed briefly here. It is not intended that this treatment be completely comprehensive, but, rather, to give some introduction to the more commonly used techniques.

Estimation Criteria.   Consider two random variables, X and Y. Assume that a value for Y can be measured during an experiment, but X is a "hidden" variable whose value cannot be observed or computed directly from a given measured value of Y. For example, Y might be the interface roughness of a given observed microstructure, and X might be its interfacial energy. According to the philosophy of modern probability, it is not possible to "know" what X is, but many ways exist for estimating X from a particular instantiation of Y. An *estimation criteria* can be used to define the *most likely* value of X given that $Y = y$ for some measured data y.

Two of the most common methods used for estimating the solution X are the *maximum-likelihood* (ML) and the *maximum a posteriori* (MAP) estimates. The MAP and ML estimates generally provide different solutions; they are only equal under certain conditions.

The posterior distribution in Equation (3.15) may be written in terms of a probability distribution, letting Y be the random variable corresponding to event A and X that corresponding to the event B:

$$p_{X|Y}(x|y) = p_{Y|X}(y|x)\frac{p_X(x)}{p_Y(y)} \tag{3.16}$$

This is the form of Bayes' rule for random variables X and Y. Each of the four functions in this equation can be either a probability mass function or a probability density function. The choice should be apparent from the problem. For example, if an absorption coefficient is to be determined, it is modeled using a density function. If a pixel classification, such as in a segmentation, is to be determined, it uses a mass function.

The ML estimate of X given Y is defined as:

$$\hat{X}_{ML} \triangleq \underset{x}{\mathrm{argmax}}\, p_{Y|X}(y|x) \tag{3.17}$$

That is, we search over all possible inverse solutions (x) to find the one that maximizes the likelihood of the observed y, given that $X = x$.

The following example will make this somewhat more concrete. Consider an intensity histogram with two peaks, which was generated from an image of a two phase mixture. If the broadening of the peaks was due to shot noise, the histogram can be modeled as a mixture of Gaussians, using two Gaussians, one representing each phase, with the separation being due to the different μ values. If we observe the intensity at a given pixel to be y, we can compute the ML estimate of the classification X at that pixel using Equation 3.17.

The two classifications we can make are α phase or β phase. So the argmax is taken over these possibilities with $p_{Y|X}(y|\alpha)$ being the Gaussian that models the α phase intensity and $p_{Y|X}(y|\beta)$

being that which models the β phase intensity. Whichever Gaussian has the higher likelihood for intensity y is the ML estimate of the classification.

The MAP estimate is a similar idea, except that it operates on the posterior likelihood:

$$\hat{X}_{MAP} \triangleq \underset{x}{\operatorname{argmax}}\, p_{X|Y}(x|y) \tag{3.18}$$

The MAP estimate is particularly suited for Bayesian problems because Bayes' rule may be directly substituted:

$$\hat{X}_{MAP} = \underset{x}{\operatorname{argmax}}\, p_{Y|X}(y|x)\frac{p_X(x)}{p_Y(y)} \tag{3.19}$$

Since $p_Y(y)$ is not a function of x, the solution of Equation (3.19) is the same as the solution to:

$$\hat{X}_{MAP} = \underset{x}{\operatorname{argmax}}\, p_{Y|X}(y|x)p_X(x) \tag{3.20}$$

To illustrate the Bayesian approach, we consider the example in which interfacial energy is to be estimated from the interface roughness of an observed microstructure. In this case, the interfacial energy is a random variable, Γ, and the interface roughness a random variable, S. The MAP estimate would be obtained using

$$\hat{\Gamma}_{MAP} = \underset{\gamma}{\operatorname{argmax}}\, p_{S|\Gamma}(s|\gamma)p_\Gamma(\gamma) \tag{3.21}$$

The problem of determining $p_{S|\Gamma}(s|\gamma)$ and $p_\Gamma(\gamma)$ is the *modeling* problem. In general, modeling is performed using a combination of relative frequencies obtained through experiments (often in the form of histograms), counting methods, and prior knowledge by domain experts. In this case, suppose that results from various experiments have suggested that Γ can be modeled as being Gaussian distributed. This means that $p_\Gamma(\gamma)$ is a Gaussian density function with some mean μ_Γ and standard deviation σ_Γ. Now for a fixed value of γ, the likelihood function for that value might be determined by drawing samples, using Monte Carlo sampling, of the microstructure with interfacial energy γ, and computing a histogram of the resulting interface roughness. To complete the modeling in this way, the likelihood would need to be evaluated for varying values of γ. Determination of the effectiveness of this approach to modeling the likelihood for this problem is part of the problem of model validation, which is not discussed further here.

A relationship between the ML and MAP estimations can be seen by comparing the right sides of Equations (3.17) and (3.18). The ML estimate maximizes the likelihood ($p_{Y|X}(y|x)$), whereas the MAP estimate maximizes the product of the likelihood and the prior, ($p_{Y|X}(y|x) \times p_X(x)$). If the prior is uniformly distributed, that is, $p_X(x)$ is constant over its interval of support, then the MAP estimate of X becomes

$$\hat{X}_{MAP} = \underset{x}{\operatorname{argmax}}\, c\,\frac{p_{Y|X}(y|x)}{p_Y(y)} \quad \text{(uniform prior)} \tag{3.22}$$

Again, since neither c nor $p_Y(y)$ is dependent upon x, this is equivalent to:

$$\hat{X}_{MAP} = \underset{x}{\operatorname{argmax}}\, p_{Y|X}(y|x) \quad \text{(uniform prior)}$$
$$\triangleq \hat{X}_{ML} \tag{3.23}$$

The human mind is notorious for making this approximation. For example, if a fisherman caught bass with overwhelming likelihood whenever he used a particular bait, he is likely to conclude that the likelihood that the bass would bite on that bait was overwhelmingly high. But, this is saying $\operatorname{argmax}_b p_{F|B}(\text{bass}|b) = \operatorname{argmax}_b p_{B|F}(b|\text{bass})$, where F is a random variable describing which kind of fish was caught and B is a random variable describing which kind of bait was used. Other examples are when a failure analyst observes a certain defect with high likelihood in the presence of a crack and concludes that that defect *causes* the crack or when a person observes that a certain crime is committed with higher likelihood by people of a certain group and concludes that a given person from that group is thus likely to commit that particular crime.

There are two general approaches to computing the MAP estimate of X once the models for $p_{Y|X}(y|x)$ and $p_X(x)$ have been determined. The first approach treats the optimization of $p_{X|Y}(x|y)$ as a classic optimization problem over the variable x, and the second approach uses Monte Carlo methods to draw samples from the posterior distribution $p_{X|Y}(x|y)$. For computational reasons, it is often easier to find the x that maximizes the log of the posterior distribution itself. Since the same x maximizes both $p_{X|Y}(x|y)$ and $\log p_{X|Y}(x|y)$, the MAP estimate can be computed as

$$\hat{X}_{MAP} = \operatorname*{argmax}_x \log \frac{p_{Y|X}(y|x)p_X(x)}{p_Y(y)} = \operatorname*{argmax}_x \log p_{Y|X}(y|x) + \log p_X(x), \qquad (3.24)$$

where, again, the term $p_Y(y)$ can be ignored because it does not depend on x.

In Bayesian image processing, the random variable Y becomes a *random field*, which is a collection of random variables defined on a 2D or 3D lattice, or possibly as a vector of random variables whose dimensionality is the number of pixels in the image. The "solution" X can be a denoised version of the image, a segmentation, a reconstruction from multiple views, or any of a number of other useful constructs. This is also typically represented as a random field. The MAP estimate is still defined as in Equation 3.18, but the X and Y are now random fields. Finding effective likelihood and prior models can be much more difficult in this case, but the signal processing community has made great progress in this area.

Likelihoods in Image Processing. A simple form of a likelihood function can be written by expressing the histogram of intensities as a *mixture of Gaussians*, where each Gaussian represents a peak in the histogram. In our earlier example, each Gaussian represents a particular phase, and the width of the Gaussian depends on the noise variance in the imaging system. A highly idealized case would be where this noise is entirely *shot noise*, or the counting statistics noise. Each peak actually represents the *Poisson distribution* of photons observed by the CCD. For more than, say, 10 or 20 photons, the Poisson distribution may be approximated by a Gaussian with μ representing the ideal intensity generated by that phase, absent the measurement noise. Geman and Geman[355] used this approach to form a likelihood for an image and a Markov random field (see next section) to obtain regularized segmentations of images.

More sophisticated likelihoods can be generated from a forward model of the image formation. From a presumed geometry, a forward model can be used to estimate what the observed signal would be in that case. This is an extremely common approach in medical imaging, where Beer's law is used as a forward model. Chapter 6 uses Beer's law for a forward model, but also adds a Gaussian to approximate the shot noise. Chapter 4 gives a considerably more accurate forward model for image formation in electron microscopy.

Priors in Image Processing. One of the most common classes of priors used is the *Markov random field* (MRF), a common form of which is familiar in physics as the equilibrium Ising or Potts model. That is, if the pixels are assumed to interact with an exchange energy that favors like neighbors and penalizes unlike neighbors, an MRF model should be appropriate for the image. In this case, a Monte Carlo algorithm that draws samples from the MRF can be used to generate a sequence of approximations of the correct solution. If this is allowed to continue for a long enough time, the system will simply be traveling through a phase space of equilibrium states. The MRF is an extremely popular prior in Bayesian image processing because of the *Hammersley–Clifford theorem*, [405, 87] which says that if the "energy" of a system can be written in terms of only local pixel interactions (of arbitrary distance), then the probability of observing a state x with energy $E(x)$ is Boltzmann distributed. That is, $P(x) = \frac{1}{Z}e^{-\beta E(x)}$, where Z is the classical partition function and $\beta \triangleq 1/kT$ is the reciprocal of the product of Boltzmann's constant and some "temperature." The prior model may then be written, based on, say, the pixels in the local neighborhood of a given pixel being classified in a segmentation. This is the approach used, for example, by Comer and Delp [219] for segmentation.

The intimate connection between the MRF and the Potts model makes this form of regularization ideal for reconstruction of structures formed by diffusional phase transformations. This was shown qualitatively by Simmons, *et al.* [934]. In Chapter 10, Tang and Luo give many examples of such interfaces that form in real systems. MRF regularized reconstructions will mimic these structures because of this connection.

Another common prior is the *sparsity prior*, in which an "energy" is assigned to the L_1-norm of the solution. This is an extremely popular prior when it is known that the sample contains a small, discrete number of constant values, for example. This is the case for a multiphase mixture at equilibrium, where the composition is a (discrete, constant) value within a "phase," or a polycrystalline sample, where the Euler angle triplets are constant over areas of the image.

Sparsity in reconstructions is discussed in greater detail by Romberg (Chapter 22). Willett uses it implicitly for denoising (Chapter 21) as well as in a limiting case of the Model Based Iterative Reconstruction (MBIR) tomographic reconstruction by Venkatakrishnan (Chapter 6).

3.3   Language of Materials Science: Thermodynamics

The mathematical methods of the language of the signal and image processing field are largely foreign to the materials science field. In the same sense, materials science has its own language that an imaging processing specialist may find impenetrable at first. The two fields that stand out are thermodynamics and crystallography. For space considerations, it was decided to focus on thermodynamics, being one of the classical engineering disciplines that is no longer taught in all fields of engineering. Chapter 10 gives structural observations that result from thermodynamic considerations. This chapter gives the salient features of the classical thermodynamics in an attempt to clarify some of the concepts that are second nature to a materials scientist. Chapter 10 deals with a specialization of this field, that of thermodynamics of interfaces.

One of the key concepts of thermodynamics, as it is applied typically in materials science, is that of a *phase*. This is a concept that is so foundational to the field that its origin is often forgotten: it becomes as a foreign word, itself. The next section reiterates the origin of this concept.

3.3.1   Thermodynamic Phases

The subject of this book, microstructure characterization, derives from the concept of *microstructure*, which is also a term that is so fundamental as to never be given a definition. But, it is a familiar concept in the more mathematical fields, being the *texture* of a material. That is, structures are stochastic and the description of the random processes by which structures are produced is the texture. In materials science, the word "texture" usually is used to describe a subset of this description: the orientation distribution of crystals. The term "microstructure" is generally used for the full description of textures.

Textures, or microstructures, in materials are often sparse. In the limiting case of thermodynamic equilibrium, they are extremely sparse because the structure is only composed of a very small number of uniform regions. These uniform regions are generally termed *phases*.

The term *phase* was originally coined by Gibbs in his classic work "On the Equilibrium of Heterogeneous Substances:" [362]

In considering the different homogenous bodies which can be formed out of any set of component substances, it will be convenient to have a term which shall refer solely to the composition and thermodynamic state of any such body without regard to its quantity of form. We may call such bodies as differ in composition or state different *phases* of the matter considered, regarding all bodies which differ only in quantity and form as different examples of the same phase.

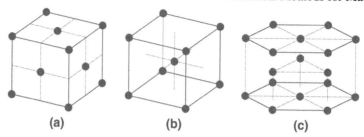

Figure 3.1: Common crystal structures found in metallic systems. (a) FCC, (b) BCC, (c) HCP .

In materials, the word "phase" is used somewhat more loosely. It retains the idea that Gibbs used, but the phase really describes an equivalency between homogeneous bodies. If, by changing the composition, the thermodynamic state of one homogenous region may be varied to where it is identical to that of another, the two regions are said to be "the same phase." If, by changing the pressure or temperature, the thermodynamic state of one region may be continuously varied until it is identical to another, it often would be referred to as the "same phase," the exception being if this only occurred along a single path. In that case, it would be referred to as a *phase change*, borrowing from the theory of phase transitions.[532] The phase change that would occur continuously along this single path is referred to as a *second-order phase change*. The paramagnetic to ferromagnetic phase change that occurs in Fe is a classic example of a second-order phase change.

Normally, in crystalline solids, phases refer to different crystal structures in which a material could form. The three classic crystal structures that form in metallic systems are *face centered cubic* (FCC), *body centered cubic* (BCC), and *hexagonal close packed* (HCP), and are shown in Figure 3.1.

Each of these involves (ideally) atoms being placed at the points of these three distinct lattices. Each lattice is an element of a group, a *space group*, each with its own characteristic group of elements. Being discrete groups, it is not possible to distort one continuously into another. Within a structure, having two different crystal structures, there are at least two phases, one for each of these crystal structures.

But, it is possible to have two separate phases, in the materials science sense. There are energetic reasons why a material crystallized in one phase may become unstable at, say, that composition. In this case, it will separate into two phases of the same crystal structure. Sterling silver is an example of this. Sterling silver is an alloy of Ag and Cu. Both of these, in pure form, crystallize in an FCC structure. But, for energetic reasons, it is not possible to vary the composition from pure Ag to pure Cu without a discontinuous change in the equilibrium composition. It is actually intentional that sterling does this, since the two different compositions of FCC have slightly different volumes, inducing a strain in the structure. This strain serves to essentially create friction when the sterling deforms: the Cu is added to harden it.

Thermodynamic Origin of Uniform Phases. In microscopy, regions of uniform composition and structure are often colloquially referred to as "phases," which fits, more or less, the Gibbs definition. The underlying, and often unstated, assumption is the system is in equilibrium or nearly so. The implications of specifying equilibrium are enormous, since that allows for a tremendous reduction in the dimensionality. *À priori*, the dimensionality of the state of a material is six times the number of atoms in the system, three being position coordinates and three being momentum coordinates. Equilibrium reduces this to a "handful" of parameters that actually need to be tracked.

The specification of equilibrium imposes an equivalence relation on all of the states that may occur at equilibrium. A classic result of statistical mechanics is that, in a large ensemble of duplicate systems, the probability that the system will be in state i, where the energy of that state is E_i will be

Gibbs (AKA Boltzmann) distributed in energy:

$$p_i = \frac{e^{-E_i/k_b T}}{Z} \tag{3.25}$$

where p_i is the probability that the system is in state i, k_b is a constant, Boltzmann's constant, and Z is the *partition function* that normalizes the probabilities to sum to 1. T is the *thermodynamic temperature*, defined as:

$$T \triangleq \left(\frac{\partial U_{eq}}{\partial S_{eq}} \right)_V \tag{3.26}$$

where U_{eq} is the macroscopic "system energy" at equilibrium, S_{eq}, is the Boltzmann entropy at equilibrium, and the subscript indicates that the path of differentiation is one of constant V.

The term $1/k_b T$ occurs so often that it is generally given the symbol $\beta \triangleq 1/k_b T$, so the Gibbs distribution is more commonly written as:

$$p_i = \frac{e^{-\beta E_i}}{Z} \tag{3.27}$$

Equation (3.27) can be rearranged to give an equivalence relation among all of the states at equilibrium:

$$p_i e^{\beta E_i} = \frac{1}{Z} \tag{3.28}$$

Since the partition function is a sum over all states of the system at equilibrium, it is independent of the population of any one state and, for a given value of Z, states i and j are equivalent if:

$$p_i e^{\beta E_i} = p_j e^{\beta E_j} = \frac{1}{Z} \tag{3.29}$$

Equation (3.29) has the properties of reflexivity, symmetry, and transitivity, making equilibrium an equivalence relationship among all of the states of the system. This forms a definition of what Gibbs described as the "thermodynamic state," from which the term *phase* was defined.

The Partition Function in Thermodynamics. The partition function is actually more than a normalization constant. It constitutes a *fundamental relation*, from which all of the thermodynamic information of the system may be derived.[161] For example, the internal energy may be determined from Z by the action of a specific operator. The partition function is defined as:

$$Z \triangleq \sum_{i \in \mathscr{I}} e^{-\beta E_i} \tag{3.30}$$

where $\mathscr{I}$ is the set of all equilibrium states of the system. In statistical mechanics, by hypothesis, the internal energy of the macroscopic system is equal to the expected value of the energy over the set of all equilibrium states:

$$U = \sum_{i \in \mathscr{I}} E_i p_i \tag{3.31}$$

Substituting Equation (3.27) into Equation (3.31) and rearranging slightly gives:

$$U = \frac{1}{Z} \sum_{i \in \mathscr{I}} E_i e^{-\beta E_i} \tag{3.32}$$

which may be derived from a term-by-term differentiation of the partition sum in:

$$U = -\frac{d \ln Z}{d\beta} \tag{3.33}$$

Equation (3.33) gives the thermodynamic internal energy as a function of the partition function. Relationships such as this exist for all thermodynamic variables of the system.[161]

Dimensionality Reduction. This gives us one important result: since the partition sum is independent of the population of any particular state, the dimensionality of the partition sum is just that of the macroscopic thermodynamic state. The number of dimensions of the thermodynamic state is the number of thermodynamic variables. In a chemical system, where the thermodynamic state is determined by the internal energy, volume (V), and the compositions ($x_i, i = 1, 2, 3, \ldots c$, where c is the number of components), the total dimensionality of the thermodynamic state is $|\{U, V, x_1, x_2, \ldots x_c\}| - 1$, since the compositions sum to one.

This represents an enormous reduction in dimensionality. For example, if we have a system that does not change volume or allow energy or chemical elements to enter or leave, there is only one thermodynamic state: it is fixed by the internal energy, volume, the compositions, *and the specification of equilibrium.* Its dimensionality is 0, existing at a specific point in the thermodynamic space. If the volume were allowed to change, but otherwise the equilibrium condition were maintained, the dimensionality would be 1. The process of changing a thermodynamic variable while maintaining the equilibrium state is the *reversible process* favored by many thermodynamic analyses and is necessary for the dimensionality reduction to apply. For a mole of material, consisting of 6.02×10^{23} atoms, specifying equilibrium amounts to a reduction in dimensionality of $6 \times 6.02 \times 10^{23} : 1$! If the system is perturbed in a way that does not maintain equilibrium (the *irreversible process*), then the dimensionality will be higher. In practical terms, materials are *annealed* when they have undergone an irreversible process in order to return them to the better behaved "near equilibrium state."

The preceding argument applies if there is only one phase of the material present. In that case, the total dimensionality of the system is $2 + c - 1$, where the 2 comes from the fact that we could change U and V reversibly and c comes from the number of components, each of which being reversibly variable, as well. The subtracted 1 is for the sum-to-one constraint on the compositions. For each additional phase added, the dimensionality will change by -1, since this adds a constraint to the equilibrium. To show this will require a discussion of the *free energy*, which is the macroscopic description of the fundamental relation and will be addressed in the next section. Essentially, the argument is that the gradients in the free energy between two phases must be equal, where the dimensions of the gradient are the thermodynamic variables. Suffice it to say, at this point, that the relationship between the dimensionality f (historically for "freedom," as in "degrees of freedom"), the number of components c, and the number of phases p that can be in equilibrium at once is:

$$f = c - p + 2 \tag{3.34}$$

This is the famous *Gibbs phase rule* that gives a very general sparsity relation: if a c-component material consists of an equilibrium mixture of p phases, the dimensionality is given by Equation (3.34), irrespective of the amount of material there.

3.3.2   Free Energies

It is nearly impossible to work with a materials scientist without the topic of "free energy" being raised at some point. This term originated from the early days of the development of thermodynamics and is, loosely speaking, the energy available to do work in a reaction. The concept itself is mathematically justified, but engineering students are not generally subjected to the full rigor. The end result is that engineers can usually apply the methods, but have difficulty explaining to others, exactly where the concepts originated and why one free energy should be used in favor of another. This section gives a brief description of free energies from a more rigorous standpoint than that needed by a practitioner in an attempt to make the literature more transparent to an outsider when concepts of chemical equilibrium are discussed.

The Helmholtz Free Energy. In its original formulation, the first and second laws of thermodynamics were enunciated by Rudolf Clausius:[211]

The energy of the world is constant.
The entropy of the world aspires to a maximum.

The second of these implies that $\delta S \leq 0$ for any system at equilibrium, where S is the *Boltzmann entropy*, which is the Shannon entropy, but with equal probability weighting to each state. Clausius' statements defines an entropy, S_{eq}, which is that limiting maximal entropy to which the classical thermodynamic texts on chemical equilibrium refer as S. This is, essentially, the entropy of the convex hull of the entropies of the set of *possible* states within which the system could exist. The field of non-equilibrium thermodynamics deals with these *non-equilibrium* states, but that area is still a specialization of the field. This section discusses the classical equilibrium thermodynamics, which are those that obey the equivalence relation in Equation (3.29).

To make Clausius' statement a bit more rigorous, if the thermodynamic state were subjected to an arbitrary perturbation, ϕ, over the manifold of equilibrium states, we could consider S to be dependent on $\alpha\phi$, where $\alpha \in [0, \infty)$. Then, by Clausius' hypothesis that S_{eq} is a maximum value if U is held constant:

$$\left(\frac{\partial S_{eq}}{\partial \alpha}\right)_U \Bigg|_{\alpha=0} = 0 \tag{3.35}$$

The subscripted parentheses is a standard notation in thermodynamics, indicating that U was held constant when performing the partial derivative. $\alpha = 0$ at equilibrium because, if it did not, there would be another thermodynamic state with greater entropy and S_{eq} would not actually refer to the equilibrium state.

An equivalent result could have been obtained by using a Lagrangian multiplier:

$$\frac{\partial S_{eq}}{\partial \alpha} - \lambda \frac{\partial U_{eq}}{\partial \alpha} = 0 \tag{3.36}$$

where λ is the Lagrange multiplier and the subscript on U also means that it is only defined on the manifold of equilibrium states.

λ, in the above expression may be shown to be the reciprocal of the thermodynamic temperature, defined in Equation (3.26). Solving for λ:

$$\lambda = \frac{\partial S_{eq}}{\partial \alpha} \left(\frac{\partial U_{eq}}{\partial \alpha}\right)^{-1} \tag{3.37}$$

Since both derivatives are over the same path, all variables other than α may be regarded as parameters and this can be simplified to:

$$\lambda = \left(\frac{\partial U_{eq}}{\partial S_{eq}}\right)^{-1} \tag{3.38}$$

If the volume is constrained during the differentiation, $\lambda = 1/T$, as defined in Equation (3.26), so that:

$$\lambda = \left(\frac{\partial U_{eq}}{\partial S_{eq}}\right)_V^{-1} = 1 / \left(\frac{\partial S_{eq}}{\partial U_{eq}}\right)_V = \frac{1}{T} \tag{3.39}$$

If Equation (3.36) is modified so that the perturbation is constrained to constant volume, so we could consider S and U to be functions of $\alpha'\phi'$, where the primes indicate constant volume, then we have:

$$\frac{\partial S_{eq}}{\partial \alpha'} - \lambda \frac{\partial U_{eq}}{\partial \alpha'} = 0 \tag{3.40}$$

as a necessary condition for equilibrium, according to Clausius' statements above. Formally, this suggests a new function, A, defined as:

$$A \triangleq U - TS \tag{3.41}$$

which would achieve a minimum at equilibrium over any constant volume, isothermal perturbation.

A is defined as the *Helmholtz free energy*, the zero derivative being a necessary condition for equilibrium for systems where the volume is constrained to be constant and the temperature is a (constant) parameter. Minimizing $A(T,V)$ with T constrained is equivalent to maximizing $S(U,V)$ with U constrained in the sense that equilibrium could be found by either approach. Minimizing $A(T,V)$ is more convenient where fixing the temperature is more convenient to use because regions in thermal equilibrium all would have the same temperature. Constraining the energy to stay in one region of a sample would be problematic.

The salient point of this section is that the equilibrium free energies represent the convex hull of all free energy states in which the system may be. In the argument above, each of these states can be modeled as being accessible by some continuous perturbation by $\alpha\phi$, where all of the α values must be zero at equilibrium. In Clausius' original statement, there are many ways the entropy may not reach a maximal state such as vibrational modes, atomic arrangements, or inhomogeneities that could be modeled this way. The free energy formulation, essentially "solves the equation" $S(U,V)$ for U to give $U(S,V)$. In this relationship, the equilibrium condition becomes the state that gives a minimum in U. But, it also transforms one of the coordinates from S to T, essentially allowing us to treat S as just another perturbation which must be minimized in order to reach equilibrium.

Independent Variables. According to this argument, the coordinates have been changed from (U,V) to $(1/T,V)$ in Equation (3.40) and to (T,V) in Equation (3.41) through a somewhat informal discussion. There was an implicit assumption that S_{eq} is bijective mapping onto U_{eq}, allowing for one variable to be "solved" in terms of the other. The Lagrange multiplier in Equation (3.35) was used to perform another change of variables from a dependency of U_{eq} on S_{eq} to a dependency on $(\partial S_{eq}/\partial U_{eq})_V$! The first of these assumptions is justified by the hypothesis that $S_{eq}(U_{eq})$ is an increasing function. The second, in a more formal treatment, is accomplished by the Legendre transform.[354]

The Legendre transform of a function $f(\xi)$ is defined as:

$$H(p) = \max_{\xi}\left[-f(\xi) + p\xi\right] \tag{3.42}$$

where $H(p)$ is the Legendre transform and $p \triangleq \frac{df}{d\xi}$. That is, it transforms the dependency on one variable to the dependency on its derivative with respect to that variable. In the above example, the dependency of S_{eq} on U_{eq} was transformed to a dependency on $1/T \triangleq (\partial S_{eq}/\partial U_{eq})_V$. By hypothesis, the thermodynamic variables are smooth,[2] so the max operator in Equation (3.42) is equivalent to a vanishing derivative of the expression.

This raises the question, "what are the independent variables of the thermodynamic variables?" By hypothesis, entropy is explicitly a function of all of the variables that scale with the size of the system, e.g., U_{eq}, V, numbers of atomic species, and any other variable that scales with the size of the system. These variables are referred to as *extensive variables*. Equivalently, energy is explicitly a function of S_{eq}, with the other extensive variables being the same as those for entropy.

The derivatives of the internal energy with respect to these extensive variables play an important role in thermodynamics, being analogous to generalized forces. This is discussed further in the next section, but, for now, it is necessary to develop some notation for describing the various derivative.

To make the discussion manageable, we restrict the set of extensive variables to be just S_{eq}, V, and the numbers of each atomic species, $n_1, n_2, n_3, \ldots, n_c$. This is the normal assumption for the fluid-based thermodynamics that is assumed to apply to solids in materials science. Other extensive variables exist and are used for specific material systems, e.g., magnetization.

Defining the set of thermodynamic extensive variables as $\mathscr{V} = \{S_{eq}, V, n_1, n_2, n_3, \ldots, n_c\}$, we can proceed to identify the thermodynamic derivatives. We have seen that the temperature is $T \triangleq \left(\frac{\partial U_{eq}}{\partial S_{eq}}\right)_{\mathscr{V} \setminus S_{eq}}$. The thermodynamic pressure is similarly defined to be $P \triangleq -\left(\frac{\partial U_{eq}}{\partial V}\right)_{\mathscr{V} \setminus V}$. The final derivative is a purely mathematical construct, the *chemical potential*, and is defined as $\mu_i \triangleq -\left(\frac{\partial U_{eq}}{\partial n_i}\right)_{\mathscr{V} \setminus n_i}$

---

[2]This is consistent with experimental observations everywhere except at critical points, where some of them develop diverging second derivatives.

The thermodynamic derivatives, also known as *intensive variables*, may be regarded as field variables, much like electric and magnetic fields. T induces an increase in S_{eq}, P induces a decrease in V, and μ_i induces an absorption of chemical species. This is made somewhat more rigorous in the next section.

Homogeneous Functions. Thermodynamics makes a number of assumptions of the functional form of the energy and entropy functions. The existence of extrema implies that they must be potentials, for example. It is generally assumed that they are differentiable almost everywhere. There is another additional function of homogeneity that is made, as well. The internal energy is assumed to be a homogeneous first degree function of its extensive variables, that is:

$$U(\alpha S, \alpha V, \alpha n_1, \alpha n_2, \alpha n_3, \ldots, \alpha n_c) = \alpha U(S, V, n_1, n_2, n_3, \ldots, n_c), \alpha \in \mathbb{R} \tag{3.43}$$

There are two implications of this. First, U may be written in the form:

$$U = \sum_{v \in \mathscr{V}} v \left(\frac{\partial U}{\partial v} \right)_{\mathscr{V} \backslash v} \tag{3.44}$$

or, specifically,

$$U = TS - PV + \mu_1 n_1 + \mu_2 n_2 + \mu_3 n_3 + \ldots \mu_c n_c \tag{3.45}$$

This is a general result and is known as Euler's homogeneous function theorem.

At equilibrium for an isolated volume of material, U assumes an extremal value for a constant value of S and V. Additionally, all of the extensive variables are conserved (they can move within the isolated volume, but cannot be created or destroyed). In order for U_{eq} to maintain its extremal value, a volume contraction in one area must be offset by a volume expansion in another. The net change in U_{eq} must be zero, but, by Equation (3.45), it must also be:

$$0 = U_{eq}^{(2)} - U_{eq}^{(1)} = -P_2 \Delta V_2 - (-P_1 \Delta V_1) \tag{3.46}$$

where the 2 subscript refers to a small subvolume of the isolated volume and the 1 subscript refers to its complement. Since $\Delta V_2 = -\Delta V_1$, this also implies that $P_2 = P_1$: the pressure is equal between the two volumes. A similar argument may be made for all of the thermodynamic derivatives, showing that the thermodynamic derivatives of all regions must be the same at equilibrium. This must also be true between any two different phases. This is the constraint alluded to above with the Gibbs phase rule.

The second implication of the homogeneous first degree assumption of U is that all of its derivatives are homogeneous, zero degree functions. That is:

$$\left(\frac{\partial}{\partial v} \right)_{\mathscr{V} \backslash v} U(\alpha S, \alpha V, \alpha n_1, \alpha n_2, \alpha n_3, \ldots, \alpha n_c)$$
$$= \left(\frac{\partial}{\partial v} \right)_{\mathscr{V} \backslash v} U(S, V, n_1, n_2, n_3, \ldots, n_c), \alpha \in \mathbb{R} \tag{3.47}$$

In other words, the thermodynamic derivatives are *independent of the size of the system*, hence their designation as "intensive variables," or "field variables."

The effect of using the Legendre transform to change the dependence on an extensive variable to one on an intensive variable is that the system can be considered in equilibrium with a *reservoir* that provides unlimited amounts of that extensive variable, so that the system can reach the equilibrium required by the intensive variable. And this is completely analogous to the Lagrange multiplier argument used above.

Generalized Free Energies.   All free energies are Legendre transformed representations of the internal energy, where the dependence upon S_{eq} has been changed to a dependence upon T. The difference in the various free energies is which other extensive variables have been Legendre transformed.

Using the Legendre transform, it is possible to make a choice between any thermodynamic extensive variable and its corresponding intensive variable. This choice is generally made on the basis of which variable is easier to control. For example, if the volume is difficult to constrain (e.g., with an incompressible phase whose volume changes as a result of a reaction), U_{eq} may be transformed to:

$$H \triangleq -\max_{V} \left[-U_{eq} + (-P)V \right] \tag{3.48}$$

$H(S_{eq}, P, n_1, n_2, n_3, \ldots, n_c)$ is known as the *enthalpy* and is a favorite of chemists who work with liquids whose volume can change during a reaction. The enthalpy reaches a minimum at equilibrium when V is constrained by losing *entropy*. While not strictly a *free energy*, the enthalpy uses the same methodology as is used in developing free energies.

For example, starting with the enthalpy, a Legendre transform may be used to produce a function that is minimum at equilibrium for constant volume, constant temperature processes. This gives the *Gibbs free energy*, $G(T, -P, n_1, n_2, n_3, \ldots, n_c)$. At equilibrium, the volume and the entropy assume the value that minimizes G. The Gibbs free energy is, by far, the most used of the thermodynamic free energies in materials science.

3.4   Concluding Remarks

This chapter provides what we hope to be a primer for the greater challenges of applying methods from the solution rich fields of signal and image processing to the problem-rich fields of materials science and microscopy. With the increase the world has experienced in the last decades in communication between experts, the explosion of data and the capability to generate even more, and an Internet that makes for the modern equivalent of the Library of Alexandria, the possibilities are endless.

Chapter 4

Forward Modeling

Marc De Graef
Carnegie Mellon University

4.1 What Is Forward Modeling?

In very broad terms, the concept of "forward modeling" refers to the ability to predict the outcome of an experiment through a physics-based simulation of all the individual components of that experiment. In the context of this book, the "experiment" will be an observation, or a series of observations, of a material, in 2-D, 3-D, or 4-D (with time typically being the fourth dimension), using any of a wide array of instruments, ranging from basic optical microscopes to scanning and transmission electron microscopes, x-ray computed tomography instruments, serial sectioning instruments, scanning probe microscopes, synchrotrons, and so on. The central idea behind the concept of forward modeling is that a complete understanding of how the observed quantities are generated should be a tremendous help in interpreting those observations in a quantitative way. All too often, the way in which experimental data is generated is not subsequently used to guide the interpretation of the data, which can lead to incomplete or even incorrect interpretations. In this chapter, we begin by formalizing the forward modeling process, and dividing it into steps that can be formally described in terms of physical models. Then, in Section 4.2, we consider a number of electron scattering modalities in more detail.

4.1.1 What Are the Unknowns in Materials Characterization?

Materials characterization refers to a broad array of experimental techniques that focus on elucidating the internal structure of an engineering material. This internal structure, usually referred to as the "microstructure," plays a central role in the behavior and performance of a material. The properties of any given material are generally defined as the proportionality factors between externally applied fields and the material's responses to those fields. For instance, thermal expansion is a material property that represents the strain (response) induced by a change in temperature (field); electrical conductivity is a property that relates an applied electric field to the resulting electrical current density. There are dozens of potential properties, and whether or not a material displays a given property depends in part on its crystal symmetry, i.e., the collection of symmetry operations that leave the crystalline lattice invariant. Among the material properties there are many that depend on the microstructure of the material; examples include yield stress, electrical breakdown voltage, creep resistance, and so on. Since these properties depend on the microstructure, often in a very sensitive way, it stands to reason that the characterization of such microstructures is an important area of interest for the materials community.

The microstructure has distinctive features at a number of different length scales, ranging from the atomistic level to the macroscopic level, and different experimental tools are used to access each length scale. Despite the wide variety of experimental techniques, the information that is to be extracted from the observations is almost always a function of the position $\mathbf{r}$ inside the sample, and typically belongs to one of four mathematically different categories:

48 Statistical Methods for Materials Science

- *discrete data:* the goal of many spectroscopic observations is to determine the atomic number, Z, as a function of position in the sample; clearly, the atomic number is a discrete variable. In atom probe tomography and mass spectroscopy, the mass or baryon number (number of particles in the atomic nucleus) is measured. In other cases, one measures which crystalline or chemical phase is present at a given location, and this is also a discrete variable. The shape of a sample or a component of a microstructure can also be represented as a discrete (binary) variable (i.e., a point is either inside or outside the region of interest) [158]. In general, we represent discrete position dependent data by the symbol $d(\mathbf{r})$.

- *scalar data:* in x-ray computed tomography experiments, mass absorption contrast is often used to reconstruct objects in 3D; in principle, the mass absorption coefficient may vary continuously with position in the sample, perhaps due to smooth chemical composition variations across the region of interest. Therefore, the mass absorption coefficient is an example of the class of scalar parameters, which we will represent by the symbol $s(\mathbf{r})$ [722]. Electron tomography observations of semiconductor structures attempt to reconstruct the charge density in the material, another example of a scalar field.

- *vector data:* In magnetic materials, the magnetic moment on individual atoms gives rise to a magnetization, i.e., a volume averaged moment, represented by an axial vector field $\mathbf{M}(\mathbf{r})$ that can vary with location in the sample. Electron microscopy and magnetic force microscopy techniques can be used to study the spatial distribution of the magnetization [244]. Similarly, the electrostatic polarization (the volume averaged atomic polarization density) is also represented by a vector field; understanding the behavior of ferroelectric materials often requires determination of the spatial variations of this polarization field. Displacement fields around lattice defects form another example of vector fields [784]. In general, we represent vector quantities by the symbol $\mathbf{v}(\mathbf{r})$. The spatial orientation of the crystal lattice inside a sample can also be considered as a type of vector quantity, since three numbers are needed to uniquely define an orientation in 3-D; in the materials community, the study of the spatial distribution of orientations is known as "texture analysis."

- *tensor data:* While scalars and vectors can be considered as special cases of tensor quantities (with tensor ranks zero and one, respectively) it is useful to distinguish quantities by their tensor rank, since the experimental difficulties associated with determining tensor fields generally increase substantially with increasing tensor rank. As examples of higher rank tensor fields, with general symbol $t(\mathbf{r})$, we mention here the local strain or stress state in a material, which can be measured experimentally by means of synchrotron x-ray scattering techniques [422].

Each of the categories $d(\mathbf{r})$, $s(\mathbf{r})$, $\mathbf{v}(\mathbf{r})$, and $t(\mathbf{r})$ above could have additional arguments, such as time t, or applied field $\mathbf{F}$ (which could itself be a multi-axial tensor field), which increase the dimensionality and complexity of the microstructure reconstruction process. In recent years there has been an increased interest in synchronous or asynchronous measurement of multiple sample characteristics, such as the local chemistry (scalar), the phase distribution (discrete), and the lattice orientation (unit vector in 4D quaternion space). The combined measurement of multiple sample characteristics is referred to as "multi-modal" materials characterization. While synchronous measurements are preferred, since they reduce the need for tedious data registration after the experiment, it is not always practical or possible to carry out synchronous measurements, so that there will always be a need for multi-modal registration algorithms.

The measurement of the spatial distribution of any material property can provide information on the 3D microstructure of that material. Many engineering materials consist of more than one *phase*; phase here refers to matter with a particular atomic arrangement, i.e., a particular crystal structure. Pure aluminum would be considered a single-phase material, even if it is made up of regions (*grains*) in which the crystal lattice has different orientations. Lead-tin solder, on the other hand, is a two-phase material; its microstructure consists of regions that are nearly pure lead, and other regions that are mostly tin. In x-ray tomography, spatial variations in the density produce an absorption profile when a uniform x-ray beam of known intensity is passed through the sample.

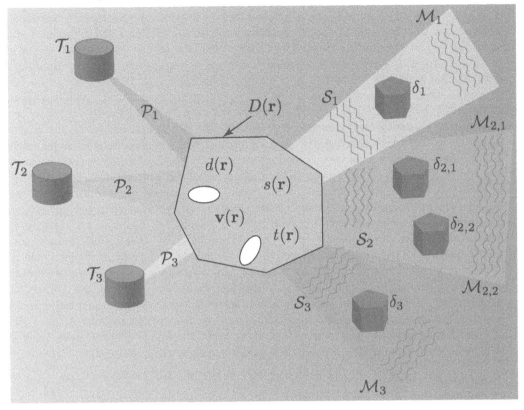

Figure 4.1: Schematic illustration of materials characterization experiments (see text for full explanation).

After recording the sinogram (the absorption profiles as a function of the sample orientation), mathematical algorithms can be used to reconstruct the microstructure, which in this case would refer to the spatial distribution of mass. In the case of a single phase material, with only one density value, such an observation may reveal internal porosity or cracks in the microstructure. In the case of the two-phase lead-tin material, the two crystal structures have very different densities, so an x-ray tomography experiment will reveal a high contrast between the two phases. The microstructure in this case is typically one in which alternating plate-like (lamellar) regions of the two phases are present, often in complex spatial arrangements. Instead of using the spatial distribution of the density, one can make use of other quantities to determine the microstructure of the material. For instance, a measurement of the orientation of the crystal lattice with respect to an external reference frame (described in Section 4.2) can produce a detailed representation of the grain microstructure of a polycrystalline material. The arsenal of instruments that is used in the materials community to determine the microstructure of a complex engineering material is quite extensive; in the following section, we introduce an abstract description of microstructure characterization and the concept of forward modeling.

4.1.2   A Schematic Description of Forward Modeling

Consider a material with an unknown 3D microstructure that we wish to determine by measuring one or more functions of the type $d(\mathbf{r})$, $s(\mathbf{r})$, $\mathbf{v}(\mathbf{r})$, and $t(\mathbf{r})$. To do so, we employ one or more characterization instruments, represented schematically by the blue cylinders in Figure 4.1. Each instrument perturbs the sample in a specific way, e.g., by illuminating it with a beam of electrons, or by sending a sound wave pulse through it, and then the response of the sample is recorded; this response may take the form of an image, a diffraction pattern, a spectrum, and so on. Each instrument $\mathcal{T}_i$ produces a *generalized forward projection*, $\mathcal{P}_i$, of the sample. Mathematically, this may be a simple linear operation, e.g., a convolution with the instrument's point spread function, or a more complex non-linear operation, e.g., a series of consecutive inelastic scattering events inside the sample. Each instrument is characterized by its point spread function $\mathcal{T}_i$, which describes the characteristics of the illumination and the signal formation process. The symbols $\mathcal{S}_i$ describe the ideal signals that are incident upon one or more detectors, $\delta_{i,j}$, represented by the red prisms in Figure 4.1; it is not unusual to have two or more different detectors inside an electron microscope. Each detector imparts its point spread function onto the signal and produces the measurement $\mathcal{M}_{i,j}$, which is typically a smoothed version of the original signal $\mathcal{S}_i$. Different detectors inside an instrument may measure different aspects of the signal $\mathcal{S}_i$; for instance, one detector could focus on backscattered electrons, whereas the second one measures the light generated when the electronic excitations caused by the incident electron beam relax, resulting in the emission of photons. The sample itself may be characterized by means of indicator functions, $D_j(\mathbf{r})$, which describe the exterior boundaries of the sample as well as any interior boundaries that may be present in the form of voids or other microstructural features. At each stage of the process, noise contributions may deteriorate the signals; noise can occur at the instrument stage (stage vibrations, sample drift, beam intensity fluctuations, …), in the sample (e.g., thermal vibrations of the atoms may blur the elastic scattering processes), and in the various detectors (counting noise, electronic noise, …).

The measurements can be written as functionals of the microstructural parameters, the instrument and detector point spread functions, and appropriate noise terms:

$$\mathcal{M} = \mathcal{P}\left[D(\mathbf{r}), d(\mathbf{r}), s(\mathbf{r}), \mathbf{v}(\mathbf{r}), t(\mathbf{r}); \mathcal{T}(\theta), \delta(\theta'); \text{noise}\right]. \tag{4.1}$$

The generalized forward projection operator, $\mathcal{P}$, describes the complete forward model, and can be linear or non-linear. The arguments θ and θ' of the instrument and detector point spread functions indicate all relevant parameters, such as beam convergence angle, lens aberrations, scintillator properties, and so on. The first group of arguments of $\mathcal{P}$ are the unknowns, to be reconstructed from the measurements $\mathcal{M}$. The second set of arguments represent the things we do know about the experiment; these could either be exact mathematical relations or realistic approximations of the interaction of the incident probe with the microstructure, and the subsequent detection process. Often, it is beneficial to obtain accurate but linearized approximations of the instrument and detector processes. The final argument represents all the noise terms of the measurement process; these are generally not known exactly, but they can often be estimated for a given noise model (Gaussian, Poisson, etc.).

The extraction of the unknowns from Equation (4.1) is in general a complex inverse problem, whereas the simulation of $\mathcal{M}$ for a given instrument and microstructure is known as the forward model. If the forward projector $\mathcal{P}$ is known, then the unknowns can often be determined via an iterative process, in which a model microstructure is updated iteratively to obtain a closer match between the forward projections and the actual measurements. The simultaneous iterative reconstruction technique (SIRT) used in tomographic reconstructions is an example of such an iterative approach. Model-based iterative reconstructions (MBIR, see Chapter 6) form another class of reconstruction algorithms that use prior information about the material as part of the reconstruction process.

Each of the measurements $\mathcal{M}_{i,j}$ provides a different piece of information about the actual microstructure of the sample. Some measurements provide chemical information, others structural or orientational information. In general, each measurement has a different spatial resolution, which complicates the process of extracting the 3D microstructure information from multi-modal measurements. Traditionally, the materials community has started from the measurements $\mathcal{M}_{i,j}$ to reconstruct various aspects of the microstructure, but it is rare in this community to find approaches that "close the loop," i.e., use relations of the type (4.1) to self-consistently solve for the microstructure while trying to accurately match the observations through the use of generalized forward projectors.

4.2   A Brief Overview of Electron Scattering Modalities

Over the last sixty years or so, the materials community has made extensive use of a large number of characterization tools and modalities, many of them introduced in the seventies. With the advent of affordable desktop computing systems in the late eighties/early nineties, materials characterization has experienced a complete change from analog observation modes to digital techniques for both instrument operation and data acquisition. This rapid change has brought with it the need for high-quality physics-based forward models, to help in the interpretation of large, often multimodal, datasets. In this section, we provide a brief overview of the most important electron-based modalities, to illustrate the rich variety of characterization techniques available to the community.

To classify characterization modalities we ask several questions: What kind of sample illumination is used? Where is the detector placed? Is the detector position sensitive? What kind of particle is detected? Do we measure the particle's momentum, position, or energy? For each question, there are several possible answers listed below, along with citations to relevant papers. For a more complete description of these techniques we refer the interested reader to [1090].

- *What kind of sample illumination is used?* The incident electron beam can have a number of different shapes. For parallel illumination, the beam is cylindrical and all electrons have parallel momentum vectors; the transmission electron microscope (TEM) employs this illumination mode for defect contrast imaging (DCI, [784]) or for high-resolution phase contrast imaging (HRTEM, [956]). For a converged electron probe, the electron momentum vectors span a conical volume, giving rise to modalities such as convergent beam electron diffraction (CBED, [997, 998, 999]) and scanning transmission electron microscopy (STEM, [497]). A converged electron beam is also the primary modality in the scanning electron microscope (SEM, [840]), where electron channeling patterns (ECP, [787]) and electron backscattered diffraction patterns (EBSD, [920]) form the main diffraction modalities, and electron channeling contrast imaging (ECCI, [787]) allows for the imaging of near-surface defects. Finally, the incident beam can illuminate a large sample region all at once, or it can be scanned point-by-point across the sample, with the detectors recording the signal synchronized with the incident beam position.

- *Where is the detector placed?* There are two major classes of modalities: in one, the detector is on the same side of the sample as the electron source, in the other detector and sources are on opposite sides (which requires a thin electron-transparent sample). The former geometry is known as the Laue geometry, the latter as the Bragg geometry. In the SEM, the sample is typically in bulk form, and the Laue geometry is used; in the TEM, the Bragg geometry is more common, although it is possible to place detectors above the sample as well, leading to the Laue geometry.

- *Is the detector position sensitive?* Detectors can be divided into two broad classes: detectors with individual pixels, in which a signal can be measured that depends on position, and integrating detectors, which do not have position sensitivity. In spectroscopic modalities, the detector has energy sensitivity, sometimes integrating, as in energy dispersive spectroscopy (EDS, [840]), sometimes position sensitive, as in energy filtered imaging (EFTEM, [512]).

- *What kind of particle is detected?* If an incident electron is scattered by the sample and subsequently detected then we refer to this as primary electron detection; if the signal is created by electrons that were knocked out of the sample, we refer to them as secondary electrons (SEs). Sometimes, the measured signal is generated by a different particle, for instance an x-ray photon emitted by a sample atom, or an optical photon (e.g., cathodoluminescence or CL).

- *Do we measure the particle's momentum, position, or energy?* When scattered electrons travel through a magnetic lens, one can acquire a diffraction pattern, i.e., a visualization of the momentum distribution, in the back focal plane of the lens, and an image in the image plane. For energy dispersive measurements, one typically applies a magnetic field normal to the electron travel direction, which gives rise to deflections that depend on the electron's energy; this leads to energy spectra (e.g., electron energy loss spectroscopy or EELS, [295]) which are analyzed to obtain local chemical compositions. Similarly, one can analyze the energy distribution of secondary particles (x-ray photons) to acquire position-sensitive compositional information.

In modern electron microscopes, several of the modalities cited above may be present in the same instrument, and often they can be operated at the same time to acquire truly multi-modal data. Extracting all possible information from such multi-modal data requires the use of accurate forward models that take into account the physics of the scattering events as well as the characteristics of all detectors used in the experiment. In the following section, we describe two case studies of forward models, one for electron backscatter diffraction (EBSD), the other for Lorentz vector field electron tomography (VFET).

4.3   Case Studies

In this section, we provide two examples of forward models that illustrate the potential complexity of physics-based models. The first model provides predictions of electron backscatter diffraction (EBSD) patterns that are routinely acquired using scanning electron microscopes (SEM); EBSD analyses are used to determine the spatial distribution of lattice orientations in polycrystalline engineering materials. The second example deals with the 3D reconstruction of the magnetic vector potential of a magnetic sample using phase-reconstructed Lorentz transmission electron microscopy (LTEM); the forward model in this case allows for the computation of the electrostatic and magnetic phase shifts experienced by an electron as it travels through or near a magnetic sample.

4.3.1   Electron Backscatter Diffraction

In a polycrystalline material, the relevant microstructural entities are known as grains, i.e., regions of the material for which the crystal lattice has very nearly a constant orientation. Knowing the distribution of orientations is an essential prerequisite for the determination of material properties; if the properties of a single grain are known, then the bulk properties can often be approximated by averaging the single crystal property over the grain orientation distribution function (ODF). Electron backscatter diffraction (EBSD) has emerged as a leading technique for the determination of the ODF; EBSD is carried out in a scanning electron microscope (SEM), in which a fine electron probe with an energy of 20-30 keV is scanned across a region of interest and electron backscatter patterns are recorded on a scintillator screen. The experimental setup is shown schematically in Figure 4.2. The electron beam exits the objective lens pole piece and enters the sample surface at a given scan point; the goal of the measurement is to determine the orientation of the crystal lattice at this point. Electrons undergo both elastic and inelastic scattering events inside the material, and a fraction of the incident electrons emerges from the inclined sample as backscattered electrons (BSEs). A portion of those BSEs is intercepted by a scintillator screen at a distance L from the illuminated point; the light intensity pattern on the scintillator is then propagated along a fiber optic cable (or viewed by a lens) and recorded by a CCD camera. An example EBSD pattern is shown in Figure 4.2(b); this pattern was obtained for a silicon single crystal sample. The EBSD pattern shown in Figure 4.2(c) is a simulated pattern, using the forward modeling approach described in the remainder of this section.

The EBSD forward model consists of three distinct components: (1) a Monte Carlo simulation to determine the energy, depth, and directional distributions of the BSEs leaving the sample; (2) a dynamical scattering simulation to determine how the BSE yield is modulated in each direction due to channeling of the BSEs on their way out of the sample; and (3) a model for the detector geometry and electron-to-photon conversion process.

4.3.1.1   BSE Monte Carlo Simulations

BSEs are generated by the process of Rutherford scattering, which is an elastic scattering process caused by point-like Coulomb potentials inside the material. The differential Rutherford scattering

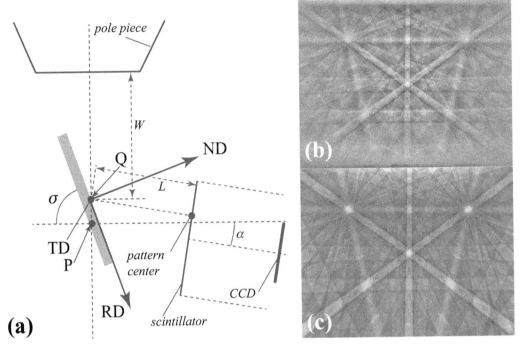

Figure 4.2: (a) Schematic representation of the EBSD geometry; electrons exit the pole piece along the optical axis and enter a crystalline sample inclined at an angle σ. A plume of backscattered electrons then reaches the scintillator screen where the electron energy is converted into photons which are subsequently detected on a CCD camera. (b) Shows a 20-keV EBSD pattern for single crystal Si; the crystallographic [111] direction intersects the scintillator screen approximately at the center of the pattern (image courtesy of D. Fullwood, BYU). The pattern in (c) is the result of a dynamical pattern simulation described in detail in the text.

cross-section for electrons of (relativistic) energy E_{kin} is given by

$$\frac{d\sigma}{d\Omega} = \left(\frac{Ze^2}{16\pi\varepsilon_0 E_{kin}}\right)^2 \frac{1}{\sin^4\frac{\beta}{2}}, \tag{4.2}$$

where $\varepsilon_0 = 8.854 \times 10^{-12}$F/m is the permittivity of vacuum, Ω the solid angle, Z the atomic number of the target material, and β the scattering angle with respect to the forward direction. For typical accelerating voltages, the pre-factor in this relation is of the order of a few barns (1 barn $= 10^{-28}$ m^2).

As the electron travels inside the material, it can undergo a large number of inelastic scattering events, including plasmon and phonon excitations and core-shell excitations; while models for the scattering cross-sections for all of these potential events exist, one can approximate the energy losses associated with these scattering events by assuming that the electron continuously loses energy. This is known as the continuous slowing down approximation (CSDA). The distance between consecutive scattering events is known as the mean free path, which can be estimated based on the material's crystal structure and composition. The overall scattering process can then be modeled in a Monte Carlo fashion by repeating the following two steps: (1) travel a random distance based on the mean free path; (2) undergo a Rutherford scattering event and travel in a new random direction extracted from the differential scattering cross-section. Before each scattering event, the electron trajectory is queried to determine whether or not it has left the sample; for those electrons that do

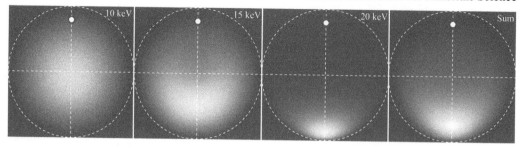

Figure 4.3: Stereographic projections of the spatial distribution of BSEs according to their exit energy, 10, 15, or 20 keV (for an incident electron energy of 20 keV); each projection represents a histogram with a bin size of 1 keV, and the dot in the upper portion of the projection represents the incident beam direction, inclined at 70° with respect to the sample surface. The rightmost projection represents the sum of all projections and represents the complete BSE plume.

leave the sample before they have lost most of their kinetic energy, the direction of travel, the depth of the last scattering event, and the electron energy are stored as histograms.

Figure 4.3 shows stereographic projections of the exit BSE distributions as a function of electron energy for a 20-keV incident electron energy. A stereographic projection is an equal-angle projection from the surface of a sphere to a 2D disk, and is used in materials science to represent functions on a sphere in a convenient 2D projection; in particular, circles on the sphere remain circles in the projection. For each energy level, the projection represents all electrons within 0.5-keV from the indicated energy. The incident beam is incident from the top, at 20° above the horizontal plane (corresponding to the standard sample tilt angle of 70°). The rightmost pattern shows the sum of all the individual patterns (which were computed for 0.5 keV step sizes between 5 and 20 keV). The sum projection clearly shows that the BSE process is a highly directional process, and that the best detector position is around the lower half of the projection.

4.3.1.2 Dynamical Scattering Simulations

The probability of a backscattered electron reaching the scintillator can be expressed as an integral over the sample depth of the electron wave function evaluated at the atom positions, multiplied by the corresponding Z^2 value to model the Rutherford scattering cross-section. The resulting integration is as follows:

$$\mathscr{P}(\mathbf{k}) = \sum_{i \in \mathscr{S}} \frac{Z_i^2 e^{-M_i}}{z_0(E)} \int_0^{z_0(E)} \mathrm{d}z \, \lambda(E, z) \, |\Psi_{\mathbf{k}}(\mathbf{r}_i)|^2 ; \tag{4.3}$$

$\mathbf{k}$ is the wave vector along the travel direction of the electron, the exponential represents the Debye–Waller factor, and the functions $z_0(E)$ and $\lambda(E, z)$ are extracted from the Monte Carlo histograms described in the previous section. The electron wave function is evaluated at the atom positions $\mathbf{r}_i$, and the sum goes over the set $\mathscr{S}$ of all atoms in the unit cell.

The wave function $\Psi(\mathbf{r})$ is computed as the solution to the Schrödinger equation for a given electron wave vector $\mathbf{k}$ [245]. There are two leading theoretical approaches to solving for the wave function: the Bloch wave approach and the scattering matrix approach. In the Bloch wave approach, the ansatz is made that the wave function is a superposition of plane waves with the periodicity of the underlying lattice, which leads to a complex non-symmetric eigenvalue problem. In the scattering matrix approach, the ansatz is made that the wave function is a superposition of plane waves traveling along the directions predicted by the Bragg equation. This leads to a system of first-order coupled differential equations for the amplitudes of the plane waves. Both methods result in the same final solution, but the scattering matrix approach is more flexible than the Bloch wave approach when it comes to incorporating lattice defects into the simulations; it is also more suitable for implementation on graphical processing units (GPUs).

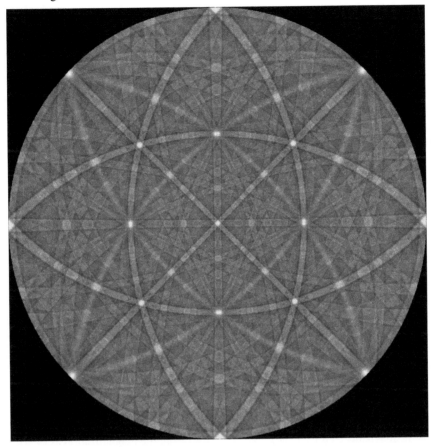

Figure 4.4: EBSD master pattern for Si at 20 keV represented as a stereographic projection; the brightness level is proportional to the BSE yield in the direction of the corresponding wave vector.

For both approaches, the final expression for the BSE yield for a given energy level E can be written as the sum (su) of all the entries in a Hadamard product ($\circ$) between two matrices [159]:

$$\mathscr{P}(\mathbf{k},E) = \mathrm{su}(S \circ L), \qquad (4.4)$$

where the matrix S depends on the atom positions and atomic numbers, and L depends on the wave function and the Monte Carlo weight factors. The total BSE yield is then obtained as a weighted sum of the yields over all energy levels. This computation is then repeated for all wave vectors covering a hemisphere above the sample surface. The spatial distribution of the BSE yield is known as the "EBSD master pattern." Figure 4.4 shows an example of an EBSD master pattern which depicts the BSE yield, $\mathscr{P}(\mathbf{k},E)$, as a function of the electron wave vector $\mathbf{k}$ for pure silicon and BSEs of energy $E = 20$ keV.

4.3.1.3 Detector Parameters

The final step in the EBSD forward model is the application of the detector geometry, so that a realistic pattern can be extracted from the master pattern obtained in the previous step. The detector geometry requires a number of parameters: scintillator pixel size, distance between scintillator and illumination point on the sample, tilt angle of the detector, as well as the so-called "pattern center" which is the point where the BSE direction is normal to the plane of the scintillator. Once these parameters are known, then for each pixel one determines the direction cosines of the line connecting the pixel to the illumination point; these direction cosines are then converted into the crystal reference frame, using the known Euler angles that represent the rotation from the sample frame to the crystal frame, and, finally, the EBSD master pattern is interpolated using bilinear interpolation to obtain the required BSE intensity. This process is then repeated for all detector pixels, and the master patterns are averaged over all energy levels to obtain the final result, which is a realistic EBSD pattern. Figure 4.2(c) shows the interpolated Si pattern for a set of detector parameters obtained from a least-squares fit to the experimental pattern. This EBSD forward model is sufficiently accurate to allow for an iterative fitting of the detector's geometrical parameters against an experimental pattern with a known crystal orientation. The model will be employed in Chapter 6 as the basis for the creation of a dictionary of EBSD patterns that will be used to index experimental datasets.

4.3.2   Lorentz Vector Field Electron Tomography

One of the core components of a model-based tomographic reconstruction algorithm is the forward model, which allows for the computation of the data, given a model for the unknown object as well as for the instrument and detector system. In Lorentz transmission electron microscopy, a technique used to study the magnetization state of a material, a high-energy electron wave is produced by the microscope, and this wave propagates through and around a magnetic object. The electron wave is described by a complex-valued wave function that satisfies the Schrödinger equation for the electron+sample system. The sample is described by two fundamental quantities, the electron density $\rho(\mathbf{r})$ and the magnetic moment density (or magnetization) $\mathbf{M}(\mathbf{r})$. According to classical electrodynamics theory, each of these densities is associated with a potential function; the electrostatic potential, $V(\mathbf{r})$, can be determined from the following convolution integral:

$$V(\mathbf{r}) = \frac{1}{4\pi\varepsilon_0} \iiint d\mathbf{r}' \, \frac{\rho(\mathbf{r}')}{|\mathbf{r} - \mathbf{r}'|}, \tag{4.5}$$

whereas the magnetic vector potential can be derived from the magnetization as [454]:

$$\mathbf{A}(\mathbf{r}) = \frac{\mu_0}{4\pi} \iiint d\mathbf{r}' \, \mathbf{M}(\mathbf{r}') \times \frac{\mathbf{r} - \mathbf{r}'}{|\mathbf{r} - \mathbf{r}'|^3}; \tag{4.6}$$

$\varepsilon_0 = 8.854 \times 10^{-21}$ F/nm is the permittivity of free space, and $\mu_0 = 4\pi \times 10^7$ N/A^2 is the permeability of vacuum. The electrostatic potential is measured in V, whereas the vector potential is expressed in units of Vs/nm.

4.3.2.1   Lorentz Forward Model

Aharonov and Bohm have shown in 1959 [16] that the phase $\varphi(\mathbf{r})$ of a plane electron wave traveling through a region of space with non-zero electrostatic and magnetic potentials is modified by these potentials; the phase shift can be written as:

$$\varphi(\mathbf{r}_p) = \varphi_e(\mathbf{r}_p) + \varphi_m(\mathbf{r}_p)$$
$$= \frac{\pi}{\lambda E} \int_{-\infty}^{+\infty} V(\mathbf{r}_p + \ell\boldsymbol{\omega}) \, d\ell - \frac{e}{\hbar} \int_{-\infty}^{+\infty} \mathbf{A}(\mathbf{r}_p + \ell\boldsymbol{\omega}) \cdot \boldsymbol{\omega} \, d\ell, \tag{4.7}$$

where $\boldsymbol{\omega}$ is a unit vector along the electron beam propagation direction, and ℓ parameterizes the electron trajectory; $\mathbf{r}_p$ is a vector in the plane normal to $\boldsymbol{\omega}$. The prefactor of the magnetic phase shift φ_m can also be written as π/ϕ_0, with $\phi_0 = h/2e$ the flux quantum ($\phi_0 = 2070$ T nm^2); λ is the (relativistic) electron wavelength, and E is the relativistic accelerating potential of the microscope. Note that both of these integrals can be regarded as tomographic projection integrals, indicating that repeated measurements along a series of directions can lead to reconstruction of both the electrostatic and magnetic vector potentials.

As the electron wave travels through the sample region, it experiences a total phase shift given by $\varphi = \varphi_e + \varphi_m$; in addition, there may be amplitude changes as well due to absorption and dynamical (Bragg) scattering. As a result, the image intensity that is measured on the detector is a non-linear function of the total phase shift and a number of other contributions. The magnetic lenses in the microscope (primarily the main objective lens) add their own phase shifts and attenuation factors; this is typically described by a point spread function, $\mathscr{T}(\mathbf{r}_p)$, which depends on phase shifts due to lens aberrations and attenuations due to chromatic aberration. The image intensity acquired by a detector is then described as (ignoring an image magnification factor):

$$I(\mathbf{r_p}) = \left| \mathscr{F}^{-1} \left[\mathscr{F} \left[A(\mathbf{r_p}) e^{-i\varphi(\mathbf{r_p})} \right] \mathscr{T}(\mathbf{q}) \right] \right|^2 \tag{4.8}$$

$\mathscr{F}$ is the Fourier transform operator, and $\mathbf{q}$ the frequency vector conjugate to $\mathbf{r}_p$ in the plane normal to the electron beam. Typically, the image intensity is computed using fast Fourier transforms. Equations (4.7) and (4.8) constitute the forward model for Lorentz transmission electron microscopy. The microscope user has control over the nature of the recorded image intensity $I(\mathbf{r})$ via the microscope transfer function $\mathscr{T}(\mathbf{q})$; in particular, defocusing of the objective lens produces image contrast at the locations of the magnetic domain walls in the sample. The defocused images can then be used in the transport-of-intensity equation (TIE) formalism to numerically extract the total electron wave phase shift $\varphi(\mathbf{r}_p)$ from the image intensities.

$\rho(\mathbf{r})$ and $\mathbf{A}(\mathbf{r})$ are the two quantities of interest, and in principle they can be reconstructed in 3D from tomographic tilt series. This requires a forward model for the efficient computation of the phase shifts, as well as a detailed characterization of the transfer function of the microscope. In the remainder of this section, we will focus on the former problem, i.e., the computation of the projection integrals in Equation 4.7.

4.3.2.2 Electron Wave Phase Shift Computations

There are three computational approaches for the magnetic phase shift simulation, all of them based on a combination of Equations (4.6) and (4.7). If the magnetization pattern $\mathbf{M}(\mathbf{r})$ is periodic and constant along the thickness of the foil, then one can use the discrete inverse Fourier transform to derive the Mansuripur expression [635] for the phase shift as:

$$\phi_m(\mathbf{r}_p) = \frac{2e}{\hbar} \sum_{m=0}^{P} \sum_{n=0}^{Q}{}' i \frac{t}{|\mathbf{q}|} G_{\mathbf{p}}(t|\mathbf{q}|)(\hat{\mathbf{q}} \times \mathbf{e}_z) \cdot (\mathbf{p} \times (\mathbf{p} \times \mathbf{M}_{mn})) e^{2\pi i \mathbf{r}_p \cdot \mathbf{q}}, \tag{4.9}$$

where the prime indicates that the term $(m,n) = (0,0)$ does not contribute to the summation, $\mathbf{q} = \frac{m}{P}\mathbf{e}_x^* + \frac{n}{Q}\mathbf{e}_y^*$ is the frequency vector, $\mathbf{e}_x^*$ and $\mathbf{e}_y^*$ are reciprocal unit vectors, $\mathbf{p}$ is the beam direction expressed in the orthonormal reference frame, t is the sample thickness, a hat indicates a unit vector, and the function $G_{\mathbf{p}}(t|\mathbf{q}|)$ is given by

$$G_{\mathbf{p}}(t|\mathbf{q}|) = \frac{1}{(\mathbf{p}\cdot\hat{\mathbf{q}})^2 + p_z^2} \operatorname{sinc}\left(\pi t |\mathbf{q}| \frac{\mathbf{p}\cdot\hat{\mathbf{q}}}{p_z} \right), \tag{4.10}$$

where $\operatorname{sinc}(x) \equiv \frac{\sin(x)}{x}$; for normal beam incidence, we have $G_{\mathbf{p}} = 1$. This expression has been extended to more complex magnetization configurations by Haug et al. [415] and has been applied to the simulation of interference fringes in coherent Fresnel domain wall images. Figure 4.5(a) shows a periodic magnetization configuration that is typically observed in finely twinned materials such as the magnetostrictive alloy Terfenol-D [281]. The vertical domain walls correspond to a 71° change in the magnetization direction between the cubic [111] and [$\bar{1}$11] directions; the zig-zag walls are 180° walls, giving rise to a so-called "12 reciprocal laminate" with domain wall planes (2$\bar{1}\bar{1}$) and (211). The field of view in Figure 4.5(a) is 512 × 512 nm^2, and the domain wall widths are $\delta_{71} = 5$ nm and $\delta_{180} = 20$ nm. Using the Mansuripur algorithm with normal incidence electrons and

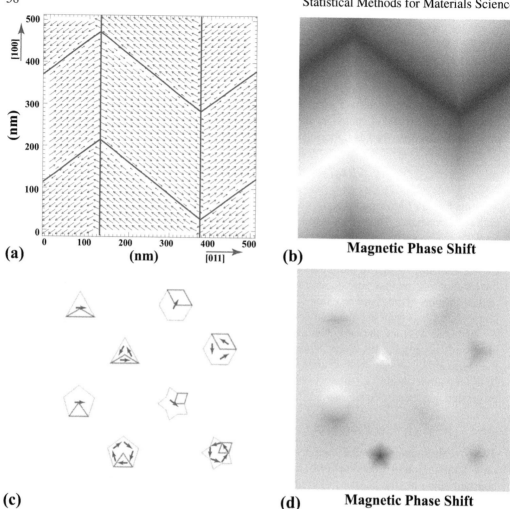

(a)

(b) **Magnetic Phase Shift**

(c)

(d) **Magnetic Phase Shift**

Figure 4.5: (a) Magnetization configuration commonly observed in finely twinned materials, Terfenol-D in this case; (b) shows the magnetic phase shift computed using the Mansuripur approach of Equation (4.9); (c) polygonal isolated particles with either a uniform magnetization state or a closure domain; (d) magnetic phase shift map for the configuration in (c), computed using the shape amplitude approach of Equation (4.11).

a saturation induction B_0 of 1.2 T we obtain the magnetic phase shift shown in Figure 4.5(b); the intensity range from black to white corresponds to a phase range of $(m)7.3\pi$. This phase shift can then be substituted into Equation (4.8) to compute the image intensity for a given set of microscope parameters.

The second approach is useful when all magnetic objects are uniformly magnetized; for that case, Beleggia and Zhu [80] have proposed an expression for the 2D Fourier transform of the magnetic phase shift for magnetized particles of arbitrary shape. If we use the shape function $D(\mathbf{r})$, equal to 1 inside and 0 outside the object, then the magnetization can be expressed as $\mathbf{M}(\mathbf{r}) = M_0\hat{\mathbf{m}}D(\mathbf{r})$ with M_0 as the saturation magnetization. This results in the following expression for the Fourier transform of the magnetic phase shift:

$$\varphi_m(\mathbf{k}) = \frac{i\pi B_0}{\phi_0}\frac{D(k_x,k_y,0)}{k_x^2+k_y^2}\left.(\hat{\mathbf{m}} \times \mathbf{k})\right|_z, \qquad (4.11)$$

where $\phi_0 = h/2e = 2070$ T nm^2 is the flux quantum, $B_0 = \mu_0 M_0$ is the saturation induction, and the shape amplitude $D(\mathbf{k})$ (the Fourier transform of $D(\mathbf{r})$) is evaluated in the plane (k_x, k_y) normal to the beam direction. This expression is essentially a magnetic version of the Fourier slice theorem [722]. For many basic shapes the shape amplitude can be computed analytically, so that a variety of nano-particle magnetization states can be handled. Additional applications to arbitrary polyhedral shapes can be found in [79]. Figure 4.5(c) shows a schematic of several uniformly magnetized polygonal particles, as well as particles with a magnetization closure state; the corresponding phase shifts are shown in Figure 4.5(d), which shows clearly that the closure domain states have no magnetic phase shift outside of the particle, whereas the uniformly magnetized particles show a phase shift extending well outside the particles, corresponding to the demagnetization field.

The third and final approach to the computation of the magnetic phase shift is capable of handling the most general magnetization configuration (i.e., not periodic, and not uniform) and is based on the fact that the projection of a 3D sphere is independent of the projection direction, $\boldsymbol{\omega}$ [449]. One of the necessary ingredients of a model-based tomographic reconstruction model is a forward projection algorithm that provides the ability to update individual object voxels without having to recompute the entire projection set. To accomplish this, we consider the magnetic phase shift for a uniformly magnetized sphere of radius R. For the shape amplitude approach from the previous paragraph, we need the sphere shape amplitude which is given by:

$$D_{\text{sphere}}(\mathbf{k}) = 3V \frac{j_1(kR)}{kR} \tag{4.12}$$

where $j_1(x) = \frac{\sin x}{x^2} - \frac{\cos x}{x}$ is the spherical Bessel function of the first order and V the volume. The magnetic phase shift can then be written in Fourier space as:

$$\varphi_m(k_x, k_y) = \mu_x \mathscr{S}_y - \mu_y \mathscr{S}_x, \tag{4.13}$$

where the factors

$$\mathscr{S}_\alpha \equiv 4\pi^2 R^2 \frac{iB_0}{\phi_0} \frac{j_1(k_p R)}{k_p^3} k_\alpha, \tag{4.14}$$

can be pre-computed once for a given sphere radius and grid sampling size. The factors μ_x and μ_y are the components of a unit magnetization vector $\boldsymbol{\mu}$, projected onto the plane normal to the electron beam direction $\boldsymbol{\omega}$. A magnetized object can now be described by means of a cubic grid of lattice paramater $2a$ by placing a sphere of radius

$$R = a \left(\frac{6}{\pi} \right)^{\frac{1}{3}} = 1.2407a \tag{4.15}$$

on each lattice node; the spheres have the same volume as the cubic grid cells ($8a^3$) and overlap slightly, as shown in Figure 4.6(a). At the center of each sphere we place a magnetic moment vector with magnitude set so that the moment density (i.e., the magnetization) becomes equal to that of the actual continuous object. It should be noted that any other choice for the shape of the elementary volume results in projection equations that contain the components of $\boldsymbol{\omega}$; the sphere is the only object for which all projections are identical. Reducing the grid size also reduce the sphere radius, as shown in Figure 4.6(a), and in the limit of $a \to 0$ the magnetization of the continuous object is recovered, along with the correct magnetic phase shifts.

Once the phase shift for a single sphere is known, then an entire 3D grid of spheres can be modeled by linear superposition, since the equations of electrodynamics are linear. If we label each sphere by its integer indices (i, j, k) with respect to an orthogonal reference frame with unit cell size $2a$, then the contribution of this sphere to the total phase shift can be written in Fourier space as:

$$\varphi_m^\omega(k_x, k_y) = \sum_{(i,j,k)\in D} e^{i\mathbf{k}_\perp \cdot \mathbf{r}_{ijk}^\theta} (\mu_{ijk,x}^\theta \mathscr{S}_y - \mu_{ijk,y}^\theta \mathscr{S}_x). \tag{4.16}$$

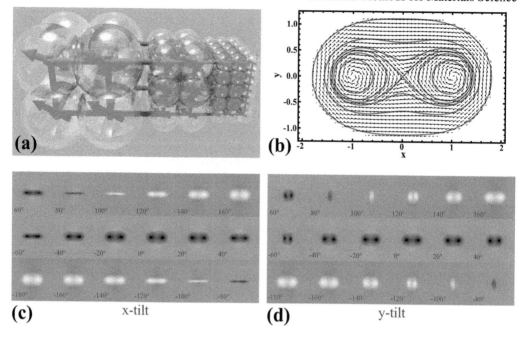

Figure 4.6: (a) Schematic illustration of the multi-grid approach for magnetic phase shift computations; (b) magnetization pattern (arrows) based on the Cassini oval (solid lines); (c) and (d) are magnetic phase shift maps as a function of sample tilt around the x and y axes, respectively. The phase shift range corresponds to the interval $[-0.645, 0.645]$ radians.

In this expression, $\mathbf{r}_{ijk}$ are the 2D components of the position vector of the sphere projected onto a plane normal to the incident beam direction, and (θ) represents the tilt of the incident beam direction $\boldsymbol{\omega}$ with respect to the untilted orthogonal reference frame. Note that the moment vector direction cosines $(\mu_{ijk,x}, \mu_{ijk,y})$ also have a superscript ω, since these components depend on the sample tilt angle; in a tomographic reconstruction, the vectors $\boldsymbol{\mu}_{ijk}$ are the unknowns.

In a VFET experiment, one records two sample tilt series, usually with respect to two orthogonal tilt axes. We represent a tilt around the x axis by a (counterclockwise) tilt angle γ, and a tilt around the y axis by δ. The position $2a(i,j,k)$ becomes $2a(i, j\cos\gamma - k\sin\gamma, j\sin\gamma + k\cos\gamma)$ for the x tilt and $2a(i\cos\delta + k\sin\delta, j, -i\sin\delta + k\cos\delta)$ for the y tilt. For the direction cosines of the magnetization unit vector we find

$$\hat{\boldsymbol{\mu}}_{ijk}^{\gamma} = (\mu_{ijk,x}, \mu_{ijk,y}\cos\gamma - \mu_{ijk,z}\sin\gamma, \mu_{ijk,y}\sin\gamma + \mu_{ijk,z}\cos\gamma); \tag{4.17}$$

$$\hat{\boldsymbol{\mu}}_{ijk}^{\delta} = (\mu_{ijk,x}\cos\delta + \mu_{ijk,z}\sin\delta, \mu_{ijk,y}, -\mu_{ijk,x}\sin\delta + \mu_{ijk,z}\cos\delta). \tag{4.18}$$

The only components that matter are the ones parallel to the detector plane, i.e., the first two components.

Summing these expressions over the entire lattice then yields expressions for the magnetic phase shift for each of the tilt series:

$$\varphi_m^{\gamma}(k_x, k_y) = \sum_{(i,j,k)\in D} v_{ijk,\gamma} \tag{4.19}$$

$$\varphi_m^{\delta}(k_x, k_y) = \sum_{(i,j,k)\in D} w_{ijk,\delta} \tag{4.20}$$

where

$$v_{ijk,\gamma} \equiv e^{2ai\left(k_x i + k_y \left(j \cos\gamma - k \sin\gamma\right)\right)} \left[\mu_{ijk,x}\mathscr{S}_y - \left(\mu_{ijk,y}\cos\gamma - \mu_{ijk,z}\sin\gamma\right)\mathscr{S}_x\right]; \tag{4.21}$$

$$w_{ijk,\delta} \equiv e^{2ai\left(k_x \left(i \cos\delta + k \sin\delta\right) + k_y j\right)} \left[\left(\mu_{ijk,x}\cos\delta + \mu_{ijk,z}\sin\delta\right)\mathscr{S}_y - \mu_{ijk,y}\mathscr{S}_x\right]. \tag{4.22}$$

These sums run over all object voxels, so that we can easily change the magnetization in a single voxel (i, j, k) and compute the result of that change on all the phase shifts for all tilt angles by means of the update equations:

$$\Delta\varphi_m^\gamma(k_x, k_y) = v'_{ijk,\gamma} - v_{ijk,\gamma}; \tag{4.23}$$

$$\Delta\varphi_m^\delta(k_x, k_y) = w'_{ijk,\delta} - w_{ijk,\delta}, \tag{4.24}$$

where the primed quantities refer to the new magnetization state of the voxel. As a result, updating the phase shifts after a single voxel magnetization change only requires the contributions of that single voxel to be computed, which leads to an efficient update algorithm. It should be noted, however, that the phase shift of each object voxel affects the intensity in each image pixel, so that the tomographic projection problem cannot be reduced to a sparse matrix description.

An example application of this forward projection model is shown in Figure 4.6(b)-(d). The magnetization pattern in (b) is constructed by taking the unit tangent vectors to a series of Cassini ovals described by the equation:

$$\left[(x - a)^2 + y^2\right]\left[(x + a)^2 + y^2\right] = b^4. \tag{4.25}$$

The ratio b/a determines whether the curves satisfying the equation are ovals, dog bone shaped, leminiscate, or two separate loops, resulting in the magnetization pattern shown in Figure 4.6(b) for a range of b/a ratios in the interval $[0, 3/2]$. The pattern has two vortices and a cross-tie wall at the center. The vortex core on the left points along the positive z-direction, and the rightmost vortex has the opposite core polarity.

The sphere-based forward projection model was applied to this configuration, using a multigrid lattice parameter of $a = 0.5$ nm, and oval disk semi-axes 72.5 and 84.5 nm for a thickness of $t = 11$ nm; the resulting object grid is made up of 117,975 voxels. Figure 4.6(c) and (d) show the resulting magnetic phase shifts for two tilt series around the x and y axes in steps of 20°. Note the reversal of the sign of the phase shift between two orientations that are 180° apart. The opposite core polarities give rise to an asymmetry in the vertical positions of the core projections. All phase maps are displayed with a common grayscale with phase shift extrema of $(m)0.645$ radians. Once the phase shift is known, the corresponding Lorentz images can be computed using Equation (4.8).

4.3.2.3 Example Lorentz Image Simulation

As a final illustration of the forward modeling of Lorentz TEM imaging we consider the magnetic phase shift map of Figure 4.5(d). The islands have a saturation magnetic induction of $B_0 = 1$ T, and a mean inner electrostatic potential of $V_0 = 20$ V, which gives rise to the electrostatic phase shift φ_e. We assume a microscope operating voltage of 200 kV, a spherical aberration constant of $C_s = 1$ m, a beam divergence angle of $\theta_c = 10^{-6}$ rad, and a defocus spread of $\Delta = 10\,\mu$m; these parameters enter the microscope transfer function $\mathscr{T}(\mathbf{r}_p)$, which is described in more detail in [245, Chapter 10]. The particles to the left of the vertical center line have a thickness of 20 nm, those on the right are 10 nm thick; the computational array has 512×512 pixels with a pixel size of 2 nm, resulting in a field of view of about 1 μm^2. The total phase shift ranges from -3.725 to $+15.91$ rad, and the particles are "supported" by an amorphous support film of 20-nm thickness and a mean inner potential of 10 V.

Combining all microscope parameters and the phase shift maps into Equation (4.8) results in the images shown in Figure 4.7, which represents a so-called "through-focus series" of Fresnel (out-of-focus) images. Such images are readily obtained experimentally by defocusing the microscope's Lorentz objective lens. The defocus values of $\Delta f = (m)0.5$ mm indicated in the figure are

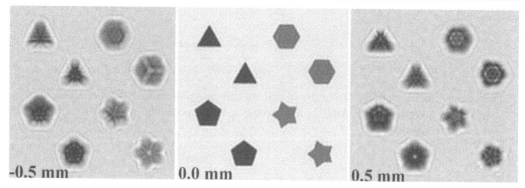

Figure 4.7: Through focus series of Fresnel images for the magnetic particles illustrated in Figure 4.5.

rather large; typically, for quantitative phase reconstructions using the transport-of-intensity equation (TIE) formalism [755], defocus values are kept smaller, of the order of a few hundreds of nanometers to a few micrometers.

4.4   Summary

In this chapter, we have reviewed the concept of forward modeling as applied to electron microscopy imaging and diffraction modalities. We have introduced a number of electron-based data acquisition modalities and provided two case studies of forward models: a three-step model for electron backscatter diffraction (EBSD) patterns, which relies on a combination of Monte Carlo modeling of the stochastic aspects of electron scattering with dynamical (elastic) scattering; and a forward model for Lorentz transmission electron microscopy (LTEM) that can be used to compute the magnetic phase shift for an arbitrary magnetization configuration projected along an arbitrary direction. All of these numerical approaches lend themselves to massive parallelization on multi-CPU and/or GPU platforms.

Acknowledgments

The author would like to acknowledge financial support from an AFOSR-MURI grant # FA9550-12-1-0458 during the preparation of this chapter. The original EBSD work was funded by an ONR grant # N00014-12-1-0075, whereas the original work on the sphere projections for Lorentz microscopy was funded by the US Department of Energy, BES # DE-FG02-01ER45893.

Chapter 5

Inverse Problems and Sensing

Charles A. Bouman
Purdue University

5.1 Introduction

Experimental scientists are taught throughout their careers that their objective is to strive towards the pure measurement of any quantity of interest. They are trained with a diverse bag of creative tricks, some of which are general, and others specific to their field. But all these tricks and innovations are designed to "reduce variation," "correct for distortion," "remove the contaminating signal," in essence to fix the problem. So for example, each pixel should measure a physical property at a point, with no distortion, and no error. Their objective is to build a perfect instrument that measures the single desired quantity of interest without contamination, noise, or distortion.

This might seem to be an unassailable goal, but on closer inspection, this approach has serious and growing flaws. In order to understand why this is true, we consider the following three trends, which make the goal of achieving pure measurements less desirable.

- *The cost of making "pure" measurements is growing exponentially.* Science seems to be on an endless march towards the purest possible measurement of a single uncontaminated physical property at a single point in space and time. However, as experimental science approaches physical limits, the cost of achieving such pure measurements is limiting progress. In fact, no measurement is ever pure, and adopting this viewpoint tends to obfuscate the required tradeoffs.

- *Computation is becoming cheap much faster than sensor hardware is improving.* While Moore's law is perennially questioned, available computation per second per dollar has continued its exponential growth curve; while sensor performance, which is often limited by fundamental laws of physics, tends to grow slowly. Leveraging inexpensive computation therefore represents a powerful strategy for improving sensor performance.

- *Making many flawed measurements may be better than making a few pure measurements.* The disadvantage of pure measurements is each one tells you so little and at such great cost. In fact, a rich and diverse set of inexpensive measurements which are "contaminated" across many unknowns may be much more informative.

5.2 Traditional Approaches to Inversion

In order to take a fresh look at the problem, it is first necessary to distinguish between the measurements, what we will call y, and the desired quantity, what we will call x. In the traditional view, x and y should be the same, except for the non-idealities of the sensor. However, in the real world, we must determine x from the measurements y.

In the simplest view of the world, the measurements might be assumed to have no noise as shown in Figure 5.1. Here the measurement vector $y \in \mathbb{R}^M$ is a deterministic function of the unknown vector $x \in \mathbb{R}^N$ given by $y = f(x)$. In practice, y and x may be very high dimensional signals, images, or even volumes. Nonetheless, in all these cases the object can be rasterized and formed into a simple

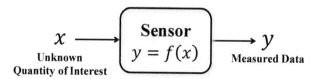

Figure 5.1: Illustration of an over-simplified sensor model in which the measurement, y, is assumed to be deterministically related to the unknown quantity, x. In this case, recovery of x from y can be viewed as an inverse problem. In simple cases, this inverse is characterized as data "pre-processing" or "calibration," but in more complex cases it is characterized as reconstruction.

column vector of data. The function $f(x)$ then maps an N-dimensional object to an M-dimensional set of measurements.

The problem of recovering the true x from the measured y is then known as an inverse problem. For simple sensor systems, we sometimes refer to this inverse problem as "pre-processing" or "calibration." For example, we may need to calibrate out gain or offset variations, or remove the effects of blurring in optical systems. However, in more complex systems, we refer to this inverse problem as reconstruction. For example, in tomographic systems, the measurements are only indirectly related to the unknown object through integral projections; so the object must be reconstructed from these projections.

In order to illustrate the challenges of inversion, we can consider the simple case of a linear forward model given by $f(x) = Ax$. In this case, we have that

$$y = Ax , \tag{5.1}$$

where y is an $M \times 1$ column vector, x is an $N \times 1$ column vector, and A is an $M \times N$ matrix. We refer to Ax as the forward model since it describes how the measurements should vary with changes in the unknown, x.

If the number of measurements is equal to the number of unknowns (i.e., $M = N$) and the matrix A is invertible, then we can solve this inverse problem to find the unknown vector; and the solution is given by

$$\hat{x} = A^{-1}y .$$

Notice we use $\hat{x}$ to denote our solution, which in this case is exactly x. However, this a very uncommon case since our estimate will typically have errors.

A more common case might be that the number of measurements exceeds the number of unknowns so that $M > N$. In this case, there may not be any solution to the inverse problem. However, we can still get an answer by solving for the least-squares inverse given by the expression

$$\hat{x} = \arg\min_x ||y - Ax||^2 . \tag{5.2}$$

Here the notation of $\arg\min_x f(x)$ simply means that we should find the value of x that achieves the minimum of $f(x)$, rather than finding the minimum value of $f(x)$ itself. So for example, if we asked for the min to the function which provides the altitude above sea level in North America, then the answer would be -85 meters. However, if we asked for the arg min to this function, then the answer would be the location of the minimum, in this case Death Valley.

This least-squares (LS) "fit" approach is widely used in scientific and engineering applications such as curve fitting through noisy data points. However, here we will treat the approach in a more abstract way and as a method for solving inverse problems, which may not be so familiar. Nonetheless, we will show through an example, how these are simply different views of the same problem.

When A has rank N, then the least-squares inverse of Equation (5.2) can be abstractly expressed with a simple closed-form expression given by

$$\hat{x} = (A^t A)^{-1} A^t y . \tag{5.3}$$

Here, the interpretation is that the least-squares inverse minimizes the average squared error between the data, y, and the hypothesized forward model values, Ax. This solution to the LS problem can be more compactly expressed using the so-called pseudo-inverse of the matrix A given by

$$A^+ = (A^t A)^{-1} A^t \ .\mathbb{E}$$

So then the solution to the LS problem can be expressed as

$$\hat{x} = A^+ y \ .$$

The pseudo-inverse exists in cases when A^{-1} does not, for example when when $M > N$ and A is not square. However, the pseudo-inverse acts like an inverse because

$$[A^+]A = [(A^t A)^{-1} A^t]A = (A^t A)^{-1}(A^t A) = I \ .$$

In other words, when $y = Ax$, then the LS solution is given by

$$\hat{x} = A^+ y = A^+ A x = x \ .$$

So when the measurements are noiseless and the matrix A has full rank, then the pseudo-inverse gives the exact answer to the inverse problem.

In order to make this more concrete, consider the problem of fitting noisy data on the interval $[0,1]$ with a 3rd-order polynomial. For this problem, the measured values at samples $m = 1, \cdots, M$ are given by

$$y_m = x_1 + x_2 \left(\frac{m-1}{M-1} \right) + x_3 \left(\frac{m-1}{M-1} \right)^2 + x_4 \left(\frac{m-1}{M-1} \right)^3 + \varepsilon_m \ ,$$

where ε represents the curve fitting error to be minimized, $t = \left(\frac{m-1}{M-1} \right)$ is the sample time, and x_n are the unknown coefficients to be estimated. We can represent this problem in the form of the LS minimization of Equation (5.2) by constructing a matrix A with entries

$$A_{m,n} = \left(\frac{m-1}{M-1} \right)^{n-1} \ ,$$

for $m = 1, \cdots, M$ and $n = 1, \cdots, N$.

Figure 5.2 illustrates the properties of the LS fit for a simple example of fitting a third-order polynomial ($N = 4$) to 20 data points ($M = 20$) in the interval $[0,1]$ with additive Gaussian noise of

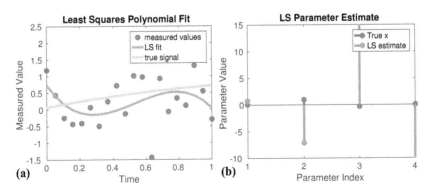

Figure 5.2: Illustration of the use of least-squares estimates for 3rd-order polynomial fit to data with additive white Gaussian noise. (a) Plot of least-squares fit to third-order polynomial model. (b) Plot of true versus estimated model parameters (i.e., the polynomial coefficients) versus the parameter index (i.e., n) for least-squares fit resulting in a normalized RMSE error of 21.7. *Notice that the least-squares estimate tends to over-fit the model, leading to very inaccurate parameter estimates.*

variance 1. Here, Figure 5.2(a) shows the LS fit of the polynomial to the data resulting from the use of the four coefficients shown in Figure 5.2(b). Notice the estimated coefficients are far from their true values and the polynomial is not a very good fit to the data. In fact, if we define the normalized root mean squared error to be given by

$$\text{NRMSE} = \sqrt{\frac{\sum_{n=1}^{N}(\hat{x}_n - x_n)^2}{\sum_{n=1}^{N}(x_n)^2}} \, ,$$

then the NRMSE for the LS estimate is an enormous 21.7. This means that the error is on average $21.7\times$ the value of the signal. One might think this problem is easy because the number of data points is large ($M = 20$) and the number of unknowns is small ($N = 4$), but we see it is actually very difficult to reliably estimate high-order polynomial coefficients from such noisy data.

While the LS estimate clearly has problems in this example, the problem is much worse when there are fewer measurements than unknowns and $M < N$. In this case, the LS estimate is not unique because we have more unknowns than we have equations, and there is an infinite number of solutions to the LS problem.

In order to better understand this issue, consider the matrix inversion $(A^t A)^{-1}$ required in Equation (5.3). Since the matrix $A^t A$ is symmetric, we know that it must have an eigen decomposition with the form

$$A^t A = E \Lambda E^t \, ,$$

where Λ is a diagonal matrix of eigenvalues and the columns of the matrix E are the corresponding eigenvectors. The symmetry also ensures that both the eigenvalues and eigenvectors are real valued and the eigenvectors are orthonormal. In this case, the required inverse is given by

$$(A^t A)^{-1} = E \Lambda^{-1} E^t \, .$$

However, since $M < N$, then we know that at least $N - M$ of the eigenvalues will be zero; so the diagonal matrix must have the form

$$\Lambda = \text{diag}\{\lambda_1, \cdots, \lambda_M, 0, \cdots 0\} \, .$$

This makes the key problem clear because the matrix Λ cannot be inverted. However, we can define the pseudo-inverse of the matrix as[1]

$$\Lambda^+ = \text{diag}\{\lambda_1^{-1}, \cdots, \lambda_M^{-1}, 0, \cdots 0\} \, ,$$

where we only invert the non-zero eigenvalues. From this we may define the pseudo-inverse of A as

$$A^+ = E \Lambda^+ E^t A^t \, . \tag{5.4}$$

This more general form of the pseudo-inverse is well defined even when the rank of A is not full and always admits a solution to Equation (5.2) given by

$$\hat{x} = A^+ y \, .$$

However, when $M < N$ and A is rank deficient, then this solution will not be unique. In fact, there will be an infinite number of solutions to this inverse problem with the form

$$x = A^+ y + z \, ,$$

where z is any vector in the $N - M$-dimensional null space of A defined by $\{z : Az = 0\}$.

So when $M < N$ and the number of measurements is less than the number of unknowns, then the uncertainty in any estimate is potentially infinite! Clearly, there must be a better alternative to the LS estimate in this case. The central problem is that, when the data is sparse or noisy, then our estimates for some parameters may be very poor, but the LS estimate does not impose any reasonable constraints to the estimates in this case.

---

[1] Here we assume that M of the eigenvalues are non-zero. Any additional zero eigenvalues are also left unchanged.

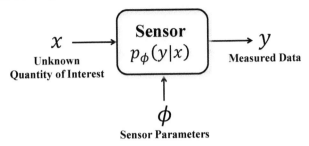

Figure 5.3: Illustration of an improved sensor model in which the measurement, y, is assumed to be stochastically related to the unknown quantity, x. Here there are usually additional parameters representing, for example, instrument calibration that must be recovered in order to accurately determine x. With this model, the recovery of x from y can be viewed as a statistical inverse problem.

5.3   Bayesian and Regularized Approaches to Inversion

In order to address the deficiencies of the LS estimate, we will need to bring in the concepts of statistical modeling for both the data, y, and the quantity being estimated, x. Figure 5.3 illustrates a much more realistic viewpoint in which the measurements are related to the unknowns, through some general probabilistic relationship, which we conveniently denote by the conditional probability density $p_\phi(y|x)$. Here we also allow for the common situation in which the sensor has some, perhaps unknown, parameters, ϕ. So for example, calibration parameters corresponding to focus, alignment, gain, offset, and any number of additional factors represent nuisance parameters, that are required to be known, but are not the quantity of direct interest. For clarity, we will drop the dependency on ϕ unless it is needed.

Chapter 3 covers the methods of probabilistic modeling in detail, but here we provide a quick overview. Mathematically, we can express the *conditional probability* of y given x as

$$p(y|x) = \frac{p(x,y)}{p_x(x)} \,,$$

where $p(x,y)$ is the joint density of the random variables X and Y, and $p_x(x)$ is the marginal density of X alone. The conditional probability $p(y|x)$ will be the *forward* model for the sensor system. Notice that our forward model of the system is not represented by a deterministic relationship of the form $y = f(x)$. Instead, it is represented by a conditional probability density function that describes the uncertainty in the vector measurements, y, given the vector unknown, x.

This abstract representation is very powerful because the simple notation can hide an enormous amount of complexity. In the traditional case, the conditional distribution $p(y|x)$ may be viewed as being peaked or having a delta function at the value $y = f(x)$. In this case, an input of x produces an output of $y = f(x)$ with certainty. However, in the more general case, the conditional distribution allows for substantial and complex variation about this value. In particular, the variation may be large in some directions and small in others, or the variation may even contain additional information when its magnitude is related to x itself.

In this new framework, our goal is to *estimate x* from the measurements y, rather than to measure x directly. Figure 5.4 shows such a sensor system, in which the measured data is used as input to a function

$$\hat{X} = T(Y) \,,$$

that estimates the unknown. In this framework, the function, $T(\cdot)$, is known as an estimator, and the result, $\hat{X}$, is known as the estimate. For example, in the least-squares problem we chose $T(x) = A^+x$, where A^+ denotes the pseudo-inverse. However, here we will introduce a broader range of estimators that are more effective when the data are noisy and sparse.

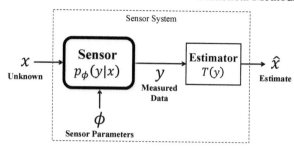

Figure 5.4: Illustration of a sensor system in which the measurement is used to estimate the unknown x. This estimation process typically requires computation and can be viewed as an inverse problem.

Computing an estimate, $\hat{X}$, as shown in Figure 5.4, can be viewed as an inverse problem because the goal is to invert the effects of the sensor to recover X. However, this inverse problem is a bit different than in the traditional deterministic inverse problem previously discussed. In this problem, we cannot hope to exactly invert the effect of the sensor due to the uncertainty introduced by noise. Nonetheless even with noise, we can often get a good estimate of X from Y by making assumptions about the distribution of the unknown X.

As it turns out, it is typically difficult, usually impossible, to identify a "best estimator." As the old saying goes, even a broken clock correctly estimates the time twice a day. However, there are a variety of estimators with desirable properties.

Generally speaking, estimators can be separated into the two categories of Frequentist and Bayesian. Frequentist estimates tend to be most useful when the number of unknowns is much less than the number of measurements; while Bayesian estimates are typically used when the problem is underdetermined, with the number of parameters being large and the data meager. We will see that the least-squares estimate of Section 5.2 is typical of a Frequentist estimate because it does not assume anything about the distribution of the unknown. In this section, we will introduce the concept of a Bayesian estimate that makes assumptions about the distribution of x in order to reduce the expected error.

We start by considering the Frequentist approach and illustrating it with the ubiquitous *maximum likelihood* (ML) estimate defined by

$$\hat{X}_{ML} = \arg\min_{x \in \mathbb{R}^N} -\log p(Y|x) \ . \tag{5.5}$$

Here we refer to the function $l(x) = p(Y|x)$ as the likelihood as a function of the vector x, which we maximize and $-\log p(Y|x)$ as the negative log likelihood that we minimize.

In order to better illustrate this approach, consider a simple sensor model given by

$$Y = Ax + W \ , \tag{5.6}$$

where the vector $W \in \mathbb{R}^M$ is assumed to be a Gaussian random vector with independent components of mean zero and variance σ_w^2. Here W represents the additive noise that typically occurs in sensor systems.

Notice that in contrast to Equation (5.1), here we use an upper case letter for $Y \in \mathbb{R}^M$ and a lower case letter for $x \in \mathbb{R}^M$. This is because in this context Y represents a random vector and x represents a deterministic one. This is a very important distinction. Intuitively, x is assumed to be unknown not deterministic.

Using this model, the conditional distribution of Y given x is given by

$$p(y|x) = \frac{1}{(2\pi\sigma_w^2)^{M/2}} \exp\left\{ -\frac{1}{2\sigma_w^2} ||y - Ax||^2 \right\}$$

From this, we can calculate the negative log likelihood of x given by

$$-\log p(y|x) = \frac{1}{2\sigma_w^2}||y - Ax||^2 + \frac{M}{2}\log(2\pi\sigma_w^2) \,.$$

The ML estimator is then given by

$$T(y) = \arg\min_{x \in \mathbb{R}^N} \left\{ \frac{1}{2\sigma_w^2}||y - Ax||^2 + \frac{M}{2}\log(2\pi\sigma_w^2) \right\} \,,$$

and the ML estimate is given by

$$\hat{X}_{ML} = T(Y) = \arg\min_{x \in \mathbb{R}^N} \left\{ \frac{1}{2\sigma_w^2}||Y - Ax||^2 \right\} \,. \tag{5.7}$$

Notice that in this case, the ML estimator turns out to be exactly the same as the LS estimate of Equation (5.2).

From the discussion of the LS estimate, we know that the ML estimate will have at least one solution given by

$$\hat{X}_{ML} = A^+ Y \,,$$

but that the solution may have very high variance when the data is noisy and sparse. If there is insufficient data and $M < N$, then the number of ML solutions can be infinite and the corresponding variance of the estimate will be unbounded. In fact, even when $A^t A$ is invertible, the matrix may have eigenvalues so close to zero that its inverse will hugely amplify any small noise or error. In this case, the inverse problem can be viewed as having a unique solution but being practically unstable.

This leads to the three conditions associated with a well posed inverse:

- The solution should exist;
- The solution should be unique;
- The solution should be stable.

In order to ensure all three conditions, Tikhonov proposed that the inverse problem be "regularized" through the inclusion of an additional stabilizing function resulting in the following modified optimization problem,

$$\hat{X}_R = \arg\min_{x \in \mathbb{R}^N} \left\{ \frac{1}{2\sigma_w^2}||Y - Ax||^2 + u(x) \right\} \,, \tag{5.8}$$

where $u(x)$ is a stabilizing function selected to regularize the inverse problem. While the question of how to best choose the regularizing function, $u(x)$, is largely left open, a traditional choice for the regularizing function is simply the weighted Euclidean norm given by $u(x) = \lambda ||x||^2$, where λ is a user selected weighting.

In the Bayesian approach, the unknown signal X is modeled as a random vector. The advantage of the Bayesian approach is that it provides a popular statistical framework for solving inverse problems that makes the inverse well posed and provides at least a reasonable method for selecting the regularizing function. However, its disadvantage is that it requires one to select a reasonable distribution for X, which can in some cases be challenging.

A popular Bayesian estimator is the minimum mean squared error (MMSE) estimator given by

$$\hat{X}_{MMSE} = T(Y) = \mathbb{E}[X|Y] = \int_{\mathbb{R}^N} x p(x|Y) dx \,,$$

where x is assumed to be represented by an N-dimensional vector. Not too surprisingly, the MMSE estimator has the potentially desirable property that it minimizes the mean squared error given by

$$\mathbb{E}\left[||X - \hat{X}_{MMSE}||^2\right] \,.$$

However, a major down side of the MMSE estimate is that it often requires the evaluation of an N-dimensional integral, which can be very computationally expensive.

The maximum *a posteriori* (MAP) estimate is also very widely used with the goal of selecting the estimate that maximizes the posterior probability, $p(x|y)$. Formally, this is written as

$$\hat{X}_{MAP} = \arg \max_{x \in \mathbb{R}^N} p(x|Y) \, .$$

Using this notation, we can derive a general expression for the MAP estimate.

$$
\begin{aligned}
\hat{X}_{MAP} &= \arg \max_{x \in \mathbb{R}^N} p(x|y) \\
&= \arg \max_{x \in \mathbb{R}^N} \log p(x|y) \\
&= \arg \max_{x \in \mathbb{R}^N} \log \frac{p(y|x)p(x)}{p(y)} \\
&= \arg \max_{x \in \mathbb{R}^N} \left\{ \log p(y|x) + \log p(x) - \log p(y) \right\} \\
&= \arg \min_{x \in \mathbb{R}^N} \left\{ -\log p(y|x) - \log p(x) \right\} \, .
\end{aligned}
$$

Notice that the MAP estimate depends on a key relationship, known as *Bayes' rule*, that also serves as the name sake for the method of Bayesian statistics.

$$p(x|y) = \frac{p(y|x)p(x)}{p(y)}$$

Usually, the physics of a real sensing system is most naturally described by the dependence of Y on X; so Bayes' rule provides a mechanism to formally invert this relationship, and express the dependence of X on Y.

While there are many good Bayesian estimators, the MAP estimate is widely used and illustrates the key issues associated with designing sensing systems.

$$\hat{X}_{MAP} = \arg \min_{x \in \mathbb{R}^N} \Big\{ \underbrace{-\log p(y|x)}_{\text{Forward Model}} + \underbrace{-\log p(x)}_{\text{Prior Model}} \Big\} \tag{5.9}$$

Intuitively, one can think of the two terms in the MAP cost function as corresponding to a forward model of the sensor and a prior model of the unknown. Minimization of the first term corresponds to improving the fit of the unknown, x, to the data, y. Alternatively, minimization of the second term corresponds to maximization of the probability $p(x)$. The resulting MAP estimate balances these two competing goals by minimizing their sum.

Again, we illustrate the use of the MAP estimate using our previous simple example where

$$Y = AX + W \, , \tag{5.10}$$

but where X is now assumed to be a Gaussian random vector with independent components of mean zero and variance σ_x^2. In this framework, the prior distribution for X is given by

$$p(x) = \frac{1}{\sqrt{2\pi\sigma_x^{2N}}} \exp \left\{ -\frac{1}{2\sigma_x^2} ||x||^2 \right\} \, ;$$

so the MAP estimate is given by

$$\hat{x}_{MAP} = \arg \min_{x} \left\{ -\log p(Y|x) - \log p(x) \right\} \, .$$

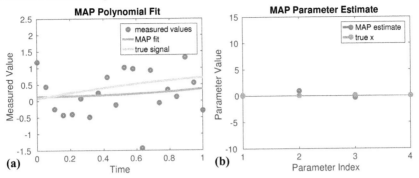

Figure 5.5: Illustration of the use of MAP estimates for 3rd-order polynomial fit to data with additive white Gaussian noise. (a) Plot of MAP fit to third-order polynomial model. (b) Plot of true versus estimated model parameters (polynomial coefficients) for MAP fit resulting in a normalized RMSE error of 1.0.

Notice that we use a lower case x in the argument of $p(x)$ since this is the deterministic vector over which we are doing the optimization, rather than the actual random vector X. Plugging in the assumed distributions for our example and dropping constant terms, we then get

$$\hat{x}_{MAP} = \arg \min_{x \in \mathbb{R}^N} \left\{ \frac{1}{\sigma_w^2} ||Y - Ax||^2 + \frac{1}{\sigma_x^2} ||x||^2 \right\} , \qquad (5.11)$$

Notice that this optimization problem has the form of the regularized inverse from Equation (5.8), but with $u(x) = \frac{1}{\sigma_x^2}||x||^2$. So we see that the MAP estimate is much like the regularized inverse, but with a particular function specified for the regularizing term.

Solving this MAP optimization problem results in the following closed form solution.

$$\hat{x}_{MAP} = \left(A^t A + \frac{\sigma^2}{\sigma_x^2} I \right)^{-1} A^t Y \qquad (5.12)$$

Notice that even when $A^t A$ is not invertible, this problem can always be solved because the additional term ensures that the eigenvalues of the matrix to be inverted are $\geq \frac{\sigma^2}{\sigma_x^2} > 0$.

Figure 5.5 again illustrates the properties of the MAP estimate for the simple problem of fitting a 3rd-order polynomial with $\sigma_w^2 = 1$ and $\sigma_x^2 = 1$. Notice that the coefficients, $\hat{x}$, are now much more accurately estimated than in the LS case of Figure 5.2 with an NRMSE of 1.0, as opposed to the NRMSE of 21.7 for the LS estimate. Also notice that the polynomial fit computed by $A\hat{x}$ is much more reasonable. The key advantage of the MAP estimate is that it does not only fit the data, it also accounts for enforcing constraints on the parameters X that ensure that the estimated coefficients of the polynomial are not unreasonable.

5.4   Why Does Bayesian Estimation Work?

The prior term of the MAP estimate is both the key to the Bayesian inverse methods and perhaps the part that is most suspicious for traditionally trained scientists. This is because it suggests that the solution should be chosen not just to fit the data, but also to fit our model of how x should behave. Since physical scientists are trained to be "unbiased" in their views, this seems very wrong.

What if the prior model of the MAP estimate is wrong? This is a good question, and a healthy suspicion of prior models in Bayesian estimation is very much warranted, but we will see that in situations where the number of unknowns is large and the data meager, unbiased estimators can have very large or even infinite variance, rendering them useless. Alternatively, Bayesian estimators offer a method which can provide reasonable estimates in these situations; so we are compelled to consider them. Moreover, real-world estimates are almost always biased because underlying physical phenomenon such as images or signals, typically have infinite dimension. In order to make signals finite dimensional, we must inevitably assume that the bandwidth is finite, the sampling rate is limited, or that some other similar approximation applies. So the Bayesian framework allows us to make these assumptions very openly, where they can be examined for their veracity.

From an inspection of Equations (5.8) and (5.9), we can see that the MAP estimate is simply a regularized inverse, in which the stabilizing function is chosen to be

$$u(x) = -\log p(x) + \text{constant} ,$$

or equivalently, we have that the MAP prior model must be given by

$$p(x) = \frac{1}{z} \exp\{-u(x)\} ,$$

where z is a normalizing constant that is given by $z = \int_{\mathbb{R}^N} \exp\{-u(x)\} dx$. In this form, we can see that the prior distribution can be viewed as a Boltzmann or Gibbs distribution with an associated energy function $u(x)$. This leads to another perspective of the MAP estimate as the solution to an energy minimization problem given by

$$\hat{X} = \arg\min_{x \in \mathbb{R}^N} \{l(x) + u(x)\} ,$$

where both $l(x)$ and $u(x)$ represent energy terms that must be balanced.

In fact, the distinction between the MAP estimate and the regularized or penalized ML estimate is really one of philosophy rather than substance. In the case of the penalized ML estimate, the solution is viewed as a regularized fit to the data; whereas, with the MAP estimate, the solution is viewed as the maximum of the posterior density. But they are both functionally equivalent. Nonetheless, the two perspectives do have practical ramifications because in the Bayesian MAP framework, the objective is to select $p(x)$ so that it empirically fits the distribution of the unknown when possible. Of course, in practice, the selection of the prior is not only done to fit the observed distribution of the data, but also to produce a useful estimate, $\hat{X}$. So practical MAP estimates tend to blend these two viewpoints.

In order to better understand the power of model-based methods, one must understand the fundamental tradeoffs that these methods exploit. Consider again a measurement problem with the form of Equation (5.10), but this time where the number of measurements, M, is much less than the number of unknowns, N.

$$Y = AX + W$$

In this case, A is an $M \times N$ matrix and the number of unknowns is much greater than the number of measurements. In some sense, this is a very common situation since in physical problems it is always possible to increase the resolution of the reconstruction until the number of unknown pixels is much larger than the available data. In this case, the variance of the reconstruction becomes very large, really infinite, because there is an infinite number of possible solutions that fit the data.

A scientist might suggest that this problem can be fixed by reducing the resolution of the reconstruction image until the value of N is less than or equal to M. Then the reconstruction is unique, and the variance is finite. However, by limiting the resolution, we are implicitly imposing a constraint on the solution that is not physical. In essence, we are unfairly biasing the solution by requiring that it be band limited to the Nyquist frequency associated with the selected sampling rate. If we are required to impose unfair biases on our solution, then this raises the question of whether these are the best biases to impose.

In the MAP problem, the variance of our estimate remains finite because the prior model regularizes the inverse problem and makes the solution to the MAP estimate unique. But this reduction in variance occurs at the cost of introducing a bias due to the prior modeling assumption.

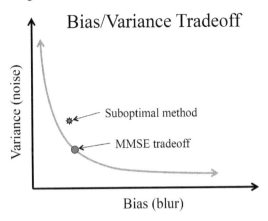

Figure 5.6: The tradeoff between bias and variance is at the core of model-based image processing. Variance typically represents noise in the image and bias typically represents excessive smoothing or blur. In many problems, the variance can become large or even infinite if the bias is driven to zero. In practice, this means it is not possible to estimate a solution with infinite resolution using a finite amount of data.

Figure 5.6 illustrates the general problem graphically. Virtually all estimation problems require a tradeoff between bias and variance. Both bias and variance contribute to the total mean squared error (MSE) as shown in the following formula.

$$\text{MSE} = (\text{Expected Squared Estimate Bias}) + (\text{Expected Estimate Variance})$$

So the total MSE is the sum of the expected errors due to bias and variance. We can break down this relationship in more detail in order to better understand why this is the case. To do this, we first define the bias which is the expected difference between the true image, X, and its estimate, $\hat{X} = T(Y)$.

$$\text{Bias}(x) = \mathbb{E}\left[X - \hat{X}|X = x\right]$$

Notice, that the bias assumes that the random variable X is known and equal to x. Intuitively, for a problem such as image reconstruction, the bias represents the systematic errors in the reconstruction due to blurring or artifacts. These errors cannot be eliminated by averaging over many datasets; so in this sense, they are more pernicious. Alternatively, the variance in the estimate is caused by the random variations due to noise in the measurements and is given by

$$\text{Var}(x) \quad = \quad \mathbb{E}\left[\left|\left|\hat{X} - \mathbb{E}\left[\hat{X}|X\right]\right|\right|^2 |X = x\right] .$$

In the Frequentist framework, it can be easily shown that the MSE in the estimate of the unknown x is given by the sum of errors due to the bias and the variance.

$$\text{MSE}(x) = \mathbb{E}\left[||X - \hat{X}||^2 \big| X = x\right] = ||\text{Bias}(x)||^2 + \text{Var}(x) .$$

Notice that in the Frequentist framework, the MSE depends on the specific image, x, that is being estimated. In a Bayesian framework, we take the expectation over x to find the single expected MSE over all cases.[2]

$$\text{MSE} = \mathbb{E}\left[||\text{Bias}(X)||^2\right] + \mathbb{E}\left[\text{Var}(X)\right]$$

---

[2]Note that the notion $\text{Var}(X)$ does not refer to the variance of the random variable X. Instead, it refers to the conditional variance of the estimator $\hat{X}$ given X.

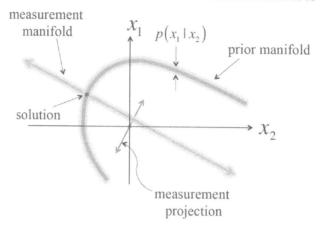

Figure 5.7: Graphical illustration of how forward and prior models can interact synergistically to dramatically improve results. The green curve represents the thin manifold in which real images lie, and the gray line represents the thin manifold defined by noisy linear measurements. If the number of measurements, M, is less than the number of unknowns, N, then classical methods admit no unique solution. However, in this case model-based approaches can still provide a useful and unique solution at the intersection of the measurement and prior manifolds.

So we see that the MSE is the sum of two terms corresponding to the bias and the variance in the estimator.

The bias and variance each have different implications in applications. Typically, bias is considered less desirable since it will not average away over many measurements. One very important point is that while there is always a tradeoff between bias and variance, different approaches may present a more favorable tradeoff for different problems. In other words, it is always possible to do poorly as is shown by the star in Figure 5.6. In this case, the star represents an estimator with increased variance at the same bias or increased bias at the same variance. It is easy to show based on randomization arguments that the optimal bias-variance curve must be convex; so, if an estimator falls at a point that lies above the convex hull of the curve, then it must have sub-optimal performance.

The great power of model-based methods is that they can provide a framework for constructing estimators that yield a much more favorable bias-variance tradeoff. This is because they allow for accurate models of both the physical measurement and the object being imaged to be built into the estimation process.

In order to better understand why dramatic improvements in reconstruction are sometimes possible, consider Figure 5.7. The green line represents the set of all possible images in the N dimensional space; for illustrative purposes the space has dimension $N = 2$ in the figure. So in this simple example, there are really only two pixels, x_1 and x_2. Notice that this set of all images is thin in its embedding N dimensional space; so we say that this set forms a thin manifold. In order to see this, imagine that we are missing the value x_1, but we have all other pixels, and we would like to estimate the missing value of x_1. In this case, the conditional distribution of the unknown pixel is given by $p(x_1|x_2)$ as is shown in the figure. In practice, a single missing pixel can almost always be accurately estimated from all the remaining pixels using even a simple interpolation algorithm. So that indicates that in the vast majority of cases, the conditional distribution $p(x_1|x_2)$ is very peaked, and the manifold is very thin.

The fact that real images almost always live on a thin manifold[3] is an indication that real images have much lower intrinsic dimension than their embedding space. Bayesian estimation can effectively exploit this low intrinsic dimension by constraining the solution of any inverse problem to be on such a manifold. By enforcing this constraint, the number of measurements required to accurately reconstruct an image, X, is only equal to the number of dimensions of the manifold, which in this case is denoted by M. In applications, it may be that $M << N$ so it is possible to achieve accurate reconstruction even when the number of measurements is much smaller than the number of unknowns.

While Bayesian reconstruction methods have great potential, there are a few important ideas to keep in mind. First, as illustrated in Figure 5.7, the sparse measurements need to be made in a direction that is along the direction of the manifold. This is represented by the thin red line. If the measurement is perpendicular to the manifold, then little additional information will be obtained. While measurements that are designed to be along the manifold are best, this requires dynamic methods since the manifold's orientation varies depending upon the specific solution to the problem [368]. Randomized measurement directions are typically not as good, but at least they avoid the worse case possibility that the measurements will be completely aligned with the manifold. In fact, conventional measurement approaches can sometimes unintentionally lead to highly structured and uninformative measurements. Another danger with Bayesian reconstruction is that a poorly constructed prior model may mis-characterize the thin manifold, and result in a reconstruction with substantial inaccuracies. In other words, if done improperly, the Bayesian estimation framework may introduce large bias into the estimator and generate a reconstruction that is substantially different from the true image.

5.5   Model-Based Reconstruction

We have seen that Bayesian methods have the potential to greatly improve reconstruction accuracy, but to be successful, they require the careful choice of an accurate prior model. In other words, a good prior model can make the reconstruction more accurate, but a bad model can also present a danger by introducing large biases that might lead to self-fulfilling prophecies. One way to manage the risk associated with prior models is to keep them simple so that their effect can be well understood.

A simple class of prior models is formed by interactions between neighboring pixel pairs. This class of models, which we call pair wise Markov random fields, has the form of a Gibbs distribution given by

$$p_x(x) = \frac{1}{z} \exp \left\{ -\sum_{\{s,r\} \in \mathscr{P}} b_{s-r} \rho \left(\frac{x_s - x_r}{\sigma_x} \right) \right\} \tag{5.13}$$

where:

- The quantities s and r are short-hand notation for 2D discrete pixel indices such as $s = (s_1, s_2)$ where s_1, s_2 are in the set $\{\cdots, -1, 0, 1, \cdots\}$;
- The scalar parameter σ_x controls the level of regularization (i.e., smaller σ_x provides more regularization and larger σ_x provides less regularization;
- $\mathscr{P}$ is the set of all neighboring pixel pairs in the image;
- The normalizing constant z creates a legitimate probability density;
- The neighborhood weights b_s are typically normalized so that $\sum_s b_s = 1$;
- The potential function $\rho(\cdot)$ is selected to model the histogram of edge detail in real images.

The pairwise MRF of Equation (5.13) provides a simple but flexible framework for modeling

---

[3]An exception would be an image formed by independent identically distributed random variables. Such an image would densely fill the N dimensional space, but this would be a very unusual image for a real application.

images with sharp edge discontinuities through the selection of the function $\rho(\cdot)$ and the regularization parameter σ_x. The selection of the potential function $\rho(\cdot)$ is as much an art as a science. The derivative of ρ, known as the influence function, gives additional insight into the choice since it is analogous to the "force" that pulls neighboring pixels together.

Figure 5.8 shows some commonly used convex potential functions along with their associated influence functions. The baseline quadratic case corresponds to a Gaussian prior. Notice that the Gaussian's influence function increases rapidly. This implies that it tends to excessively smooth the sharp discontinuities that commonly occur at image edges.

Alternatively, potential functions such as the total variation [88, 871], Huber function [970], generalized Gaussian [119], and Q-GGMRF [1038] have influence functions which increase more slowly. This reduces the tendency to smooth across image edges, and therefore produces MAP estimates that better preserve edges and textures in images.

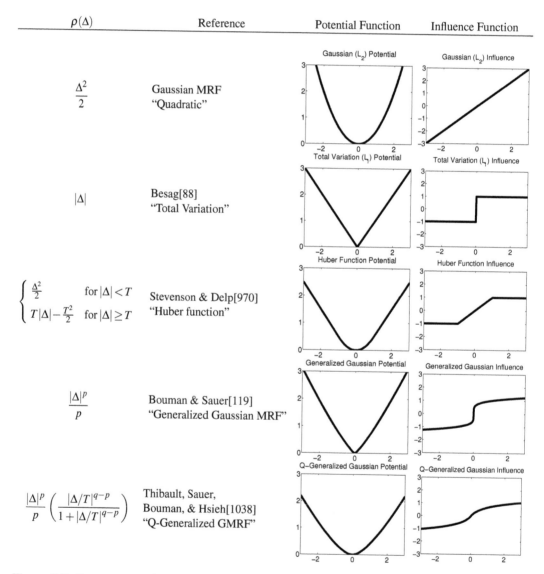

$\rho(\Delta)$	Reference	Potential Function	Influence Function
$\dfrac{\Delta^2}{2}$	Gaussian MRF "Quadratic"		
$\lvert\Delta\rvert$	Besag[88] "Total Variation"		
$\begin{cases} \dfrac{\Delta^2}{2} & \text{for } \lvert\Delta\rvert < T \\ T\lvert\Delta\rvert - \dfrac{T^2}{2} & \text{for } \lvert\Delta\rvert \geq T \end{cases}$	Stevenson & Delp[970] "Huber function"		
$\dfrac{\lvert\Delta\rvert^p}{p}$	Bouman & Sauer[119] "Generalized Gaussian MRF"		
$\dfrac{\lvert\Delta\rvert^p}{p}\left(\dfrac{\lvert\Delta/T\rvert^{q-p}}{1+\lvert\Delta/T\rvert^{q-p}}\right)$	Thibault, Sauer, Bouman, & Hsieh[1038] "Q-Generalized GMRF"		

Figure 5.8: Some commonly used convex potential functions, $\rho(\cdot)$, along with their associated influence functions, $\rho'(\cdot)$ plotted for $T = 1$ and shape parameters $p = 1.2$ and $q = 2$.

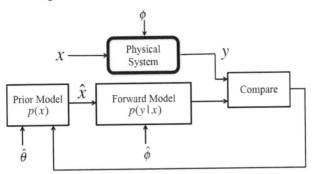

Figure 5.9: Graphical illustration of how model-based methods are typically used to solve inverse problems in imaging applications. The measurements Y of a physical system depend on an unknown image X. In addition, there are often unknown system-model parameters, ϕ, and prior-model parameters, θ. Model-based inversion works by searching for an input $\hat{x}$ that matches the measured output y, but also balances the need for the solution to be consistent with a prior model of the data.

Using this basic prior model, the MAP estimate for the simple forward model of Equation (5.6) is given by

$$\hat{x} = \arg\min_{x \geq 0} \left\{ \frac{1}{2\sigma_w^2} ||y - Ax||^2 + \sum_{\{s,r\} \in \mathscr{P}} b_{s-r} \rho \left(\frac{x_s - x_r}{\sigma_x} \right) \right\}, \qquad (5.14)$$

where we have also incorporated the non-negativity constraint, $x \geq 0$, which is often applicable in practical imaging problems. Notice, that if the potential function, $\rho(\cdot)$, is convex, then the overall MAP optimization problem is also convex. Strict convexity, which typically holds for such a problem, implies that the MAP estimate exists and is unique.

Computing the MAP estimate of Equation (5.14) requires numerical optimization of the MAP cost function. Over the years, a wide variety of numerical optimization methods have been studied for efficiently computing the solution to this optimization problem. Commonly used methods include gradient descent, conjugate gradient descent, preconditioned gradient descent, and iterative coordinate descent.

While a detailed review of these optimization techniques goes beyond the scope of this chapter, there are a number of key ideas to keep in mind. First, all these methods require iterative optimization of the cost function as illustrated in Figure 5.9. Each iteration is typically based on computing some kind of gradient to both the log likelihood and prior terms, and then moving in the direction of that gradient in a manner that reduces the cost function. Second, it is critical to understand that, ideally, the optimization technique has no effect on the quality of the reconstruction. Instead, it only affects the speed of the MAP reconstruction algorithm. This is because the quality of the reconstruction is defined by the model and the resulting optimization problem of Equation (5.14). This separation of the MAP reconstruction into the image quality, defined by the model, and the computational performance, defined by the optimization algorithm, is perhaps a unique feature of the model-based approach. It allows these two aspects of the problem to be studied separately, but it also can cause confusion when investigating alternative approaches.

5.6 Successes and Opportunities of Bayesian Inversion

Figure 5.10 shows the practical value of model-based methods [1038]. This result compares state-of-the-art direct reconstruction results and model-based iterative reconstruction (MBIR) applied to data obtained from a General Electric 64 slice Volumetric Computed Tomography (VCT) scanner. The dramatic reduction in noise could not simply be obtained by post-processing of the image.

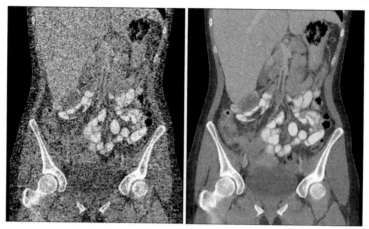

Figure 5.10: Comparison of state-of-the-art direct methods on the left to model-based iterative reconstruction (MBIR) on the right for the 3-D reconstruction of medical images from data collected from a General Electric 64 slice Volumetric Computed Tomography (VCT) scanner. Notice that by using an accurate model of both the scanner and the image being reconstructed, it is possible to recover fine detail of the anatomy while simultaneously reducing noise.

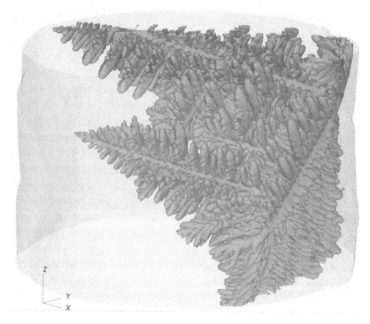

Figure 5.11: 3D rendering of a snapshot in time of a dendrite imaged using time-interlaced MBIR (TIMBIR). The image is possible because the data was reconstructed using $1/16^{th}$ of the data normally required for conventional reconstruction, which resulted in $16x$ the temporal resolution of 1.8 sec per frame. The dramatic increase in time resolution is a result of the fact that TIMBIR constrains the reconstruction to live a lower dimensional thin manifold, so that the number of measurements required for accurate reconstruction is greatly reduced.

The increased resolution and reduced noise is the result of the synergistic interaction between the measured data and the prior model.

Figure 5.11 shows an example of an important scientific imaging problem which benefits from dramatic reduction in data that can result from Bayesian reconstruction with correctly designed measurements [697, 361]. The objective in this problem is to reconstruct a volume of solidifying aluminum in both time and space so that the dendritic structures formed during solidification can be measured and better understood. Traditional reconstruction methods require the full set of 1536 views in order to compute a single 3D volume. At 60 frames per second, obtaining this many views takes approximately 26 sec, resulting in a sampling frequency of 0.039 Hz, which is much too slow to observe the physical process of solidification. However, by using interlaced samples in time it is possible to reconstruct using only $1/16^{th}$ the number of views per time sample. This results in a temporal sampling rate of approximately 1.625 sec or equivalently 0.62 Hz. Figure 5.11 shows how by using TIMBIR (time-interlaced MBIR) it is possible to resolve fine details of the 3D dendritic structure that are essential to obtain a broader understanding of grain and crystal structure in metals.

Part III

Inverse Methods for Analysis of Data

As a preliminary to the book, we thought it best to give some examples of successful microstructure analysis using inverse methods. To that end, we asked "Venkat" Venkatakrishnan, Qiu "Emma" Wang and Peter Doerschuk, and Song Wang to give examples of successful approaches.

Venkatakrishnan's chapter gives a development of model-based iterative reconstruction (MBIR), as it is applied to electron tomography. MBIR has been successfully implemented for medical imaging applications and Venkatakrishnan's model incorporates crystal effects for high angle annular dark field (HAADF) scanning electron microscopy (SEM).

Wang and Doerschuk's chapter gives their method for reconstructing models of repeated structures (e.g., viruses) from many randomly oriented instances as occur in cryo-electron microscopy. Their approach is to develop statistical signal processing algorithms that account for both the noise inherent in real data and the random orientations of each of the structures.

Finally, Song Wang et al.'s chapter is on a classical method of tracking observations through image sequences. His chapter is an introduction to the Kalman filter with applications to continuously reinforced fiber composites. Kalman filtering has been extensively used in other fields, particularly radar. What is unique about a materials science application is that the objects to be tracked are typically packed in close proximity to one another and are nearly identical to one another. This chapter gives a development with some practical examples and is intended to be tutorial in nature.

Chapter 6

Model-Based Iterative Reconstruction for Electron Tomography

S.V. Venkatakrishnan
Oak Ridge National Laboratory

Lawrence F. Drummy
Air Force Research Laboratory

6.1 Introduction

Driven by rapid advances in nanoscience, the past decade has witnessed an unprecedented interest in the development of new materials to solve key problems in areas ranging from energy to medicine. A key component in the development of advanced materials is to have the ability to quantitatively characterize samples at the nanometer and angstrom scale. Transmission electron microscopy (TEM) is a popular imaging method used in the physical and biological sciences for characterizing samples at the nanometer scale in 2D. The TEM has also been modified to characterize samples in 3D using the principles of tomography. Typically, tomography is carried out by using a sample holder that can tilt the sample about an axis and obtaining a sequence of images formed by measuring the scattered electrons which can then be inverted (see Figure 6.1). Because of sample motion during tilting, the images are aligned prior to the tomographic inversion. Due to mechanical limitations, it is usually not possible to obtain a full set of tilts, leading to a limited view dataset. Furthermore, in several applications, in order to limit beam damage to the sample (low-dose imaging) or acquire the data rapidly, the measurements need to be made rapidly, resulting in a dataset that is extremely noisy. Additionally, the measurements can be complex to model (and hence invert) due to dynamical diffraction effects such as Bragg scatter from crystalline materials. As a result, tomographic inversion of electron microscope data can be challenging. For a detailed review of electron tomography, the reader can refer to [680, 330, 309].

While there has been significant progress in enhancing the optics used in a TEM to dramatically improve 2D image resolution, the widely used algorithms for tomographic reconstruction have not fully exploited the statistical information in the data and the properties of the material being imaged; leading to a sub-optimal 3D reconstruction quality. The typical algorithms used for tomographic inversion are either the filtered back projection (FBP) or the simultaneous iterative reconstruction technique (SIRT) [473]. These techniques implicitly assume a linear relationship between the measurement (or some transformed version of it) and the unknown volume to be reconstructed. Furthermore, they do not treat the presence of noise in the measurement in a principled manner. The FBP, based on inverting the radon transform, is a fast algorithm but is not suited to electron tomography due to the noisy, and limited view nature of the measurement. The SIRT algorithm, based on iteratively solving a system of linear equations, is not stable to noise and the limited angular sampling since the system of equations does not have a unique solution. In summary, the widely used approaches are not effective because of the inherently ill-posed nature of the inversion and the implicit assumption of a linear model.

Incoming electron beam

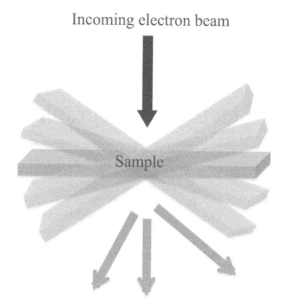

Scattered electrons used for various imaging
modalities

Figure 6.1: Schematic of the general process used for electron tomography. The sample is repeat-edly tilted and imaged to obtain a sequence of images. This "tilt-series" is then inverted using a reconstruction algorithm.

In contrast to traditional approaches, we will present a fundamentally different approach to the problem — model-based iterative reconstruction (MBIR), that can incorporate sophisticated probabilistic models for the measurement as well as the unknown volume and cast the reconstruction as an inference problem. The organization of the rest of this chapter is as follows. In Section 6.2 we will outline the MBIR paradigm; in Section 6.3 and Section 6.4 we will present MBIR methods for different modalities in electron microscopy; and in Section 6.5 we will present some important future directions for MBIR approaches.

6.2   Model-Based Iterative Reconstruction

Model-based iterative reconstruction (MBIR) is a systematic approach to solving inverse problems in imaging. Model-based approaches to imaging combine the physics of data formation with a noise model for the sensors and a probabilistic model for the underlying image to formulate a posterior probability distribution for the image given the measurements. Using this description we can esti-mate the unknown image/volume in several ways, the most popular being maximum *a posteriori* (MAP) estimation. Due to the complexity of the estimation for high-dimensional problems, the estimation is often done in an iterative manner.

MBIR approaches have been shown to dramatically improve the quality of reconstructions in a variety of imaging applications [1184, 893, 894, 1038, 973, 812, 319, 478, 803, 1116, 565, 110, 1174, 257] compared to traditional methods. A key benefit of MBIR approaches is that we can in-clude the estimation of calibration parameters (for example, detector offsets, gains, etc.) associated with the measurement as a part of the reconstruction. A popular method of formulating the recon-struction problem in the presence of unknown calibration parameters is given by the joint-MAP estimate [696]

$$(\hat{f}, \hat{\phi}) \leftarrow \underset{f, \phi}{\operatorname{argmin}} \{ -\log p(g|f, \phi) - \log p(f) \} \qquad (6.1)$$

where g is the data typically organized into a $M \times 1$ vector, f is the image/volume to be reconstructed organized into a $N \times 1$ vector, ϕ refers to a vector of unknown calibration parameters, $p(g|f, \phi)$ is the probability density function of the data given values of the voxels and calibration parameters (*forward model*), and $p(f)$ is the probability density function for the image/volume (*prior model*). The reconstruction is obtained by solving the (*optimization problem*) (6.1). The MBIR framework can also be interpreted as solving a regularized inverse problem where the first term is a data-fidelity term and the second term enforces regularity/constraints on the image/volume to be reconstructed. In such a context, the task is that of finding the reconstruction that best matches the measurements while preserving essential features of the reconstructed image.

The forward model/data-fidelity term is derived based on the physics of image formation, the noise statistics of the measured signal and the characteristics of the detector. For example, in X-ray tomography the physics of propagation is typically modeled by Beer's law while the measurements are assumed to follow a Poisson distribution. Prior models control properties of the image such as sharpness/smoothness of the edges, the presence of textures, etc. Choosing the "best" prior model for images/volumes in MBIR techniques is challenging and the appropriate choice depends on the application. However, a simple and powerful family of prior models is based on Markov random fields (MRF). MRFs model the probability density function in terms of interactions between pairs of neighboring pixels/voxels. The popular total-variation regularizer [873, 893] and Tikhonov regularizer belong to this family of priors. More advanced priors include dictionary-based sparsity models [15] and Gaussian mixture models of patches [1198]. A more complicated prior-model involves a larger number of parameters and can make the MBIR optimization computationally more expensive. The assumption of an image prior model is a reasonable assumption because most images of interest can be modeled using a thin manifold, i.e., if we are given all the pixels in a neighborhood except one, we can estimate that pixel with a reasonably high degree of accuracy. As a result, iterative image reconstruction techniques have also become very popular in the context of "compressed sensing" [270, 607] problems.

The design of fast mathematical optimization methods that can exploit modern parallel computing platforms plays a vital role in making model-based methods useful in practice for the large datasets that are typically encountered in electron tomography applications. The large size of the problems also dictates that these methods are largely iterative in nature wherein we start with an initial guess for the reconstruction and iteratively update the values of the voxels (and other parameters) to find a desirable minimum of the optimization problem in (6.1). The choice of optimization algorithm depends on the specific properties of the cost-function that has to be minimized. Mathematical properties such as convexity, differentiability, presence of constraints, etc., restrict the options available to the algorithm designer to find a desirable solution to the reconstruction problem. For a detailed discussion of optimization techniques, readers may refer to [202, 124]. We note that while a rich literature exists in the field of mathematical optimization, directly applying these techniques to the tomographic reconstruction problem is typically not desirable because it does not fully exploit the structure specific to tomography applications and can be undesirably slow for practical applications. Hence, several algorithms have been explored for model-based tomographic reconstruction in the context of specific applications. These range from sequential update algorithms like the iterative-coordinate descent (ICD) that minimize the cost-function with respect to each unknown [894], parallel update techniques like nonlinear conjugate gradient [320] that update all the variables simultaneously, and methods like block-coordinate descent that loosely is a hybrid of the two techniques [318, 321]. The parallel update techniques have been shown to require a larger number of iterations for tomographic reconstruction than the sequential update techniques in order to attain convergence [246]. However, the former are easy to parallelize on modern computing platforms like graphics processing units (GPUs) and hence each iteration can require a smaller amount of time.

In summary, the main components of MBIR are the *formulation of tractable forward and prior models* and the subsequent *development of convergent optimization algorithms for large datasets*. For a detailed review of model-based image processing, see [121]. In the next two sections, we will demonstrate the development of MBIR for different modalities in an electron microscope.

6.3 High-Angle Annular Dark-Field STEM Tomography

There is a growing interest in the use of HAADF-STEM tomography to study materials in 3D (e.g., [567, 146, 1127, 796, 881, 375]). This modality, which is based on measuring electrons scattered by the material, is relatively free from diffraction effects such as Bragg scatter. The HAADF-STEM signal contains information about the atomic number of the region being imaged and hence is useful for extracting chemical information about the material. As a result HAADF-STEM tomography is also referred to as *Z*-contrast tomography [680].

A typical HAADF-STEM acquisition involves focusing an electron probe at a point on the material for a short duration and integrating all the electrons scattered into an annular detector as shown in Figure 6.2(a). The electron beam is raster scanned and at each point a measurement is made to obtain a projection image of the object. The object is then tilted along a single axis and the process is repeated. At the end of the acquisition, a set of projection images is obtained corresponding to each tilt of the object. Therefore, HAADF-STEM tomography can be classified as a parallel beam, limited angle tomography modality. More details of HAADF-STEM acquisition and pre-processing can be found in [680, 681].

In this section, we develop a MBIR method for HAADF-STEM tomography. In Section 6.3.1 we develop a forward model for the measurement process in HAADF-STEM tomography. In Section 6.3.2 we present details of the prior model for the 3D volume. In Section 6.3.3 we use the developed models to formulate the MBIR cost-function and briefly outline an algorithm to find a minimum. In Section 6.3.4 we present results from a simulated dataset and follow it up with results on a real dataset.

6.3.1 HAADF-STEM Forward Model

In HAADF-STEM tomography, a single measurement is obtained by focusing an electron beam at a point on the surface of the material of interest. As the beam propagates through the material, the electrons which are scattered through high angles, approximately in the range $50 - 300$ mrad, are detected by an annular detector (see Figure 6.2). The number of electrons detected depends on the value of the HAADF scatter coefficients along the region being probed which is related to the type and number of atoms per unit volume at each location, the beam energy used as well as the inner and outer angle of the detector. The total number of electrons detected by the annular detector is typically assumed to be proportional to the projected value of the scatter coefficients through the region being probed. The electron beam is raster scanned and at each point we get a single measurement. The measurements resulting from a single raster scan of the material constitute a single electron microscope image. The sample is then tilted around a single axis (y axis in Figure 6.2) and the process is repeated. Thus we get 2-D parallel beam tomography data which we need to invert in order to reconstruct the HAADF scatter coefficients.

The goal of HAADF-STEM tomography is to reconstruct the HAADF scatter coefficients (units of nm^{-1}) denoted by $f(x, y, z)$ at every point in space. If (x, y, z) is the frame of reference of the object and (u, v, w) is the reference frame for the electron source (see Figure 6.2), then any function of space can be reparameterized so that $f_k(u, v, w) = f([u, v, w]R_{\theta_k})$ where R_{θ_k} is an orthonormal rotation of the spatial coordinates by an angle θ_k. For tomographic reconstruction we require measurements of the projection integral $\int_{-\infty}^{+\infty} f_k(u, v, w)dw$ through the object for every tilt θ_k and every point (u, v). We begin by describing how this measurement can be obtained from the HAADF-STEM signal.

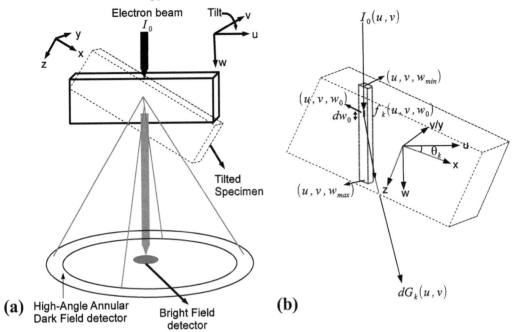

Figure 6.2: Illustration of the measurement process in a HAADF-STEM acquisition system. (a) A single measurement is obtained by focusing the electron beam I_0 on the material and integrating the signal from all the electrons scattered into the annular detector. This corresponds to a single pixel in the HAADF-STEM image. (b) Equivalent model for the measurement process. The scattered beam from an infinitesimal strip at location w_0, $dG_k(u,v)$, is the total signal deflected into the range of $50 - 300$ mrad, detected by the annular detector.

Let $I_0(u,v)$ be the source electron flux in units of electrons per nm^2 corresponding to the beam at location (u,v). Under the assumptions that attenuation effects are minimal and the sample is relatively thin, the total number of electrons scattered into the HAADF detector at location (u,v) at tilt k, $G_k(u,v)$, is given by [941]

$$G_k(u,v) \;=\; I_0(u,v)\int_{w_{min}}^{w_{max}} f_k(u,v,w)\,dw. \qquad (6.2)$$

Thus we get

$$\begin{aligned}\frac{G_k(u,v)}{I_0(u,v)} \;&=\; \int_{w_{min}}^{w_{max}} f_k(u,v,w)\,dw \\ &=\; \int_{w_{min}}^{w_{max}} f\left([u,v,w]R_{\theta_k}\right)dw. \qquad (6.3)\end{aligned}$$

So the normalized quantity $G_k(u,v)/I_0(u,v)$ is an estimate of the tomographic projection at angle θ_k and at position (u,v). We note that the above model, though widely used for HAADF-STEM tomography, does not account for the attenuation effects observed during the imaging of thick specimens of heavy elements [880]. While Van den Broek et al. [257] and Venkatakrishnan et al. [1098] have addressed some of these nonlinear effects, the topic of developing simple tractable models that describe all the experimentally observed characteristics of the HAADF-STEM signal including contrast reversal [310] remains an open problem. We use the linear model because it is still widely applicable for a range of materials and thicknesses studied in STEM tomography including all the cases considered in this chapter.

Next, we model the process by which the detector converts the incident electron flux to a measured signal. The HAADF-STEM detector typically consists of a photomultiplier tube (PMT) which converts the detected electron flux to a current which is then converted to a voltage using a preamplifier and read out using an A/D converter [548]. Since the HAADF-STEM detector has a limited dynamic range, the gain associated with the PMT (referred to as contrast) and offset associated with the preamplifier [548, 860] (referred to as brightness) can be adjusted so a wide range of materials and thicknesses can be imaged. We denote the PMT gain associated with the measurements at tilt k by α_k and the additive offset associated with the measurements at tilt k by d_k. Therefore, we model the ith measurement (corresponding to the electron source at (u_i, v_i)) at tilt k by a Gaussian random variable $g_{k,i}$, with mean

$$\mathbb{E}[g_{k,i}] \quad = \quad I_0 \alpha_k \int \int G_k(u,v) h_i(u,v) \, du dv + d_k \tag{6.4}$$

where $h_i(u,v)$ is a kernel which averages the electron flux over the area corresponding to the ith measurement and turns the continuous domain quantity to the discrete domain. A typical example of h_i is a truncated 2-D Gaussian-shaped kernel that is normalized so that it behaves like an averaging operator and models a Gaussian-like profile for the electron beam. Let f be a discretized version of $f(x,y,z)$ organized as a $N \times 1$ vector, where N is the total number of voxels, A_k is a $M \times N$ projection matrix for tilt k, and M is the number of measurements per tilt. Then substituting (6.2) in (6.4) we get

$$\mathbb{E}[g_{k,i}] \quad \approx \quad \alpha_k \int \int \left(\int_{w_{min}}^{w_{max}} f_k(u,v,w) dw \right) h_i(u,v) du dv + d_k$$
$$= \quad I_k [A_k f]_i + d_k \tag{6.5}$$

where $[A_k f]_i$ is the ith entry of the vector $A_k f$, $I_k = I_0 \alpha_k$ is the product of the source electron dose (counts) and the PMT gain at tilt k. For simplicity we will call I_k the gain associated with the measurements at tilt k. The variance of each measurement is modeled by

$$\mathrm{Var}[g_{k,i}] \quad = \quad \sigma_k^2 \mathbb{E}[g_{k,i}] \tag{6.6}$$

where σ_k^2 is a parameter used to model the noise variance at tilt k and $\mathbb{E}[g_{k,i}]$ accounts for the Poisson characteristics of the measurement. We assume that all the measurements are conditionally independent. If $\mathbb{E}[g_{k,i}] \approx g_{k,i}$, $\Lambda_k = \mathrm{diag}\left(\frac{1}{g_{k,1}}, \cdots, \frac{1}{g_{k,M}}\right)$, $g_k = [g_{k,1}, \cdots, g_{k,M}]^t$, $g = [g_1^t \cdots g_K^t]^t$, $I = [I_1, \cdots, I_K]$, $d = [d_1, \cdots, d_K]$ and $\sigma^2 = [\sigma_1^2, \cdots \sigma_K^2]$ then using (6.5) and (6.6) we get

$$p(g|f, I, d, \sigma^2) = \left(\prod_{k=1}^{K} \frac{1}{(2\pi\sigma_k^2)^{\frac{M}{2}} |\Lambda_k|^{-\frac{1}{2}}} \right)$$
$$\exp\left\{ -\frac{1}{2} \sum_{k=1}^{K} \frac{1}{\sigma_k^2} \|g_k - I_k A_k f - d_k \mathbb{1}\|_{\Lambda_k}^2 \right\} \tag{6.7}$$

where K is the total number of tilts.

6.3.2   Prior Model

We use a q-GGMRF [1038] model for the probability density of f. The parameters of this model can be adjusted to account for sharp or diffuse interfaces between materials. If $\mathcal{N}$ is the set of all pairs of neighboring voxels (e.g., a 26-point neighborhood in 3D), w_{ij} is a weighting kernel which is inversely proportional to the distance between voxel i and voxel j, normalized so that $\sum_{j \in \mathcal{N}_i} w_{ij} = 1$,

$\mathcal{N}_i$ is the set of all neighbors of voxel i, then the density function corresponding to the q-GGMRF prior is given by

$$p(f) = \frac{1}{Z}\exp\left\{-\sum_{\{i,j\}\in\mathcal{N}} w_{ij}\rho(f_i - f_j)\right\} \qquad (6.8)$$

$$\rho(f_i - f_j) = \frac{\left|\frac{f_i - f_j}{\sigma_f}\right|^q}{c + \left|\frac{f_i - f_j}{\sigma_f}\right|^{q-p}}$$

where Z is a normalizing constant and p, q, c, and σ_f are q-GGMRF parameters. Typically $1 \le p \le q \le 2$ is used to ensure convexity of the function $\rho(.)$, thereby simplifying the subsequent MAP optimization. The value of σ_f is typically set to achieve a balance between resolution and noise in the reconstructions. In this chapter we fix $q = 2$. When $p = 1$ the prior model corresponds to strong edge preserving reconstructions while $p = 2$ corresponds to smooth reconstructions (bearing some similarities to the Cahn–Hilliard phase field model [153]). We note that when c is zero, the $p = 1$ case of the q-GGMRF corresponds to the total variation prior [893]. Thus, the q-GGMRF provides a flexible prior model framework enabling us to model a range of possible materials from those with very sharp interfaces to those with smooth interfaces.

6.3.3 Cost Function Formulation and Optimization Algorithm

We use the MAP estimate to reconstruct the values of the HAADF scatter coefficients. In the course of a typical experiment the gains (I) and offsets (d) associated with the detector are adjusted for the specific sample being viewed. However, these gains (I), offsets (d), and variance parameters (σ^2) are typically not explicitly recorded and hence we treat them as nuisance parameters and jointly estimate them in the MBIR framework. The reconstruction is given by

$$(\hat{f}, \hat{I}, \hat{d}, \hat{\sigma}^2) = \underset{f\ge 0, I\in\Omega, d, \sigma^2}{\operatorname{argmin}} \left\{-\log p(g|f, I, d, \sigma^2) - \log p(f)\right\}$$

where $\Omega = \left\{I \in \mathbb{R}^K : \frac{1}{K}\sum_{k=1}^{K} I_k = \bar{I}\right\}$, $p(I, d, \sigma^2)$ is uniformly distributed, f is conditionally independent of (I, d, σ^2). We impose positivity constraints on the voxels ($f \ge 0$) as it is physically meaningful to have positive values of the HAADF scatter coefficients. The constraint Ω forces the average value of the gains to be equal to a value $\bar{I}$ to prevent our algorithm from diverging to unreasonable values of the HAADF scatter coefficients. The choice of $\bar{I}$ is arbitrary but affects the scaling of f and hence the choice of the prior model parameter, σ_f. Hence if $\bar{I}$ is set to the product of detector gain and source electron dose, then the reconstructions will be quantitative.

Using (6.7) and (6.8) we obtain the MAP estimate by minimizing the cost

$$c_{\text{HAADF}}(f, I, d, \sigma^2) = \frac{1}{2}\sum_{k=1}^{K} \frac{1}{\sigma_k^2}\|g_k - I_k A_k f - d_k \mathbb{1}\|_{\Lambda_k}^2$$

$$+ \frac{1}{2}\sum_{k=1}^{K} \log\left((2\pi\sigma_k^2)^M |\Lambda_k|^{-1}\right) + \sum_{\{i,j\}\in\mathcal{N}} w_{ij}\rho(f_i - f_j) \qquad (6.9)$$

In general the cost function $c_{\text{HAADF}}(f, I, d, \sigma^2)$ is convex in f but not jointly convex in (f, I, d, σ^2). We adapt the ICD algorithm [894] to minimize the cost function (6.9). In ICD the parameters are updated one at a time such that each update results in a lower value of the cost function. The basic structure of our algorithm is to repeatedly perform the following steps until convergence is achieved.

(i) For each voxel j, $\hat{f}_j \leftarrow \underset{f_j \ge 0}{\operatorname{argmin}} c_{\text{HAADF}}(f, I, d, \sigma^2)$

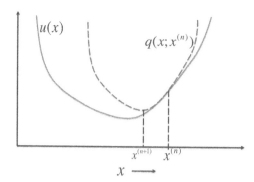

Figure 6.3: Illustration of the surrogate function approach. At each iteration, a simple surrogate is constructed and minimized in order to decrease the original function.

(ii) $(\hat{I}, \hat{d}) \leftarrow \underset{I \in \Omega, d}{\mathrm{argmin}} \, c_{\mathrm{HAADF}}(\hat{f}, I, d, \sigma^2)$

(iii) $\hat{\sigma}^2 \leftarrow \underset{\sigma^2}{\mathrm{argmin}} \, c_{\mathrm{HAADF}}(\hat{f}, \hat{I}, \hat{d}, \sigma^2)$

In Step (i) we update each voxel to lower the original cost function. The original cost-function with respect to a single voxel has a complicated non-quadratic form due to the prior model used for the reconstruction. As a result, finding the minimum of this function can be computationally expensive often requiring a line-search. Hence we use a surrogate function approach [1184] to lower the cost-function with respect to a single voxel. This approach, sometimes, also referred to as majorize-minimize in the optimization literature, involves the repeated construction and minimization of a simpler function such that minimizing the surrogate leads to a lower value of the original cost-function (Figure 6.3). The form of the surrogate function is chosen so that it results in simple closed form updates for the voxels, speeding up the implementation of our method. It has also been shown [1184] that the surrogate function approach speeds up the overall convergence of the algorithm to the minimum. Minimizing the surrogate function does not minimize the original cost but it results in voxel updates that lower the original cost. In Step (ii) we find the minimum of the cost function with respect to the gain I and offset parameters d by turning the constrained optimization to an unconstrained problem by using a Lagrange multiplier. Finally in (iii) we minimize the cost function with respect to σ^2. Thus each of the above updates lowers the value of the original cost function. The algorithm is terminated if the relative change in the total magnitude of the reconstruction is less than a preset threshold.

To further accelerate the algorithm, we use a multiresolution initialization strategy [1174]. The central idea behind multiresolution approaches is to run the entire algorithm detailed above using large-voxels and then use that as an initialization for a finer resolution (smaller voxels) reconstruction by up-sampling the coarse-scale reconstruction. Because the reconstruction using large voxels (coarse-scale) takes less time and the finer scale reconstruction is initialized with a better starting point the overall algorithm takes less run-time. For example, in order to reconstruct a $1024 \times 1024 \times 1024$ voxel volume with a voxel side of size 1nm, a multi-resolution algorithm would first reconstruct a $512 \times 512 \times 512$ voxel volume with a voxel size of 2nm and then up-sample this and use it as an initialization for the finer scale reconstruction.

Further details of the overall algorithm can be found in [1099].

6.3.4 Experimental Results

6.3.4.1 Simulated Dataset

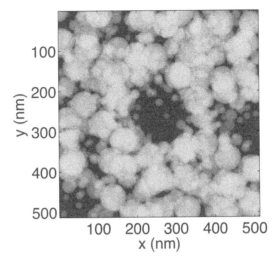

Figure 6.4: Simulated HAADF microscope image of aluminum spheres in vacuum when the object is not tilted.

Table 6.1: Comparison of the root mean square error (RMSE) of the reconstruction with respect to the original phantom for various scenarios. SIRT and FBP have higher RMSE than MBIR, indicating that MBIR can produce quantitatively accurate reconstructions.

Method	RMSE ($\times 10^{-4}$ nm^{-1})
FBP	1.278
SIRT	0.988
MBIR ($p = 1$)	0.284
MBIR ($p = 1.2$)	0.324
MBIR ($p = 2$)	0.496

We begin by studying the performance of the MBIR algorithm on a simulated dataset produced from a phantom consisting of spheres in vacuum. Our objective is to compare the results of our method to the most widely used algorithms for HAADF-STEM tomography, FBP, and SIRT. The FBP and SIRT reconstructions are performed using a popular electron microscopy package, IMOD [651]. Since the scaling is arbitrary and no positivity is enforced in IMOD [651], we clip the reconstructions to be positive and perform a least squares fit to scale the reconstructions to a similar range as the true phantom. We use the visual quality as well as the root mean square error (RMSE) between the reconstruction and the original phantom to evaluate the reconstructions.

Figure 6.4 shows the simulated HAADF data corresponding to the phantom at zero tilt. The images have been displayed by scaling them using the minimum and maximum count in the data. The spheres in the phantom have a scattering coefficient of 4.132×10^{-4} nm^{-1} corresponding to aluminum at 300 kV with detection angles $50 - 250$ mrad (value obtained using the Monte-Carlo simulator, CASINO [256]). The sphere diameters vary (up to 100 nm) and the sample thickness is about 128 nm. The object is tilted from $+70°$ to $-70°$ in steps of $1°$ and the projection images are obtained using (6.2) with an electron flux of 50000 counts per nm^2 for every measurement and the detector gain set to 1. The offset d_k is set to 9000 counts for each view. To each HAADF projection

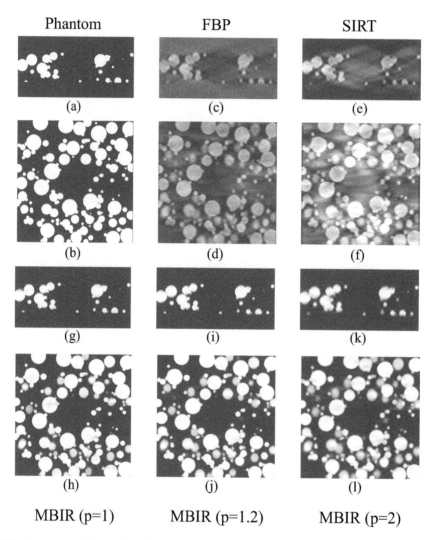

Phantom   FBP   SIRT

(a)   (c)   (e)

(b)   (d)   (f)

(g)   (i)   (k)

(h)   (j)   (l)

MBIR (p=1)   MBIR (p=1.2)   MBIR (p=2)

Figure 6.5: Reconstructions using FBP, SIRT, and MBIR with different prior models. (a), (c), (e), (g), (i), and (k) show a $x - z$ slice and (b), (d), (f), (h), (j), and (l) show a $x - y$ slice. (a) and (b) show ground truth corresponding to a single slice. (c) and (d) show reconstructions from FBP, (e) and (f) show SIRT reconstructions using IMOD. (g) - (l) show MBIR reconstructions with different values of the prior model parameter p. FBP and SIRT reconstructions are noisy and have streaking artifacts in the $x - z$ plane while MBIR significantly suppresses these artifacts and can produce sharper reconstructions. Increasing the value of the prior model parameter p from 1 to 2 produces smoother MBIR reconstructions. Thus the value of p can be chosen to best match the type of material being imaged.

measurement Gaussian noise is added with variance parameter σ^2 set so that the noise variance increases with tilt. The values of σ_k^2 corresponds to a minimum SNR $\left(\min_{k,i} 10 \log \left(\frac{g_{k,i}}{\sigma_k^2} \right) \right)$ of 34.471 dB. The projection images are acquired at a pixel size of 1nm $\times$ 1nm. All the reconstructions are performed with $\sigma_f = 4.1 \times 10^{-5}$ nm^{-1}, $q = 2$, and $c = 0.01$ using a 3-stage multiresolution initialization. The interpolation between resolutions is performed using pixel replication. The value of $\bar{I}$ is set to 50000, the value of the source electron dose - detector gain product, so that the reconstructions are quantitative. The stopping threshold is set to 0.1%.

Figure 6.5 (a) and (b) show a single $x-z$ and $x-y$ slice from the original phantom. Figure 6.5 (c) and (d) show the corresponding slices from the reconstruction obtained using the FBP algorithm implemented in IMOD [651]. The algorithm results in blurry reconstructions with significant streaking artifacts in the $x-z$ plane. In the $x-y$ plane the reconstructions are noisy. Figure 6.5 (e) and (f) show the reconstructions using the SIRT algorithm. While some of the noise appears suppressed compared to FBP we still observe the streaking artifacts in the $x-z$ plane and noise in the $x-y$ plane. Moreover in regions where the phantom has no material, we observe that the reconstructions still contain material with nonzero scatter coefficients. The RMSE of the SIRT reconstructions are lower than those of the FBP reconstructions as shown in Table. 6.1.

Figure 6.5 (g) - (l) shows the reconstructions when we apply MBIR algorithm to the simulated dataset with different values of the q-GGMRF parameter p. This phantom has discontinuous boundaries, so as we would expect, the total-variation prior ($p = 1$ in Figure 6.5 (g) and (h)) is well matched to its behavior and produces the lowest RMSE (see Table. 6.1) reconstruction. Figure 6.5 (i) and (j) show the MBIR reconstruction when we set $p = 1.2$. This produces results with slightly more smooth edges than the $p = 1$ case. This prior has been found to be useful [1038] since it provides a good balance between preserving edges and modeling the smooth regions in the reconstruction. Finally, Figure 6.5 (k) and (l) show the MBIR reconstruction when $p = 2$. In this case the edges are most diffuse and hence the RMSE is higher than the $p = 1$ and $p = 1.2$ case for this phantom. In all cases the streaking artifacts in the $x-z$ plane are significantly suppressed and noise in the $x-y$ slices is effectively reduced compared to FBP and SIRT. The RMSE of the MBIR reconstructions (see Table 6.1) are lower than those of the SIRT and FBP suggesting that the MBIR reconstructions are quantitatively more accurate than FBP and SIRT. Thus this experiment illustrates that MBIR is superior to FBP and SIRT and furthermore the parameters of the algorithm can be chosen to model a range of interfaces from very sharp to diffuse.

6.3.4.2   Experimental Dataset

In order to evaluate our approach on real data, we compare our algorithm with FBP and SIRT from IMOD [651]. The data acquired is of a $\approx$ 150-nm-thick sample of polystyrene functionalized titanium dioxide nanoparticle assembly [1007]. The TEM used was a FEI Titan operating in STEM mode with 300 kV accelerating voltage, spot size 7, +/- 70° with 2° increments for +/- 54° and 1° increments for 54° to 70° and $-54°$ to $-70°$. The exposure time was 12.6 seconds, magnification was set to 225 kX, the frame size set to 2048 $\times$ 2048, with 0.34 nm pixel size, and STEM dynamic focus activated. A Fischione model 3000 HAADF photomultiplier tube detector was used at camera length of 130 mm. We use a $\approx$ 350 nm $\times$350 nm section of the projection images for reconstruction. Figure 6.6 shows a single projection image acquired when the object is at zero tilt, displayed by scaling it to the range of the data.

The FBP and SIRT reconstructions are performed with voxels of size 0.343 nm $\times$ 0.343 nm $\times$ 0.343 nm. The filter parameters for FBP are chosen to produce the most visually appealing results. The particles of interest in this dataset are approximately cylindrical with a diameter of 18 nm and a height of 40 nm [1007]. Thus in order to reduce computation, MBIR is performed with voxels of size (3×0.343) nm $\times (3 \times 0.343)$ nm $\times (3 \times 0.343)$ nm. The parameter σ_f is chosen for the best visual quality of reconstruction. The value of c is set to 0.01. We use a three-stage multiresolution initialization for the reconstruction. The interpolation between resolutions is performed using pixel

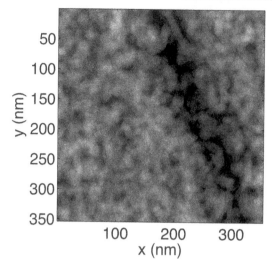

Figure 6.6: Acquired HAADF data for a titanium dioxide nanoparticle dataset at zero tilt. The dark regions represent a void in the material.

replication. Since the source electron dose and detector gains are unknown in this case, we set $\bar{I} = 20000$ and jointly estimate the gains along with the offsets and variance parameters as a part of the reconstruction. The stopping threshold is set to 0.9%. The dimensions of the reconstructed volume are set so as to account for all the voxels contributing to the projection data. In presenting the results we only show voxels that can be reliably reconstructed from the projection data, i.e., at every tilt there is a projection measurement corresponding to those voxels. Additionally while displaying the results, we use a scaling window ranging from the minimum to the maximum value in the reconstructed volume.

Figure 6.7 shows a single $x-z$ and $x-y$ slice from FBP, SIRT, and the MBIR reconstruction. We observe that in SIRT and FBP there are streaking artifacts in the $x-z$ plane of reconstruction while MBIR significantly suppresses these artifacts. In the $x-y$ plane the effects of noise are effectively suppressed in MBIR clearly showing the titanium dioxide nanoparticles against the background support material. This demonstrates the effectiveness of the method even for this particularly limited tilt dataset.

6.4 Bright-Field Electron Tomography

While HAADF-STEM tomography has become popular for studying samples in the physical sciences, bright field (BF) electron tomography (ET) has been widely used in the life sciences to characterize biological specimens in 3D [59] using either a transmission electron microscope (TEM) or a scanning transmission electron microscope (STEM) [955]. BF-ET typically involves focusing an electron beam on a sample, acquiring images of transmitted electrons corresponding to various sample tilts, and using an algorithm on the acquired "tilt-series" to reconstruct the object. In most cases due to the geometry of the acquisition and mechanical limitations of the tilting stages, BF-ET is a limited angle parallel beam transmission tomography modality.

While there are a few instances where BF-ET has been used in the physical sciences [959, 510, 53], it has generally been avoided [680, 47], due to the occurrence of contrast reversals [311] from dynamical diffraction effects such as Bragg scatter [245]. Bragg scatter occurs when the crystal lattice is oriented in such a manner that the incident electrons are elastically scattered away from the direct path [245] leading to a measurement uncharacteristic of attenuation due to thickness alone. We refer to measurements which are not well modeled by attenuation due to thickness alone

FBP   SIRT   MBIR (p=1.2)

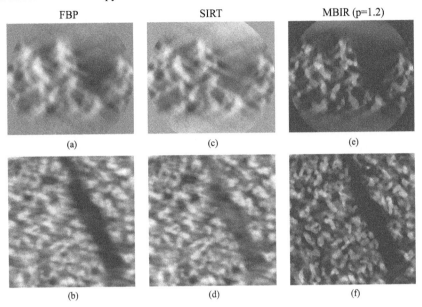

     (a)   (c)   (e)

     (b)   (d)   (f)

Figure 6.7: Comparison of MBIR with FBP and SIRT on a real HAADF-STEM dataset. The top row shows a $x - z$ reconstructed slice and the bottom shows a $x - y$ reconstructed slice. (a) and (b) FBP reconstruction, (c) and (d) SIRT with 20 iterations, (e) and (f) MBIR with $p = 1.2$. MBIR produces images with no streaks in the $x - z$ plane and significantly suppresses noise in the $x - y$ plane.

as anomalous measurements. These anomalies make interpretation of the individual BF images complicated and result in strong artifacts if the BF tilt-series is used for tomographic reconstruction using standard reconstruction algorithms such as FBP [47]. Thus BF-ET has generally been avoided in the physical sciences due to the complicated nature of the data and the inability of the standard reconstruction algorithms like FBP to handle such data.

In this section, we present an MBIR algorithm for accurate reconstruction of BF-ET data containing anomalous measurements that typically result from crystalline samples. Our approach is based on a novel generalized Huber function that is used in the forward model (i.e., log likelihood) to account for the anomalous measurements due to Bragg or other errors. The generalized Huber function is parameterized so that it can model the heavy tailed distribution of the errors that are present in anomalous measurements. Using this forward model, we derive an MBIR cost function which allows for joint estimation of both the unknown image, f, and a key parameter of the generalized Huber function. This approach allows for adaptive parameter estimation in the reconstruction process, which is important in practical applications.

We demonstrate via results from the simulated and a real dataset that MBIR with anomaly modeling can significantly improve the reconstruction quality compared to FBP and conventional MBIR, suppressing the artifacts that arise due to the anomalous measurements.

6.4.1   BF-TEM Forward Model and Cost Function Formulation

The goal of BF-ET is to reconstruct an attenuation coefficient at every point in the sample. The attenuation coefficient is related to the ability of the material to scatter the incident beam away from the direct path which is dependent on the differential cross-section, geometry of the detector, density of the material, and incident electron energy. An electron beam is focused on the material and the electrons that are scattered by the sample through small angles are captured by a BF detector to obtain a single image. The sample is then tilted along a fixed axis and the process is repeated. Thus,

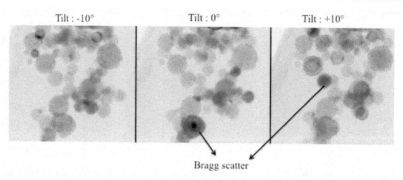

Figure 6.8: Illustration of the "anomalies" present in a real BF-TEM dataset of aluminum nanoparticles. The figure shows BF images corresponding to three different tilts of the specimen. Note that certain spheres turn dark (fewer counts) and then again turn bright due to Bragg scatter (contrast reversal). These effects make it challenging to directly apply standard analytic tomographic reconstruction algorithms to the data.

at the end of the acquisition, we obtain a collection of BF images that can be used for tomographic reconstruction of the attenuation coefficients.

In order to develop a forward model for BF-ET that accounts for the anomalous effects, we start with the simple case when there are no anomalies. Let $\lambda_{k,i}$ be the electron counts corresponding to the ith measurement at the kth tilt and $\lambda_{D,k}$ be the counts that would be measured in the absence of the sample at that tilt (blank scan). We model the attenuation of the beam through the material using Beer's law. Thus, an estimate of the projection integral corresponding to the ith measurement at the kth tilt is given by $\log\left(\frac{\lambda_{D,k}}{\lambda_{k,i}}\right)$. Notice that unlike in high-angle annular dark field electron microscopy [1099], the BF case requires a log operation to be applied to a normalized version of the measurement. There can be cases in which the blank scan, $\lambda_{D,k}$, is not measured, and we can include it as an unknown parameter in the MBIR framework and estimate it as a part of the reconstruction. If g_k is a $M \times 1$ vector with $g_{k,i} = -\log(\lambda_{k,i})$, f is an $N \times 1$ vector of unknown attenuation coefficients of the material and, $d_k = -\log(\lambda_{D,k})$, then, assuming $\lambda_{k,i}$'s are conditionally independent Poisson random variables it has been shown that [120] the conditional mean of $g_{k,i}$ can be approximated by $A_{k,i,*}f + d_k$ and the conditional variance is proportional to $\frac{1}{\mathbb{E}[\lambda_{k,i}]}$, where A_k is a $M \times N$ forward projection matrix and $A_{k,i,*}$ is the ith row of A_k. Modeling the conditional density of the measurements as a Gaussian distribution [1197] results in a probability density function (pdf),

$$p(g|f,d,\sigma) = \frac{1}{Z_l}\exp\left\{-\frac{1}{2}\sum_{k=1}^{K}\sum_{i=1}^{M}\left(g_{k,i} - A_{k,i,*}f - d_k\right)^2\frac{\Lambda_{k,ii}}{\sigma^2}\right\} \tag{6.10}$$

where K is the total number of tilts, $g = [g_1^t \cdots g_K^t]^t$ is a $KM \times 1$ data vector, Λ_k is a diagonal matrix with entries set so that $\frac{\sigma^2}{\Lambda_{k,ii}}$ is the variance of the measurement $g_{k,i}$ with σ^2 being a proportionality constant, $d = [d_1 \cdots d_K]$ is a vector containing the offset parameters, and Z_l is a normalizing constant. For such a transmission tomography model it has been shown that $\Lambda_{k,ii} = \mathbb{E}[\lambda_{k,i}] \approx \lambda_{k,i}$ [893]. We note that our formulation can account for more sophisticated physics models as introduced in [564], but in this chapter we focus on using Beer's law as it has been found to be accurate for a class of materials and thickness typically studied using BF-ET [564].

6.4.1.1 Generalized Huber Functions for Anomaly Modeling

Bragg scatter from crystalline material can cause the BF-ET measurements to vary substantially from the model of Equation (6.10). Figure 6.8 shows an example of three tilts from a BF tilt series with regions having significant anomalies.

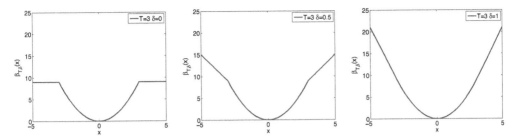

Figure 6.9: Illustration of the generalized Huber function $\beta_{T,\delta}$ used for the likelihood term with $T = 3$ and $\delta = 0, 0.5$ and 1. When $\delta = 1$ the function reduces to the Huber function. Large model mismatch errors are penalized by restricting their influence on the overall cost function.

A precise way of accounting for these anomalies would require identifying 3D regions of the object that consist of a single crystal, and modeling the associated crystal structure. While possible, this would be a highly ill-posed inverse problem to recover from a single 2D tilt series due to the unknown 3D orientation of the crystals. Furthermore, modeling other classes of anomalies such as Fresnel fringes [1147] and extinction contours involves more complex physics making the data more difficult to invert. Therefore, instead of modeling the complicated physics of dynamical diffraction that leads to anomalies, we will use an alternate approach.

In order to account for anomalous effects like Bragg scatter, we use a modified likelihood function that models the anomalies as outliers of a pdf

$$p(g|f,d,\sigma) \quad = \quad \frac{1}{Z}\exp\left\{-\frac{1}{2}\sum_{k=1}^{K}\sum_{i=1}^{M}\beta_{T,\delta}\left((g_{k,i}-A_{k,i,*}f-d_k)\frac{\sqrt{\Lambda_{k,ii}}}{\sigma}\right)\right\} \quad (6.11)$$

where $\beta_{T,\delta}: \mathbb{R} \to \mathbb{R}$ such that

$$\beta_{T,\delta}(x) = \begin{cases} x^2 & |x| < T \\ 2\delta T|x| + T^2(1-2\delta) & |x| \geq T \end{cases}$$

and Z is a normalizing constant. We call $\beta_{T,\delta}$ the generalized Huber function. Figure 6.9 shows the generalized Huber function for three different values of δ. Notice that δ controls the tail behavior of the density function while T controls the threshold beyond which a measurement is considered anomalous. When $\delta = 0$, $\beta_{T,\delta}$ corresponds to the weak-spring potential [107] used for image modeling and results in a function with the heaviest tails among the three cases. However, when $\delta = 0$ we cannot jointly estimate the calibration parameters because the likelihood is not a valid density function since it does not integrate to 1. When $\delta = 1$, $\beta_{T,\delta}$ reduces to the Huber function [970] which is a convex function and corresponds to a pdf with the lightest tail among the three cases. When T is very large then $\beta_{T,\delta}$ is effectively a quadratic function and the likelihood reduces to the standard transmission tomography model in (6.10). Thus the generalized Huber function can be adjusted to have heavier tails than the density function in (6.10) to account for the various anomalies in the dataset.

Restricting $0 < \delta \leq 1$ and using the fact that

$$\int p(g|f,d,\sigma)dg = 1$$

we can show that the normalizing constant has the form

$$Z = \sigma^{MK} \times \text{Constants}.$$

Hence, the modified log-likelihood function for the BF-ET case is given by

$$-\log p(g|f,d,\sigma) = \frac{1}{2}\sum_{k=1}^{K}\sum_{i=1}^{M}\beta_{T,\delta}\left((g_{k,i}-A_{k,i,*}f-d_k)\frac{\sqrt{\Lambda_{k,ii}}}{\sigma}\right)+MK\log(\sigma)$$

$$+\text{Constants} \quad (6.12)$$

Each term in the summation corresponds to a penalty on the ratio of the data mismatch error ($g_{k,i} - A_{k,i,*}f - d_k$) to the noise standard deviation $\left(\frac{\sigma}{\sqrt{\Lambda_{k,ii}}}\right)$. Thus T has the interpretation that if the data fit error is greater than T times the noise standard deviation then that measurement is likely to be an anomaly. Notice that typically σ is not known and hence we will jointly estimate it as a part of the reconstruction.

6.4.1.2 MBIR Cost Formulation

Combining the statistical model in (6.12) with a prior model of the form

$$p(f) \;=\; \frac{1}{Z_f}\exp\{-s(f)\} \quad (6.13)$$

where Z_f is a normalizing constant, the reconstruction is obtained by minimizing the cost

$$c_{\text{BF}}(f,d,\sigma) = \frac{1}{2}\sum_{k=1}^{K}\sum_{i=1}^{M}\beta_{T,\delta}\left((g_{k,i}-A_{k,i,*}f-d_k)\frac{\sqrt{\Lambda_{k,ii}}}{\sigma}\right)+MK\log(\sigma)+s(f). \quad (6.14)$$

We note that the MRF type prior models introduced in the context of MBIR for HAADF-STEM tomography can be used here. If we define $e_{k,i} : \mathbb{R}^{N+K+1} \to \mathbb{R}$ to be a function such that

$$e_{k,i}(f,d,\sigma) = (g_{k,i}-A_{k,i,*}f-d_k)\frac{\sqrt{\Lambda_{k,ii}}}{\sigma}$$

the cost function can then be written as

$$c_{\text{BF}}(f,d,\sigma) = \frac{1}{2}\sum_{k=1}^{K}\sum_{i=1}^{M}\beta_{T,\delta}(e_{k,i}(f,d,\sigma))+MK\log(\sigma)+s(f). \quad (6.15)$$

Additionally, we will constrain $f \geq 0$, as it is physically meaningful to have positive values of the attenuation coefficients. Thus, the MBIR BF-ET reconstruction is given by

$$(\hat{f},\hat{d},\hat{\sigma}) \leftarrow \operatorname*{argmin}_{f \geq 0, d, \sigma} c_{\text{BF}}(f,d,\sigma)$$

The cost function (6.15) is non-convex in general, and thus finding the global minimum is computationally expensive. Furthermore, since the likelihood term of (6.15), is not differentiable, gradient-based methods cannot be directly applied. Therefore we use an alternating minimization approach based on surrogate functions as in the HAADF-STEM MBIR case. Further details of the optimization can be found in [1097].

6.4.2 Results

In this section we compare four algorithms for BF-ET: FBP, conventional model-based iterative reconstruction (MBIR), the proposed MBIR with anomaly modeling and known parameter values (MBIR-AM), and the proposed method with anomaly modeling and parameter estimation (MBIR-AM-PE). We apply the methods to simulated data as well as real data. For the simulated data we will

Table 6.2: Comparison of the root mean square error (RMSE) of the reconstruction with respect to the original phantom for various scenarios. MBIR with anomaly modeling produces quantitatively more accurate reconstructions.

Algorithm	RMSE ($\times 10^{-4}$ nm^{-1})
FBP	13.90
MBIR	4.95
MBIR-AM ($T = 3, \delta = 0.5$)	4.30
MBIR-AM-PE ($T = 3, \delta = 0.5$)	4.31

compare results from all four methods while in the real data, since we do not know the parameters, we will not consider the MBIR-AM case.

The FBP reconstructions are performed in MATLAB® using the *iradon* command and the output is clipped to be positive. For the MBIR reconstructions with anomaly modeling, we set $T = 3$, $\delta = 0.5$, and $p = 1.2$. The value of σ_f is chosen to obtain the lowest root mean square error (RMSE) for the simulated dataset and is chosen to obtain the best visual quality of reconstruction for the real dataset. The offset parameter for each tilt, d_k, is initialized to the mean value of the log of the measurements from a void region in the sample. The variance parameter, σ^2, is initialized as the ratio of the mean value of the log measurements to the mean value of the measurements from a void region in the sample.

6.4.2.1 Simulated Dataset

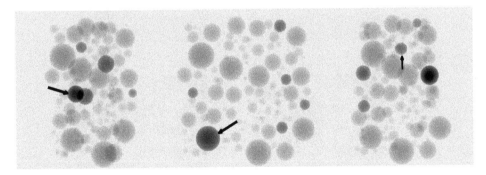

Figure 6.10: Simulated BF data corresponding to a 3D phantom of spheres for three successive tilts. The arrows in the figure show example locations with the simulated Bragg scatter obtained by increasing the attenuation coefficient of a few spheres in the phantom. We model these as anomalies in the projection data as they can cause artifacts in the reconstructions produced using the standard reconstruction techniques.

Furthermore, since the likelihood term of (6.15), is not differentiable, gradient-based methods cannot be directly applied. Therefore we use an alternating minimization approach based on surrogate functions as in the HAADF-STEM MBIR case. Further details of the optimization can be found in [1097].

We use a 3-D cubic phantom containing spheres of varying radii with an attenuation coefficient of 7.45×10^{-3} nm^{-1} to generate the simulated dataset. The phantom has a dimension of 256 nm $\times$ 512 nm $\times$ 512 nm ($z - x - y$ respectively). It is projected at 36 tilts in the range of $-70°$ to $+70°$ in steps of $4°$ about the y axis with a dosage $\lambda_{D,k} = 1865$ counts using the Beer's law model with Gaussian noise having variance equal to the mean of the signal. The value of σ is set to 1. At certain

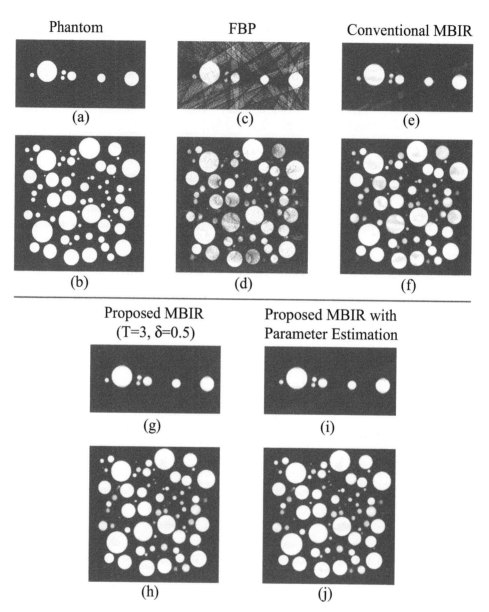

Figure 6.11: Comparison of BF reconstructions for a dataset with Bragg scatter like anomalies. (a) and (b) show a single $x - z$ and $x - y$ cross-section from the phantom used. The horizontal direction represents the x axis. (c) and (d) show the corresponding cross-sections from an FBP reconstruction. (e) and (f) show the conventional MBIR reconstruction. The reconstruction has streaks because of Bragg scatter but much less compared to FBP. (g) and (h) show the cross-section from MBIR with anomaly modeling ($T = 3$ and $\delta = 0.5$). The method effectively suppresses the artifacts in (c) - (f), and produces a more accurate reconstruction. Finally (i) and (j) show the reconstruction using MBIR with anomaly modeling and nuisance parameter estimation. The reconstructions are comparable to the MBIR-AM case showing that the algorithm can work well despite the unknown parameters. All images are scaled in the range of $0 - 7.45 \times 10^{-3}$ nm^{-1}.

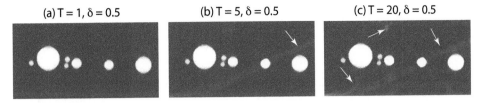

(a) T = 1, δ = 0.5   (b) T = 5, δ = 0.5   (c) T = 20, δ = 0.5

Figure 6.12: Illustrates the impact of varying anomaly threshold T on the proposed MBIR reconstructions. (a) shows a $x - z$ cross-section from the 3-D reconstruction when $T = 1$. (b) and (c) shows the corresponding slices when $T = 5$ and $T = 20$. Notice that for (b) and (c) there are visible streaking artifacts. A value of $T = 3$ as shown in Figure 6.11 produces an accurate reconstruction.

Table 6.3: Comparison of the root mean square error (RMSE) ($\times 10^{-4}\text{nm}^{-1}$) of the reconstruction with respect to the original phantom when varying T and δ. A value of $T = 3$ and $\delta = 0.5$ produces the lowest RMSE reconstruction.

$T \parallel \delta$	0.1	0.5	1
1	4.50	4.42	4.44
3	4.40	**4.31**	4.60
5	4.33	5.09	5.14
20	5.06	5.06	5.06

tilts the attenuation of a fraction of the spheres is adjusted to simulate Bragg scatter like effects (Figure 6.10) as in a real dataset.

Figure 6.11 (a) and (b) shows a single $x - z$ and $x - y$ cross-section from the original phantom. Figure 6.11 (c) - (j) shows the corresponding cross-section from the reconstructed volume using the different algorithms. The FBP reconstruction (Figure 6.11(c), (d)) has strong streaking artifacts in the $x - z$ cross-section and noise in the $x - y$ cross-section.

The conventional MBIR (Figure 6.11 (e), (f)) shows prominent streaking artifacts in the $x - z$ cross-section even though there are fewer artifacts than in FBP. Furthermore, there is some underestimation at the center of the spherical particles.

However, MBIR with anomaly modeling (MBIR-AM) (Figure 6.11(g)-(h)) produces a reconstruction which effectively suppresses these artifacts. In the $x - y$ cross-section, we notice that the MBIR reconstructions are less noisy as compared to FBP and that the anomaly modeling significantly improves the reconstruction. Next, we evaluate the performance of the proposed MBIR algorithm with parameter estimation (MBIR-AM-PE). Figure 6.11 (i) and (j) show that the MBIR-AM-PE can accurately reconstruct the 3D volume suppressing the artifacts despite the unknown nuisance parameters. In addition to the qualitative improvements shown in Figure 6.11, Table 6.2 shows that MBIR with the anomaly modeling (MBIR-AM and MBIR-AM-PE) significantly improves the quantitative accuracy of the reconstruction compared to FBP as well as conventional MBIR.

Finally, we study the sensitivity of the MBIR reconstructions to the choice of parameters T and δ. Figure 6.12 shows an $x - z$ cross-section from the reconstructions for different values of T when $\delta = 0.5$. Notice that as T increases, streak artifacts start to appear in the reconstruction. This is because some of the anomalous measurements are misclassified. Figure 6.13 shows the binary classifier mask, b', corresponding to three successive tilts from simulated data upon completion of the reconstruction. This variable indicates which measurements are classified as anomalous based on the generalized Huber function used for the reconstruction. Notice that when $T = 1$, several non-anomalous measurements are classified as anomalous (false alarms). When $T = 5$ the algorithm misses certain anomalies. When $T = 20$, all the measurements are classified as non-anomalous

Data and anomaly classifier as T is varied

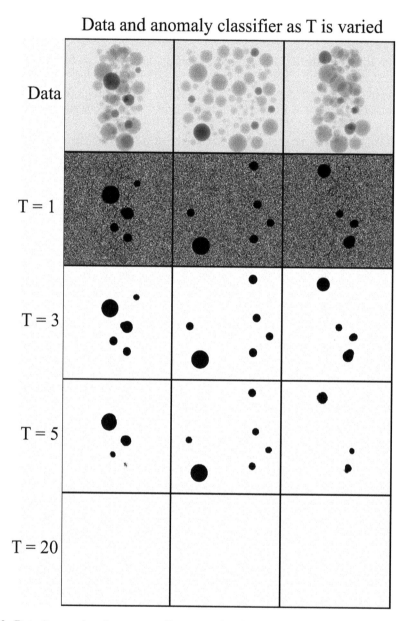

Figure 6.13: Data (top row) and corresponding anomaly classifier upon termination of the algorithm for three tilts from the phantom dataset corresponding to different values of T. The white regions indicate areas classified as non-anomalous and the black regions correspond to the anomalies identified by the algorithm. As the value of T increases the algorithm starts to misclassify anomalies. A value of $T = 3$ provides a good tradeoff between the false alarms and missed detections.

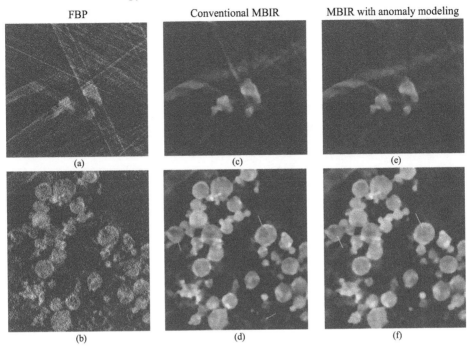

Figure 6.14: A single $x - z$ and $x - y$ cross-section reconstructed using different algorithms from a BF-TEM dataset of aluminum sphere nanoparticles. The horizontal direction represents the x axis. The FBP reconstruction (a)-(b) has very strong streaking artifacts in the $x - z$ cross-section, and noise in the $x - y$ cross-section suggesting why it has been avoided for BF-ET. The MBIR algorithm with the anomaly modeling and parameter estimation ($T = 3$ and $\delta = 0.5$) (e)-(f) is superior to the conventional MBIR (c)-(d), suppressing the streaking artifacts seen in (c). In the case of MBIR, the circular cross-section of the spherical particles is clearly visible compared to FBP. All images are scaled in the range of $0 - 6.0 \times 10^{-3}$ nm^{-1}.

leading to large errors in the reconstruction. A value of $T = 3$ provides a good tradeoff and is intuitively appealing because this implies that if the data fit error for a measurement is less than 3 times the standard deviation of the noise, then that measurement is non-anomalous. Thus the trade off between false positives and missed detection of anomalies can be varied via the parameter T in the algorithm. Table. 6.3 shows the RMSE when we vary δ for the different values of T. The value of δ controls the influence of the anomalous measurements on the reconstruction. A value of δ close to 0 implies the anomalous measurements are weighted less in the cost function, while $\delta = 1$ implies the anomalies are weighted significantly. For this particular simulation, we get the lowest RMSE for the $T = 3$ and $\delta = 0.5$ case.

6.4.2.2 *Real Dataset*

In order to evaluate our approach on real data, we use a dataset of approximately 700-nm-thick crystalline aluminum nanoparticles in a carbon support. We used an FEI Titan TEM with a 300 kV accelerating voltage, and a spot size[1] of 5. The exposure time was set to 1 second, magnification was set to 100 kX, the frame size set to 2048 × 2048, with a pixel size of 0.83 nm × 0.83 nm. The detector used was a CCD with a 30 μm objective aperture resulting in a detector which captures electron scattered in the $0 - 15$ mrad range. The BF-TEM data consists of 33 tilts in the range of

---

[1]The spot size is a manufacturer dependent unit-less parameter that refers to the size of the condenser aperture and controls the electron flux on the sample.

Data and anomaly classifier (MBIR-AM-PE)

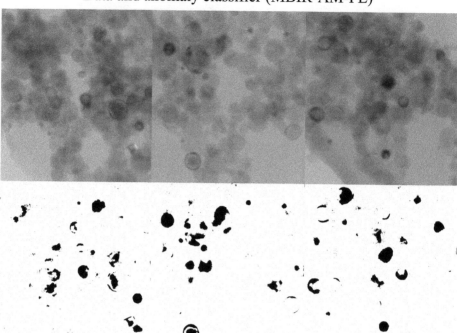

Figure 6.15: Data and corresponding anomaly classifier upon termination of the algorithm for the real dataset corresponding to three different tilts. The white regions in the classifier correspond to non-anomalous measurements and the black regions indicate an anomaly. While the classifier selects certain non-anomalous regions notice that the regions in the data with anomalies are accurately classified by the algorithm.

$-70°$ to $+70°$. We use a ≈ 580 nm $\times 580$ nm section of the projection images for reconstruction. The dimensions of the reconstructed volume are set so as to account for all the voxels contributing to the projection data. We reconstructed the dataset using our algorithm (MBIR-AM-PE), FBP and conventional MBIR without anomaly modeling. All reconstructions are performed with voxels of size 0.83 nm $\times$ 0.83 nm $\times$ 0.83 nm.

Figure 6.14 (a) and (b) show a $x-z$ and $x-y$ cross-section reconstructed from the data using FBP. The reconstruction has strong streaking artifacts in the $x-z$ plane and noise in the $x-y$ plane similar to the simulated dataset. The reconstruction using the conventional MBIR algorithm (Figure 6.14 (c)-(d)), also has streaking artifacts in the $x-z$ plane that are similar to those in the simulated dataset of Figure 6.11. However, the conventional MBIR result also significantly reduces streaking artifacts as compared to FBP. This is likely due to the fact that MBIR reduces the weighting of the highly attenuated projections corresponding to measurements with anomalous Bragg scatter. Figure 6.14 (e) shows that using the anomaly modeling and parameter estimation reduces streaking in the $x-z$ plane. The arrows in Figure 6.14 (d,f) indicate regions where the MBIR with anomaly modeling reduces the under-estimation as well as other artifacts in the $x-y$ cross-section compared to the conventional MBIR. The wall-clock time taken for the proposed MBIR reconstruction ($844 \times 4516 \times 1008$ voxels) using a node with two Intel Xeon-E5s (total of 16 cores) was approximately 9 hours and 40 minutes.

Figure 6.15 shows the binary classifier mask b' along with 3 successive tilts from the real data upon termination of the reconstruction algorithm (MBIR-AM-PE). Notice that most of the anomalous measurements are successfully identified by the generalized Huber function at the end of the reconstruction. Similar to the simulated dataset a few of the noisy measurements are also classified as anomalous but this does not affect the final quality of the reconstruction significantly.

6.5 Future Directions

(a) BF and HAADF STEM images acquired simultaneously

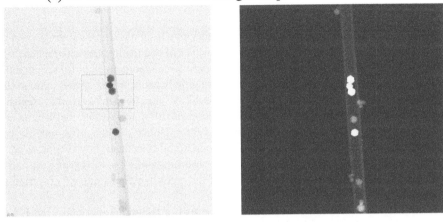

(b) BF and HAADF STEM reconstructed cross-section

Figure 6.16: (a) Multi-modal dataset consisting of a HAADF and BF image simultaneously acquired. (b) Reconstructed $x - y$ cross-section. Note that the BF image corresponds to the attenuation coefficient while the HAADF corresponds to the scatter coefficient.

In the previous two sections, we discussed MBIR approaches the two popular modalities: HAADF and BF electron tomography. Both these methods are primarily used for morphological studies of various materials though they reconstruct different properties. HAADF, when collected at sufficiently high angles, is a Rutherford scattering modality which can be related to the atomic number of the material. BF scattering intensities, when phase contrast is minimized at near zero defocus, is correlated with the material density. A less explored area lies in being able to reconstruct both quantities simultaneously. In general, being able to form multi-modal reconstructions can extract multiple material properties from a given STEM scan. The relationship between density and atomic number is nonlinear and not single valued, therefore extracting 3D relationships from a variety of single element and alloyed materials is challenging. These concepts aim to broaden the capabilities of electron tomography beyond a 3D imaging tool, to a 3D materials measurement tool through the use of multi-modal data.

Experimentally, it is possible to acquire both the dark and bright field signals in STEM mode simultaneously and perform independent HAADF and BF MBIR reconstructions. For example in

Table 6.4: Table comparing expected and measured z-contrast ratios.

Material	Expected ratio	Z-ratio	Measured ratio
$\frac{Al}{C}$	3.94	$\left(\frac{Z_{Al}}{Z_C}\right)^{1.7}$	3.72
$\frac{Au}{Al}$	21.92	$\left(\frac{Z_{Al}}{Z_C}\right)^{1.7}$	21.49
$\frac{Al}{C}$	86.53	$\left(\frac{Z_{Al}}{Z_C}\right)^{1.7}$	80

Figure 6.16 we collected both the signals simultaneously from a gold-aluminum dataset adsorbed on a curved thin strip of carbon. For the simultaneous STEM acquisition case shown here, co-alignment of the data is not necessary; however for other experimental geometries, or for sequential acquisition of the data, registration of the data can be challenging because of the spatial anticorrelation of the image intensities. The mutual information approach [793] has been successful for registration of the datasets. We then used the data to reconstruct the attenuation and scatter coefficients. Thus at each voxel location we reconstructed a vector of two values. Notice that the cross-sections from the two reconstructions (Figure 6.16, bottom panels) appear to be correlated; however there are differences which have not yet been fully investigated. If this information is known a priori, we can incorporate it into the prior model to improve the quality of the reconstruction since we constrain the solution space.

Potentially more important in the near term, such a technique can make the HAADF reconstructions quantitative by collecting the normalizing beam (in the simultaneous BF measurement) as discussed in Section 6.3. The region of sample in Figure 6.16 contains two types of nanoparticles, gold and aluminum, on a thin strip of carbon. With simultaneous HAADF and BF STEM acquisitions, and subsequent normalization of the HAADF reconstruction using the BF data quantitative reconstruction showed good agreement between the quantitative theoretical ratios of atomic numbers for Au, Al and C as shown in Table 6.4.

More generally, collecting the full diffracted intensity in a STEM imaging geometry (instead of integrating over selected angular regions as is done in HAADF and BF imaging) can potentially provide a more detailed description of the material, albeit at a significant increase in data handling overhead. With the advent of direct electron detectors, and other types of fast acquisition pixel array detectors [308] the diffraction pattern can be captured at every pixel in a large raster scanned area. Area detectors can also be used to improve the resolution of images using the concept of ptychography [450]. In the future, the key to inverting more general TEM data will rely on being able to use the multi-slice method [496] effectively to solve inverse problems.

One can also combine tomography techniques with modalities such as electron energy loss spectroscopy (EELS) and x-ray energy dispersive spectroscopy (XEDS) to form a more complete description of the sample under study, although at this stage the current state of the practice is SIRT reconstructions of the resulting EELS or XEDS intensity maps. Significant improvements in the reconstruction methods will be needed for quantitative, accurate, and informative 3D reconstructions for these types of 4D datasets, and it will become increasing critical to address this as advances in detectors continue to drive improvements in the time and ease for data collection. If it still takes a full-time graduate student several months to analyze a dataset such as this, advances in the hardware for tomography will have limited impact on materials and biological sciences.

6.6 Conclusion

In this chapter, we introduced the MBIR framework and demonstrated how model-based approaches to electron tomography can dramatically improve reconstruction quality by reducing noise and missing wedge artifacts while being robust to outliers and missing calibration parameters. Some key directions to increasing the applicability of **MBIR** techniques is the development of algorithms to combine complex forward models with novel priors, the automatic selection of regularization parameters and the development of fast optimization methods on heterogeneous computing platforms. The source code for the work presented in this chapter along with a **GUI** application implementing our method is publicly available at the website, *www.openmbir.org*.

Chapter 7

Statistical Reconstruction and Heterogeneity Characterization in 3-D Biological Macromolecular Complexes

Qiu Wang*
Cornell University

Peter C. Doerschuk
Cornell University

7.1 Introduction

The structure and function of biological macromolecular complexes is currently a topic of great interest in biology. The primary contribution of this chapter is the mathematical description of a specific problem in this area of biology, the development of algorithms and software to solve the problem, and the demonstration of the relevance of the solution to biology.

Recent success with 3-dimensional reconstructions of biological macromolecular particles employing single-particle cryo electron microscopy (cryo EM) has been remarkable. Subnanometer icosahedral virus structures are virtually routine and protein and nucleo-protein structures, without symmetry, are appearing more frequently at comparable resolution. Structures of icosahedral viruses at near-atomic resolution have been achieved with this technology in recent years [585, 43, 1200]. Cryo EM data captures biological macromolecular particles that are trapped in one of a smooth continuum of conformations at the moment of vitrification in liquid ethane. The amount of conformational change accessible to the particle is presumably space dependent, but there are limited tools available for assessing the global amount of conformational change let alone creating a spatial map of the amount of conformational change occurring. Heterogeneity among a set of particles can be detected by methods such as cross-common lines residuals [340]. In this chapter, maximum likelihood estimation is used not to estimate a single reconstruction or to find a homogeneous subset of particles but rather to estimate the statistics of an entire ensemble of reconstructions where the statistics of the images predicted by the statistics of the ensemble of reconstructions match the statistics of the experimental images.

Figure 7.1 shows a visualization of the 3-D electron scattering intensity of Flock House Virus (FHV) [324] and four cryo electron microscopy images of FHV. Each image is a 2-D projection of the 3-D electron scattering intensity of the virus particle where the projection direction is not known. Only one image of a particular instance of the virus particle is recorded, so these four images show four different instances. The goal of this chapter is not to compute just a nominal reconstruction, as is shown in Figure 7.1(a), but rather to compute a nominal reconstruction and to characterize the heterogeneity of the instances of the particle. The approach is statistical and the nominal reconstruction and characterization of heterogeneity are the first- and second-order statistics, respectively, of the electron scattering intensity. An alternative method, electron tomography, is described in Chapter 6.

*Presently at Google, New York.

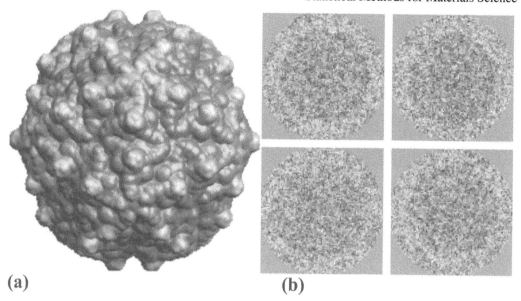

(a)   **(b)**

Figure 7.1: Flock House Virus: Panel (a): A contour-surface plot of the 3-D electron scattering intensity of the particle. Panel (b): Four example projection images from four different instances of the virus particle. Panel (a) is from http://viperdb.scripps.edu/info_page.php?VDB=4ftb.

In electron tomography, multiple images of a single particle are recorded at a controlled sequence of tilts of the microscope stage corresponding to projection directions. The full range of tilts, $-90°$ to $+90°$, cannot be attained. Among the advantages of electron tomography is that a unique particle can be studied and the reconstruction process is simplified because the projection directions are known. However, the total electron dose required in electron tomography is generally greater than the dose used for the one image of single-particle cryo electron microscopy so that, when the particle is sensitive to the dose which is common in biological particles, there is greater damage to the particle in electron tomography. The greater damage has limited the achievable spatial resolution of the reconstruction.

Viruses are parasites. While viruses take advantage of molecular machinery of the infected host cell, the evolutionary pressures on the cell and its machinery are in the direction of hindering rather than aiding viruses. One evolutionary direction is simple viruses where the virus genome only encodes two or three peptides and where the production of virus progeny in an infected host cell has two steps: (1) an assembly step in which a complete virus assembles from constituent molecules in an essentially reversible non-covalent chemical reaction where the reversibility increases the yield of correctly formed virus particles followed by (2) a maturation step in which the structure of the virus particle is reorganized in a non-reversible chemical reaction where non-reversibility leads to a virus particle that is durable outside of the cell. In Section 7.3.2 [1119], the method of Section 7.2 is used to analyze time-resolved single-particle cryo electron microscopy (cryo-EM) images of Nudaurelia Capensis Omega Virus (NωV) [655] which is a simple virus of the type described in the preceding sentences. Each virus particle contains 240 copies of the same capsid protein molecule in four different geometric positions. During maturation, each copy undergoes an autocatalytic cleavage reaction. Molecules in different geometries have different kinetics for this reaction which range from minutes to hours. Using time-resolved cryo-EM images (i.e., sets of images taken at specific times following the initiation of maturation), the method of Section 7.2 is able to demonstrate that capsid protein copies in different geometric positions have different time-varying trajectories of heterogeneity as measured by variance and that the heterogeneity tracks the kinetics previously measured by different methods [655].

7.2 Statistical 3-D Signal Reconstruction of Macromolecular Complexes

7.2.1 Introduction

Single-particle cryo electron microscopy of multiple instances of a biological object in the 10^2–10^3Å spatial scale, such as a virus or a ribosome, is used to determine the object's 3-D structure, i.e., the spatial variation of the object's electron scattering intensity [331]. Each image is roughly a 2-D projection of the 3-D scattering intensity modified by the so-called contrast transfer function (CTF) of the microscope. Because the electron beam of the microscope rapidly damages the objects, most studies with spatial resolution goals of less than 25Å limit the dose in at least two ways. First, the dose per image is minimized, which implies computing reconstructions from images with low SNR, e.g., less than 0.1. Second, only one image is recorded per instance of the object, rather than a series of images at different projection orientations, so that reconstructions are computed by fusing information from one image of each of many instances of the object. Because the orientation in the microscope of the instances of the object are not controlled, the projection direction that results in a particular image is not known *a priori*. Because the image SNR is low, it is difficult from an individual image to determine either the projection direction that created that image or the projected location in the image of the center of the object. Most estimation and computation approaches assume that each instance of an object has identical 3-D structure. Exceptions include [913, 555, 1180, 265] which assume that there are a few (e.g., 2–5) classes of objects and all instances within a class are identical. However, there are important situations in which these assumptions are not valid. For example, at high spatial resolution (e.g., < 5Å), most biological objects of this size have some element of flexibility and therefore instances in the same class are not identical even if each instance is composed of an identical number of identical chemical constituents. Furthermore, there exist many objects (e.g., [1000]) where there is variability in the chemical constituents such as variability in the number of copies of a macromolecule. For the situation where each instance within a class has a different 3-D structure, this chapter (see also [1204, 1118]) describes a statistical model, estimator, and algorithm for determining the structure of each class as a statistical distribution based on generalizations of Gaussian mixture models, maximum likelihood (ML) parameter estimation, and generalized expectation maximization (EM) algorithms.

The problem considered in this section concerns data that is a linear transformation with structured stochasticity of a vector from a Gaussian mixture distribution with unknown parameters plus a second vector from a Gaussian distribution with known parameters. The goal is to estimate the unknown parameters of the Gaussian mixture distribution. By exploiting the Gaussian assumptions, the problem is equivalent to a Gaussian mixture problem with stochastically-structured mean and covariance matrices. Gaussian mixture problems with structured covariances have been studied in a wide range of application areas (e.g., remote sensing [84], speech [933, 1043, 259], etc.) and from a large number of points of view (e.g., trees [1043], constrained bases [933], a broad range of linear structures such as covariances arising from stationary time series [951], eigen decompositions with partial sharing between classes [84], subspaces with rank constraints where the subspaces are shared or unshared between classes [259], etc.). However, the problem considered in this chapter appears to be dominated by the complicated structure of the stochasticity of the linear transformation, which includes a projection from 3 to 2 dimensions, and therefore the approach taken is a development of the approach of [265].

7.2.2 Statistical Model

Let $\eta \in \{1,\ldots,N_\eta\}$ index the possible classes of an object where the number of classes, denoted by N_η, is known but the *a priori* probabilities of each class, denoted by q_η, are not known. Assume that the electron scattering intensity of an instance in the ηth class is represented as a linear combination of $N_c(\eta)$ basis functions, denoted by $\phi_\tau^{(\eta)}(\mathbf{x})$, with support $S^{(\eta)} \subset \mathbb{R}^3$. Let $S = \cup_{\eta=1}^{N_\eta} S^{(\eta)}$. Therefore, the electron scattering intensity (denoted by $\rho(\mathbf{x})$) of an instance is

$$\rho(\mathbf{x}) = \sum_{\tau=1}^{N_c(\eta)} c_\tau \phi_\tau^{(\eta)}(\mathbf{x}) \tag{7.1}$$

where the unknown weights are denoted by c_τ. Since $\rho(\mathbf{x}) \in \mathbb{R}$, it is sufficient to consider systems where $c_\tau \in \mathbb{R}$ and $\phi_\tau^{(\eta)}(\mathbf{x}) \in \mathbb{R}$. Let $c \in \mathbb{R}^{N_c(\eta)}$ be a vector with components c_τ. Heterogeneity among instances within one class is described by making c a Gaussian random vector with mean $\bar{c}^\eta$ and covariance V_η. Reconstruction of the object is equivalent to estimation of $\bar{c}^\eta$ and V_η for $\eta \in \{1, \ldots, N_\eta\}$. Assume that the image is discretized and the samples are the components of a vector. Because the image formation process is linear, the unknown weight vector c and the image vector are related by a matrix denoted by L. The matrix depends on unknown parameters: the class of the instance (η) and the projection direction of the image and the location in the image coordinate system of the projection of the center of the object (the projection direction and center location are collectively denoted by θ).

The measurement noise in the image is described by an additive Gaussian zero-mean model ($N(\mu, \Sigma)(x)$ denotes the Gaussian pdf with mean μ and covariance Σ evaluated at location x). When the pixel noises are grouped into a vector, the covariance of this vector is denoted by Q where Q is to be estimated from the image data. The index $i \in \{1, \ldots, N_v\}$ indicates which instance of the object. The possibility of recording several images at known relative projection directions, a so-called tilt series, is included by adding an index $j \in \{1, \ldots, N_T\}$. Therefore the image formation model for the jth tilt of the ith instance is

$$y_{i,j} = L_{i,j}(\theta_i, \eta_i) c_i + w_{i,j} \tag{7.2}$$

where $\{\theta_i\}$, $\{\eta_i\}$, $\{c_i\}$, and $\{w_{i,j}\}$ are independent stochastic sequences; θ_i are i.i.d. with probability density function (pdf) $p(\theta)$ which is known (e.g., uniform over all rotations for projection orientation, i.e., Haar measure on the group $SO(3)$, times uniform on a disk of known radius for center location); η_i are i.i.d. with probability mass function (pmf) q_η which is unknown; c_i are independent but not identically distributed; for a fixed i, the pdf of c_i conditional on η_i is $N(\bar{c}^{\eta_i}, V_{\eta_i})$ where $\bar{c}^{\eta_i}$ and V_{η_i} are unknown; and $w_{i,j}$ is i.i.d. (jointly in i and j) with pdf $N(0, Q_{i,j})$ where $Q_{i,j}$ is unknown. To include the possibility of a tilt series of images, define $y_i = \left[y_{i,1}^T, \ldots, y_{i,N_T}^T \right]^T$, $L_i(\theta_i, \eta_i) = \left[L_{i,1}^T(\theta_i, \eta_i), \ldots, L_{i,N_T}^T(\theta_i, \eta_i) \right]^T$, $w_i = \left[w_{i,1}^T, \ldots, w_{i,N_T}^T \right]^T$, and $Q_i = \mathrm{diag}(Q_{i,1}, \ldots, Q_{i,N_T})$. Then,

$$y_i = L_i(\theta_i, \eta_i) c_i + w_i \tag{7.3}$$

where w_i is i.i.d. with pdf $N(0, Q_i)$ and Q_i is unknown. Denote the conditional mean and covariance of y_i by

$$\mu_i(\theta_i, \eta_i, \bar{c}^{\eta_i}) \doteq \mathrm{E}[y_i | \theta_i, \eta_i, \bar{c}^{\eta_i}, V_{\eta_i}, Q_i] \tag{7.4}$$

$$= L_i(\theta_i, \eta_i) \bar{c}^{\eta_i} \tag{7.5}$$

$$\Sigma_i(\theta_i, \eta_i, V_{\eta_i}, Q_i) \doteq \mathrm{Cov}[y_i | \theta_i, \eta_i, \bar{c}^{\eta_i}, V_{\eta_i}, Q_i] \tag{7.6}$$

$$= L_i(\theta_i, \eta_i) V_{\eta_i} L_i^T(\theta_i, \eta_i) + Q_i \tag{7.7}$$

In this notation, the conditional pdf for y_i is

$$p(y_i | \theta_i, \eta_i, \bar{c}^{\eta_i}, V_{\eta_i}, Q_i) = N\left(\mu_i(\theta_i, \eta_i, \bar{c}^{\eta_i}), \Sigma_i(\theta_i, \eta_i, V_{\eta_i}, Q_i)\right)(y_i). \tag{7.8}$$

Collected in this paragraph is all of the notation used to describe the size of a problem. Some of this notation has not yet been used. The number of classes of object is N_η. The number of coefficients used to describe the electron scattering intensity of the ηth class (Equation 7.1) is $N_c(\eta)$. The number of objects imaged is N_v. The number of images taken of each object is N_T. The number of pixels in each image is N_y.

7.2.3   Relationship between the Moments of the Weights and the Moments of the Electron Scattering Intensity

By Equation 7.1, it follows that the mean of the electron scattering intensity for a particular class is

$$\bar{\rho}_{\eta_0}(\mathbf{x}) \doteq E[\rho(\mathbf{x})|\eta = \eta_0] = \sum_{\tau=1}^{N_c(\eta_0)} \bar{c}_{\tau}^{\eta_0} \phi_{\tau}^{(\eta_0)}(\mathbf{x}) \tag{7.9}$$

and the autocorrelation is

$$r_{\eta_0}(\mathbf{x},\mathbf{x}') \doteq E[[\rho(\mathbf{x}) - \bar{\rho}_{\eta_0}(\mathbf{x})][\rho(\mathbf{x}') - \bar{\rho}_{\eta_0}(\mathbf{x}')]|\eta = \eta_0] \tag{7.10}$$

$$= \sum_{\tau=1}^{N_c(\eta_0)} \sum_{\tau'=1}^{N_c(\eta_0)} (V_{\eta})_{\tau,\tau'} \phi_{\tau}^{(\eta_0)}(\mathbf{x}) \phi_{\tau'}^{(\eta_0)}(\mathbf{x}'). \tag{7.11}$$

Let $\hat{\bar{\rho}}_{\eta_0}(\mathbf{x})$ and $\hat{r}_{\eta_0}(\mathbf{x},\mathbf{x}')$ be Equations 7.9 and 7.11 evaluated at the estimated values of $\bar{c}$ and V rather than the true values. For biological purposes, the natural quantities to visualize are $\hat{\bar{\rho}}_{\eta_0}(\mathbf{x})$ and $\hat{r}_{\eta_0}(\mathbf{x},\mathbf{x}')$, especially the case $\hat{r}_{\eta_0}(\mathbf{x},\mathbf{x})$. The estimators used in this chapter are maximum likelihood (ML) estimators (Section 7.2.4). For such estimators, if $y = f(x)$ and the ML estimate of x is $\hat{x}$ then the ML estimate of y is $f(\hat{x})$ [180, Theorem 7.2.10, p. 320]. Therefore, $\hat{\bar{\rho}}_{\eta_0}(\mathbf{x})$ and $\hat{r}_{\eta_0}(\mathbf{x},\mathbf{x}')$ are ML estimates of $\bar{\rho}_{\eta_0}(\mathbf{x})$ and $r_{\eta_0}(\mathbf{x},\mathbf{x}')$, respectively.

7.2.4   Estimation Criterion

The goal of the estimation problem is to determine $\omega = \{q_{\eta}, \bar{c}^{\eta}, V_{\eta}, Q_{i,j} : \eta \in \{1,\dots,N_{\eta}\}, i \in \{1,\dots,N_v\}, j \in \{1,\dots,N_T\}\}$ for which no *a priori* pdfs are available. The parameters that describe the class and the projection orientation and coordinate system, $\Omega = \{\eta_i, \theta_i : i \in \{1,\dots,N_v\}\}$, are of less biological interest and have *a priori* pdfs. The measurement noise, $\{w_{i,j} : i \in \{1,\dots,N_v\}, j \in \{1,\dots,N_T\}\}$, is not of biological interest. The approach used in this chapter to estimate ω is maximum likelihood (ML) estimation. Once the estimate of ω, denoted by $\hat{\omega}$, is computed, it is sometimes useful to estimate Ω and that is done via $\hat{\Omega} = \text{argmax}_{\Omega} p(y|\hat{\omega}, \Omega)$ where $y = \{y_i : i \in \{1,\dots,N_v\}\}$. Define $q = \{q_{\eta} : \eta \in \{1,\dots,N_{\eta}\}\}$ and similarly for $\bar{c}$ and V. Define $Q = \{Q_{i,j} : i \in \{1,\dots,N_v\}, j \in \{1,\dots,N_T\}\}$ and $Q_i = \{Q_{i,j} : j \in \{1,\dots,N_T\}\}$. By direct calculation, the log likelihood is

$$\ln p(y|\bar{c},V,q,Q) = \sum_{i=1}^{N_v} \ln \left[\sum_{\eta_i=1}^{N_{\eta}} \int_{\theta_i} p(y_i|\theta_i, \eta_i, \bar{c}^{\eta_i}, V_{\eta_i}, Q_i) q_{\eta_i} p(\theta_i) d\theta_i \right] \tag{7.12}$$

where $p(y_i|\theta_i, \eta_i, \bar{c}^{\eta_i}, V_{\eta_i}, Q_i)$ is given in Equation 7.8 and $p(\theta_i)$ is the *a priori* pdf on θ_i.

In order to compute the ML estimate, a generalized expectation-maximization (EM) algorithm is used. The sense in which the algorithm is a generalized EM versus standard EM algorithm is that at any particular iteration, only a subset of the variables to be estimated is updated. Specifically, only $(q,\bar{c})$, (q,V), or (q,Q) is updated. As will be described in Section 7.2.4.1, the update for q is independent of the updates for the other variables but shares some of the same computations which motivates combining updates of q with the updating of each of the other variables. In the following paragraph, the expectation of the EM algorithm is derived and then, in Sections 7.2.4.1 and 7.2.4.2, the updates for q and for $\bar{c}$, V, and Q are derived.

The expectation in the EM algorithm is to compute

$$
\mathcal{Q}(\bar{c}, V, q, Q | {}_{0}\bar{c}, {}_{0}V, {}_{0}q, {}_{0}Q, y)
$$

$$
= \int_{\theta} \sum_{\eta} \left[\ln p(y, \theta, \eta | \bar{c}, V, q, Q) \right] p(\theta, \eta | {}_{0}\bar{c}, {}_{0}V, {}_{0}q, {}_{0}Q, y) d\theta \tag{7.13}
$$

$$
= \sum_{i=1}^{N_v} \int_{\theta_i} \sum_{\eta_i=1}^{N_\eta} \left[\ln p(y_i | \theta_i, \eta_i, \bar{c}^{\eta_i}, V_{\eta_i}, Q_i) + \ln p(\theta_i) + \ln q_{\eta_i} \right]
$$

$$
\times p(\theta_i, \eta_i | y_i, {}_{0}\bar{c}^{\eta_i}, {}_{0}V_{\eta_i}, {}_{0}q, {}_{0}Q_i) d\theta_i
$$

$$
\tag{7.14}
$$

where $\bar{c}, V, q, Q$ $({}_{0}\bar{c}, {}_{0}V, {}_{0}q, {}_{0}Q)$ are the new (old) values of the parameters. The conditional pdf on the nuisance parameters is

$$
p(\theta_i, \eta_i | y_i, \bar{c}, V, q, Q) = \frac{p(y_i | \theta_i, \eta_i, \bar{c}^{\eta_i}, V_{\eta_i}, Q_i) p(\theta_i) q_{\eta_i}}{\sum_{\eta'=1}^{N_\eta} \int_{\theta'} q_{\eta'} p(\theta') p(y_i | \eta', \theta', \bar{c}^{\eta'}, V_{\eta'}, Q_i) d\theta'}. \tag{7.15}
$$

7.2.4.1   q as a Function of ${}_{0}\bar{c}$, ${}_{0}V$, ${}_{0}Q$

Maximizing Equation 7.14 with respect to q subject to the two constraints

$$
\sum_{\eta'=1}^{N_\eta} q_{\eta'} = 1 \tag{7.16}
$$

$$
q_{\eta'} \geq 0 \quad \forall \eta' \in \{1, \dots, N_\eta\} \tag{7.17}
$$

is equivalent to maximizing

$$
\mathcal{Q}_3(q | {}_{0}\bar{c}, {}_{0}V, {}_{0}q, {}_{0}Q, y) \doteq \sum_{i=1}^{N_v} \int_{\theta_i} \sum_{\eta_i=1}^{N_\eta} [\ln q_{\eta_i}] p(\theta_i, \eta_i | y_i, {}_{0}\bar{c}^{\eta_i}, {}_{0}V_{\eta_i}, {}_{0}q, {}_{0}Q_i) d\theta_i \tag{7.18}
$$

subject to the same two constraints. As is standard in the derivation of ML parameter estimates for Gaussian mixture models by EM, e.g., [835, 670, 96], the optimization problem is solved by ignoring Equation 7.17, combining Equation 7.16 with Equation 7.18 by using a Lagrange multiplier, solving the resulting optimization problem, and verifying that the solution satisfies Equation 7.17. The result is that

$$
q_{\eta'} = \frac{1}{N_v} \sum_{i=1}^{N_v} \int_{\theta_i} p(\theta_i, \eta' | y_i, {}_{0}\bar{c}^{\eta'}, {}_{0}V_{\eta'}, {}_{0}q, {}_{0}Q_i) d\theta_i \quad \forall \eta' \in \{1, \dots, N_\eta\} \tag{7.19}
$$

where the computation of $p(\theta_i, \eta' | y_i, {}_{0}\bar{c}^{\eta'}, {}_{0}V_{\eta'}, {}_{0}q, {}_{0}Q_i)$ is from Equation 7.15.

7.2.4.2 $\bar{c}$, V, and Q as a Function of ${}_{0}\bar{c}$, ${}_{0}V$, ${}_{0}Q$

Maximizing Equation 7.14 with respect to $\bar{c}$, V and Q subject to

$$
V_\eta = V_\eta^T \tag{7.20}
$$

$$
V_\eta > 0 \tag{7.21}
$$

is equivalent to maximizing

$$\mathcal{Q}_1(\bar{c}, V, Q | {}_0\bar{c}, {}_0V, {}_0q, {}_0Q, y)$$

$$\doteq \sum_{i=1}^{N_v} \int_{\theta_i} \sum_{\eta_i=1}^{N_\eta} [\ln p(y_i | \theta_i, \eta_i, \bar{c}^{\eta_i}, V_{\eta_i}, Q_i)] p(\theta_i, \eta_i | y_i, {}_0\bar{c}^{\eta_i}, {}_0V_{\eta_i}, {}_0q, {}_0Q_i) d\theta_i \qquad (7.22)$$

$$= -\frac{N_y}{2} \ln(2\pi) N_v$$

$$- \frac{1}{2} \sum_{i=1}^{N_v} \int_{\theta_i} \sum_{\eta_i=1}^{N_\eta} \ln \det(\Sigma_i(\theta_i, \eta_i, V_{\eta_i}, Q_i)) p(\theta_i, \eta_i | y_i, {}_0\bar{c}^{\eta_i}, {}_0V_{\eta_i}, {}_0q, {}_0Q_i) d\theta_i$$

$$- \frac{1}{2} \sum_{i=1}^{N_v} \int_{\theta_i} \sum_{\eta_i=1}^{N_\eta} (y_i - \mu_i(\theta_i, \eta_i, \bar{c}^{\eta_i}))^T \Sigma_i^{-1}(\theta_i, \eta_i, V_{\eta_i}, Q_i)(y_i - \mu_i(\theta_i, \eta_i, \bar{c}^{\eta_i}))$$

$$\times p(\theta_i, \eta_i | y_i, {}_0\bar{c}^{\eta_i}, {}_0V_{\eta_i}, {}_0q, {}_0Q_i) d\theta_i \qquad (7.23)$$

(N_y is the number of pixels in an image) with respect to $\bar{c}$, V, and Q with the same two constraints.

7.2.4.3 $\bar{c}$ as a Function of V, Q, $_0\bar{c}$, $_0V$, $_0Q$

The gradient of $\mathcal{Q}_1$ with respect to $\bar{c}$ concerns only μ in the third term of Equation 7.23.

Setting the gradient equal to zero gives a linear system of equations for the new value of $\bar{c}^{\eta'}$, specifically, $F_{\eta'} = g_{\eta'}$ where

$$F_{\eta'} = \sum_{i=1}^{N_v} \int_{\theta_i} L_i^T(\theta_i, \eta')$$

$$\times \Sigma_i^{-1}(\theta_i, \eta', V_{\eta'}, Q_i) L_i(\theta_i, \eta') p(\theta_i, \eta' | y_i, {}_0\bar{c}^{\eta'}, {}_0V_{\eta'}, {}_0q, {}_0Q_i) d\theta_i \qquad (7.24)$$

$$g_{\eta'} = \sum_{i=1}^{N_v} \int_{\theta_i} L_i^T(\theta_i, \eta')$$

$$\times \Sigma_i^{-1}(\theta_i, \eta', V_{\eta'}, Q_i) y_i p(\theta_i, \eta' | y_i, {}_0\bar{c}^{\eta'}, {}_0V_{\eta'}, {}_0q, {}_0Q_i) d\theta_i \qquad (7.25)$$

Please note that $F_{\eta'}$ and $g_{\eta'}$ both depend on $V_{\eta'}$ which is not known.

7.2.4.4 V as a Function of $\bar{c}$, Q, $_0\bar{c}$, $_0V$, $_0Q$

Define

$$N_i(y_i, \theta_i, \eta_i, \bar{c}^{\eta_i}) \doteq (y_i - \mu_i(\theta_i, \eta_i, \bar{c}^{\eta_i}))(y_i - \mu_i(\theta_i, \eta_i, \bar{c}^{\eta_i}))^T \qquad (7.26)$$

and use the trace operator to rewrite Equation 7.23 in the form

$$\mathcal{Q}_1(\bar{c}, V, q, Q | {}_0\bar{c}, {}_0V, {}_0q, {}_0Q, y)$$

$$= -\frac{N_y}{2} \ln(2\pi) N_v$$

$$+ \frac{1}{2} \sum_{i=1}^{N_v} \int_{\theta_i} \sum_{\eta_i=1}^{N_\eta} \ln \det(\Sigma_i^{-1}(\theta_i, \eta_i, V_{\eta_i}, Q_i)) p(\theta_i, \eta_i | y_i, {}_0\bar{c}^{\eta_i}, {}_0V_{\eta_i}, {}_0q) d\theta_i$$

$$- \frac{1}{2} \sum_{i=1}^{N_v} \int_{\theta_i} \sum_{\eta_i=1}^{N_\eta} \text{tr} \left[\Sigma_i^{-1}(\theta_i, \eta_i, V_{\eta_i}, Q_i) N_i(y_i, \theta_i, \eta_i, \bar{c}^{\eta_i}) \right]$$

$$\times p(\theta_i, \eta_i | y_i, {}_0\bar{c}^{\eta_i}, {}_0V_{\eta_i}, {}_0q, {}_0Q_i) d\theta_i. \qquad (7.27)$$

It is not possible to follow the standard derivation of ML parameter estimates for Gaussian mixture models by EM, which involves computing derivatives with respect to Σ^{-1}, because it is necessary to account for the dependence of Σ on V.

Therefore we maximize $\mathcal{Q}_1$ directly. To limit the complexity of the calculation, the most general convariance we consider is $V_\eta = \text{diag}(v_\eta)$ and we optimize $\mathcal{Q}_1$ with respect to v_η by MATLAB®'s fmincon function with symbolic formulas for the gradient and Hessian.

An alternative to the method of the previous paragraph is to approximate $\mathcal{Q}_1$ so that the approximation is concave in V and then use convex optimization methods. The log of the Gaussian pdf is jointly concave in the mean and inverse covariance and linear combinations of concave functions are concave so the expectation integral makes $\mathcal{Q}_1$ a concave function of $\bar{c}^\eta$ and Σ_i^{-1}. However, we seek to optimize $\mathcal{Q}_1$ with respect to $\bar{c}^\eta$ and V_η and Σ_i^{-1} is a complicated function of V_η. Since if $f(x)$ is concave in x then $g(y) = f(a+By)$ is concave in y, a linear approximation for Σ_i^{-1} as a function of V_η would lead to a convex optimization problem. Iteration application of a linearization of Σ_i^{-1} as a function of V_η around the current value of V_η followed by solution of the resulting convex optimization problem provides a complete algorithm for updating V_η. For the case of $V_\eta = \text{diag}(v_\eta)$, the linearization of Σ_i^{-1} as a function of V_η around $V_{\eta 0}$ has, in abbreviated notation, the simple form $\Sigma^{-1} = (LV_0L^T + Q)^{-1} - T(V - V_0)T^T$ where $T = Q^{-1}L(L^TQ^{-1}L + V_0^{-1})^{-1}V_0^{-1}$. This approach has not been implemented but will be reconsidered for larger problems.

7.2.4.5   Q as a Function of V, $\bar{c}$, $_0\bar{c}$, $_0V$, $_0Q$

Only the case of $Q_{i,j} = \lambda I_{N_y}$ is considered so there is only a single scalar parameter. An initial condition is available by computing the sample variance of the image in an annulus outside of the virus particle [1180, Section 2.8]. Simple formulas are available for the first and second derivatives of $\mathcal{Q}_1$ in the case of $V = 0$ (not shown) but the corresponding formulas for $V \neq 0$ are complicated. Therefore, numerical optimization was done using MATLAB [654] function fminbnd with a limit of 8 iterations and bounds of 0 and ∞.

7.2.5   Relationship with Other Results

If $V_\eta = 0$ then the results of [1180] are regained. If $Q_i = 0$ and $L_i(\theta_i, \eta_i) = I_{N_c(\eta_i)}$ (I_n is the $n \times n$ identity matrix) then the standard results [835, 670, 96] on ML parameter estimation for Gaussian mixture problems by EM are regained.

7.2.6   Algorithm

The generalized EM algorithm operates by updating either $(q, \bar{c})$, (q, V), or (q, Q). Because of the historical focus on $\bar{c}$ in biology, the particular pattern of updating that is used focuses on $\bar{c}$. In particular, updating of $(q, \bar{c})$ is performed until $\bar{c}$ converges in the sense of small quadratic norm of the difference between the current and the immediately previous value of $\bar{c}$. Then (q, V) and (q, Q) are updated. Then $(q, \bar{c})$ is updated. If the change in $\bar{c}$ relative to its immediately previous value is sufficiently small to satisfy the convergence criteria then the algorithm terminates. Otherwise, the algorithm continues to update $(q, \bar{c})$ until convergence followed by another update of (q, V) and (q, Q), etc. Joint optimization of Q and V was tested but works poorly possibly due to the difference in the sizes of these two covariance matrices. Pseudocode is given in Algorithm 7.1.

7.2.7   Performance

The standard measure of performance in structural biology is the Fourier Shell Correlation (FSC) [1088, Equation 2][407, Equation 17][44, p. 879] between a pair of reconstructions computed from disjoint sets of images. The FSC is the spherical average of the correlation in the frequency domain between the two reconstructions normalized by the square root of the product of the

---

Algorithm 7.1: The generalized EM algorithm

1: set the initial condition from a homogeneous (i.e., $V = 0$) calculation
2: **while** true **do**
3:     **while** $\bar{c}$ not converged **do**
4:         update q (Equation 7.19) and $\bar{c}$ (Equation 7.25)
5:     **end while**
6:     update q (Equation 7.19) and Q (Section 7.2.4.5)
7:     update q (Equation 7.19) and V (Newton's method for $V_\eta = \mathrm{diag}(v_\eta)$ in Section 7.2.4.4)
8:     update q (Equation 7.19) and $\bar{c}$ (Equation 7.25)
9:     **if** $\bar{c}$ converged **then**
10:        break
11:    **end if**
12: **end while**

---

spherical average of the magnitude squared of the individual reconstructions. The resolution of the reconstruction is defined to be the magnitude of the spatial frequency vector at the first crossing of a threshold where the threshold is typically taken to be $1/2$.

Because existing reconstruction tools are based on having homogeneous objects (i.e., $V_\eta = 0$), the natural way in which to apply FSC to the heterogeneous (i.e., $V_\eta \neq 0$) work described in this chapter is to measure performance in terms of the mean, i.e., $\hat{\bar{\rho}}_\eta(\mathbf{x})$.

Since this chapter concerns maximum likelihood (ML) estimators, the standard theory for the performance of ML estimators [291] can be applied. The basic result is that the difference between the true and the estimated values of the parameters has a Gaussian probability distribution with mean zero and covariance equal to the negative of the inverse of the Hessian of the log likelihood evaluated at the estimated parameter values. We have used these ideas on simpler problems where $V_\eta = 0$ including methods related to the errors in the parameters with FSC. With V_η as additional parameters, the situation is more complicated and, with respect to the V_η parameters, it may be more natural to use the Cholesky factors as parameters so that the constraint that V_η be a positive definite matrix is automatically imposed.

7.2.8 Estimation of the a priori Probability Distribution on the Nuisance Parameters

Some objects, especially objects such as tailed bacteriophages that are far from spherical in shape, adopt a particular range of orientations relative to the air-water interface during the preparation of the cryo electron microscopy specimen. Therefore, the *a priori* probability distribution on the projection orientation nuisance parameters in θ_i is not uniform over all rotations (i.e., not Haar measure on $SO(3)$). Due to space limitations, please refer to detailed discussion on this topic in [1118].

7.2.9 Pre- and Post-Processing

Pre-processing.   Different types of preprocessing are required for different datasets. Our calculations [1119, 1205, 266, 1001] start with so-called boxed images which are small subimages from the micrograph where each subimage contains exactly one centered particle. In some datasets, some boxed images may show a defective, e.g., broken, particle. We have removed such boxed images [1119] by computing the average boxed image and computing the difference between each boxed image and the average boxed image. Then a boxed image is retained only if the l_2 norm of the difference is sufficiently small. In some datasets, boxed images come from many micrographs and need to be scaled. Because our algorithms use both first- and second-order statistics, we scale by $y^{new} = ay^{old} + b$ where a and b are chosen so that the sample mean and variance of the region outside of the particle have fixed values, e.g., 0 and 1, respectively.

Post-processing. The type of post-processing needed depends on the questions that the study is designed to answer. The spatially-dependent mean and variance for each class of particle are key products of our calculations and can be computed via

$$\bar{\rho}_{\eta'}(\mathbf{x}) \doteq E[\rho(\mathbf{x})|\eta = \eta'] \tag{7.28}$$

$$= \sum_{j=1}^{N_c(\eta')} (\bar{c}^{\eta'})_j \phi_j^{(\eta')}(\mathbf{x}) \tag{7.29}$$

which is Equation 7.9 and

$$v_{\eta'}(\mathbf{x}) \doteq E[[\rho(\mathbf{x}) - \bar{\rho}_{\eta'}(\mathbf{x})]^2|\eta = \eta'] \tag{7.30}$$

$$= \sum_{j=1}^{N_c(\eta')} \sum_{j'=1}^{N_c(\eta')} (V_{\eta'})_{j,j'} \phi_j^{(\eta')}(\mathbf{x}) \phi_{j'}^{(\eta')}(\mathbf{x}) \tag{7.31}$$

which is Equation 7.11, respectively. Let $\hat{\bar{\rho}}_{\eta'}(\mathbf{x})$ and $\hat{v}_{\eta'}(\mathbf{x})$ be Equations 7.29 and 7.31 evaluated at the estimated values of $\bar{c}$ and V rather than the true values. For biological purposes, the natural quantities to visualize are $\hat{\bar{\rho}}_{\eta'}(\mathbf{x})$ and $\hat{v}_{\eta'}(\mathbf{x})$, especially the standard deviation $s_{\eta'}(\mathbf{x}) = \sqrt{v_{\eta'}(\mathbf{x})}$ ($\hat{s}_{\eta'}(\mathbf{x}) = \sqrt{\hat{v}_{\eta'}(\mathbf{x})}$). Many problems benefit from the ability to compute the spherical average of the variance map as a function of radial position [1119, 372]. This calculation can be done exactly because of the use of spherical harmonics in our problem formulation. In particular, using a formula analogous to [59, Equations 22-25], the spherical average of the variance map is

$$\bar{v}_{\eta'}(x) = \frac{1}{4\pi} \int v_{\eta'}(x) d\Omega \tag{7.32}$$

$$= \frac{1}{4\pi} \sum_{l=0}^{L} \sum_{p=1}^{P} \left[\sum_{n=0}^{N_l-1} v_{l,n,p}^{(\eta')} \left(\sum_{m=-l}^{+l} |b_{l,n,m}|^2 \right) \right] h_{l,p}^2(x) \tag{7.33}$$

where $(V_{\eta'})_{j,j'} = v_{l(j),n(j),p(j)}^{(\eta')} \delta_{j,j'}$, $h_{l,p}(\cdot)$ is the radial basis function [1180], $\int d\Omega$ is integration over the sphere, and

$$I_{l,n}(\theta, \phi) = \sum_{m=-l}^{+l} b_{l,n,m} Y_{l,m}(\theta, \phi) \tag{7.34}$$

where $I_{l,n}(\cdot, \cdot)$ is the (l,n)th icosahedral harmonic [1203] and $Y_{l,m}(\cdot, \cdot)$ is the (l,m)th spherical harmonic. Applying Equation 7.33 to the results of multiple calculations on different datasets indexed by $\delta \in \{1, \ldots, \Delta\}$ gives $\bar{v}_{\eta',\delta}$. Then, $\bar{s}_{\eta',\delta} = \sqrt{\bar{v}_{\eta',\delta}}$.

7.3 Biological Examples

7.3.1 Flock House Virus (FHV)

Flock House Virus (FHV) [324, 535, 1078] is an insect virus, hence a eukaryote virus, that has been intensively studied including both single-particle cryo electron microscopy [535] and x-ray crystallography [324] structures of the entire particle. We have performed a reconstruction using the methods of Section 7.2 [1180, Section 3.1] and the input images and output results are summarized in Figure 7.2.

A key part of the x-ray crystallographic structure of FHV is that a part of the RNA genome of the particle can be seen in the structure as a dodecahedral cage just below the protein shell ("capsid") of the structure. After removing the capsid by using the UCSF Chimera [781] visualization software, we also can see the dodecahedral cage as a low variance (blue color) region in Figure 7.2(d). Thus we both confirm the x-ray crystallographic result and increase our confidence in the meaningful nature of our variance maps.

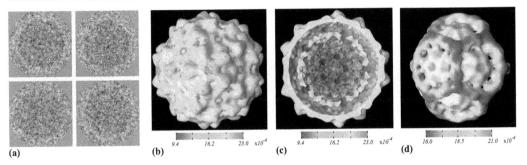

Figure 7.2: Reconstruction of FHV from experimental images. Panel (a): Example boxed experimental images (same color map). Panel (b): Surface plot of the mean $\hat{\bar{\rho}}_{\eta_0}(\mathbf{x})$ pseudo-colored by the variance $\hat{v}_{\eta_0}(\mathbf{x})$. The variance is highest near the 5-fold symmetry axes which is consistent with the idea that the binding of the particle to a new host cell and possibly later events concerning RNA translocation occur around a 5-fold axis [197, 141, 438]. Panel (c): Cross-section of the mean though the center of the particle perpendicular to a 5-fold symmetry axis pseudo-colored by the variance. Panel (d): The RNA core after removal of the protein capsid. Surface plot of the mean pseudo-colored by the variance. The ordered dodecahedral RNA cage [1046, 324] is detected and, as expected, the variance of the cage is low (blue) while that of the surrounding less well-ordered material is high (red). Visualizations in panels (b–d) by UCSF Chimera [781].

7.3.2   Nudaurelia Capensis ω Virus (NωV)

Many viruses self-assemble from constituent macromolecules and then undergo a series of irreversible so-called maturation reactions before exiting the host cell. Nudaurelia Capensis omega virus (NωV) [655] has such a production mechanism that is amenable to time-resolved study because maturation of the assembled particles can be triggered by dropping the pH to 5.0. NωV is a $T = 4$ icosahedral RNA virus so there are 60×4 copies of the capsid peptide each in one of four different chemical environments. The four environments are referred to as subunits A, B, C, and D. Decreasing the pH first causes the particle to decrease in diameter, which is reversible, but then, at the smaller diameter, a self-catalyzed peptide cleavage occurs in each of the 60×4 peptides. The kinetics of the cleavage reaction are different in the four different subunits. The different cleavage kinetics were studied by Matsui, Lander, Khayat, and Johnson [655] by computing difference maps between pairs of reconstructions based on sets of particles that have been exposed to low pH for different times (3 min, 30 min, 4 hr, and 3 day). This analysis approach based on difference maps takes advantage of the unusual fact that the particle does not change shape as the cleavage reactions occur.

Using the data from [655], we have performed separate reconstructions at each of the four time points. Sufficient data is available to make four independent reconstructions for each of the four time points. Each of the 16 reconstructions provides mean and variance maps. As is shown in Figure 7.3, the four mean maps (which define the shapes) are very similar but the four variance maps (which define the colors) are very different. Overall, as maturation proceeds, the variance decreases dramatically. Therefore, cleavage causes variance reduction.

The sites of the cleavages are known from the x-ray crystallographic structure of the particle. Averaging the variance map at the C_α carbons that fall within a sphere of radius 10 Å gives a local measure of variance. For each of the four subunits, the time variation of this variance, along with standard deviations computed from the four repetitions, are shown in Figure 7.4. The similarity between Figure 7.4 and the work of Matsui, Lander, Khayat, and Johnson [655, Figure 3B] is striking. Thus we both confirm the result of Matsui et al. and increase our confidence in the meaningful nature of our variance maps. Note that our calculation does not require that the structure of

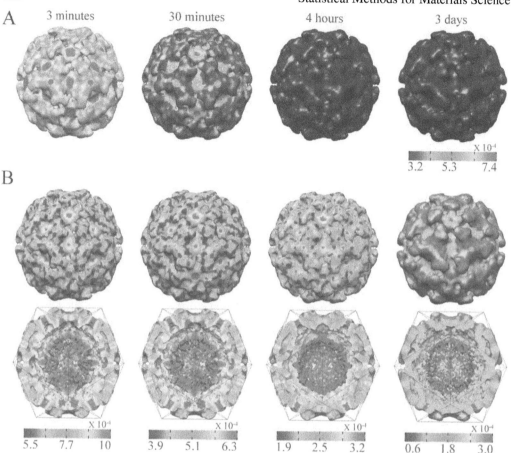

Figure 7.3: The four time-resolved reconstructions. Panel A: Surface of each of the four reconstructions colored by the square root of the variance map (i.e., the standard deviation map) and displayed using the VIPERdb [1079] convention. The same color map is used in all images. Panel B: The surface and a cross-section perpendicular to a 2-fold axis of each of the four reconstructions colored by the standard deviation map. The surface and cross-section visualizations at a particular time point share the same color map. Different color maps are used at different time points. Visualization by UCSF Chimera [781].

the maturing particles does not change since the algorithm never simultaneously uses images from multiple time-points.

In order to achieve a more global view of the variance map, Figure 7.5 shows a ribbon diagram for each of the subunits at each of the four times where the variance is encoded in the color of the ribbon diagram. This presentation emphasizes the spatial variation in the variance as well as the localized major differences in the geometry of the four subunits.

7.3.3 Summary

In both the FHV and the NωV examples, the calculations described in this section reproduce important results from previous calculations. In the case of NωV, the reproductions are more quantitative than the original results and do not require the special situation that the particle undergoes the 60×4 cleavage reactions without a change in the particle's shape.

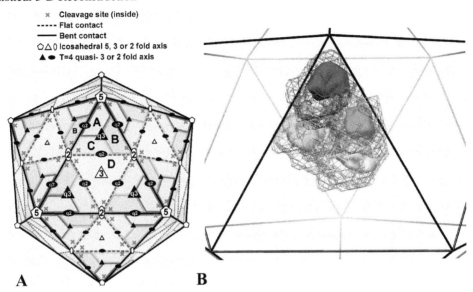

Figure 7.4: Part I of region-specific variability analysis of the NwV protein capsid in different stages of maturation. Panels A and B: Variance analysis around the cleavage sites of subunits A, B, C, and D that form the asymmetric unit of the NwV protein capsid. Both panels show the $T = 4$ surface lattice with the subunits' locations. The total volume occupied by each subunit is rendered as a mesh in panel B. The variance was calculated over a smaller region, enclosing the cleavage site, which is shown as a solid volume within the subunit density. As is described in Section 7.2.9, the smaller region is essentially the region occupied by $C\alpha$ atoms within 10Å of the active site. This is the same region analyzed by Matsui et al. [655] using difference maps.

7.4 Discussion

7.4.1 Challenges and Future Directions

The work described in this chapter is focused on diagonal covariance matrices V for the description of heterogeneity in the unknown weights c_τ of Equation 7.1. In order to fully describe the weights within the assumption of Gaussian statistics on the weights, it is necessary to consider full covariance matrices. This leads to two coupled challenges and directions. First, estimating large covariance matrices from data is a challenging and much studied task. Incorporation of methods from this literature into the reconstruction problem is an important unresolved issue. Second, estimating full covariance matrices increases the computational requirements of the reconstruction problem and these challenges must also be addressed if a practical tool is to be provided to biologist users. In the following paragraphs, both of these challenges are described in greater detail.

Sparse Full-Matrix Covariance for the Coefficients

The estimation of large matrices from limited data is an important current topic in statistics [728, 849, 156, 531, 92, 823, 673, 93]. Methods for covariance matrices include [156] banding, tapering, thresholding, penalties, and regularization. An appropriate method for V_η depends in part on the choice of basis functions. For instance, using the voxel basis functions, every element (i, j) of V_η corresponds to a pair of triple-integer indices $(\mathbf{m}, \mathbf{n})$ for the corresponding voxels and it is natural that the elements of V_η decay as a function of $\|\mathbf{m} - \mathbf{n}\|_2$ where $\|\cdot\|_2$ is the Euclidean norm. On the other hand, using harmonic basis functions also leads to triple-integer indices but now two of the indices describe the linear combination of spherical harmonics and one describes the radial function and so $\|\mathbf{m} - \mathbf{n}\|_2$ no longer has a geometric interpretation. Possibly the most promising

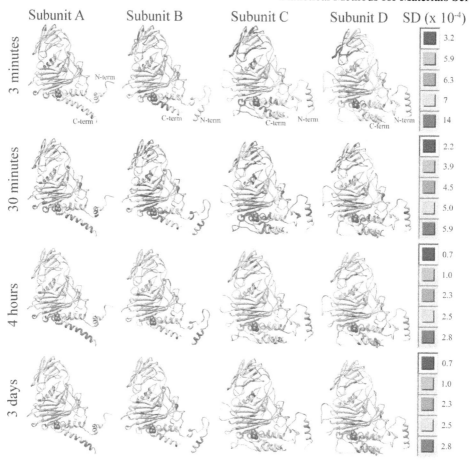

Figure 7.5: Ribbon diagrams of the four subunits at the four times colored by the square root of the variance map (i.e., the standard deviation map) with the asparagine at the self-catalytic site (Asn 570) shown as a ball-and-stick model. Each time point has its own color map analogous to the second row of Figure 7.3. For instance, red at the 3-minute time point is $14 \times 10^{-4}/5.9 \times 10^{-4} = 2.4$ times higher than red at the 30-minute time point.

approach is to focus on regularizers [823, 93], transform the regularizer into an *a priori* pdf on V_η (e.g., square of the Euclidean norm of a vector transforms into a Gaussian *a priori* pdf), replace the maximum likelihood criteria by a maximum *a posteriori* (MAP) criteria, and continue to use expectation maximization to compute the solution.

Faster Computation

In order to have an impact on the biological community, better software is required. First of all, it is important to change to a programming language and user interface that are familiar to the biological community. The current implementation is done in MATLAB [654], which is less familiar to the biological community compared to python [1072]. Therefore a python implementation, which can include MPI [1073, 1074, 1075] type message-passing parallel functionality, is attractive. Such an implementation could be integrated into the Appion system [533, 1076] which would immediately provide a familiar user interface. A second sense in which the software needs to be improved is through better algorithms to implement the estimators. Several areas of effort are described in [1118]. In particular, the use of *a priori* information as described in [1118] seems promising.

7.5 Conclusion

The primary conclusions of this chapter are that maximum likelihood estimators that solve the heterogeneous particle reconstruction problem can be posed, solved, and implemented in practical software and that the resulting heterogeneous reconstructions are relevant to biology. We describe the relevance in terms of giving two examples, which are FHV (Section 7.3.1) and NωV (Section 7.3.2). However, these ideas have also been useful in a second NωV maturation problem [1001] and in studying the maturation of the bacteriophage Hong Kong 97 [372].

In one sense the estimators are substantial generalizations of previous homogeneous reconstruction estimators [913, 555, 1180, 265] because the new estimators treat the problem as a stochastic signal in noise problem rather than a deterministic signal in noise problem. In a second sense the estimators are substantial generalizations of previous pattern analysis estimators [835, 670, 96] because the new estimators do not measure the random vector with a Gaussian mixture pdf directly but rather through a stochastic linear transformation with the addition of noise. The relevance of the results to biology is most clearly demonstrated in the work on the maturation of NωV which is described in Section 7.3.2. Even the homogeneous case of these estimators has proven useful in the characterization of nanoscale synthetic particles [609].

Chapter 8

Object Tracking through Image Sequences

Song Wang
University of South Carolina

Hongkai Yu
University of South Carolina

Youjie Zhou
University of South Carolina

Jeffrey P. Simmons
Air Force Research Laboratory

Craig Przybyla
Air Force Research Laboratory

Fast and accurate characterization of micro-structures plays an important role for materials scientists to analyze physical properties of materials and can help facilitate the new material design and development. In materials science, this is usually achieved by cross-sectioning study [801] – continuously cross-sectioning a 3D material sample for a sequence of 2D microscopic images, followed by image analysis to detect, segment, and reconstruct the 3D structures of interest from the obtained image sequence. Many micro-structures, such as fibers in continuous fiber reinforced composite (FRC) materials, can be efficiently and accurately extracted by detecting and tracking their cross-sections along the obtained microscopic image sequence [801]. In this chapter, we will study the use of tracking for the micro-structures from the material image sequence.

Object tracking has been well studied in radar signal processing and video analysis [106]. In this chapter, we introduce the use of object tracking, particularly the recursive Kalman-filter tracking algorithm [477], to extract the 3D micro-structures from cross-sectioned material image sequences. Specifically, we will take fiber tracking from continuous FRC materials as an example, where the tracking states are simply the fiber location on each slice and its location change between adjacent slices. We will then focus on addressing several specific issues in large-scale fiber tracking. First, the tracking objects, e.g., the fibers, are of large scale, crowded, and show similar appearance and to track all of them simultaneously, it is important to associate them correctly from slice to slice. Second, to speed up the cross-sectioning study for the underlying micro-structures, sparse sampling along the sequence, i.e., using larger inter-slice distance in cross-sectioning, is usually desired. We will study the effect of sampling sparsity to the tracking accuracy. Third, compared with the radar or video tracking over a long sequence, material image sequences may contain a much smaller number of slices, especially under sparse sampling. As a result, we need to construct more accurate initialization of the tracking states to make the tracking converge quickly. Finally, objective evaluation of the proposed tracking is a challenging issue given the large number of the objects (e.g., fibers) and the difficulty to construct all the ground truth.

The intention of this chapter is to give a tutorial style presentation of tracking algorithms and will focus on the Kalman filter based tracking algorithm, since it is widely used and intuitive.

8.1   Tracking and Kalman Filters

The Kalman filter is one of the most widely used tracking models [477]. It is a probabilistic model that recursively updates the tracking target/object's state through the image sequence. Let $\mathbf{s}^i$ be the state of the target (e.g., location, velocity, acceleration) at slice (or time for temporal tracking) i. In the Kalman filter, we first define the state transition from slice $i-1$ to slice i by

$$\mathbf{s}^i = A\mathbf{s}^{i-1} + \mathbf{w}^{i-1}, \tag{8.1}$$

where A is the $n \times n$ state transition matrix, with n being the dimension of the tracking state $\mathbf{s}$, and $\mathbf{w}$ is the n-dimensional noise for the state transition. For example, if one fiber is characterized by its center location (x and y) in a particular slice, with a velocity of v_x and v_y, the fiber state would be a 4-dimensional vector describing the fiber at this slice. The action of A on this fiber state is to project it to the next slice, making an initial prediction of the state of this fiber in the next slice.

We further define the observation model by

$$\mathbf{o}^i = H\mathbf{s}^i + \mathbf{r}^i, \tag{8.2}$$

where $\mathbf{o}^i$ is the observation of the target on slice i with dimension m, H is the $m \times n$ observation matrix, and $\mathbf{r}$ is the m-dimensional noise for observation. For example, the observation for one fiber is its detected center location (x and y) in a particular slice. Please refer to Section 8.2.2 for the detailed state transition and observation models.

Typically, Gaussian noise models are used for the transition noise $\mathbf{w}$ and the observation noise $\mathbf{r}$, i.e.,

$$\mathbf{w}^i \sim \mathcal{N}(\mathbf{0}, Q) \tag{8.3}$$
$$\mathbf{r}^i \sim \mathcal{N}(\mathbf{0}, R). \tag{8.4}$$

As in many other tracking applications, here we assume that the noise models are independent of each other.

The Kalman filter is a recursive Bayesian filter that carries out tracking slice by slice [477]. Based on the observation and tracking on slices $1, 2, \cdots, i-1$, a recursive Bayesian filter first makes a *prediction* of the state of the tracking target on slice i by following

$$p\left(\mathbf{s}^i \mid O^{i-1}\right) = \int p\left(\mathbf{s}^i \mid \mathbf{s}^{i-1}\right) p\left(\mathbf{s}^{i-1} \mid O^{i-1}\right) d\mathbf{s}^{i-1} \tag{8.5}$$

where $O^{i-1} = \left\{\mathbf{o}^1, \mathbf{o}^2, \ldots, \mathbf{o}^{i-1}\right\}$ are the target observations on slices $1, 2, \cdots, i-1$. The distribution in Equation (8.5) can be considered as a *prior* distribution for $\mathbf{s}^i$ before observation $\mathbf{o}^i$ is collected. After that, by considering the observation on the slice i, a *correction* step is taken to update the predicted state of the target to accomplish the tracking on slice i by

$$p\left(\mathbf{s}^i \mid O^i\right) = p\left(\mathbf{s}^i \mid \mathbf{o}^i, O^{i-1}\right) = \frac{1}{Z} p\left(\mathbf{o}^i \mid \mathbf{s}^i\right) p\left(\mathbf{s}^i \mid O^{i-1}\right) \tag{8.6}$$

where Z is the normalization constant.

By defining $p\left(\mathbf{s}^i \mid \mathbf{s}^{i-1}\right)$ and $p\left(\mathbf{o}^i \mid \mathbf{s}^i\right)$ using Equations (8.1) and (8.2), respectively, The Kalman-filter tracking algorithm can be written in a matrix form [477]. The matrix form contains three main steps, i.e., *prediction*, *association*, and *correction*, for executing the Kalman filter, which is shown as follows. "-1" denotes matrix inverse.

Kalman-filter tracking from slice $i-1$ to slice i

1. Prediction: estimate the target state at slice i as

$$\mu^i = A\mathbf{s}^{i-1} \tag{8.7}$$

and calculate the prior estimate error covariance as

$$\hat{P}^i = AP^{i-1}A^T + Q, \tag{8.8}$$

where $\mathbf{s}^{i-1}$ and P^{i-1} are the posterior state estimate and the posterior estimate error covariance at slice $i-1$.

2. Association: associate the predictions with the detections.

3. Correction: compute the posterior state estimate at slice i as

$$\mathbf{s}^i = \mu^i + K^i(\mathbf{o}^i - H\mu^i) \tag{8.9}$$

and compute the posterior estimate error covariance at slice i as

$$P^i = (I - K^iH)\hat{P}^i, \tag{8.10}$$

where I is an identity matrix and K^i is the Kalman gain at slice i:

$$K^i = \hat{P}^iH^T\left(H\hat{P}^iH^T + R\right)^{-1}. \tag{8.11}$$

8.2   Fiber Tracking Using the Kalman Filter

Continuous fiber reinforced composite (FRC) materials have been widely used in modern industry [499, 989] because their strength and stiffness are far better than those of traditional materials [989]. These superior properties of FRC materials are largely dependent on the micro-structure of the reinforced fibers. For example, the strength of an FRC material is generally much higher along the direction of reinforced fibers than the direction that is perpendicular to the fibers. Fast and accurate characterization of the underlying fiber micro-structure can substantially speed up the design and development of new composite materials. In the following, we describe the use of Kalman filtering for tracking fibers through the FRC image sequences.

Figure 8.1 shows two adjacent slices (microscopy images) from a S200 FRC material image sequence. S200 is an amorphous SiNC matrix reinforced by continuous Nicalon fibers [1208].

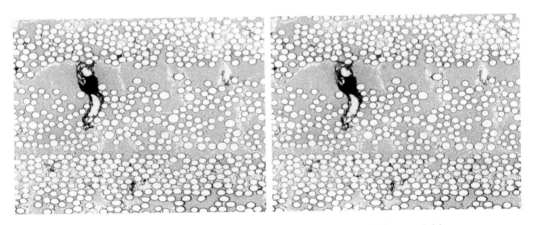

Figure 8.1: Two adjacent slices (microscopy images) from an S200 FRC material image sequence. The inter-slice distance between these two slices is $1\mu m$.

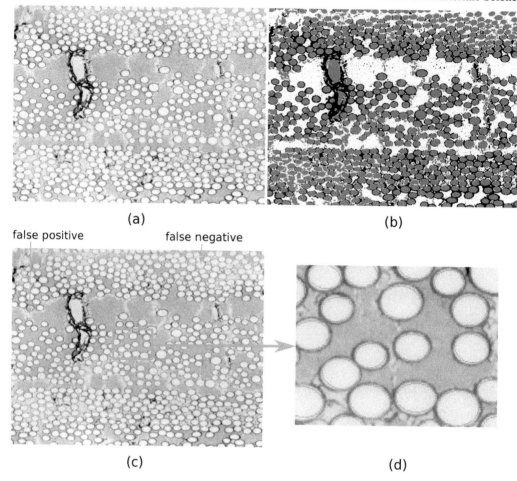

false positive   false negative

(a)   (b)

(c)   (d)

Figure 8.2: An illustration of fiber detection: (a) Original image, (b) EM/MPM segmentation, (c) ellipse fitting, and (d) enlarged view of the ellipse-fitting results.

8.2.1 Fiber Detection

When Kalman-filter tracking moves to a new slice, we need to first detect the tracking targets on this slice as the observations, which are allowed to contain noise as modeled by $\mathbf{r}$ in Equation (8.2). In the computer vision and image processing communities, a large variety of the object detection methods have been developed for detecting different kinds of targets, based on their colors, textures, shape, and other features. For the fiber tracking, from Figure 8.1 we can see that the 2D cross-sections of fibers are usually of an ellipsoidal shape. Therefore, we can detect them using an ellipse detection algorithm.

We first apply the EM/MPM algorithm [219] to segment the image slice into fiber (foreground) and non-fiber (background) regions. As shown in Figure 8.2(b), three segments are obtained using EM/MPM, where the red segment indicates the fiber region and the black/white segments indicate the non-fiber regions, including void, coating and SiC matrix. EM/MPM is a Markov random field (MRF) based pixel labeling technique that minimizes the expected number of mis-segmented pixels. The pixel labeling problem is formularized as the maximization of the posterior marginal (MPM) problem, and model parameters are estimated by the EM algorithm. EM/MPM has been found to be very effective in segmenting various materials-science images [219, 1113]. From the fiber region, we extract a set of connected components, each of which is treated as a candidate of a fiber.

We apply a Sobel operator to detect the boundary of each connected component, which is then fitted by an ellipse using a set of ellipse geometry constraints [1163]. We then take the locations (the center coordinates) of the fitted ellipses as the fiber observations on this slice and use them for fiber tracking.

Note that fiber detection is not perfect. Other than the possible inaccuracies in detecting the fiber (ellipse) centers and boundaries, both false positives and false negatives may occur in the fiber detection, as indicated in Figure 8.2(c), given the image noise and blurs. In addition, 3D fibers may be cropped by the image boundary when extending from one slice to another. As a result, a fiber tracked from previous slices may not have a corresponding observation on a new slice and a detected fiber on the new slice may not have a corresponding fiber track from previous slices.

8.2.2 Model Parameters

The first step in using the Kalman filter for tracking is to choose appropriate model parameters, including the definition of the tracking states, the transition matrix, the observation matrix, and both the transition and the observation noise model parameters. For the proposed fiber tracking, we assume that the 3D fibers to be tracked are highly smooth without quick turnings. This way, we define the state $\mathbf{s} = (x, y, v_x, v_y)^T$ to describe the corresponding fiber in 2D slices, where $\mathbf{z} = (x, y)^T$ is the fiber location (e.g., ellipse center) and (v_x, v_y) is the fiber velocity (e.g., fiber location change between neighboring slices), in horizontal and vertical directions, respectively.

For simplicity, we define the transition matrix

$$A = \begin{bmatrix} 1 & 0 & 1 & 0 \\ 0 & 1 & 0 & 1 \\ 0 & 0 & 1 & 0 \\ 0 & 0 & 0 & 1 \end{bmatrix}, \tag{8.12}$$

from which we can see that the new location of the fiber can be estimated by adding the velocity (offset) to the location of the fiber in the previous slice. The change of the velocity from one slice to the next slice is modeled by the transition noise $\mathbf{w}$.

From Section 8.2.1, the observation of a fiber is its location, i.e., the center of the ellipse. Therefore, we define the observation matrix

$$H = \begin{bmatrix} 1 & 0 & 0 & 0 \\ 0 & 1 & 0 & 0 \end{bmatrix}. \tag{8.13}$$

For the transition and observation noises, we set them to be zero-mean Gaussian as in Equation (8.4). Typically, in Kalman-filter tracking the covariance matrices of the two noise models should be very small if the underlying tracking path is very smooth and there is little observation noise. Based on this consideration, we set the covariance matrices Q and R to be diagonal matrices with constant diagonal element of 10^{-3} for fiber tracking.

8.2.3 Multiple Fiber Association

The Kalman filter described above is developed for tracking a single target (fiber) – each fiber is tracked by a Kalman filter with sequential steps of prediction and correction using its observation in the new slice. From Figures 8.1 and 8.2, we can see that we must simultaneously track hundreds or thousands of fibers. This is a large-scale multiple target tracking problem and an important step in multiple target tracking is *association* – deciding the correspondence between the fiber predications and fiber observations. Different from many other tracking problems, all the fiber observations share similar appearance and shape and it is difficult to distinguish them using their appearance and shape features. The problem addressed here is similar to the multi-vehicle tracking on a freeway, except that there are large-scale crowded similar-appearance fibers. Instead, we examine the spatial proximity to associate the fiber predictions and their observations.

When tracking moves into a new slice i, we have a set of N fiber predictions $\{\hat{\mathbf{s}}_p^i\}_{p=1}^N$ derived from the previous slices and a set of M fiber observations $\{\mathbf{o}_q^i\}_{q=1}^M$ detected on the new slice. For simplicity, we drop the superscript i and denote the predictions and observations as $\{\hat{\mathbf{s}}_p\}_{p=1}^N$ and $\{\mathbf{o}_q\}_{q=1}^M$, respectively, when it does not introduce any ambiguity. The goal of association is to build a correspondence between them, i.e., for each prediction, find its corresponding observation.

If the inter-slice distance is small, the 3D shapes of fibers are smooth, and the 2D fiber detections (i.e., fiber observations) contain low noise, we expect that the prediction of a fiber exhibits a good spatial proximity to its corresponding observation in the new slice. In this section, we introduce two algorithms, a greedy search algorithm and a global matching algorithm, for constructing the fiber association.

The greedy search algorithm can be summarized as:

Greedy search algorithm for fiber association

1. Initialize the pool of available observations as $O = \{\mathbf{o}_q\}_{q=1}^M$.

2. Initialize $p = 1$.

3. For the fiber prediction $\hat{\mathbf{s}}_p$, find from the current observation pool a fiber observation $\mathbf{o}_{q^*}$ with the smallest Euclidean distance, i.e.,

$$q^* = \arg\min_{\mathbf{o}_q \in O} \|\hat{\mathbf{z}}_p - \mathbf{o}_q\|_2, \qquad (8.14)$$

where $\hat{\mathbf{z}}_p$ is the location (first two dimensions) of the fiber prediction $\hat{\mathbf{s}}_p$.

4. If $\|\hat{\mathbf{z}}_p - \mathbf{o}_{q^*}\|_2 < T_L$, we associate $\hat{\mathbf{s}}_p$ and $\mathbf{o}_{q^*}$ and remove the observation $\mathbf{o}_{q^*}$ from the pool O, i.e.,

$$O = O \setminus \mathbf{o}_{q^*}, \qquad (8.15)$$

where T_L is a preset threshold. Otherwise, there is no observation associated to the fiber prediction $\hat{\mathbf{s}}_p$.

5. Increase $p = p + 1$ and go back to Step 3, until $p > N$.

The operation in Equation (8.15) ensures that no two observations can be associated to the same prediction because in the proposed fiber tracking, different fibers do not merge or spatially overlap with each other. From this algorithm, we can also see that it is allowed that a prediction have no associated observation because the observations may contain false negatives (missing detections in Section 8.2.1). Similarly, the observation pool O may not be empty when the algorithm completes since the observations may contain false positives in the fiber detection described in Section 8.2.1. The threshold T_L plays a critical role in identifying such false negatives and false positives.

This greedy search algorithm has one problem – the results are dependent on the processing order of the fiber predictions. For example, if an observation is the closest one to two different predictions and its association will be the prediction that is processed earlier. In multiple target tracking, it is usually desired that the association results are independent of the processing order of the targets. One way to address this problem is to formulate a global association cost, and then find a globally optimal solution that minimizes the association cost. To ensure the one-on-one matching between the predictions and observations, we can formulate it as a bipartite perfect matching problem.

More specifically, we add M dummy predictions and N dummy observations to the original fiber predictions and observations, respectively, such that the total number of predictions and the total number of observations are all extended to $M + N$. We then find the bipartite matching [517] – one-on-one matching between $M + N$ predictions and $M + N$ observations to construct the association. As illustrated in Figure 8.3, this global matching algorithm for association can summarized as follows:

Global matching algorithm for fiber association

1. Extend the N predictions to $M+N$ predictions $\{\hat{\mathbf{s}}_p\}_{p=1}^{M+N}$ by adding M dummy predictions $\{\hat{\mathbf{s}}_p\}_{p=N+1}^{M+N}$.

2. Extend the M observations to $M+N$ observations $\{\mathbf{o}_q\}_{q=1}^{M+N}$ by adding N dummy observations $\{\mathbf{o}_q\}_{q=M+1}^{M+N}$.

3. Between each prediction and each observation, define a matching cost $\phi(\cdot,\cdot)$. If $p \leq N$ and $q \leq M$, we define

$$\phi(\hat{\mathbf{s}}_p,\mathbf{o}_q) = \|\hat{\mathbf{z}}_p - \mathbf{o}_q\|_2,$$

which describes the matching cost between a real prediction and a real observation using their Euclidean distance. For the matching costs involving dummy predictions/observations, their definitions are illustrated in Figure 8.3. More specifically, if $p > N$ and $q > M$, we define

$$\phi(\hat{\mathbf{s}}_p,\mathbf{o}_q) = 0$$

and if $p > N$ or $q > M$ but not both, we define

$$\phi(\hat{\mathbf{s}}_p,\mathbf{o}_q) = \begin{cases} T_G & \text{if } q = p-N \text{ or } p = q-M \\ \infty & \text{otherwise,} \end{cases}$$

where T_G is a preset (dummy) cost. Similar to the threshold T_L in the greedy search algorithm, T_G provides a threshold of the Euclidean distance for identifying the false-positive and false-negative detections in the observations.

4. Find a bipartite (one-on-one) matching between $M+N$ predictions and $M+N$ observations with the minimum total matching cost

$$\sum_{p=1}^{M+N} \phi(\hat{\mathbf{s}}_p,\mathbf{o}_{q(p)}),$$

where $(\hat{\mathbf{s}}_p,\mathbf{o}_{q(p)})$, $p = 1,2,\cdots,M+N$ are the matching pairs of predictions and observations, which provide us the final associations. Real predictions that are matched to dummy observations indicate the cases of false negatives in fiber detection (missing detections) or fibers extending out of the image perimeter. Real observations that are matched to dummy predictions indicate the cases of false positives in fiber detection on this slice or a new fiber getting into the perimeter of the slice.

There are many bipartite matching algorithms that can find the global optimal solution in Step 4. The most widely used one is the Hungarian algorithm [517], which finds the optimal bipartite matching by iteratively performing vertex labeling and path augmenting to the underlying bipartite graph. The Hungarian algorithm has a time complexity of $O((M+N)^3)$.

Both the greedy-search algorithm and the global matching algorithm identify the false positives and false negatives in the fiber detection. In using the Kalman filter for fiber tracking, we use different strategies to handle the cases of false positives and false negatives when the tracking moves into a new slice. For false positives in the observations that are not associated to any fiber prediction, we simply discard them. For false negatives in the observations, we actually have tracked fibers without corresponding observations in the new slice. For such fibers, we can divide them into two cases: if the predicted location of such a fiber is outside the image perimeter, we stop tracking this fiber into the new slice; if the predicted location of such a fiber is still inside the image perimeter, we take its predicted location and velocity as the tracking state on the new slice, without the step of correction, and continue tracking into the next slice.

8.3 Tracking Performance Evaluation

To achieve objective and quantitative performance evaluation, we manually annotate fibers on the tested image sequences as the ground truth, which include the annotation of the fiber centers on each

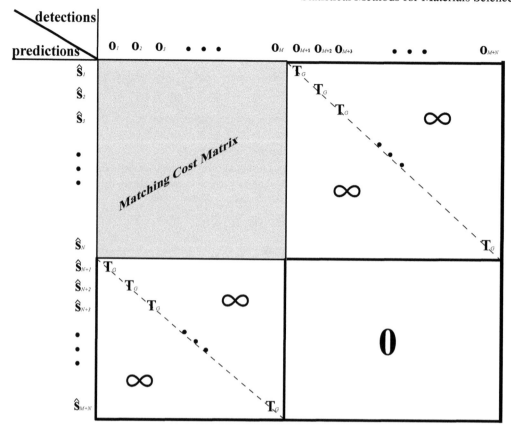

Figure 8.3: An illustration of the global-matching based association, with matching cost defined between the real/dummy fiber predictions and observations.

2D slice and the linking of annotated 2D fiber centers between slices. Given the tracked fibers and the ground-truth fibers, we can then examine the coincidence of them on each slice – the better the coincidence, the better the tracking performance. In the following, we focus on introducing several metrics that have been used for other multi-target tracking tasks and can be used for evaluating the proposed large-scale fiber tracking.

On each slice i, we have the 2D locations of the tracked fibers and the 2D locations of the ground-truth fibers. We match them by using the same global bipartite matching algorithm that is developed in Section 8.2.3 for association; here we also use the Hungarian algorithm. Note that the number of tracked fibers and the number of ground-truth fibers may be different. In this case, we introduce dummy tracked fibers and dummy ground-truth fibers as described in Section 8.2.3. Based on the bipartite matching results, we can compute fp_i, the number of false positives, and fn_i, the number of false negatives, on each slice i.

In addition, based on the matching results on two neighboring slices $i-1$ and i, we can find the fiber tracks that switch their identities. More specifically, if a same tracked fiber is matched to two different ground truth fibers on slice $i-1$ and slice i (note that the global matching is performed on each slice independently), we count it as one identity switch on slice i. We denote the number of identity switches on slice i as mm_i.

Multiple Object Tracking Accuracy (MOTA). By combining the number of false positives (fp), false negatives (fn) and the identity switches/mismatches (mm) on each slice, we can define the multiple object tracking accuracy (MOTA)

$$MOTA = 1 - \frac{\sum_i (fp_i + fn_i + mm_i)}{\sum_i g_i}, \tag{8.16}$$

where g_i is the number of ground-truth fibers on slice i. We can see that MOTA is a comprehensive metric and the higher the MOTA, the better the tracking performance.

Multiple Object Tracking Precision (MOTP). For the tracked fibers that are successfully matched to corresponding ground truth on each slice, we can also measure their matching precision using the Euclidean distance between their locations. Let d_i be the sum of the Euclidean distances between the matched pairs of tracked fibers and ground-truth fibers on slice i, excluding the false positives and false negatives. We can define the multiple object tracking precision (MOTP) as

$$MOTP = \frac{\sum_i d_i}{\sum_i c_i}, \tag{8.17}$$

where c_i is the number of matched pairs of the tracked fibers and ground-truth fibers on slice i. The lower the MOTP, the better the tracking performance. Note that, MOTP only reflects the precision of the tracked fibers that are well aligned with the ground-truth fibers and does not reflect the number of false positives and false negatives.

Identity Switch (IDSW). By summing up the number of identity switches over all the slices, we can define a metric

$$IDSW = \sum_i mm_i. \tag{8.18}$$

This metric reflects the occurrences of the tracked fibers that switch from one ground-truth fiber to another between neighboring slices. The lower the IDSW, the better the tracking performance.

Mostly Tracked (MT) and Mostly Lost (ML). Based on the matching results on each slice i, we examine each ground-truth fiber whether it is tracked or not. If it is matched to a tracked fiber, we consider this ground-truth fiber is tracked on slice i. Otherwise, we consider this ground-truth fiber is lost on slice i. For each 3D ground-truth fiber, if it is tracked on more than 80% of slices, we consider this 3D fiber is mostly tracked. The mostly tracked (MT) metric is defined as the total number of fibers that are mostly tracked over all the slices. Similarly, if a 3D ground-truth fiber is lost on more than 80% of slices, we consider this 3D fiber is mostly lost. The mostly lost (ML) metric is defined as the total number of fibers that are mostly lost over all the slices.

In practice, the construction of the ground truth is not a trivial problem. It is usually difficult to manually annotate all the fibers in the image sequence since i) the number of fibers is of large scale, e.g., thousands to hundreds of thousands in an image sequence, ii) some 3D fibers may not be present in all the slices – they may get out of image perimeter when moving from one slice to another, iii) fibers are crowded and the inter-slice linking may be difficult for some of them, especially when the inter-slice distance is large and the location of a fiber from one slice to another may show poor continuity. In this work, we construct the ground truth by annotating as many full 3D fibers as possible from an image sequence. Here full 3D fibers mean that such fibers are present in all the slices in the image sequence.

This introduces a problem in using the above metrics for quantitative performance evaluation. The counted fp_i false positives may not be all false positives. Some of them may be correctly tracked fibers but they are not annotated in the ground truth. This may undervalue the real performance of the fiber tracking. To address this problem, we first prune the tracked fibers that are far away from the ground-truth fibers and then use the remaining tracked fibers for computing the above five metrics against the ground-truth fibers. Specifically, a tracked fiber $\{\mathbf{z}^i\}_{i=1}^I$, where $\mathbf{z}^i$ is the tracked-fiber center at slice i, is pruned if

$$\frac{\sum_{i=1}^I d(\|\mathbf{z}^i - \mathbf{z}_g^i\|_2 < T_P)}{I} < t_o \tag{8.19}$$

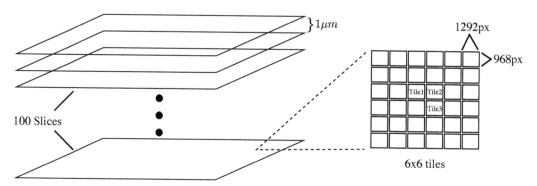

Figure 8.4: An illustration of the test data.

where $\mathbf{z}_g^i$ is the center of the ground-truth fiber that matches the tracked fiber $\mathbf{z}^i$ on slice i, and $d(\cdot)$ is an indicator function which equals 1 if the condition is satisfied and 0 otherwise. We set $T_p = 20$ pixels and $t_o = 50\%$ and find that this is not a very strict condition as it does not overly prune the tracked fibers and reduce the recall much. Specifically, we find out that, after applying this step of pruning, the remaining number of tracked fibers is usually similar to the number of the annotated ground-truth fibers.

8.4   Testing Data and Sparse Sampling

We tested the above Kalman-filter tracking algorithm on a set of material image sequences. Specifically, these images are collected in Air Force Research Laboratories (AFRL) using the RoboMet.3D automated serial sectioning instrument [1066]. The tested material is S200, which is an amorphous SiNC matrix reinforced by continuous Nicalon fibers. One hundred serial sections are produced with the dense inter-slice distance of $1\mu m$. Because each serial section of the tested S200 sample is too large to be covered by a single shot of the microscopic image, it is divided into $6 \times 6 = 36$ tiles. In the following experiments, we test the performance on three center tiles, which we named Tile1, Tile2, and Tile3, respectively, and their locations in the original 36 tiles are shown in Figure 8.4. For each tile, we have an image sequence consisting of 100 slices and the resolution of each slice is 1292×968. Two sample slices of a tile are shown in Figure 8.1, which contains hundreds of crowded fibers.

While dense sampling with small inter-slice distance ensures high spatial continuity between neighboring slices and can facilitate the tracking of the fibers through the image sequence, it substantially increases the cross-sectioning and imaging time. Therefore, sparse sampling in cross-sectioning, i.e., using large inter-slice distance, is highly desirable in practice. One important problem is whether the proposed Kalman-filter tracking algorithm can work on a sparsely sampled image sequence. To address this problem, we study the fiber-tracking performance through image sequences with different sampling sparsities.

To test the tracking performance under sparsely sampled image sequences, we downsample the original image sequence. In particular, we skip $C \geq 0$ slices before taking the next slice in the original sequence, until the end of the original sequence is reached, to construct such sparsely sampled image sequences. For convenience, we name parameter C the *sparsity*: The larger the parameter C, the sparser the constructed image sequence. One issue is that the constructed image sequences with large C are much shorter than the original image sequence, e.g., if the original sequence has 100 slices, a sparse sequence with $C = 5$ only contains around 20 slices, as shown in Figure 8.5(a). The tracking performance obtained on such a short sequence may not be statistically reliable. To alleviate this issue, for a given sparsity C we construct $C+1$ image sequences, starting from original slice $1, 2, \cdots, C+1$, respectively. These $C+1$ image sequences do not share any slice. We perform track-

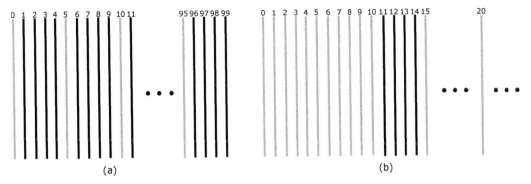

Figure 8.5: An illustration of downsampling for sparse image sequences. (a) Downsampling the image sequence with sparsity $C = 5$. (b) Sparsely sampled image sequence started with several densely sampled slices, with which we can make a better initial fiber velocity estimation. Lines indicate slices and the sampled slices are highlighted in red.

ing on each of them independently and then average their performances, e.g., MOTA and MOTP, as the performance of tracking under the sparsity C. Note that when $C = 0$, tracking is directly performed and evaluated on a single image sequence: the original densely sampled image sequence without any downsampling. In our experiments, we continuously vary the sparsity C from 0 to 19 and examine the tracking performance under different sparsities.

Based on the Kalman filter algorithm described in Section 8.1, we also need to set the initial tracking state on the first slice to start the recursive predictions and corrections. This state consists of the initial location and velocity of each fiber on the first slice. The initial locations can directly use the centers of the detected ellipses. The major issue is the setting of the initial velocity.

When the sparsity C is low, we expect the velocity is also low and we can set a small or even zero initial velocity for each fiber. The transition model and its noise as shown in Equation (8.1) can quickly adapt the velocity to the correct value in the slice-by-slice tracking. However, when the sparsity C is high, an inaccurate initial velocity may lead to incorrect fiber predictions that are far from their underlying observations and result in incorrect fiber association. Meanwhile, with high sparsity C, the image sequence is much shorter. In this case, the velocity of each fiber may not get adapted to the correct ones before reaching the end of the sequence.

To address this issue, we propose to start with several densely-sampled image slices, e.g., the first 5 to 10 slices in the original image sequence, to estimate the initial velocity of each fiber. In the cross-sectioning study, this just means that we start with several slices with small inter-slice distance to estimate the initial velocities of the fibers and then sample using the desired larger sparsity for the remaining image sequence. Note that the first several densely sampled slices only add a little bit of overhead to the number of cross-sectioning slices and the imaging time. From these few densely sampled slices, we can carry the proposed Kalman filter tracking to estimate the velocity of each fiber. We can then multiply this velocity by C to take it as the initial velocity of this fiber for tracking through the remaining sparsity-C image sequence, as shown in Figure 8.5(b).

8.5 Experiment Results

Figures 8.6 and 8.7 show the MOTA and MOTP performance of the proposed Kalman-filter tracking method using the greedy-search and global-matching algorithms for fiber association, through three image sequences (tiles) under different sparsity C. In these results, "Kalman-greedy" and "Kalman-greedy-PV" indicate the use of the greedy-search algorithm for association, without and with the use of several densely sampled slices for initial predicted velocity, respectively. "Kalman-global" and "Kalman-global-PV" indicate the use of the global-matching algorithm for association, without and with the use of several densely sampled slices for initial predicted velocity, respectively. In all the experiments, we set $T_L = 100$ pixels for the greedy-search association algorithm and $T_G = 40$ for the global-matching association algorithm.

We can see that, when the sparsity C is low, the proposed Kalman-filter tracking methods produce satisfactory fiber tracking with very high MOTA and very low MOTP. However, the performance of tracking decreases with the increase of the sparsity C, since the spatial continuity between neighboring slices gets poorer. On the tested S200 image data, we can still get very high MOTA performance with $C = 8$. This indicates that for this kind of FRC materials, we can actually perform a sampling with an inter-slice distance as large as $9\mu m$ without affecting the fiber micro-structure tracking and reconstruction very much. This is much sparser than the original $1\mu m$ inter-slice sampling and can substantially save the cross-sectioning and imaging time.

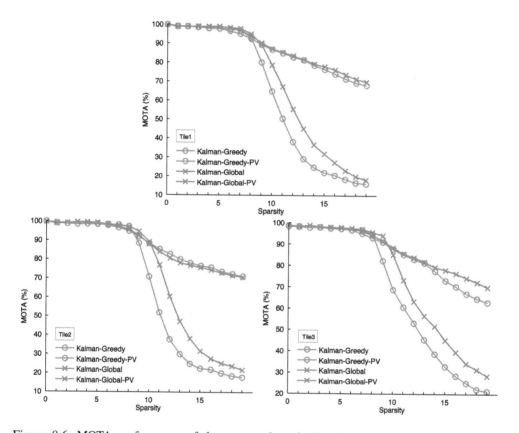

Figure 8.6: MOTA performance of the proposed methods using the greedy-search and global-matching algorithms for fiber association, through three image sequences (tiles) under different sparsity C.

Tables 8.1, 8.2, and 8.3 show the IDSW, MT, and ML metrics of the proposed tracking methods under different sparsities. The performance shown in this table is the average of all three test image sequences. We can see that all the methods show approximately increased IDSW, decreased MT, and increased ML with the increase of the sparsity C. Figure 8.8 shows the tracking results on several slices in the Tile2 image sequence, where the fibers with the same color/number are on the same fiber track.

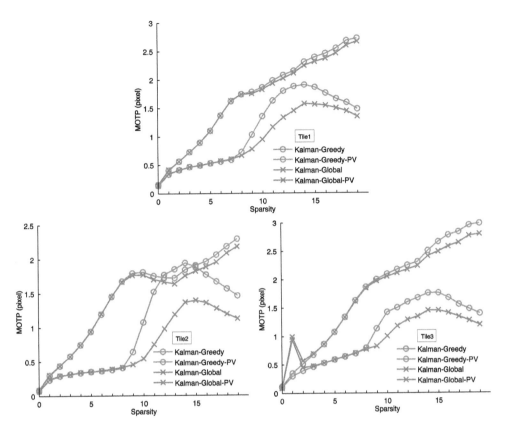

Figure 8.7: MOTP performance of the proposed methods using the greedy-search and global-matching algorithms for fiber association, through three image sequences (tiles) under different sparsity C.

Table 8.1: IDSW performance of the proposed methods under different sparsity C. The performance is the average over all three image sequences.

IDSW	$C = 0$	$C = 5$	$C = 10$	$C = 15$	$C = 19$
Kalman-Greedy	9.0	6.3	596.6	1162.4	1100.5
Kalman-Global	6.3	3.3	209.9	937.9	999.1
Kalman-Greedy-PV	9.0	6.1	57.2	126.4	186.6
Kalman-Global-PV	6.3	3.3	54.2	89.8	133.0

Table 8.2: MT performance of the proposed methods under different sparsity C. The performance is the average over all three image sequences.

MT	$C = 0$	$C = 5$	$C = 10$	$C = 15$	$C = 19$
Kalman-Greedy	376.3	371.1	309.2	268.2	307.5
Kalman-Global	377.0	374.9	337.6	271.9	297.5
Kalman-Greedy-PV	376.7	371.6	357.2	337.7	339.2
Kalman-Global-PV	377.0	375.3	362.5	344.3	339.8

Table 8.3: ML performance of the proposed methods under different sparsity C. The performance is the average over all three image sequences.

ML	$C = 0$	$C = 5$	$C = 10$	$C = 15$	$C = 19$
Kalman-Greedy	0.3	1.6	3.9	8.8	1.5
Kalman-Global	0.3	1.8	5.4	11.2	1.7
Kalman-Greedy-PV	0.3	1.8	4.4	9.2	3.7
Kalman-Global-PV	0.3	1.6	6.4	12.8	7.3

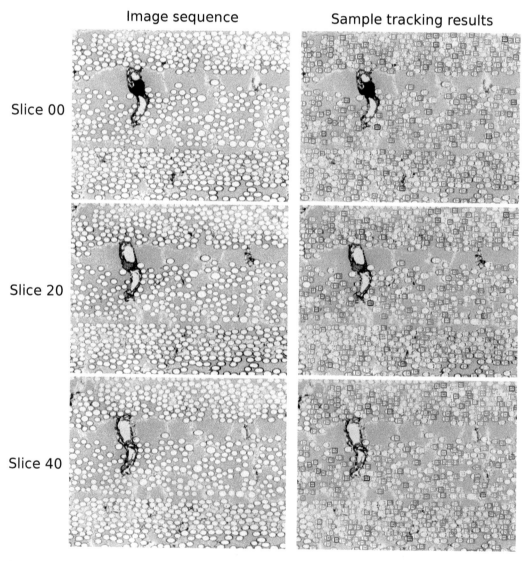

Figure 8.8: The tracking results on several slices in the Tile2 image sequence, where the fibers with the same color/number are on the same fiber track.

8.6   Other Tracking Methods

In this section, some other tracking algorithms that are more complex than Kalman filters are briefly introduced for completeness. Existing multi-target tracking algorithms are mainly developed for tracking multiple persons or vehicles from videos, or multiple cells from biomedical image sequences. They can be generally categorized as recursive and non-recursive methods.

Recursive tracking methods estimate the state of the target in a new slice (frame for video tracking) only using the information from previous slices that have been processed. Typical recursive tracking methods include Kalman filters [477, 105, 839], particle filters [129, 750], and non-parametric mixture particle filters [1101]. When tracking moves to a new slice, these recursive methods first make a prediction of the state for each target, such as target location and velocity, using the information from previous slices. They then detect targets in this new slice as observations, followed by an association step that matches the predictions and observations of multiple targets. Finally, the posterior distribution of each tracker is corrected using the associated observation. These three steps of prediction, association, and correction are recursively performed on each slice along the image sequence. The recursive methods are particularly applicable to online tracking tasks and are usually desirable in materials science applications. We can adaptively tune the inter-slice distance for cross-sectioning the next slice based on the tracking results from previous slices. This may prevent the over-dense sampling in the cross-sectioning study and substantially shorten the required imaging time for reconstructing and characterizing the underlying micro-structures.

Non-recursive tracking methods assume the availability of the whole image sequence before tracking multiple targets over this sequence. In these methods, observations of multiple targets are first detected on all the slices and then linked across slices along the image sequence for the final tracks. Graph model and algorithms are usually employed for non-recursive tracking. That is, each graph node represents an observation, each graph edge indicates a possible linking of the observations across two slices, and the tracking along the sequence can be reduced to find optimal paths in the constructed graph, given appropriately defined cost functions, such as maximum a posteriori (MAP) [461, 1196, 83, 791]. In [260], motion dynamics similarity is incorporated into the cost function, resulting in a non-recursive SMOT tracking method for multi-target video tracking. In [616], a KTH tracking method is developed by searching the shortest path in the constructed graph model and this method was successfully used to track living cells along microscopy biomedical image sequences. In [685], a CEM tracking method was proposed by optimizing a continuous cost function that considers the detection, appearance, motion priors, and physical constraints of the targets.

As mentioned in Section 8.2.3, one important component in the multi-target tracking methods is the association. For recursive methods, association is usually formulated as an explicit step which finds the correspondence between the predictions and the observations, as we did in Section 8.2.3. For non-recursive methods, the association is implied in the cost function and the optimization algorithm – the extracted paths find the associations between the observations across neighboring slices. In many multi-target tracking tasks, such as human/vehicle tracking, the targets to be tracked are of small number, and they are spatially scattered or of different appearance [461, 1196, 83, 791, 260, 685]. However, in materials science, the goal is to track large-scale (hundreds or thousands), crowded fibers with very similar appearance and shape in each slice. Therefore, the association in fiber tracking is a much more challenging problem than that in the other tracking applications.

8.7   Summary

In this chapter, we introduced Kalman-filter based methods for tracking micro-structures through microscopic image sequences of cross-sectioned material samples. In particular, we introduced the algorithms to track large-scale fibers from image sequences of continuous fiber reinforced composite materials. The noisy observations of the fibers are first detected on each slice by combining image segmentation and ellipse fitting operations. Each fiber is tracked by a Kalman filter consisting of recursive fiber predictions and corrections, following linear transition and observation models.

We introduced two algorithms for associating the predictions and observations of the large-scale, crowded fibers when the tracking moves into a new slice – a greedy search algorithm and a global matching algorithm. Both association algorithms can handle false positives and false negatives in the fiber detection. Quantitative performance was reported on three sequences of an S200 material sample, under different sampling sparsity. We showed that the proposed Kalman-filter based methods can accurately track the large-scale fibers in this kind of material when the sampled inter-slice distance is up to $9\mu m$.

Part IV

Structure Formation in Materials

In microstructure characterization, it is generally desired to understand what is in a particular image, not just a homogeneous average of a number of observations. A consequence of this is that the researcher is perennially in a "data poor" situation, having more unknowns than experimental values to unambiguously determine their values. In Chapter 5, the point is made that, in the data poor regime, it is necessary to apply some sort of regularization in order to stabilize the equations to be solved. This regularization biases the results towards what is reasonable. Since regularization is a bias, it should be based on well-established facts, independent of the actual observation being analyzed. Thus, it is valuable to understand some key aspects of material structure. We included this chapter in order to provide some foundation, for the algorithm developer, for knowing what would be reasonable and what would not.

A perfect understanding of material structures would, of course, require more than a lifetime of study. But, it can be said that material structures are generally viewed in terms of features which can be recognized at specific length scales. To this end, we decided to illuminate three important length scales in materials science: (1) the size of crystalline regions, or *grains*, which typically occur over a scale of a hundred microns or so, (2) interfaces, which can be viewed as "thin manifolds," having tens or hundreds of microns in two dimensions, but are on the order of atomic distances in the third, and (3) nanostructures, which, as the name implies, apply over a scale of nanometers.

To illustrate the grain scale, we asked Beladi and Rohrer to give a background chapter on grains and grain structures. It may be said that grain structures are the "poster child for sparsity," because, within each grain is a uniform orientation, with the boundaries between being effectively modeled as discontinuities between regions of differing orientations.

To illustrate interface behavior, we asked Tang and Luo to prepare a chapter on the various phenomena that occur at interfaces between grains or different phases. One of the most common forms of regularization in image science involves applying an "energy penalty" on interface length. In real material imagery, this is not just a mathematical construct: there is an energy cost associated with the existence of a boundary. Tang and Luo's chapter illustrates the interplay between this energy and the various other energies of the system to produce sometimes dramatic structural results.

Finally, the emerging field of nanoscience involves many complexities, not just of structure, but of the instrumentation used for measurements, as well. To this end, we asked Nepal and colleagues to prepare a chapter illuminating the realm of nanoscience, in which we participated. Material behavior at the nanoscale being much less mature than those at the other two scales, this chapter needed to cover a wide range of both structural and physical phenomena.

Chapter 9

Grain Boundary Characteristics in Polycrystalline Materials

Hossein Beladi
Deakin University

Gregory S. Rohrer
Carnegie Mellon University

9.1 Introduction

There is an ongoing drive among research groups around the world to develop novel materials to meet the growing need for higher performance at lower cost. A major strategy to achieve this goal is to engineer the microstructure constituents of polycrystalline materials according to their contributions to the particular property of interest. Polycrystalline materials can be thought of as composite structures consisting of grain interiors and grain boundary regions, where two grains are joined. The grain size is not a sufficient factor to evaluate the polycrystalline mechanical property, as grain refinement may or may not lead to increased strength [1131]. In addition, the grain size only provides limited information about the grain boundary (i.e., density and spacing) rather than the types of grain boundaries and their characteristics.

   Grain boundaries are active structural elements as certain properties of polycrystalline materials are controlled to a large extent by their characteristics. The grain boundary type influences the atomic structure of the interface (i.e., defect density). The nature and crystallography of the boundary network is therefore a critical parameter governing the performance (e.g., diffusion, strength). The characterization of grain boundary network requires three-dimensional knowledge of microstructure. New advances in both experimental instruments and computational approaches have made it possible to fully characterize the three-dimensional structure at a variety of length scales. This enables the full analysis of the characteristics of the grain boundary networks developed in different polycrystalline materials subjected to distinct processing routes. This chapter will explore different aspects of grain boundaries in single-phase polycrystalline materials and review recent results relevant to the structure of interfacial networks. Firstly, grain boundary crystallography is described. In the next section, experimental methods used to resolve the three-dimensional interfacial networks structure will be discussed. Novel approaches to measure both the grain boundary plane character and energy distributions will also be described. This is followed by a summary of recent findings related to the parameters affecting the grain boundary character distribution (i.e., type and population). Recently developed methods to analyze the connectivity of grain boundary networks are reviewed. The final part summarizes the most important points and highlights the current challenges relevant to the grain boundary research.

9.2   Grain Boundary Representation

Five macroscopic parameters are required to fully describe the crystallography of a grain boundary; three parameters to define the orientation relationship across the grain boundary (i.e., the crystallographic lattice misorientation) and two parameters specify the boundary plane orientation [856]. The following describes each of these characteristics.

9.2.1   The Crystallographic Lattice Misorientation

Each single grain orientation in a polycrystalline material is often described by three Eulerian angles, $(\varphi_1, \Phi, \varphi_2)$, which specify the transformation that brings the reference frame and the crystal frame into coincidence. The lattice misorientation between two neighboring grains with distinct orientations (i.e., g_A and g_B) is, therefore, defined by two passive rotations to bring two adjacent crystals into coincidence: i) back rotation from orientation A to the reference frame, g_A^{-1}, and ii) the rotation of the reference frame to coincide with crystal B, g_B. Therefore, the net rotation, Δg, can be described by a new set of Eulerian angles $(\varphi_1', \Phi', \varphi_2')$, ranging from zero to 2π, π and 2π, respectively. The lattice misorientation, Δg, can be also expressed by a rotation angle, ω, and a rotation axis $\langle hkl \rangle$, which is common for both lattice coordinate systems. However, the rotation of one crystal to coincide with the other is an arbitrary choice (i.e., $\Delta g = g_B \cdot g_A^{-1} = g_A \cdot g_B^{-1}$). Therefore, there would be $2 \times N^2$ equivalent measurements with N representing the number of symmetry operators. In case of materials with cubic structure, there are 24 symmetry operators, yielding $1,152$ equivalent misorientations. Hence, a selection criterion is required to choose a single misorientation, which represents all equivalent misorientations. This criterion is defined by the fundamental zone, which refers to a region containing only one representative for any misorientation [990, 705].

The fundamental zone can be computed for the misorientation in the Euler angles space, but it has a complex shape and angle-axis pairs sharing the same axis appear as a curved line rather than a straight line in this space. In addition, there are no simple rules to combine two sets of Euler angles and calculate the misorientation [424]. These issues are resolved by using the misorientation in the Rodrigues–Frank space (R-F), where grain boundaries with the same misorientation axis occur on a straight line, with the rotation angle given by the distance from the origin [329].

In the R-F vector space, the lattice misorientation is a vector, $\mathbf{R} = n_i \tan(\omega/2)$, where n_i $(i = 1 \ldots 3)$ are the unit vector components representing the misorientation axis, ω is the misorientation angle and $\mathbf{R}$ has three components, i.e., (R_1, R_2, R_3). Considering all possible $\mathbf{R}$ vectors, the misorientations in R-F space for a cubic crystal structure lie within a truncated cube. Symmetry reduces the volume of this space by a factor of $1/48$, when only those R-F vectors equivalent to rotation axes within the standard stereographic triangle are considered ($n_1 \geq n_2 \geq n_3 \geq 0$). Consequently, the fundamental zone for grain boundaries in cubic materials appears as a truncated pyramid [424]. The three lowest index axes, i.e., [100], [110], and [111], lie along the edges of the fundamental zone (Figure 9.1). The R-F space is often sectioned parallel to the [100] − [110] plane for graphical representation. The representation of the lattice misorientation within the fundamental zone has the minimum misorientation angle, and this is often called the disorientation angle. The disorientation angle varies from $0°$ to $62.8°$ for materials with cubic crystal structures.

9.2.2   Grain Boundary Plane Orientation

A grain boundary plane is described by two orientation parameters (i.e., spherical angles of θ and Φ, Figure 9.2a), which defines the plane normal, $\mathbf{n}$. However, the boundary plane has two indistinguishable normal options, resulting in inversion symmetry in the space of grain boundary planes for a specific misorientation. The plane normal must be expressed in the reference frame of one of the two grains. While the choice is arbitrary, it can be done consistently by selecting the reference from the first (non-transposed) crystal in the calculation of misorientation. Here, the fundamental zone is defined so that the third component of the unit vector, which is parallel to the boundary plane nor-

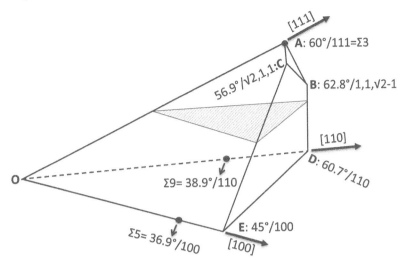

Figure 9.1: The schematic representation of the fundamental zone in Rodrigues–Frank space in cubic structure, including a planar section perpendicular to the [001] axis (i.e., R_3). The misorientation angle/axis pairs are shown at the corners of the fundamental zone. The position of Σ5 and Σ9 are shown at the base, i.e., $R_3 = 0$.

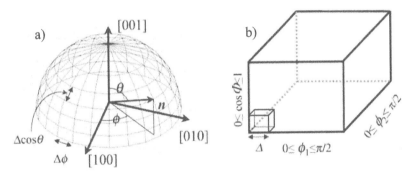

Figure 9.2: Schematic representation of $\lambda(\Delta g, \mathbf{n})$ into a) two boundary parameters of plane orientation and b) three parameters of lattice misorientation, i.e., Euler angles (reproduced with permission from [856]).

mal, becomes positive [898]. Hence, the general fundamental zone of the space of boundary planes is a hemisphere with θ ranging from 0 and $\pi/2$ and Φ ranging 0 to 2π. At certain special points, symmetry reduces the size of the fundamental zone, but for comparisons, it is most convenient to consider the entire hemisphere.

9.3   Representation of the Grain Boundary Character Distribution

Various representations are used to describe the grain boundary character distribution depending on the number of macroscopic parameters employed. The most common representation is the misorientation angle distribution, which only represents one parameter of the complete grain boundary character distribution. Here, the grain boundary population is plotted as a function of disorientation angle (i.e., the minimum misorientation angle of each boundary, Figure 9.3a) and ignores other macroscopic parameters (i.e., misorientation axis and grain boundary plane orientation). Another

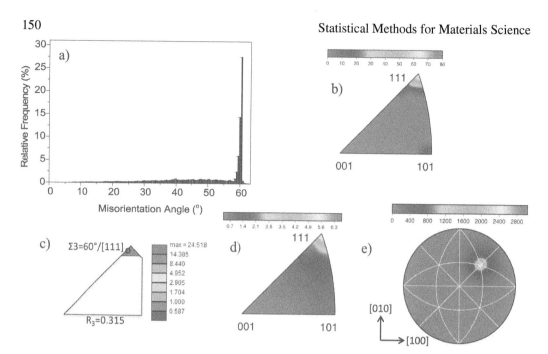

Figure 9.3: a) One-parameter misorientation angle distribution of grain boundaries in a Ni-30Fe austenitic alloy, showing a single peak at 60° misorientation angle, along with b) a two-dimensional section of the three-parameter misorientation distribution in angle-axis pair format at 60°, presenting a pronounced peak at the position of [111] misorientation axis; c) a planar section of three-parameter Rodrigues–Frank space at $R_3 = 0.315$ and R_1 and R_2 span throughout the fundamental zone, revealing a maxima at 60°/[111] misorientation; d) two-parameter representation of grain boundary planes distribution ignoring misorientation angle, showing that the boundaries largely terminated at (111) orientation, and e) five–parameter representation of grain boundary planes character distribution at 60°/[111] misorientation, displaying a single peak at (111) orientation suggesting that $\Sigma 3$ boundary has pure twist character. d and e are plotted in stereographic projection along [001]. The scale bars represent multiples of random distribution.

grain boundary representation is the misorientation distribution function (MDF), which requires three-parameters consisting of the misorientation axis along with disorientation angle. This distribution is represented using sections of angle-axis space (Figure 9.3b) or Rodrigues–Frank space (Figure 9.3c). Figure 9.3b shows the MDF for all boundaries with a misorientation angle of 60° in the fundamental zone for an austenitic Ni-30Fe (in wt%) alloy. In the R-F space, the $R_1 - R_2$ plane for $R_3 = 0.315$ reveals a maximum at the [111] position (Figure 9.3c), similar to the misorientation axis representation (Figure 9.3b).

The grain boundary plane distribution, ignoring misorientation, is a two-parameter representation of the grain boundary character distribution. The distribution is plotted on a stereogram viewed along the [001] crystal direction, and has the symmetry of the crystal. For instance, the grain boundary plane distribution for all boundaries for an austenitic Ni-30Fe alloy shows a strong preference for (111) planes (Figure 9.3d).

The full representation of the grain boundary character distribution requires all five crystallographic parameters. Here, the distribution is plotted for a given disorientation angle/axis, similarly presented as a stereogram. Therefore, the grain boundary orientations can be defined for the boundaries with a specific orientation relationship. For grain boundaries with a 60°/[111] misorientation, the distribution reveals a strong maximum at the position of (111) planes (Figure 9.3e).

9.4   The Measurement of the Grain Boundary Plane Distribution in Polycrystals

Grain boundaries in polycrystals appear as lines on planar sections using the conventional electron backscattered diffraction (EBSD) mapping technique. This approach only resolves four of the
five crystallographic parameters of grain boundary (i.e., misorientation angle/axis and also one of
the two parameters for the inclination of the grain boundary plane). The full recovery of all five-
parameters can be achieved using advanced characterization techniques including transmission electron microscopy, X-ray diffraction microscopy, and serial sectioning.

Novel X-ray diffraction techniques are recently developed to non-destructively measure all five
microscopic grain boundary parameters, $\lambda(\Delta g, \mathbf{n})$, in polycrystalline materials: i) high energy X-
ray diffraction microscopy [579], ii) X-ray diffraction contrast tomography [669, 995], and iii) differential aperture X-ray microscope using polychromic Laue transmission approach [536]. These
techniques have a significant advantage to potentially monitor a given process as a function of time
(e.g., kinetics). However, the spatial resolution is limited to a few microns, which makes them only
suitable for coarse microstructures at this stage.

Transmission electron microscopy (TEM) can also be used to measure $\lambda(\Delta g, \mathbf{n})$ as the grain
orientations are resolved on either side of boundaries by the selected area diffraction patterns and
the boundary inclination is defined by the image projection through the sample [584]. However, this
technique is restricted to the thin area close to the perforation/hole, which can only accommodate
a small number of grain boundaries. Therefore, significant time and effort are required to obtain
statistically sound measurements using the TEM approach.

In the serial sectioning approach, a thin layer of material is removed at each stage and subsequently the EBSD scan is performed on the region of interest on the surface. The EBSD map
provides sufficient crystallographic information for each layer and the grain boundary inclination
(i.e., normal direction, $\mathbf{n}$) can be measured through reconstruction of the three-dimensional data.
The grain boundary network is reconstructed from skeletons of the grain boundary traces on each
layer; two successive layers are superimposed so that the triple junctions are connected one at a
time through the nearest vertex position (Figure 9.4b-d). Next, the grain boundaries are assembled using a triangular meshing algorithm (Figure 9.4a). This process is repeated for all subsequent
layers to construct the grain boundary network three-dimensionally [902]. Figure 9.5 shows a reconstructed image of 100 slices of EBSD for an austenitic high-Mn TWIP steel. The grain boundary
normal/orientation can be measured by computing the cross product of the triple lines connecting
adjacent layers and the corresponding grain boundary line trace.

The discrete nature of the data in the serial sectioning process results in two uncertainty sources:
i) the relative in-plane (horizontal) and between plane (vertical) resolution and ii) the horizontal
alignment of the layers [854]. The former can be decreased through the connection of the triple
lines between alternate layers, which makes the vertical discretization coarser than the horizontal
discretization. The latter is diminished through the employment of a sub-pixel alignment procedure, which rigidly moves layers to obtain the mean of the triple line direction distribution normal
to the sample surface. This modification usually provides less than one pixel spacing [854]. This
approach is extensively used to measure the relative grain boundary plane character and energy
distributions using all five crystallographic parameters in a wide range of polycrystalline materials.
Recently, this approach was coupled with newly developed Dream 3D software, which reconstructs
the microstructure three-dimensionally and creates 3D mesh of boundaries using EBSD data. The
Dream 3D software is a digital representation environment for the analysis of microstructure three-
dimensionally [393].

9.4.1   The Relative Grain Boundary Character Distribution

The grain boundary character distribution (GBCD) function, $\lambda(\Delta g, \mathbf{n})$, is described as the relative areas of internal interfaces having a misorientation of $\Delta g = g_A \cdot g_B^{-1}$ and an interface normal of $\mathbf{n}$ [856].
The distribution $\lambda(\Delta g, \mathbf{n})$ is parameterized into discrete cells using three Euler angles $(\varphi_1, \Phi, \varphi_2)$

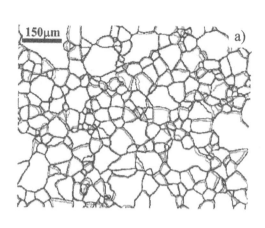
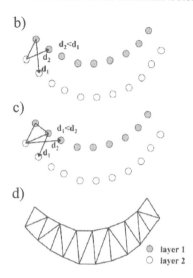

Figure 9.4: a) Positioning of the two adjacent layers through the alignment process. b)-d) Schematic representation of the grain boundary meshing algorithm. b) The projection of the second layer pixels onto the first layer plane, followed by the connection of triple junctions and calculating the distance to the next vertex position for each layer (i.e., d_i). c) The vertex related to the shortest distance is employed to form the first element and repeats the process. d) This procedure is continued to connect all pixels on both layers. (Figure reproduced with permission from [902].)

representing the lattice misorientation and two spherical angles of θ and Φ representing all possible interface normals (Figure 9.2). For the misorientation parameterization, φ_1, $\cos\Phi$, and φ_2 are employed to subdivide the Euler angle space into equal volumes (Figure 9.2b). The quantities $\cos\theta$ and Φ are used to discretize the domain of interface normals into equal areas (Figure 9.2a). This subdivision creates volumes that have an equal probability of being populated by a random grain boundary. The GBCD is quantified in units of multiples of a random distribution (MRD). For an MRD value of 1, all cells in $\lambda(\Delta g, \mathbf{n})$ would have the same values, representing a random GBCD.

9.4.2   The Relative Grain Boundary Planes Energy Distribution

Here, emphasis is given to the capillarity vector reconstruction method established by Morawiec using three-dimensional EBSD datasets, where the grain boundary energy distribution is reconstructed over the complete space of macroscopic boundary parameters [704]. Similar to other measurements, the method uses the interfacial geometry observations and assumes that the triple junction is locally in thermodynamic equilibrium. Under these circumstances, the energy of grain boundary is associated to the interfacial geometry through the Herring equation, $(\boldsymbol{\zeta}^1 + \boldsymbol{\zeta}^2 + \boldsymbol{\zeta}^3) \times \mathbf{1} = 0$. Here, $\boldsymbol{\zeta}^i$ ($i = 1\ldots3$) represent the capillarity vectors related to the three boundaries at a triple junction and $\mathbf{1}$ represents the triple line [436, 154]. Each vector has two components. One is normal to the boundary, having a length equal to the relative grain boundary energy; the other is tangential to the boundary and its magnitude denotes the differential of the energy with respect to a right-handed rotation about $\mathbf{1}$.

9.5   Grain Boundary Plane Anisotropy in Polycrystalline Materials

New developments in advanced characterization and computational techniques make it possible to readily measure the grain boundary character distribution of a wide range of materials including ceramics (e.g., MgO [903], SrTiO$_3$ [900]), metals (e.g., Ni [568], austenitic [76] and ferritic [77] steels, and Al [901]). Based on these measurements, three important conclusions can be drawn

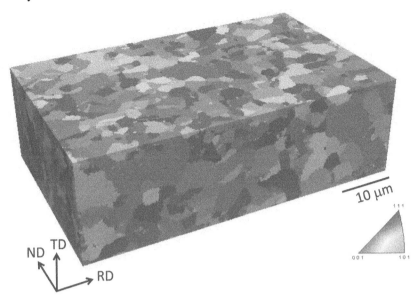

Figure 9.5: Reconstructed serial sections of electron backscatter diffraction data of the austenitic high-Mn TWIP steel, containing 100 slices. RD, ND, and TD are rolling direction, normal direction, and transverse direction, respectively. The colors are the orientations referred to the normal direction. (Figure reproduced with permission from [76].)

[852]. Firstly, the grain boundary plane distributions, $\lambda(\mathbf{n})$, reveal a significant anisotropy for all systems, though the preferred planes strongly depend on the crystal structure. For instance, {111} planes are preferred in the case of austenitic steel with the fcc crystal structure (Figure 9.6a). The second important finding is that the preferred habit planes of polycrystalline grains match well with the identical low-energy, low-index planes (Figure 9.6) [899]. In addition, there is a correlation between the summation of two surface energies making up a boundary and the grain boundary energy, $\gamma(\Delta g, \mathbf{n})$ [903, 900]. These findings suggest that the grain surface relationships are more significant than the crystal lattice orientation, which was initially assumed. Furthermore, the surface energy anisotropy is enough to predict the energy of the grain boundary. Lastly, there is a strong inverse correlation between the grain boundary population and energy [903, 900]. In other words, low-energy boundaries are observed more frequently in the distribution than high-energy boundaries (Figure 9.7). Although it is now evident that the presence of the anisotropy in boundary properties (e.g., energy) impacts the distribution of grain boundary planes, there is no clear understanding of the role of different parameters (i.e., extrinsic and intrinsic) on the distribution formed during material processing. The following section classifies these variables and highlights their impact on the grain boundary planes character distribution.

9.6 Influence of Different Parameters on the Grain Boundary Character Distribution

The parameters affecting the grain boundary character distribution can be classified into two main categories, namely intrinsic and extrinsic. The former refers to the alloy composition and crystal structure, which are inherited by the material and cannot be altered. The latter is mostly related to the processing routes, which can be controlled to alter the characteristics of grain boundaries in polycrystalline materials.

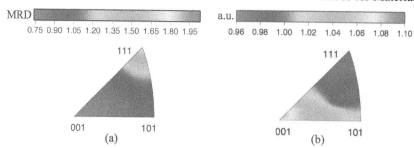

Figure 9.6: The distribution of grain boundary planes (a) and energy (b) of the TWIP steel. MRD and a.u. are multiples random distribution and arbitrary units, respectively. (Figure reproduced with permission from [76].)

9.6.1 Intrinsic Parameters

9.6.1.1 Alloy Composition

The grain boundary energy is associated with the interface structure, the temperature, and chemical composition. In pure materials, there is a strong correlation in the grain boundary energy distribution in isostructural materials (e.g., fcc) [820]. The range of grain boundary energies, however, is different for different pure materials. Atomistic simulations reveal that the grain boundary energy in Cu is greater than that of Al for a wide range of boundaries [1061]. The addition of alloying elements may enhance the grain boundary energy either through the change in the interface thermodynamics and/or segregation. The former is widely used to control the grain boundary population through the boundary dopant segregation (e.g., Y doping in alumina [199]). In a specific thermodynamic condition (i.e., temperature and pressure), a grain boundary can transit from one distinct grain boundary

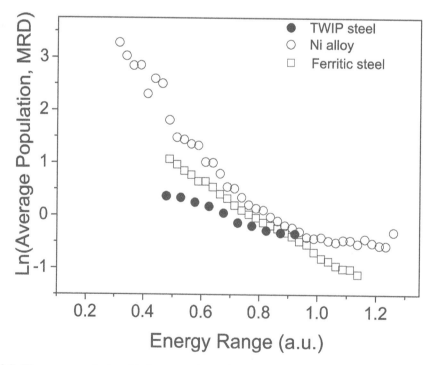

Figure 9.7: The average relationship between the grain boundary population and energy for different materials. (Figure reproduced with permission from [76].)

structure and/or composition to another if the solute concentration is changed. This is known as a complexion transition and is observed in some doped polycrystalline systems. This boundary complexion transition may significantly change the grain boundary properties (e.g., mobility) [175].

The chemical composition can strongly affect the grain boundary character distribution as it influences the stacking fault energy (SFE) and other interface energies. In pure materials, the population of coincident site lattice (CSL) boundaries (e.g., $\Sigma 3$) is generally governed by the SFE of material. The SFE is associated with the atomic bonding of material, determining the extent of dislocation dissociation into the partial dislocations. In general, a material with a lower SFE reveals a greater CSL boundary population (e.g., the $\Sigma 3$ population in pure Al, SFE=166 mJ/m^2, is three times lower than that in pure Ni, SFE=128 mJ/m^2) [815]. The addition of alloying elements may alter the SFE differently. The addition of Mn to the austenitic TWIP steel decreases the SFE, though C and Al additions reveal opposite influence on SFE in TWIP steels [879]. The SFE remains nearly unchanged through the addition of Mn to Cu alloys [305]. The addition of small amounts of impurities may also influence the grain boundary character distribution [762].

9.6.1.2   Crystal Structure

The grain boundary characteristics are significantly altered with the crystal structure [440]. For example, the materials with the bcc crystal structure reveal remarkably different grain boundary character and energy distributions compared with fcc materials. For the $\Sigma 3$ boundary, $\{111\}$ planes, for instance, show the minimum energy and highest population for materials with fcc crystal structure. This is entirely opposite for bcc materials, where the $\{112\}$ planes appear as the most populous and being the lowest energy planes (Figure 9.8). This is mostly due to the presence of distinct low energy grain boundary planes in different crystal structures [77, 76].

Interestingly, there is, on average, a strong inverse relationship between the grain boundary population and energy for both bcc and fcc materials, similar to simulations [395, 853]. In other words, the boundaries with the maximum population have the minimum energy and vice versa. However, the relationships show a prominent difference between the different materials (Figure 9.7). The ferritic steel relationship is nearly linear [77], while there is curvature for materials with fcc crystal structures (i.e., Ni [568] and TWIP steel [76]). This difference can be partly explained through multiple twinning[816], which is frequently observed in fcc materials with low stacking fault energy. This changes the misorientation distribution, selectively altering the population in specific energy ranges.

9.6.2   Extrinsic Parameters

Polycrystalline materials are mostly subjected to one or more processing routes depending on their properties and the required applications. The microstructure evolution occurring during different processing routes results in distinct preferred crystallographic texture, which significantly influences the grain boundary character distribution. The most common processing routes are solidification, thermomechanical processing (i.e., deformation and annealing), thin film growth, magnetic field processing, and phase transformation.

9.6.2.1   Solidification

The solidification is the prime production stage of most metals and alloys. In a conventional solidification process, three main zones are developed, which have distinct textures and grain structure morphologies: chill, columnar, and central zones. The chill zone is formed at the surface in contact with the mould at an early stage of solidification. This region has comparatively small volume with relatively fine equiaxed grain morphology having random or preferred orientations. The grain morphology of the second zone has a columnar morphology originated from oriented dendrite growth, resulting in a sharp texture, for example (001) texture in cubic materials (Figure 9.9a). The last zone is formed at the late stage of the solidification process and is located in the middle of the mould.

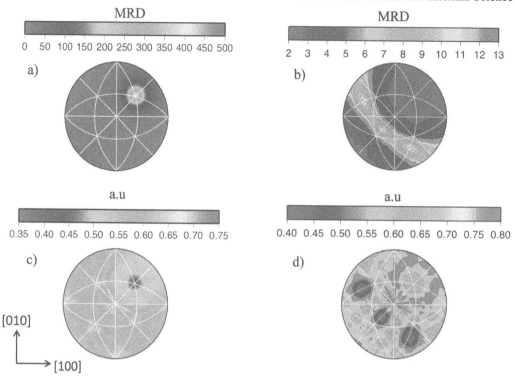

Figure 9.8: (a,b) The distribution of grain boundary planes and (c,d) the corresponding grain boundary energy distribution at the fixed misorientation of $\Sigma = 60°/[111]$ for the austenitic TWIP steel and fully ferritic steel ([77] and [76]), respectively, plotted in stereographic projection along [001]. MRD and a.u. represent multiples of a random distribution and arbitrary units, respectively.

The central zone has equiaxed grain morphology with consistently random texture. The extent of each zone differs depending on the casting condition (e.g., mould type).

The development of distinct texture at different zones results in a significant change in the population of different boundaries. For instance, the $\Sigma 3$ boundary population is remarkably reduced from $\approx 25\%$ in a thermomechanically-processed Ni-30Fe austenitic alloy to less than 5% in the as-cast condition. However, the texture developed through the solidification did not alter the grain boundary plane character of the $\Sigma 3$ boundary revealing a maxima at the (111) position for both conditions (Figures 9.3d and 9.9b).

9.6.2.2   Thermomechanical Processing

The thermomechanical processing is widely used to alter the grain boundary populations in polycrystalline materials, having moderate-to-low stacking fault energy. Essentially, two main thermomechanical routes are manipulated to change the grain boundary populations. The first approach employs a single or multiple small to moderate strain levels (e.g., $2-6\%$) followed by annealing after each pass at a relatively low temperature. At this temperature regime, no recrystallization takes place, though the temperature is high enough to reorient the grain boundaries to a low energy configuration through the recovery process [817, 921]. Despite an increase in the fraction of the CSL boundaries, this approach requires long annealing times of up to 20 h, which are not industrially viable. In addition, the prolonged annealing time leads to undesirable grain growth.

To overcome these obstacles, iterative recrystallization treatments were widely employed using a moderate strain level of up to 30% and annealing temperature in a range of $0.6-0.8T_m$ (T_m is the melting temperature in Kelvin). Here, the post-deformation annealing time is significantly

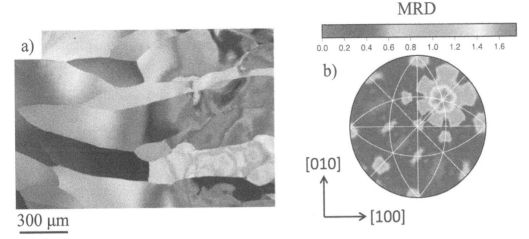

Figure 9.9: a) The AsB-SEM image of columnar microstructure formed through casting of Ni-30Fe austenitic alloy and b) the corresponding grain boundary planes distribution at the fixed misorientation of $\Sigma = 60°/[111]$, plotted in stereographic projection along $[001]$. MRD is multiples of a random distribution.

reduced to a few minutes, leading to very limited grain growth. The recrystallization takes place during annealing, resulting in an enhancement of the CSL boundary population to more than 60% [921]. The extent of the CSL boundary development is significantly influenced by the deformation mode. A recent study showed that shear strain is much more effective in the development of CSL boundaries during annealing compared with strain induced by the rolling process, where both shear and compression deformation modes operate simultaneously [322].

9.6.2.3   Thin Film

The polycrystalline thin films have a wide range of applications such as in electronic, magnetic, semiconducting, photovoltaic, and micromechanical devices. They are often produced through the nucleation of enormous numbers of isolated crystals on the surface of a substrate. Once the nuclei form, they grow laterally as well as perpendicular to the plane of substrate interface. The former results in the impingement and coalescence of crystals, leading to the formation of grain boundaries. The latter forms a columnar grain morphology having a preferred crystallographic orientation in which the crystals mostly display specific crystallographic directions normal to the film plane. For instance, the thin film with face centered cubic crystal structure, fcc, (e.g., Cu, Al) mostly reveals a strong $\langle 111 \rangle$ fiber texture, having predominantly tilt boundaries with a common $\langle 111 \rangle$ axis [62]. This results in a significant change in the grain boundary population, and can also alter the grain boundary plane character. The extent of film texture strength is governed by three energies, namely the surface energy, the grain boundary energy, and the elastic strain energy, which can be controlled during both film formation and its post-processing treatment (i.e., annealing). The reduction of these energies is the main driving force for the growth of specific grain orientation/s resulting in the development of specific film texture [1040]. For example, the presence of annealing twins with low grain boundary energy in thin films having relatively low stacking fault energy (e.g., Cu) changes the overall texture of the thin film due to the formation of new grain orientations.

9.6.2.4   Magnetic Field

The application of a magnetic field during materials processing leads to significant changes in the texture, recrystallization (i.e., nucleation and growth), phase transformations, and grain boundary migration. The atomic structure of a grain boundary is different from the grain interiors. Thereby, it would be expected to have distinct electronic structure compared with adjacent grain interiors. Therefore, grain boundary migration can be treated as the movement of an electric conductor in a magnetic field, leading to the formation of an electric field [380]. The energy dissipation because of the induced electromotive force is significantly influenced by the mobility of grain boundary.

Mullins is among the first to examine the role of a magnetic field on grain boundary migration in bismuth bicrystals and polycrystals, though no explicit boundary motion was detected [713]. Later, Molodov et al. successfully revealed that the migration direction of a planar $90°\langle 112 \rangle$ tilt boundary in bismuth bicrystals can be reversed by altering the bicrystal's position in the magnetic field [698]. They also showed that the grain boundary character influences the boundary mobility. For a given bicrystals crystallography, the boundary mobility becomes faster when the direction of motion was along with the c axis (i.e., $\langle 111 \rangle$) of the growing grain in comparison with the one having the c axis perpendicular to the direction of motion. This phenomenon makes it possible to preserve the grain boundaries with the minimum mobility and produce a grain structure with high stability.

9.6.2.5 *Transformation Path*

Most technologically important metals such as steel and titanium alloys undergo different phase transformation paths, which are the most effective ways to tailor the microstructure and properties. The composition, prior grain size, density of various defects, and cooling rate are the most important thermomechanical parameters for the control of the different phases at room temperature. In steels, the transformation of austenite to ferrite occurs at a relatively high temperature regime (i.e., slow cooling) where both nucleation and growth processes are controlled by the diffusion/reconstructive mechanism. Alternatively, the displacive shear mechanism takes place during the austenite to martensite phase transformation on rapid cooling.

The interface plane orientation distribution in transformed microstructures is mostly governed by the mechanism of transformation rather than the relative energy [75, 78]. For polygonal ferritic microstructure, the plane distribution reveals maxima at the positions of {112} symmetric tilt planes having the lowest energy for the $\Sigma 3$ misorientation (Figures 9.8b,d) [77]. By contrast, the martensitic microstructure shows the maxima at {110} symmetric tilt planes (Figure 9.10) [78]. This is consistent with the crystallographic constraints associated with the martensitic transformation rather than low energy configuration as {110} has a relatively higher energy compared with {112} orientation (Figure 9.8d). A similar observation was reported for GBCD of martensite in a Ti-6Al-4V alloy [75]. The sensitivity of the grain boundary type to the phase transformation path can potentially lead to a new route to engineer the grain boundary network for specific application/s.

9.7 Grain Boundary Network

Simulations [578, 492] and experiments [1124] have demonstrated that the grain boundary network characteristics (i.e., type, frequency, and connectivity) have a significant effect on the performance of polycrystals. The contribution of each grain boundary type to a property of interest is different. However, the characterization of the grain boundary network is a long-standing challenge since the introduction of the grain boundary engineering field. Different approaches were developed to measure the grain boundary network in polycrystalline materials.

9.7.1 *Grain Boundary Correlation Number*

The simplest approach to describe the grain boundary network is by the type and frequency of specific boundaries (e.g., random or CSL), which is relevant to the property of interest. These criteria cannot, however, define how the boundary of interest is joined throughout the microstructure. Watanabe introduced a new criterion to describe the connectivity of grain boundary network, the so-called grain boundary correlation number [1124]. This measure depicts the number of specific boundary type connected to each other in a given microstructure. He then demonstrated that a decrease in the correlation number of random boundaries for two microstructures with similar random

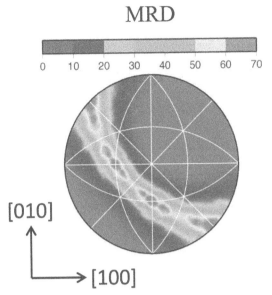

Figure 9.10: The distribution of grain boundary planes at the fixed misorientation of $\Sigma = 60°/[111]$ for the lath martensitic steel, plotted in stereographic projection along [001]. MRD is multiples of a random distribution. (Figure reproduced with permission from [78].)

and total boundary populations is correlated to improved ductility, assuming that the crack initiation and propagation take place on the random boundaries [1124]. A low grain boundary correlation number suggests less connectivity between random grain boundaries, which delays fracture by interrupting the crack propagation path.

9.7.2   Percolation Model

The percolation theory is a natural mathematical approach to specify the connection probability of a group of points, distributed in space [389]. In a polycrystalline material, there is usually a non-random correlation between different boundary types. The classification of boundaries to random and special types makes it possible to define open and closed bonds in the grain boundary network, which is used in the percolation model [358, 521]. Initially, the percolation theory employed a hexagonal lattice, reflecting the triple junctions, to represent the grain boundary network [358]. It was shown that a local correlation exists on the triple junction distribution, where the triple junctions are classified as $J_{i=0...3}$ (i denotes the number of a specific boundary type at the triple junction). For instance, J_0 and J_3 represent the triple junctions consisting of no special and three special boundaries, respectively. The local correlation at the triple junctions is mostly related to the crystallography constraints [358]. For instance, a triple junction having two special boundaries most likely has a third special boundary too, according to Σ-product rule [693]. This can be further extended to define other percolative parameters such as a grain boundary cluster size (i.e., connected path) and the percolation threshold. The cluster is composed of interconnected boundaries that are either solely random or entirely special. The clusters can be measured based on their size and dimension as a topological scalar to describe the grain boundary network [918]. There is an obvious difference in the behavior of clustering and percolation in most grain boundary engineered microstructures in comparison with what is expected from a random percolation. The percolation approach has shown that the grain boundary network structure and materials performance are linked [918]. The extent of percolation directly links to the fraction of boundaries of interest. In fact, the population of a specific type of boundary that forms a percolated network defines another characteristic of grain boundary network, the so-called percolation threshold. There is, therefore, a percolation threshold, above which the special boundaries strongly percolate, enhancing the material property of interest (e.g., long-range resistance to intergranular damage). However, this information is limited to the nearest-neighbor correlations in two-dimensional grain boundary networks, which may not necessarily represent long-range correlations in three-dimensional microstructure with a different percolation threshold [439]. In addition, the percolation threshold depends on the lattice geometry used to model the grain boundary network [342, 332], definition of boundary type, and the microstructure texture characteristics [332].

9.7.3 Homology Metrics

The homological approach is another mathematical method to define the connectivity properties of complicated microstructures, which is discussed in detail in Chapter 13 of this book. In short, the homology metrics are sensitive measures of the characteristics of grain boundary networks, which ultimately provide quantitative information about arbitrary datasets. This approach employs two topological parameters known as Betti numbers (i.e., β_0 and β_1). These numbers are positive integers and measure the connectivity of boundary networks. In a two-dimensional microstructure, β_0 measures the number of independent pieces of the network in the microstructure (i.e., separate boundary segments, which are not joined to the rest of network) and the number of continuous and closed loops (i.e., enclosed paths of boundaries) is referred to as β_1 [1122, 855]. The ratio β_0/β_1 is a metric representing the *inverse connectivity* of the boundary network structure, which can be measured as a function of varying grain boundary crystallography parameter/s threshold (i.e., dis/misorientation angle and/or axis). The effect of grain boundary network characteristics (i.e., disorientation threshold) on the inverse connectivity is schematically shown in Figure 9.11. The smallest value of β_0/β_1 refers to a fully equiaxed structure. The β_0/β_1 ratio gets larger as the disorientation threshold increases. This approach is initially employed to characterize the thermal-elastic response fields in polycrystals. Later, the connectivity of grain boundary networks was successfully measured as a function of disorientation angle and/or misorientation axis in a number of materials (i.e., martensitic steel [78], Ni, and $SrTiO_3$ [855]). It appears that the homology metrics are sensitive indicators of the grain boundary network structure.

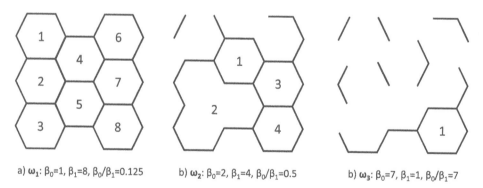

a) ω_1: β_0=1, β_1=8, β_0/β_1=0.125   b) ω_2: β_0=2, β_1=4, β_0/β_1=0.5   b) ω_3: β_0=7, β_1=1, β_0/β_1=7

Figure 9.11: a-c) Illustration of the inverse connectivity (β_0/β_1) measurement as a function of disorientation angle threshold (ω) where $\omega_1 < \omega_2 < \omega_3$. β_0 measures the number of boundaries not connected to the network and would equal 1 when the network is complete (i.e., "a"). β_1 measures the number of continuous, closed paths of grain boundaries. (Figure reproduced with permission from [78].)

9.8   Summary and Current Challenges

Recent advances in characterization techniques and computational approaches provide a great opportunity to three-dimensionally examine the grain boundary networks of a wide range of polycrystalline materials. These observations reveal a strong anisotropy in the grain boundary planes distribution, which is mostly driven by the grain boundary energy. In other words, there is a strong inverse relationship between the grain boundary population and the energy. However, the relative population of grain boundaries can be significantly altered by different intrinsic and extrinsic parameters such as texture and phase transformation mechanisms. Among different parameters, phase transformations can play a significant role to not only alter the grain boundary population, but also to change the grain boundary plane characteristics. These changes may not necessarily lead to the lowest energy population of grain boundaries. This is an area which requires further work as it significantly influences the grain boundary character distribution and consequently the final material performance.

The ultimate goal of grain boundary characterization is to measure the mesoscale structure of the grain boundary networks in polycrystals. This requires two main microstructure characteristics: a) the relative areas of the different types of grain boundaries within the microstructure and b) the way that they are connected. The current approaches use planar sections of microstructure (i.e., two dimensional datasets) to measure the relative areas of specific grain boundary types based either on the boundary lattice misorientation angle or the classification of the number of special boundaries in the triplet. Despite significant progress in the characterization of the grain boundary network, these approaches do not fully characterize the boundary network based on all five crystallographic characteristics of the boundary (i.e., the disorientation and the grain boundary plane orientation). This is an outstanding challenge, which can be potentially resolved as three-dimensional datasets are readily obtained through advanced characterization techniques.

Acknowledgments

Hossein Beladi acknowledges the grants provided by the Australian Research Council. Gregory Rohrer acknowledges an ONR-MURI program (grant no. N00014-11-0678).

Chapter 10

Interface Science and the Formation of Structure

Ming Tang
Rice University

Jian Luo
University of California at San Diego

10.1 Introduction

Interface science is a very important branch of materials science and has undergone explosive growth in the last few decades. Rather than attempting the impossible mission of capturing all the aspects of this field within this chapter, we limit the scope to a succinct discussion on interface thermodynamics and its implication for microstructures. To this end, focus will be given to the dependence of interface energy on thermodynamic state variables such as temperature, chemical potentials and interface orientation, the effects of interface energy on materials microstructures, and how interface energy and its variation can be measured or inferred from experimental data such as micrographs. Readers can find excellent books, such as Sutton and Balluffi's *Interface in Crystalline Materials* [990] and *Materials Interfaces: Atomic-Level Structure and Properties* edited by Wolf and Yip [1151], which provide a comprehensive treatment of various topics in interface science that are either not covered or only briefly mentioned here.

Triple junctions provide a crucial link between interfaces and microstructures, since the relative interface (or interfacial) energies, as well as the torque terms associated with the anisotropy in interface energies, control the geometry of triple junctions, which in turn influences the morphology of microstructures. Experimentally, such thermodynamic relations are often used to determine the relative interface energies from observed microstructures. The relation between interface energies and triple junction configurations is discussed in Section 10.2. Interface energies in crystalline materials are often anisotropic; the effects of interface energy anisotropy on crystal shape and interface faceting are discussed in Section 10.3. Interface energy is also at the heart of wetting phenomena, which is the subject of Section 10.4. In Section 10.5, we provide a discussion of some fundamental aspects of interface thermodynamics including the dependence of interface energy and adsorption of solute atoms to interfaces on thermodynamic state variables and interface transitions, and also a brief discussion of the implications of interface transitions for the kinetics of microstructural evolution.

10.2 Effect of Interface Energy on Triple Junction Geometry

A triple junction is a line where three interfaces meet. The intersecting interfaces could be either of the same or dissimilar type. An example for the former case is a triple junction in polycrystalline materials formed by three grain boundaries. The latter case is exemplified by thermal grooves, where an interface intersects the material surface, or contact lines formed by liquid droplets on substrate. Triple junctions play an important role in the measurement of interface energy because the relative

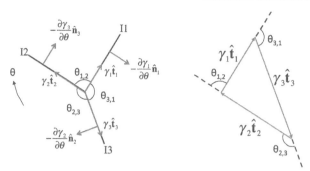

Figure 10.1: Schematics of the forces acting on a triple junction formed by interfaces $I1$, $I2$, and $I3$, which include three interface tension terms along the tangential directions of the interfaces ($\gamma_1\hat{t}_1$, $\gamma_2\hat{t}_2$ and $\gamma_3\hat{t}_3$) and three torque terms along the interface normal directions ($-(\partial\gamma_1/\partial\theta)\hat{n}_1$, $-(\partial\gamma_2/\partial\theta)\hat{n}_2$ and $-(\partial\gamma_3/\partial\theta)\hat{n}_3$). When the torque terms can be neglected, i.e., interface energy is independent of boundary plane orientation, the equilibrium dihedral angles of the triple junction, $\theta_{1,2}$, $\theta_{2,3}$, and $\theta_{3,1}$, satisfy the Young's equation, or that the vectors $\gamma_1\hat{t}_1$, $\gamma_2\hat{t}_2$, and $\gamma_3\hat{t}_3$ form a closed triangle.

energies of the intersecting interfaces could be inferred from the junction geometry, which is characterized by the three dihedral angles between the interfaces (Figure 10.1). Assuming that excess energy of a triple junction itself is negligibly small, when the system is in thermodynamic equilibrium, the three dihedral angles should take values that minimize the total interface energy. This is equivalent to requiring that the net interfacial "force" acting on the triple junction is zero. If interface energy is isotropic (or independent of the plane orientation), such as in liquid systems, the forces on the triple junction consist of the interface energies along the tangential directions of the three interfaces, as illustrated in Figure 10.1. The force balance $\sum_{i=1}^{3}\gamma_i\hat{t}_i = 0$ (where γ_i and $\hat{t}_i$ are the magnitude and the unit vector that specify the i-th interface energies, as shown in Figure 10.1) leads to the well-known Young's equation:

$$\frac{\gamma_1}{\sin\theta_{2,3}} = \frac{\gamma_2}{\sin\theta_{3,1}} = \frac{\gamma_3}{\sin\theta_{1,2}}. \tag{10.1}$$

Equation 10.1 shows that the interface energies are proportional to the sines of the dihedral angles, which can be measured from micrographs.

However, solid-solid or solid-liquid interfaces usually have anisotropic energy that depends on the plane orientation. As illustrated in Figure 10.1, under such circumstances the interfaces meeting at a triple junction experience additional forces along the interface normals, $-(\partial\gamma_i/\partial\theta)\hat{n}_i$ where θ is the rotation angle around the triple junction line and $\hat{n}_i$ is the unit normal vector. These forces are often referred to as the torque terms and their role is to rotate the interfaces towards the orientations with lower energy. The three torque terms hence need to be added to the force balance equations that dictate the equilibrium condition of a triple junction — this is known as the Herring relation [427]:

$$\sum_{i=1}^{3}\left(\gamma_i\hat{t}_i - \frac{\partial\gamma_i}{\partial\theta}\hat{n}_i\right) = 0. \tag{10.2}$$

With Equation 10.2, it is no longer possible to determine the relative magnitudes of the interface energies from the dihedral angles of a single junction (Equation 10.2 consists of two individual equations along two orthogonal directions, but a total of six unknowns).

It should be pointed out that there are circumstances under which the Herring relation and its special case Young's equation do not hold. One such case is when the orientation dependence of the interface energy has steep cusps. As shown in Figure 10.2, $\partial\gamma_i/\partial\theta$ is not continuous and jumps

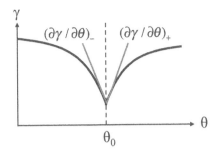

Figure 10.2: When the orientation of an interface is located at a cusp of $\gamma(\theta)$, $\partial\gamma\partial\theta$ is discontinuous at this orientation and the torque acting on the interface has a value between $(\partial\gamma\partial\theta)_-$ and $(\partial\gamma\partial\theta)_+$.

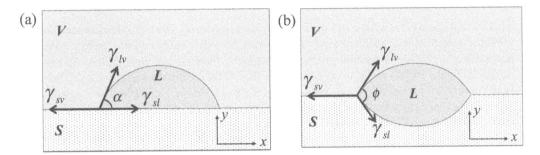

Figure 10.3: A liquid droplet sitting on an initially flat solid substrate forms a triple junction between the liquid (L), solid (S), and vapor (V) phases. (a) If the solid substrate is inert, force balance at the triple junction is only satisfied along the x axis. (b) If the solid and liquid phases can equilibrate with each other (e.g., a solid and its melt at the melting temperature) and the solid/liquid interface is mobile, then the force balance is satisfied along both x and y axes.

from $(\partial\gamma\partial\theta)_-$ to $(\partial\gamma\partial\theta)_+$ at a cusp. When the interface is at such an orientation, there is no well-defined value for $\partial\gamma_i/\partial\theta$ in Equation 10.2. Instead, the equilibrium torque acting on the interface has a value between $(\partial\gamma\partial\theta)_-$ and $(\partial\gamma\partial\theta)_+$. Applying the Herring relation in such a case implicitly assumes that the derivative of $\partial\gamma_i/\partial\theta$ is continuous and will miss the presence of the cusp, leading to incorrect interpretation of experimental measurement. Another frequently encountered situation that causes the breakdown of the Herring relation is when the equilibrium of triple junction geometry is kinetically inhibited. This is exemplified by a liquid droplet on a solid substrate, forming a contact line with the vapor and solid phases, Figure 10.3. Often, the substrate is inert and rigid, and remains flat within observable time scale. Consequently, the three interface energies (tensions) at the contact line do not cancel each other along the surface normal (y direction) of the substrate. The contact angle of the liquid droplet is determined by force balance in the x direction only, i.e.,

$$\gamma_{sv} = \gamma_{sl} + \gamma_{lv}\cos\alpha. \tag{10.3}$$

Another type of triple junctions often seen in materials are thermal grooves, which are V-shaped surface depressions formed at locations where internal interfaces (e.g., grain boundaries) intersect the free surface, as illustrated in Figure 10.4. The name "thermal groove" comes from the fact that this feature is prominent in polycrystals annealed at high temperatures. Two surface segments meet an internal interface at a thermal groove. Like other triple junctions, the reason for the surface to groove is that the minimization of the total interface energy requires the two surface segments to have an angle less than 180° between them. Assuming that the groove junction is in local equilibrium and that the surface energy is isotropic, the ratio of the internal interface and surface energies can

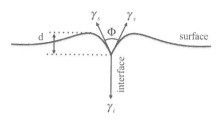

Figure 10.4: Schematics of a thermal groove.

be evaluated from the groove's dihedral angle (Φ):

$$\gamma_i/\gamma_s = 2\cos\frac{\Phi}{2}. \tag{10.4}$$

where γ_i and γ_s are the interface and surface energies, respectively. With advances in scanning probe microscopy technology, the dihedral angle Φ can now be conveniently determined by measuring the surface height profiles of thermal grooves in an atomic force microscope (AFM). An example is shown in Figure 10.5. It needs to be pointed out that while the dihedral angle of a thermal groove is an equilibrium feature, the depth and width of the groove are not and will continuously increase during annealing, known as thermal grooving. The driving force for the grooving process is the curvature difference between surfaces inside and outside the groove. The surface curvature within the groove is controlled by the dihedral angle and is always larger than that of the flat surface away from the groove. According to the Gibbs–Thomson relation [46, pp. 611], atoms in the groove region have an excess chemical potential of $\Omega\gamma_s\kappa$ (Ω—atomic volume, κ—surface curvature). Materials are hence continuously removed from the groove region to reduce the chemical potential, causing the groove to grow. Mass transport in this process may be achieved by various mechanisms, including evaporation-condensation, surface diffusion, and volume diffusion. Mullins's classic theory of thermal grooving [711, 712] shows that the surface profile and growth rate of a groove are mechanism-dependent. For surface-diffusion-dominant grooving, the surface profile exhibits characteristic ridges next to a groove (see Figure 10.4) and the groove depth d varies with annealing time as

$$d = \frac{0.973}{\tan(\Phi/2)}\left(\frac{D_s\gamma_s\Omega^2 vt}{kT}\right)^{\frac{1}{4}}, \tag{10.5}$$

where D_s is the surface diffusion coefficient, v is the number of atoms per unit volume, and T is the temperature. Measurement of the time-dependent groove dimensions can thus reveal not only the relative energies of the interface and free surface, but also the grooving mechanism and other material properties such as surface diffusivity.

Now we use grain boundary as an example to discuss the experimental measurement of interface energies. Most engineering materials are polycrystalline materials and contain grain boundaries, the structure and properties of which frequently control material performance such as strength, fracture and corrosion resistance. A grain boundary is classified by five macroscopic crystallographic parameters, with three describing the misorientation of the two crystal lattices and two describing the boundary plane orientation. Grain boundary energy usually has a non-trivial dependence on the five parameters and can display orders of magnitude variation. As the grain boundary energy value is a good indicator of the underlying boundary structure and other properties and its anisotropy could exert considerable influence on the microstructural evolution of polycrystals, it is important to determine the grain boundary energy as a function of misorientation and boundary plane orientation.

The absolute or relative energy of a grain boundary can be calculated from a triple junction using the Young's equation (Equation 10.1) if the energy of another interface is either known or chosen as the reference and the orientation dependence of interface energy can be neglected. This approach is the basis of various types of grain boundary energy measurement, such as:

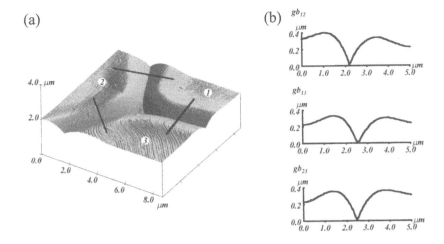

Figure 10.5: (a) Contact AFM image of a MgO polycrystal where three thermal grooves meet. Underneath the grooves are three grain boundaries formed by grain 1, 2, and 3. (b) Surface height profiles of the thermal grooves along the black lines indicated in the AFM image. From work by Saylor and Rohrer (reprinted from [895] with permission).

1. *Tri-crystal method:* A tri-crystal, which is usually grown from three crystal seeds with controlled crystal lattice and boundary plane orientations, contains a triple junction where three grain boundaries meet. This method has been applied to both metals [412] and ceramics [699].

2. *Thermal grooving method:* This method is applied to bicrystals or polycrystals by annealing samples to produce thermal grooves on sample surfaces. For polycrystal samples, electron backscattering diffraction (EBSD) is often used to determine the misorientations and plane orientations of individual boundaries. The surface energy γ_s is typically assumed to be isotropic and serves as the reference for calculating the relative energies of grain boundaries. This method has been used to measure grain boundaries in metals [365] and ceramics [926, 895].

3. *Grain boundary precipitation method:* Mori et al. [706] measured the dihedral angles of SiO_2 particles precipitated at grain boundaries after annealing in an oxidizing environment in a series of Cu-0.06 wt. % Si bicrystal samples. By assuming SiO_2/Cu interface energy to be isotropic, they were able to compare the relative energies of different Cu grain boundaries.

The traditional approaches to grain boundary energy measurements using bicrystal or tricrystal samples are very time-consuming and limit data collection to a small number of boundaries. In addition, the torque terms in the Herring relation Equation 10.2 are usually neglected in the interpretation of the measurement, which could lead to inaccurate grain boundary energy values. These limitations are addressed by recent developments in automated measurement techniques [896, 897, 851, 850], e.g., by combining automated EBSD with serial sectioning. These techniques can characterize the geometry and crystallographic orientation of a large number of grain boundaries and triple junctions in one polycrystalline sample. The availability of such large datasets facilitates the determination of grain boundary energy as a function of grain boundary character via the full Herring relation without ignoring the torque terms. Readers are referred to Chapter 13 of this book and the review articles by Rohrer [851, 850] for a detailed account of recent research findings on the structure and energy of grain boundary networks enabled by the 3D microstructure reconstruction techniques.

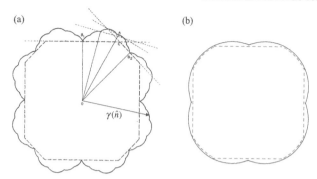

Figure 10.6: (a) Example of 2D gamma plot (solid line) and the equilibrium crystal shape (long dashed line) obtained from Wulff construction. Adapted from [426] with permission. (b) Example of anisotropic surface energy (solid line) that gives rise to both facets and curved surfaces in the Wulff shape (dashed line).

10.3   Effect of Interface Energy Anisotropy on Crystal Shape and Interface Faceting

In general, the energy of the interface between a crystalline phase and a vapor, liquid, or another solid phase is anisotropic and a function of the crystallographic orientation of the interface plane $\hat{n}$. The orientation dependence of interface energy can be represented by a polar plot of the interface energy $\gamma(\hat{n})$ as illustrated in Figure 10.6, which is known as the gamma plot. Because of the interface energy anisotropy, a crystalline particle that is free-standing or embedded in the matrix of another crystalline material usually exhibits anisometric shape at equilibrium with the surrounding. Mathematically, the equilibrium particle shape should minimize the total interface energy at constant volume. (Note that the shape of a crystallite in a matrix can also be affected by the elastic energy generated by the misfit strain between the matrix and particle. As the elastic energy scales with particle volume, interface energy may have a dominant effect on crystallite shape only at sufficiently small particle sizes.) The equilibrium crystal shape can be determined from the gamma plot $\gamma(\hat{n})$ using the Wulff construction method. As illustrated in Figures 10.6 and also 10.7(b), the procedure of the Wulff construction is as follows: first draw a line connecting the origin to a point on the gamma plot, then draw a plane (thin dashed lines in Figure 10.6(a) or thick solid line in Figure10.7(b)) that is perpendicular to the line and intersects the gamma plot at this point, and repeat the steps for each point on $\gamma(\hat{n})$. The inner envelope of all of these planes forms the equilibrium crystal shape, or the Wulff shape. As shown in Figure 10.6, if $\gamma(\hat{n})$ has deep cusps, then the Wulff shape may have a faceted morphology that is dominated by planes corresponding to the lowest interface energy orientations. Depending on the depth of the cusps, the equilibrium shape may consist of only facets (Figure 10.6(a)) or facets separated by smooth curved surface segments (Figure 10.6(b)). What is common between the two cases is that there are interface inclinations missing from the equilibrium crystal shape.

Cahn and Hoffman [436, 152] introduced another method to determine the equilibrium crystal shape from $\gamma(\hat{n})$ based on the capillarity vector $\boldsymbol{\xi}$. The capillarity vector is defined as

$$\boldsymbol{\xi} = \nabla(r\gamma(\hat{n})), \tag{10.6}$$

where $r = \sqrt{x_1^2 + x_2^2 + x_3^2}$ is the radial distance between the origin and a point at $[x_1, x_2, x_3]$, and $\hat{n} = [x_1, x_2, x_3]/r$ is the unit direction vector. It can be proven that the $\boldsymbol{\xi}$ vector is a function of $\hat{n}$ only, i.e., $\boldsymbol{\xi} = \boldsymbol{\xi}(\hat{n})$, and has the following properties:

$$\begin{cases} \boldsymbol{\xi} \cdot \hat{n} = \gamma(\hat{n}); \\ \hat{n} \cdot d\boldsymbol{\xi} = 0. \end{cases} \tag{10.7}$$

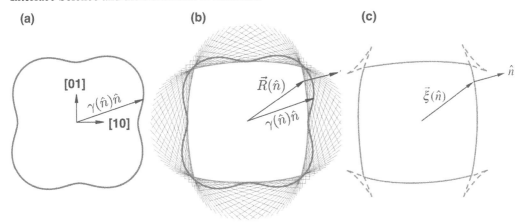

Figure 10.7: (a) A schematic surface energy plot $\gamma(\hat{n})$; (b) using $\gamma(\hat{n})$ (dashed line) and the Wulff construction to determine the equilibrium crystal shape or Wulff shape, which is the inner envelope of the thin solid lines. c) Using the ξ-plot to determine the equilibrium crystal or Wulff shape. The dashed lines represent the unstable part of the ξ-plot and are not part of the Wulff shape.

A ξ-plot is formed by connecting the end points of ξ vectors for all orientations $\hat{n}$. As illustrated in Figure 10.7(c), Equation 10.7 states that the surface normal of the ξ-plot at a point specified by $\xi(\hat{n})$ is parallel to the direction $\hat{n}$, and the projection of the $\xi(\hat{n})$ vector onto the surface normal direction is equal to $\gamma(\hat{n})$. On the other hand, let $R = R(\hat{n})$ represent the Wulff shape obtained from the Wulff construction, where R connects the origin to a point on the Wulff shape. Based on the Wulff construction procedure, it is easy to see that the Wulff shape satisfies the same criteria: $R \cdot \hat{n} = \gamma(\hat{n})$ and $\hat{n} \cdot dR = 0$. Therefore, the ξ-plot is identical to the Wulff shape, and so an alternative approach to obtaining the equilibrium crystal shape from $\gamma(\hat{n})$ is to use Equation 10.6. Figure 10.7(b) and (c) compare the crystal shapes determined by the Wulff construction method and the capillarity vector plot, which are identical. It needs to be pointed out that the "ears" of the ξ-plot (dashed lines in Figure 10.7(c)) correspond to the unstable interface inclinations and are excluded from the Wulff shape.

The reverse problem of Wulff construction, i.e., calculating $\gamma(\hat{n})$ from the equilibrium crystal shape, provides a means to experimentally determine the interface energy anisotropy. We see from Equation 10.7 that the projection of the radial vector from particle center to a point on the equilibrated particle surface onto the surface normal $\hat{n}$ gives the relative interface energy in that direction. Figure 10.8 showcases the work by Chatain et al. [188] on evaluating the surface energy anisotropy of 99.999% pure copper from crystallites equilibrated at 1240 K using this relation. It can be seen that pure Cu exhibits weak but distinguishable anisotropy in surface energy. For multi-component systems, solute elements of even minute concentrations can often have strong influence on the interface energy and its anisotropy due to their preferential segregation to interfaces. The solute segregation effect can also be studied by examining the equilibrium crystal shape. For example, Figure 10.9 compares the equilibrated shapes of 99.999% pure Pb [429] and a Pb-0.5 at. % Bi-0.08 at. % Ni alloy [1161] annealed at 523K. While the shape of pure Pb particles is dominated by curved surfaces with small facets, solute segregation at the surfaces of the Pb alloy significantly enhances the surface energy anisotropy and causes the Wulff shape to be dominated by facets. Readers are referred to the review paper by Wynblatt and Chatain [1161] for a detailed discussion on the anisotropy of segregation at interfaces.

Interface faceting is another phenomenon related to interface energy anisotropy. When a material changes its thermodynamic state (e.g., temperature or composition), an initially flat interface in the system is sometimes replaced by a "zig-zag" structure consisting of alternating facets shown in Figure 10.10. The reason for an interface to develop facets is that the change in $\gamma(\hat{n})$ or the Wulff

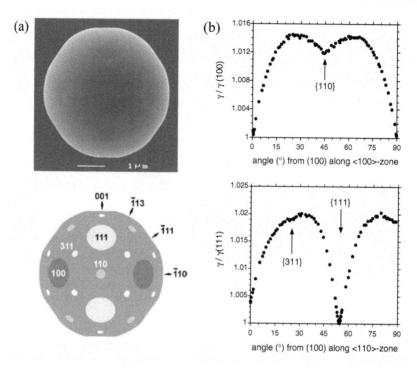

Figure 10.8: (a) Cu crystal annealed at 1240K, viewed from $\langle 110 \rangle$ direction (top), and the schematic Wulff shape (bottom). (b) Relative surface energies of Cu along $\langle 100 \rangle$ zone (top) and $\langle 110 \rangle$ zone (bottom) measured from the equilibrated Cu crystal shape. Reprinted from [188] with permission.

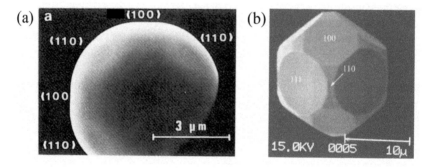

Figure 10.9: (a) Pb crystal equilibrated at 523 K (reprinted from [429] with permission). (b) Pb-0.5 at. % Bi-0.08 at. % Ni equilibrated at 523 K (reprinted from [1161] with permission).

plot causes the original interface inclination to become unstable and no longer appear in the equilibrium shape. Faceting provides a way to reduce the interface energy through local adjustment of interface structure while still maintaining the average interface orientation. As illustrated in Figure 10.10(b), a smooth surface may be replaced by two or three (but not more) facets with lower interface energies, depending on the local form of the Wulff plot. Interface faceting typically takes place with decreasing temperature [443] or increasing interface segregation [317], which tends to enhance the interface energy anisotropy.

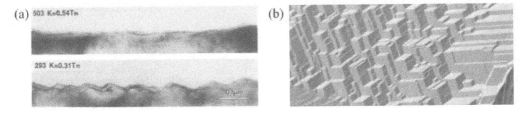

Figure 10.10: (a) A $\Sigma 3$ asymmetric $\langle 111 \rangle$ tilt grain boundary in Al is flat at 503K but faceted at 293K (reprinted from [443] with permission). (b) An originally flat interface replaced by two facet inclinations (right region of the interface) and three facet inclinations (left region) upon faceting.

10.4   Effect of Interface Energy on Wetting

10.4.1   Interface Wetting

The relative interface energies, which may change with varying thermodynamic state variables, can affect the topological and microstructural features. Specifically, a necessary (but not always sufficient) condition for the occurrence of (perfect or complete) wetting, where an α–β interface is replaced by a continuous layer of ω phase of an arbitrary thickness, is given by:

$$\gamma_{\alpha\omega} + \gamma_{\beta\omega} < \gamma_{\alpha\beta}^{(0)}. \tag{10.8}$$

Moreover, a wetting transition (from "non-wetting" to "wetting") may occur as either a first-order (characterized by a discontinuous change in one or more state variables of the system such as entropy and volume) or continuous (i.e., no discontinuity in state variables upon transition) transformation upon the change of a thermodynamic state variable [115]. Here, several notes should be made. First, the term "perfect wetting" or "complete wetting" is sometimes used to avoid ambiguity because the term "partial wetting" (which is rigorously "non-wetting") is used practically for a case where the contact angle (the angle α in Figure 10.3(a)) is less than 90°. Second, the superscript "(0)" in $\gamma_{\alpha\beta}^{(0)}$ refers to a hypothetical α–β interface (without the wetting ω phase or any adsorption) but not the equilibrium $\gamma_{\alpha\beta}$; at a thermodynamic equilibrium, however, this hypothetical α–β interface is replaced by an α–ω interface and an ω–β interface; by definition, the equilibrium interface energy for the α–β interface is given by:

$$\gamma_{\alpha\beta} = \gamma_{\alpha\omega} + \gamma_{\beta\omega}. \tag{10.9}$$

Third, it is important to emphasize that the wetting phase ω should be arbitrarily thick for a case of (complete) wetting. In contrast, if the α–β interface is continuously covered by an ω-like interfacial complexion [174] of a thermodynamically determined "equilibrium" thickness (see, e.g., Figure 10.11(a)) [599, 598], it is not a case of (perfect) wetting; such a case may be considered as "prewetting" [150] in a broad definition and was called "moist" (intermediate to "wet" and "dry") by the late Dr. Rowland Cannon [173]. The related interfacial phenomena will be discussed in the next section.

Let us consider a special case of grain boundary wetting by a liquid phase ($\omega = l$), where α and β are the same crystalline phase with different orientations. Here, a necessary (but not sufficient) condition for the occurrence of (perfect or complete) grain boundary wetting is:

$$-\Delta\gamma \equiv \gamma_{\text{gb}}^{(0)} - 2\gamma_{\text{cl}} > 0, \tag{10.10}$$

where γ_{cl} is the solid-liquid interface energy. One example of the wetting of Mo grain boundaries by a Ni-rich liquid (a binary Ni-Mo liquid with composition on the liquidus line) is shown in Figure 10.11(b,d) [929]. Other examples of grain boundary wetting are studied by Straumal et al. for Cu-In [977], Al-Sn [978], Al-Zn and Zn-Al [979] systems, where the primary phases are underlined.

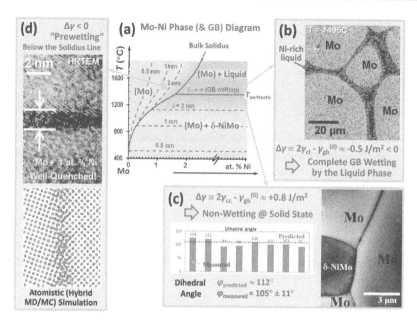

Figure 10.11: (a) (top) Grain boundary prewetting (coupled with premelting) in the single-phase region below the bulk solidus line where Ni-enriched, nanometer-thick, liquid-like, complexions are stabilized at Mo grain boundaries when the liquid phase is not yet a stable bulk phase. Reprint from [929, 931] with permission. (bottom) The hybrid molecular dynamics and Monte Carlo simulation result from [1169]. (b) Complete grain boundary wetting in the solid-liquid two-phase region above the bulk peritectic temperature, (c) non-wetting in the solid-solid two-phase region below the bulk peritectic temperature, and (d) the Mo-Ni binary phase diagram, where the solid blue lines represent bulk phase boundaries (while the red dotted lines represent estimated width of the liquid-like grain boundary complexions; see [602] for elaboration).

A couple of points should be further discussed. First, Equation 10.10 is valid only for simplified cases of isotropic interfacial energies; in reality, both grain boundary energy and crystal-liquid interfacial energy are anisotropic. Thus, wetting of different grain boundaries can occur at different conditions and the wetting transition can also trigger faceting or vice versa. Second, the dispersion interaction between the two adjoining grains is the longest-range interaction and always attractive for grain boundaries with a symmetrical configuration. Its existence limits the liquid thickness between the grains and thus prevents true complete wetting. However, long-range interactions are often weak and nearly perfect wetting can hence be achieved at least for metals, as shown in Figure 10.11(b).

In principle, grain boundary wetting can also occur in solid state, if $-\Delta\gamma \equiv \gamma_{gb}^{(0)} - 2\gamma_{cc} > 0$. However, crystal-crystal interfacial energies (γ_{cc}) are typically large, so that solid-state grain boundary wetting rarely occurs. For example, in case of $\underline{Mo}$-Ni, where complete grain boundary wetting occurs in the solid-liquid two-phase region above the bulk peritectic temperature (Figure 10.11(b,d)), the crystalline secondary phase (δ-NiMo) does not wet the Mo grain boundaries below the bulk peritectic temperature, where the dihedral angles (φ) are generally larger than $100°$ with significant boundary-to-boundary variation due to anisotropy (Figure 10.11(c)) [929]. Presumably, torque terms may also be significant in determining the configurations at the triple lines for such anisotropic cases.

Finally, it is imperative to note Equation 10.8 and Equation 10.10 are not sufficient conditions for the occurrence of complete wetting; this is because the interfacial potentials (or the interface

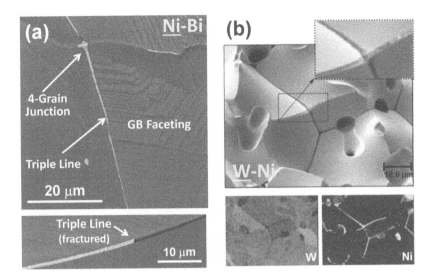

Figure 10.12: Wetting of the triple-grain junctions by liquid phases in (a) Ni-Bi and (b) W-Ni. The two bottom images in panel (b) are Auger compositional maps for W and Ni, respectively; all other images are scanning electron micrographs of (intergranularly) fractured specimens. Reprint/adapted from [605, 401] with permission.

energy as a function of adsorption or interfacial width) can be non-monotonic so that it is possible that the equilibrium $\gamma_{\alpha\beta} < \gamma_{\alpha\omega} + \gamma_{\beta\omega} < \gamma_{\alpha\beta}^{(0)}$, which leads to non-wetting; this is true for general grain boundaries of Bi-saturated Ni (where the adsorption of Bi reduced γ_{gb} to $\approx 1/4$ of $\gamma_{gb}^{(0)}$ with the formation of bilayers [605] that leads to: $\gamma_{gb} < 2\gamma_{cc} < \gamma_{gb}^{(0)}$ [37]). Other well-known cases are represented by many equilibrium-thickness intergranular and surficial films in ceramic systems [600, 598] that formed in the solid-liquid co-existence regions (where the film thickness is limited by attractive London dispersion forces), which will be discussed further subsequently. In fact, Equation 10.9 is a necessary and sufficient condition for the occurrence of complete wetting (or the necessary and sufficient condition for the non-wetting is that the equilibrium $\gamma_{\alpha\beta}$ is less than the sum of $\gamma_{\alpha\omega}$ and $\gamma_{\beta\omega}$.

10.4.2   Triple-Junction Wetting

For a polycrystalline solid with a small fraction of a liquid phase (e.g., in a case of liquid-phase sintering) where the interface energies can be assumed to be isotropic for simplicity, Equation 10.10 shows that the occurrence of grain boundary wetting implies that the $\gamma_{gb}^{(0)}/\gamma_{cl}$ ratio is greater than two (and the equilibrium $\gamma_{gb}/\gamma_{cl} = 2$ by definition). Moreover, triple-grain junctions may be wetted by the liquid phase if

$$\gamma_{gb}/\gamma_{cl} > \sqrt{3}, \tag{10.11}$$

which has been observed in several systems such as Ni-Bi [605] and W-Ni [401] (Figure 10.12). It should be noted that the above simplified condition for triple-grain junction wetting ignores the anisotropy in interface energies and the linear tensions of the triple lines.

In an extreme case, triple-grain junctions can also be wetted by a gas (vapor) phase if

$$\gamma_{gb}/\gamma_s > \sqrt{3}, \tag{10.12}$$

which leads to triple-line instability or the formation of holes along triple-grain junctions. This unusual phenomenon has recently been observed for electrodeposited Ni (with S impurities) annealed

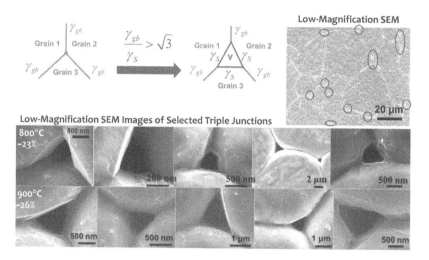

Figure 10.13: Observation of triple-line instability for electrodeposited Ni annealed in Bi vapor, which can be explained as wetting of the triple-grain junctions by a vapor phase. Reprint from [1201] with permission.

in the Bi vapor (Figure 10.13) [1201]. Theoretically, a polycrystal may spontaneously pulverize into particles (or the corresponding powder is un-sinter-able) if $\gamma_{gb}/\gamma_s > 2$, which has not yet been observed experimentally.

10.5 Change of Interface Energy and Structure with Thermodynamic State Variables

Here we will further discuss three related topics on how variations in thermodynamic potentials may influence microstructures via changing interface energies and kinetics. First, interface energies can vary with changing thermodynamic state variables, such as temperature (T), pressure (P), and chemical potentials (μ_i), which will subsequently induce changes in various microstructural features. Second, such changes can occur as either continuous or first-order phase-like transitions [174, 1002, 1003, 479, 54, 261]. Third, the occurrence of interfacial transitions can significantly influence the microstructural development via kinetic effects, particularly via substantially changing the grain boundary diffusivity and mobility [174, 602, 599].

Since interfaces can exhibit phase-like transitions, the term "complexion" was introduced to represent the relevant interfacial thermodynamic state(s) [1002, 174]. A complexion is essentially an interfacial phase that is two dimensional (2D) in thermodynamics and has no identifiable volume. Because it cannot exist by itself without abutting bulk phase(s), strictly speaking a complexion is not a phase according to Gibbs definition.

10.5.1 Variation of Interfacial Energy with Thermodynamic State Variables

In a multicomponent system, interface energy is the interfacial excess of grand potential ($\Phi_g = U - TS - \sum_i \mu_i N_i$; but not the interfacial excess of free energy [348, pp. 87], despite the common usage of the term "interfacial free energy") in the Gibbs dividing plane approach [1161]:

$$\gamma = e^{xs} - Ts^{xs} - \sum_i \mu_i \Gamma_i, \qquad (10.13)$$

where e^{xs} (internal energy), s^{xs} (entropy), and Γ_i (adsorption) are the interfacial excess quantities per unit area that depend on the selection of the Gibbs dividing plane in general (noting that the interfacial excess volume $v^{xs} \equiv 0$ in the Gibbs convention). For an interface between two different phases, one common convention is to select the Gibbs dividing plane so that the interfacial excess of one species vanishes, e.g., $\Gamma_1 = 0$. However, for a grain boundary where two abutting grains are the same phase, the interfacial excess quantities are independent of the position of the Gibbs dividing plane.

The well-known Gibbs adsorption equation states:

$$d\gamma = -s^{xs}dT - \sum_i \Gamma_i d\mu_i. \tag{10.14}$$

For example, for a specific grain boundary (with the fixed misorientation and inclination) in a one-component system that has two degrees of freedom (T, P), the grain boundary energy (γ_{GB}) varies with temperature or pressure:

$$\begin{cases} \left(\frac{\partial \gamma_{GB}}{\partial T}\right)_P = -s^{xs} - \Gamma\left(\frac{\partial \mu}{\partial T}\right)_P = -\left(s^{xs} - \Gamma\bar{S}\right) \equiv -\Delta S \\ \left(\frac{\partial \gamma_{GB}}{\partial P}\right)_T = -\Gamma\left(\frac{\partial \mu}{\partial P}\right)_T = -\Gamma\bar{V} \equiv \Delta V \end{cases}. \tag{10.15}$$

where $\bar{S}$ is the molar entropy and $\bar{V}$ is the molar volume; here Γ is typically negative ($\Gamma < 0$ because the grain boundary region usually has a lower local density) so that ΔV ($\equiv -\Gamma\bar{V}$) is essentially the grain-boundary "free volume." Consequently, the GB energy decreases with increasing temperature or decreasing pressure in most cases, since the ΔS and ΔV defined in the above equation are typically positive for a unary grain boundary.

A specific grain boundary in a binary A-B system has three degrees of freedom (T, P and μ_B or $X_B^{(bulk)}$, where $X_B^{(bulk)} = 1 - X_A^{(bulk)}$ and $X_A^{(bulk)}$ or $X_B^{(bulk)}$ is the molar fraction of A or B component in the bulk phase.) At constant T and P, the famous Gibbs isotherm is given as:

$$d\gamma_{GB} = -\Gamma_A d\mu_A - \Gamma_B d\mu_B. \tag{10.16}$$

Combining it with the Gibbs–Duhem equation [348, pp. 216] at constant T and P ($X_A^{(bulk)}d\mu_A + X_B^{(bulk)}d\mu_B = 0$) produces:

$$\left(\frac{\partial \gamma_{GB}}{\partial \mu_B}\right)_{T,P} = -\Gamma_B + \left(\frac{X_B^{(bulk)}}{X_A^{(bulk)}}\right)\Gamma_A, \tag{10.17}$$

Or:

$$\left(\frac{\partial \gamma_{GB}}{\partial X_B^{(bulk)}}\right)_{T,P} = -\left(\frac{\partial \mu_B}{\partial X_B^{(bulk)}}\right)_{T,P}\left[\Gamma_B - \left(\frac{X_B^{(bulk)}}{X_A^{(bulk)}}\right)\Gamma_A\right]. \tag{10.18}$$

If the bulk phase is a dilute solution where Henry's law applies, the above equation can be simplified to:

$$\left(\frac{\partial \gamma_{GB}}{\partial X_B^{(bulk)}}\right)_{T,P} = -\frac{RT}{X_B^{(bulk)}}\left[\Gamma_B - \left(\frac{X_B^{(bulk)}}{X_A^{(bulk)}}\right)\Gamma_A\right] = -RT\left[\frac{\Gamma_B}{X_B^{(bulk)}} - \frac{\Gamma_A}{X_A^{(bulk)}}\right]. \tag{10.19}$$

An exemplar of a γ_{gb} vs. $X_B^{(bulk)}$ curve and the associated Γ_B are shown in Figure 10.14.

Finally, a binary phase boundary (with fixed orientations) has one degree of freedom at a constant pressure according to the Gibbs phase rule [348, pp. 399], because the interface is in equilibrium with two bulk phases. Thus, we cannot vary temperature and composition (or chemical potential) of the solute independently. Specifically, let us consider a simple case of a solid-liquid interface in a binary A-B system (with a fixed crystalline surface orientation), which is important in

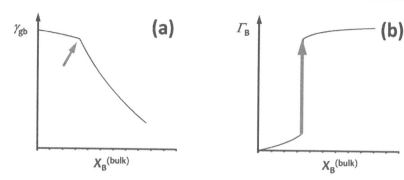

Figure 10.14: An example of the variations in (a) grain boundary energy (γ_{gb}) and (b) grain boundary excess/adsorption of solute (Γ_B) as functions of the bulk composition ($X_B^{(bulk)}$). A first-order of grain boundary adsorption transition corresponds to a slope change in the γ_{gb} vs. $X_B^{(bulk)}$ curve and a finite jump in Γ_B, as indicated by the arrows.

determining the wetting configurations. From the Gibbs adsorption equation, we can obtain:

$$\left(\frac{\partial \gamma_{S\text{-}L}}{\partial T}\right)_P \approx -s^{xs} - \Gamma_A \left(\frac{\partial \mu_A}{\partial T}\right)_{P,solidus} - \Gamma_B \left(\frac{\partial \mu_B}{\partial T}\right)_{P,solidus}$$

$$= -\left[s_{(B)}^{xs} - \Gamma_{A(B)}\left(\frac{\partial \mu_A}{\partial T}\right)_{P,solidus}\right], \qquad (10.20)$$

where we select the Gibbs dividing plane so that $\Gamma_B = 0$ and the "(B)" is added in the subscripts of the excess quantities to refer to this reference. Interested readers are referred to a review article by Eustathopoulos [312] and references therein for further discussion and examples.

In summary, the Gibbs adsorption equation dictates how the interface energy can be varied by changing temperature and chemical potentials, which can subsequently influence the microstructural features. Reversibly, measurement of interface energy as a function of temperature and chemical potentials may provide information on interface properties, e.g., interfacial adsorption and excess entropy.

10.5.2 Interface Complexions and Transitions: Thermodynamics

Following the Gibbs adsorption theory discussed above, an accelerated change in the interface energy can take place with the occurrence of a first-order interfacial transition [174, 1002, 1003, 479, 54, 261] with a discontinuous change in adsorption. Figure 10.14 illustrates one example of a first-order grain boundary adsorption transition, which corresponds to a slope change in the γ_{gb} vs. $X_B^{(bulk)}$ curve (Figure 10.14(a)) and an associated finite jump in Γ_B at the same bulk composition (Figure 10.14b)). Figure 10.15 further shows aberration-corrected high-angle annular dark field (HAADF) scanning transmission electron microscopy (STEM) evidence of the possible occurrence of such a first-order transition from a nominally "clean" grain boundary to a bilayer adsorption at an Au-doped Si twist grain boundary [611], where an abrupt change in Γ_{Au} is evident along the grain boundary. Evidence for the occurrence of first-order complexion transitions was also found for a grain boundary in (CuO + SiO_2)-doped TiO_2 [612] and free surfaces of V_2O_5-doped TiO_2 nanoparticles [804, 805].

In the physical metallurgy community, Hart first proposed that grain boundaries can be considered as 2-D interfacial phases in 1968 and this concept is further elaborated by Hondros, Seah, Cahn, Clarke, and many others (interested readers can find the historical perspective and references in two recent review articles [174, 479]). Such 2-D interfacial phases were recently named

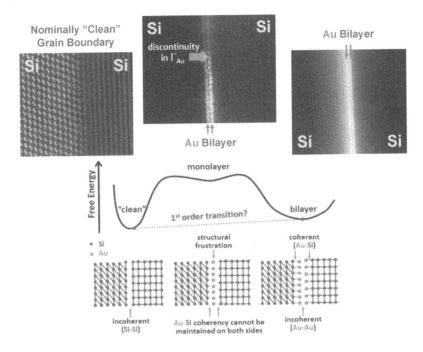

Figure 10.15: HAADF STEM micrographs indicating an abrupt (first-order) transition between the bilayer and the nominally "clean" grain boundary in an Au-doped Si twist grain boundary and associated schematic illustration of bimodal free-energy states that may lead to this first-order transition. Replotted after [611] with permission.

as "complexions" [174, 1002, 1003, 479, 54, 261] because they are not "phases" according to the rigorous Gibbs definition as they have no identifiable volume and cannot exist without abutting bulk phases. Complexions entail nanoscale structural, chemical, and morphological features in the interface region. Complexions are thermodynamically 2-D objects because the spatial variations in the composition and structure in the third dimension perpendicular to the interface are thermodynamically determined; thus, they often have so-called "equilibrium" thickness, even if their composition and structure are intrinsically non-uniform in the through-thickness direction.

As discussed in Section 10.3, an interface may undergo a faceting transition upon changing thermodynamic conditions. The reverse process is known as the roughening transition. In light of the discussion in this section, grain boundary roughening/faceting transitions could be viewed as one type of complexion transition, which was discussed by Cahn [151] in a related context. Grain boundary roughening/faceting occur in many materials, such as Zn, Cu, Nb, Cu-Bi, $SrTiO_3$, Ag, Al_2O_3, and ZrO_2 (specific references are not included here due to the length limit; but interested readers can find references in a recent review [174]). We should also note that the interfacial morphological (roughening and faceting) transitions can often occur in conjunction with atomic-level structural and adsorption transitions discussed below

Perhaps another universal type of complexion transition is the generalized prewetting transitions, which refer to transitions in wetting configuration under conditions where the wetting layer is not yet a stable bulk phase [115]. Phenomenologically, a nanometer-thick ω-like complexion may appear and cover an α–β interface continuously when the ω phase is not yet a stable bulk phase [115], if

$$-\Delta\gamma \equiv \left[(\gamma_{\alpha\omega} + \gamma_{\beta\omega}) - \gamma_{\alpha\beta}^{(0)} \right] > \Delta G_{\text{vol}}^{(\omega)} \times h, \tag{10.21}$$

where $\Delta G_{\text{vol}}^{(\omega)}$ is the volumetric free-energy penalty for forming the unstable (metastable) ω phase

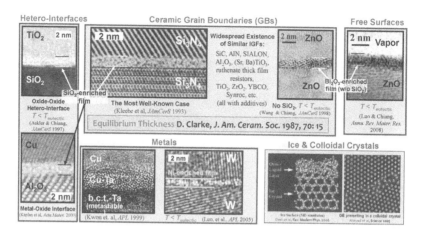

Figure 10.16: Representative nanoscale, impurity-based, quasi-liquid, interfacial films (also called "nanolayer" complexion) as a class of widespread complexions. Figure adapted from [598] with permission.

and h is the effective thickness (interfacial width) of the ω-like complexion. It should be emphasized again that this is not a case of (perfect) wetting despite the continuous coverage of the ω-like complexion at the α–β interface because the ω phase is not yet a stable bulk phase (so it cannot grow into macroscopic thickness). Prewetting adsorption transitions in binary de-mixed liquids were first predicted by the famous critical point wetting theory by Cahn in 1977 [150]. Figure 10.11(a) can be considered as a special case of grain-boundary prewetting occurring in the single-phase region below the bulk solidus line, where the interface is a Mo grain boundary; in this case, prewetting adsorption occurs in conjunction with interfacial disordering (premelting).

Yet another common type of complexion transition is represented by "premelting" or "surface melting" [236], which was rigorously defined for unary systems. To some extent, premelting can be considered as a special case of prewetting, where the (pre)wetting phase is the unstable liquid below the bulk melting temperature. Sometimes, the term "premelting" is used to emphasize interfacial structural transitions (disordering), while the term "prewetting" is used to emphasize chemical transitions (adsorption).

Tang et al. [1002] and Mishin et al. [692] used diffuse-interface (phase-field) models to show that coupled grain boundary premelting and prewetting transitions can occur in binary alloys, leading to the formation of complexions with character similar to those nanoscale, impurity-based, disordered (glass-like or liquid-like) intergranular films (IGFs) that have been widely observed in ceramic materials [598] as well as in some metallic alloys [929, 401, 601] (Figure 10.16). Clarke originally proposed that such IGFs have an "equilibrium" thickness on the order of one nanometer as a balance of attractive and repulsive interfacial interactions acting on the films [209]. Cannon further pointed out that these impurity-based IGFs can be equivalently considered as disordered multilayer adsorbates with average film composition [173]. Similar impurity-based, equilibrium, nanometer-thick IGFs, which were also called "nanolayer" complexions in a most recent overview article [174], were also observed at hetero-phase boundaries (Figure 10.16) [598, 54] and on free surfaces (Figure 10.16) [599]. It is interesting to note that these equilibrium-thickness, liquid-like interfacial complexions can often persist above the bulk solidus lines/curves, coexisting in a thermodynamic equilibrium with non-wetting bulk liquid drops, which is somewhat in contrast to the original literal meanings of the terms "premelting" and "prewetting" discussed above (since the wetting phase is already a stable bulk phase).

Here, a particular confusing point for these nanolayer complexions is that there are two distinct "interfaces" in high-resolution transmission or scanning electron micrographs (e.g., Figure 10.16). However, these "two interfaces" are not independent because of the interfacial interactions acting across the films; consequently, the interface energy for the nanolayer complexions differs from the sum of two independent interfaces, which was demonstrated directly for both IGFs at an Au/Al$_2$O$_3$ interface [54] and Bi$_2$O$_3$-enriched surficial films on the ZnO (11$\bar{2}$0) surfaces [599, 806]. Thus, there is only one interface thermodynamically for the nanolayer complexion, with a thermodynamically determined "equilibrium" thickness (as well as equilibrium through-thickness gradients in compositions and structures).

In 2007, Dillon and Harmer further reported the discovery of a series of six discrete grain boundary complexions in doped Al$_2$O$_3$ (Figure 10.17) [261, 410]. Further studies revealed the existence of this series of Dillon–Harmer complexions in metals (Figure 10.17) [174, 929, 605, 611, 523]. To some extent, this series of Dillon-Harmer complexions can be considered as derivatives of IGFs with discrete thickness of $0, 1, 2, 3, x$, and ∞ atomic layers, respectively (Figure 10.17).

10.5.3   Effects of Complexion Transitions on Microstructure Formation and Evolution: The Kinetic Aspect

Beyond controlling triple junction geometry, interfacial morphology, and wetting configuration *thermodynamically*, the occurrence of interface (complexion) transitions can also impact the microstructural evolution *kinetically* via both (1) varying the thermodynamic driving forces that scale with the interface energies and (2) changing, often drastically, interfacial diffusivity and/or mobility. Here, let us discuss how complexion formation and transition may affect materials processing and microstructural development via changing the interface diffusivity or mobility, with two sets of examples.

First, the formation of liquid-like grain boundary complexions can enhance grain boundary diffusion. The discoveries of premelting-like grain boundary complexions and the accelerated mass transport in such disordered complexions revealed the physical origin of solid-state activated sintering [602, 213]. Moreover, a new type of grain boundary λ diagram has been developed to forecast the thermodynamic tendency for average general grain boundaries to disorder [602, 1207] (see Figure 10.11(d) for an example). Although they are not yet rigorous complexion diagrams (which should have well-defined transition lines and critical points), it has been demonstrated that they can predict useful trends in activated sintering [602, 1207], as well as systematic trends in how grain boundary diffusivity depends on temperature and overall composition [931] and a counterintuitive phenomenon of decreasing grain boundary diffusivity with increasing temperature predicted that was subsequently verified experimentally [930]. Subsequently, such information can help explain the formation of polycrystalline structures during fabrication.

Second, another important area of transport kinetics during materials processing concerns the grain boundary mobility and grain growth. Here, the classic Cahn solute-drag theory suggested that grain boundary adsorption (segregation) should reduce grain boundary mobility and grain growth rates [149]. However, experiments showed that adsorption of certain impurities or dopants can drastically promote grain growth in both metallic systems like Al-Ga [1139] and ceramics like Al$_2$O$_3$-SiO$_2$-Y$_2$O$_3$ [613]; this phenomenon can be explained by the occurrence of complexion transitions since the adsorption can induce interfacial disordering to increase grain boundary mobilities (by overcoming the solute-drag effect). More recently, studies by Dillon and Harmer suggested that adsorption of impurities or dopants can lead to the formation of five different types of impurity-based complexions, in addition to the intrinsic grain boundaries (Figure 10.17); four of them are structurally more disordered and have higher mobilities than the intrinsic (nominally "clean") grain boundaries [261, 410]. Moreover, since grain boundaries have five macroscopic degrees of freedom, the complexion transitions can occur at different temperatures or chemical potentials within the same polycrystalline specimen. Consequently, the coexistence of two (or more) grain boundary complexions with distinctly different mobilities can result in abnormal grain growth, as shown in Figure 10.18 [174, 261, 410].

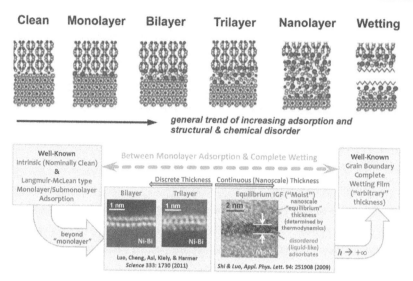

Figure 10.17: Schematic illustration of the origin of a series of Dillon–Harmer complexions. Re-plotted with permission by combing figures in [261, 410].

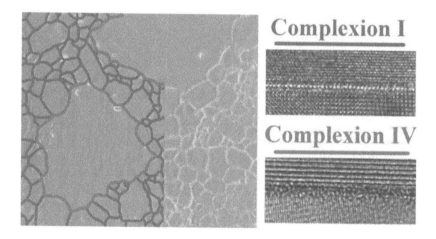

Figure 10.18: Coexistence of two different grain boundary complexions with drastically different mobilities can lead to abnormal grain growth, as shown by Dillon and Harmer; reprinted from [261] with permission.

Finally, we should also point out that the existence and transitions of complexions can affect a variety of other physical properties, including mechanical properties, electrical, thermal and ionic conductivities, and performance of lithium-ion batteries and photocatalysts. Interested readers are referred to two recent comprehensive reviews [174, 479], as well as a few more specialized reviews on the roles of interface complexions in controlling solid electrolytes [603] and liquid metal corrosion/embrittlement [604] and the development of interfacial "phase" (complexion) diagrams as a new materials science tool [602, 599], for further discussion.

10.6 Summary

Interface science for surfaces, grain boundaries, and hetero-phase interfaces provides the underlying physics for the formation of structures and microstructural development. Thermodynamically, interface energies dictate the triple-junction geometry and wetting configuration, and the anisotropy in interface energies produces faceting and controls crystal morphology. Moreover, interface energies, as extensive thermodynamic variables, vary with changes in thermodynamic state variables such as temperature, pressure and chemical potentials, which will subsequently result in changes in various microstructural features. A variety of first-order and continuous faceting/roughening, wetting and complexion transitions may occur at interfaces, leading to accelerated changes in interfacial energy, structure, chemistry, and morphology as well as related microstructural features. Furthermore, interfaces can control microstructural evolution via variation in thermodynamic driving forces and, often more importantly, drastic changes in interfacial transport kinetics (e.g., grain boundary diffusivity and mobility), particularly with the occurrence of first-order interface transitions.

Acknowledgments

MT acknowledges the support from the Department of Energy, Office of Basic Energy Science, Physical Behavior of Materials Program, under contract No. DE-SC0014435 monitored by Dr. Refik Kortan. JL acknowledges the support from several past and current research projects in different relevant scientific and technological areas that were/are supported by the Office of Naval Research (N00014-11-1-0678 for 2011-2016 and N00014-15-1-2863 for 2015-2020), the National Science Foundation (CMMI-1436976 and CMMI-1436305), the Air Force Office of Scientific Research (No. FA9550-10-1-0185 for 2010-2013 and FA9550-14-1-0174 for 2014-2019), the Department of Energy (DE-FG02-08ER46511 for 2008-2012 and DE-FE0011291 for 2013-2016), and a Vannevar Bush Faculty Fellowship sponsored by the Basic Research Office of the Assistant Secretary of Defense for Research and Engineering and funded through the Office of Naval Research (N00014-16-1-2569), from which he has obtained critical knowledge and drawn useful results for writing this review chapter.

Chapter 11

Hierarchical Assembled Structures Based on Nanoparticles: Structure-Property Relations and Advanced Three-Dimensional Characterization

Dhriti Nepal
Air Force Research Laboratory

Sushil R. Kanel
Air Force Institute of Technology

Lawrence F. Drummy
Air Force Research Laboratory

11.1 Fundamentals of Nanostructure Assembly

Gold, silver, or copper based nano-particles (NPs) with sizes smaller than (or comparable to) the wavelength of light show strong dipolar excitations in the form of localized surface plasmon resonances (LSPR) (Figure 11.1(a)). Two major characteristics of LSPR from these plasmonic NPs are a) high enhancement of electric fields near the particle's surface, and b) distinctive display of colors due to its optical absorption, which has a maximum at the plasmon resonant frequency. These features are extremely sensitive to size, shape, composition of these plasmonic NPs, and their local dielectric environment. These unique characteristics are extremely valuable in a wide range of applications including optical energy transport, chemical and biological sensors, surface-enhanced Raman scattering (SERS), near-field scanning optical microscopy, and nanoscale optical devices [359, 1157]. When these plasmonic NPs form organized assemblies or lattices, it offers another degree of freedom in tuning their optical properties as a result of the plasmon coupling (Figure 11.1(b)). The near field coupling between these plasmonic nanostructures could produce an amplified enhancement of the local field as "hot spots" in the junctions. Depending on the gap (< 100 nm) between the particles, configuration of the assembly, or orientation, these coupled plasmons induce novel optical response. This is the foundation of the next generation technology enabling superior device performance and miniaturization of devices. Hence, designing complex and hierarchical nano-particle assembly systems currently in a very high demand in wide research field including electronics, optics, energy and biotechnology.

11.2 Light Scattering and Surface Plasmons

As electromagnetic wave interacts with the discrete particle (e.g., metal NPs), the electron orbits within the particles are perturbed periodically with the same frequency as the electric field of the incident wave. The oscillation or perturbation of the electron cloud results in a periodic separation of charge with the particle, which is commonly termed plasmon. This plasmon as a source of elec-

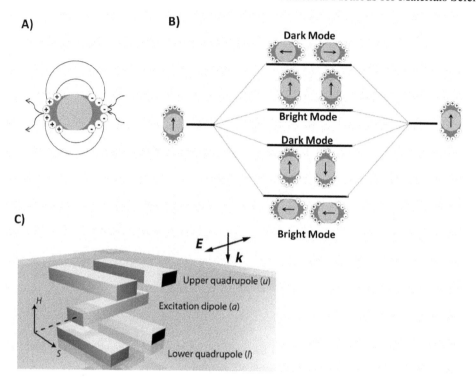

Figure 11.1: A) Light incident onto a metal NP creates charge density oscillations on the surface (left), rerouting of the incident light towards the NP (right). Left box: key formulas determining the resonance behavior of the metal NP. (B) Plasmon hybridization between strongly coupled NPs leads to the formation of dark and bright modes, depending on the relative orientation of the NP dipoles (indicated by arrows). C) Three-dimensional plasmonic ruler, the central dipole antenna excites subradiative quadrupole modes in the two pairs of nanorods. Copyright 2012 American Chemical Society [243].

tromagnetic radiation results in light scattering, and it is the fundamental for color of metal NPs. Application of the light absorption/scattering by NPs to induce dichroic effect, i.e., shifting colors depending on the angle of view, dates back to 4th century. For example, dichroic effects were applied in "The Lycurgus Cup (Romans)" or old church windows by embedding with gold and silver-based NPs. However, physical understanding of light scattering by small particles was introduced by Lord Raleigh in 1871.

The breakthrough in understanding light scattering by spherical structures came from the work of Mie in 1908. On the basis of electromagnetic theory, Mie obtained a general rigorous solution for the optical scattering by a homogeneous sphere with arbitrary size in a homogeneous medium. This theory is still valid as long as the distance between particles is large enough so that there are no coherent phase relations among the scattered light from different particles [113, 313]. To accommodate other shaped particles, especially for the spheroids, this theory modified by Richard Gans in 1912. In Gans' theory, the aspect ratio of the particles is the predominant factor to calculate the absorption. As a result, it is widely used to get the exact solution for gold/silver nanorods (NRs). Discrete dipole approximation (DDA) method is one of the most popular alternatives to model the exact shape of the particles without employing physical approximations. DDA is based on an approximation of the continuum target by a finite array of polarizable points, where the points acquire dipole moments in response to the local electric field.

Nevertheless, 100 years of history Mie theory, research in exciting physics associated with the light scattering of NPs is still a very active research area today. One of the challenging areas includes incorporating higher-order scattering modes such as quadrupole and octupole in addition to conventional electric dipole concept (Figure 11.1(c)) [243, 1157]. Nevertheless, new phenomena are being explored including fano resonances, electromagnetically induced transparency, negative dielectric susceptibility, unique scattering pattern with complicated near field structures, and unusual frequency dependencies [243, 759].

11.2.1 Plasmon Coupling

The LSPR of plasmonic NP are distinctly different when they are densely packed compared to that of isolated system as a result of electronically coupled plasmon resonance (plasmon coupling). Depending on the geometry of the configuration and inherent characteristic of the particles, complicated extinction spectra can be obtained (Figures 11.2 and 11.3) [103, 101, 102, 759]. Here the inter-particle coupling gets stronger than the coupling within the surrounding medium. As a result, Mie theory developed for isolated particles fails to describe the optical absorption spectrum on such system. Maxwell Garnett effective-medium theories have been successfully applied to this kind of assembled system. This theory is strictly valid in the quasistatic limit $(2R, i)$ along with very small interparticle distances but can be generalized to various shapes of the particles. Here, the optical properties of the small particles are determined by plasmon hybridization as a result of both contributions, isolated individuals as well as the whole ensemble (Figure 11.1). An explicit knowledge of the statistical variation of the positions and pairwise distances of all particles necessary to predict the influence of inter-particle coupling.

Plasmon hybridization in sub-wavelength systems result in the combination of strongly radiative bright modes (super-radiant) and spectrally sharp dark plasmons (subradiant) (11.1(b)). Bright modes result from dipole-active character, and they efficiently decay in the far field. On the dark modes are quadrupolar and higher-order eigenmodes and they cannot interact classically with light.

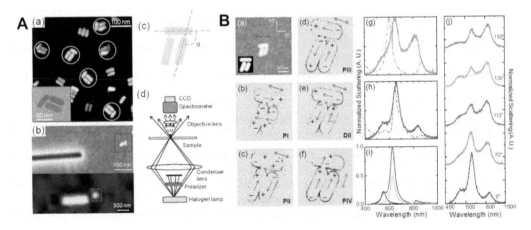

Figure 11.2: A) SEM characterization of dolmens or π structures of gold nanorods, with the help of a fiducial marker, each particle tracked both on SEM and dark field microscope, which enabled extracting individual spectra of each of the assembled particles. (B) Two major polarization directions parallel to the long axis of cap and dimer base, and corresponding FDTD charge density profiles (b-f). The results were compared (g,h) with experimental data (solid) and FDTD modeling (dashed). (i) Analytical scattering profile for the polarization direction, and (j) evolution of the dark-field scattering spectra at different polarization directions (0150). ([101], Copyright American Chemical Society 2013).

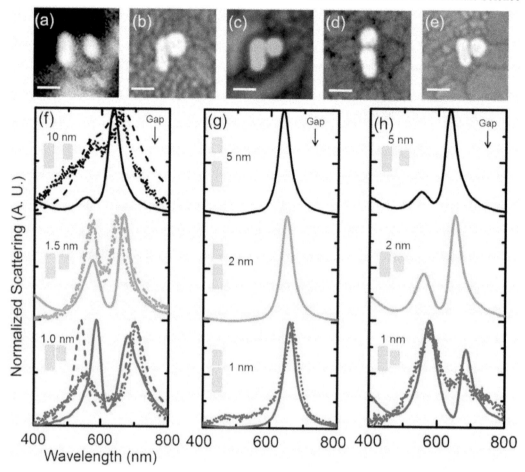

Figure 11.3: Influence of heterodimer geometry on scattering, (ae) SEM images (scale bars 50 nm). (f) to (h) are individual spectra (inset shows the corresponding structure) obtained from experiments and modeling (dotted: expt., solid: FDTD, dashed: analytical). ([102], Copyright American Chemical Society 2013.)

Around the spectral position of a dark mode, there will be an abrupt suppression of the band, resulting in an asymmetric line-shape, which is termed fanoresonance [759].

Recently, several research groups have demonstrated the influence of plasmon hybridization based on NP assembly. For example, Nordlander and co-workers have analyzed the plasmon coupling of a dimer, in which dipoles are oriented parallel or orthogonally to the inter-axes between the NP. The key aspect of this study is variation on the plasmon energies based on dimer orientations and a function of NPs inter-distance. Depending on the orientation of dipoles two hybridized configurations of bright and dark modes are shown. As the inter-particle distance decreases, the bonding and anti-bonding dimer plasmons begin to split asymmetrically (Figure 11.1(b)).

Recently, the numerical and computational models are further verified by controlled experiments using single-particle spectroscopy. For example, Vaia et al.[102] have successfully demonstrated plasmon coupling and resulting fanoresonance in wide range of assembled structure based on colloidal AuNR (Figure 11.2). As the architecture of the assembly becomes complex, more complicated spectroscopy signature can be achieved. The spectra achieved from two AuNRs are distinctly different based on the assembled orientation and spacing. Examples include spectra of AuNR dolmen structures (π like structure) (Figure 11.2) and AuNR pair (Figure 11.3) based on its assembly and the spacing. Because of the plasmon coupling, multiple radiative (bright) and confining (dark or subradiant) modes are seen (Figure 11.2). Subradiant or dark modes in plasmonic systems are especially interesting due to their inherent small line-width, high-quality factor (Q) and reduced radiative losses, which have special significance for nanolasing, waveguiding, and sensing. This is possible by bottom-up fabrication assembly where precision on gap (0.5-2.0 nm) is essential for such coupling. With the scattering data obtained by using single particle spectroscopy [101], the estimations of the scattering spectra, charge distributions and insight to mode coupling were done by analytical formulation as well as finite difference time domain (FDTD) modeling.

11.2.2 Plasmon-Exciton Coupling

Exciton is an electrically neutral quasiparticle, an existence of electron and hole bound together by electrostatic Coulomb force. Typically it exists in semiconductors and insulators. The semiconductor nanocrystals (SNC) have long-lived excitations and higher emission yield [101, 1179]. For the plasmonic nanostructures, LSPR enables the concentration of electromagnetic energy and enhance optical field and nonlinearities. If a plasmonic NP is present in the vicinity of such SNC, it can strongly influence oscillator strength of excitons, which can create either weak or strong coupling [945, 3, 745, 396]. That is called plasmon-exciton coupling, where the strength of this interaction determines both the absorption and emission properties of the hybrid system and opens up new avenue for light manipulation. Depending on the choice of the material and the architecture of the hybrid structure, it can redistribute the optical electric field, alter the de-excitation pathways, change the emission polarization, redirect the emission intensity [370], enhance the emission [231], quench the emission [690], redirect the emission intensity [690], and modulate the emission spectral profile [231, 1202, 546].

The strength of the coupling is influenced by multiple factors and it is usually realized in two ways [45] Figure 11.4. First the excitation rate can be increased by the local electric field enhancement resulting from the excitation of the localized plasmon resonance. The SNC material situated in the local field enhancement regions essentially feels an enhanced excitation light intensity. Second, in the vicinity of the plasmonic nanocrystals, the SNC material at the excited state can undergo nonradiative decay, emit a photon directly to the far field, or relax rapidly by exciting the localized plasmon resonance of the plasmonic nanocrystals via energy transfer. Once the localized plasmon is excited, it can either decay non-radiatively owing to internal damping or re-radiate into the far field. The nonradiative decay and radiative decay of the plasmonic nanocrystals are basically determined by its absorption and scattering cross-sections at the emission wavelength respectively. As a result fluorescent emission intensity is jointly affected by both the excitation enhancement and emission modification.

$$I = \gamma_{ex} \left(\frac{\gamma_{f,r}}{\gamma_{f,r} + \gamma_{f,nr} + \gamma_{f,ET}} + \frac{\gamma_{f,ET}}{\gamma_{f,r} + \gamma_{f,nr} + \gamma_{f,ET}} \frac{C_{scatt}}{C_{abs} + C_{scatt}} \right) \varepsilon_{coll}. \qquad (11.1)$$

The enhancement of local electric field is higher in anisotropic plasmonic NPs such as AuNR than that of spherical particles. The electric field intensity enhancement occurs mostly in the regions close to the two ends of AuNR. In these regions, the electric field intensity is significantly enhanced (> 1 order of magnitude). These regions will be very effective for enhancing the excitation rate of fluorophore molecules. In general, the field enhancement decays nearly exponentially away from the metal surface. The sizes of the local field enhancement region and the field enhancement factors vary with the geometry and size of plasmonic nanocrystals. For a given nanocrystal they also change as a function of the wavelength at which the localized plasmon is excited [45].

Figure 11.4 shows some of the key parameters to control the plasmon-exciton coupling especially taking example of anisotropic plasmonic NPs. For effective plasmon-exciton coupling precise control of not only the positioning of the emitter at high field enhancement region (rod ends) but also its location and orientation is very important. A plasmonic nanostructure acts as an analogue

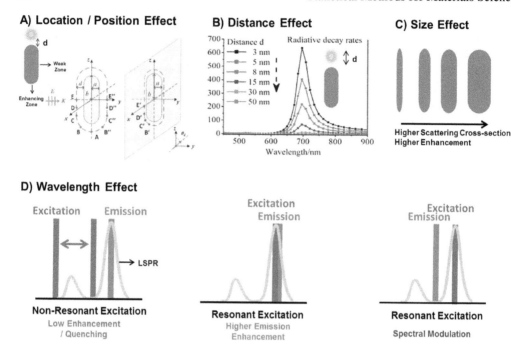

Figure 11.4: Plasmon-exciton coupling enabled modulation of emission of emitter close to anisotropic gold nanorods. A) Schematics showing effect of position/location /orientation of emitter on gold nanorods ([575] Copyright Springer 2012). B) A theoretically simulated plot showing effect of distance between an emitter and gold nanorods on radiative decay. C) A cartoon showing four different AuNR of same length but increasing diameter, which corresponds to increasing volume. Higher enhancements are expected for higher volume AuNR due to higher scattering cross-section. D) A schematic showing effect of wavelength for emission spectral modification.

of an antenna working at the optical frequency. Liaw et al. [575] systematically studied the effect of location, orientation of the emitter close to AuNR surface using semi-analytical method (multiple multipole). A schematic on positioning of emitter at different location of AuNR, which means one has to take account of its exact location as well as it orientation (Figure 11.3A). Precisely positioning emitter at the tip of AuNR (not the sides) is very crucial to achieve high fluorescence enhancement because AuNR ends (highest electric field region) are the highest enhancing zone and AuNR sides are the weakest zones. It is clear by now that precise gap control is very critical it is especially very important and experimentally challenging to fabricate anisotropic plasmonic structure with emitter at its tip with precisely controlled gap. To understand its effect, Liaw et al. simulated a plot (multiple multipole) which clearly shows the effect of distance between an emitter and AuNR on radiative decay [575] red (Figure 11.4B). The highest decay was observed when the emitter is 3 nm. Larger plasmonic nanostructures with higher scattering cross-section are beneficial for plasmon enhance fluorescence. A cartoon shows four different AuNR of same length but increasing diameter, which corresponds to increasing volume (Figure 11.4C). Higher enhancements are expected for higher volume AuNR due to higher scattering cross-section. Similarly, irrespective of the three dimensional structure of the hybrid material, the wavelength used for excitation has much stronger impact. If lower excitation wavelength is used than LSPR of AuNR, i.e., a condition of non-resonant excitation, is less likely that plasmon will be excited. Under such situation very low fluorescence enhancement or quenching of fluorescence observed. If the excitation is used with the same energy of LSPR, i.e. a condition of resonant excitation, plasmons with strongly localized enhanced fields can excite the emitter. Hence very high emission enhancement is expected. On a third example, if emission wavelength is lower than the LSPR of AuNR and wavelength of excitation is similar to LSPR of AuNR, this is still a resonant excitation but modulation of the emission spectra expected. Under such situations new peaks close to the LSPR of AuNR are expected to be seen.

11.2.3 Assembly of NPs, Thermodynamics, DLVO Theory, and Extended DLVO

Self and directed assemblies of NPs are a spontaneous assembly process. It is driven by specific interactions such as van der Waals, hydrogen-bonding, columbic interactions, etc. Based on thermodynamics theory, it is an equilibrium process driven by the minimization of Gibbs free energy where the assembled components are in equilibrium with the individual components [575]. Generally, the lower free energy is a result of a weaker intermolecular force between self-assembled moieties and is essentially enthalpic in nature. The organization is accompanied by a decrease in entropy and in order for the assembly to be spontaneous the enthalpy term must be negative and in excess of the entrophy term [753]. Key highlights of this theory include consideration of Lewis acid-base and Lifshift-van der Waals force. However, it may not be valid in the system as it assumes reversible reaction, distance independence, and does not include the effect of surface charge of the surrounding media.

The second commonly used theory is Derjaguin, Landau, Verwey, and Overbeek (DLVO), which combines the attractive Lifshitz–van der Waals (VDW) interaction and the electrostatic double-layer (EDL) interaction to model interactions in aqueous colloidal suspensions and the respective aggregation rates [753]. Here, too, van der Waals forces are almost always present, and they result from interactions of the rotating or fluctuating dipoles of atoms and molecules. While van der Waals forces are normally attractive, the double layer forces can be attractive, repulsive, or both (Figure 11.5). Similarly, aggregation or assembly of particles could be homoaggregation, which refers to aggregation of two similar particles that are successfully correlated with DLVO theory. The assembly could also be heteroaggregation, which refers to particles with different shapes and size. Classical DLVO may no longer valid due to unique NP shapes and compositions. Consequently, some other limitation of this theory is oversimplification of the resulting interactions based on only van der Waals and electrostatics. However, there are other major interactions which might play a critical role which includes acid-base, steric, magnetic, and osmotic forces, etc. Recently, other interactions are being incorporated as an extended DLVO (XDLVO) (Table 11.1). Recently, Hotze et al. summarized some of the interaction forces of nanoparticles in terms of its aggregation and its implication of its transport and reactivity [441].

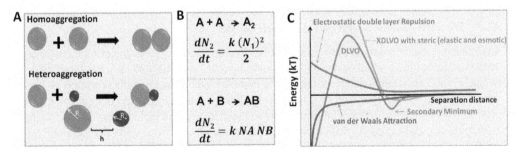

Figure 11.5: Schematic showing homoaggregation and heteroaggregation of particles where rate of dimer formation is given by the rate law van der Waals force, electrostatic double layer (EDL) force, total Derjaguin–Landau–Verwey–Overbeak (DLVO) forces, extended DLVO (XDLVO), and elastic force plotted together to find total potential as a function of separation distance.

Table 11.1: Summary of key interacting forces in nanoparticles aggregation: thermodynamics, DLVO and X-DLVO.

Force	Model	Equations[1]	Refs.
Thermodynamic	Second Law	$\Delta G_{SA} = \Delta H_{AS} - T_{SA}$	[441]
Repulsive and attractive force – London dispersion forces	Lennard–Jones model	$V_{LJ} = 4\varepsilon \left[\left(\frac{\sigma}{r}\right)^{12} - \left(\frac{\sigma}{r}\right)^6 \right]$ $= \varepsilon \left[\left(\frac{r_m}{r}\right)^{12} - 2\left(\frac{r_m}{r}\right)^6 \right];$ $\Phi = -4\varepsilon \left(\frac{\sigma}{\gamma}\right)^6$	[198, 468]
Net electrostatic effect	Debye length	$\kappa^{-1} = \left(\frac{k_B T \varepsilon \varepsilon_0}{2N_A I q^2}\right)^{1/2} = \frac{0.3\,\mathrm{nm}}{\sqrt{I}}$	[404]
van der Waals attraction	DLVO	$\frac{V_{VDW}}{\kappa T} = -\frac{A}{6\kappa T}\left(\frac{2a^2}{s(4a+s)} + \frac{2a^2}{2a+s} + \ln\frac{s(4a+s)}{(2a+s)^2}\right)$	[112]
Electrostatic double layer	DLVO	$\frac{V_{EDL}}{\kappa T} = \frac{64\pi\eta\kappa T}{\kappa}\frac{(a+\delta)^2}{2a+s}\tanh^2\left(\frac{ze\Psi_d}{4\kappa T}\right)e^{-\kappa(s-2\delta)}$	[984]
Magnetic attraction	XDLVO	$V_M = -\frac{8\pi\mu_0 M_s^2 a^3}{9(\frac{s}{a}+2)^3}$	[255]
XDLVO	Hydrophobic (Lewis acid-base)	$\frac{V_{AB}}{\text{surface area}} = \Delta G_{S_0}^{AB}\exp\frac{S_0-s}{\lambda}$	[782, 434]

[1][in order of appearance] ΔG is the Gibbs free energy; ΔH the total enthalpy; T absolute temperature; Φ L-J potential; ε depth of the potential well, σ finite distance at which the inter-particle potential is zero, r distance between the particles; DLVO: Derjaguin-Landau-Verwey-Overbeak; XDLVO: extended DLVO; A Hamaker constant; a average particle radius; d thickness of "brush" polymer layer ($\approx \delta$); δ Stern layer thickness; ε relative permittivity of water (= 78.5 C/Vm); ε_0 permittivity of a vacuum (= 8.854×10^{-12} F/m); e elementary charge of an electron; κ inverse Debye length; k Boltzmann constant; λ decay length for acidbase interactions (0.6 nm); M_s saturation magnetization; μ_0 vacuum permeability; MW molecular weight of polyelectrolyte; n number concentration of ion pairs; ρ_p density of polymer; Ψ_d diffuse potential ($\approx$ zeta potential); s distance between interacting surfaces; s_0: minimum equilibrium separation distance (= 0.158 nm); κ^{-1} Debye screening length, I Ionic strength of the electrolyte solution $I = 1/2\sum_{k=0}^n z^k a^{n-k}$; q elementary charge; F force acting between two particles; R_{eff} the effective radius; W the interaction free energy per unit surface area in the plate-plate geometry.

11.3 Directed Assembly and Self-Assembly of NPs

Complex and structured nano-particle assemblies systems are a key to engineering innovative optical response due to their novel linear and nonlinear optical properties. There is a very high demand scalable fabrication approaches to create the complex architectures with precision at nanoscale; nevertheless, it is very challenging. Most approaches for fabricating these complex nanostructures rely on top-down approaches. The limitations of these techniques include very low throughput and very high cost, and extremely difficult to reach nanoscale precision due to the fundamental limitation (diffraction limited). Alternatively, self and directed assembly of NPs have been very promising for building functional nanomaterials and nanostructures, as they provide the ultimate solution for nanoscale precision architecture with the added benefit of cost and throughput. As a result, several complex architectures based on spherical NPs are demonstrated in solution as well on the surfaces. The current challenges include expanding it to other anisotropic building blocks, and controlling their position, spacing and orientation at nanoscale precision (< 5 nm), and achieve the assembly across the multiple length scale. The key requirement to achieve this includes designing building blocks, understanding physics governing directed and self-assembly, and implementation to direct the assemblies precisely at various length scales. Significant progress has been made over the years for the assembly of NPs (wet chemically synthesized) through various assembly techniques, which can be broadly divided into two categories, a) assembly in solution, and b) assembly on surface. We will discuss recent progress made in each of these categories and discuss each of their limitation and future prospect.

11.3.1 Assemblies in Solution

Directed and self-assembly of NPs in solution is one of the most efficiently used techniques to spontaneously connect and integrate NPs using direct (covalent), and/or indirect interactions (secondary interactions, such as van der Wals interactions, electrostatic interactions, molecular surface forces, entropic effects, etc.) (Figure 11.5). In contrast to conventionally used lithographic techniques, this technique use colloidal particles which offer tremendous benefits including: a) single crystals and atomically smooth surfaces, which are essential to improve the optical performance, and b) continuous fabrication needed for large-scale production. Depending on the surface functionality present in the NPs system and type of solvent used, typically self-limiting growth process is governed by a balance between electrostatic repulsion and van der Waals attraction. The flexibility of the interactions opens up possibility for wide range of the composition, size and shape of the constituent NPs, and leads to a large family of self-assembled structures over macroscopic and microscopic scales. The architectures may have defined size, molecular weight, and stoichiometric ratio of different components [1160]. This technique has been utilized for creating Janus particles [1191] as well as nanoparticle cluster architectures based on DNA origami frames [1194] Similarly, it has been applied for series of AuNR-based architectures [1042, 1120, 526, 767] and has tremendous promise to create precisely controlled three-dimensional structures and geometries in scalable quantities if reproducibility and yield problems can be overcome.

Directed assembly of spheroid particles such as nanorods offers added advantages due to its anisotropy relative to that of spherical particles for creating hierarchical structure. Density-driven colloidal assembly, such as by solvent evaporation, produces some intriguing structures, e.g., particle chains; however, controllability and post-processibility of the final architecture is limited. Also there is still a challenge to produce homogeneous products in terms of assembly size and type. Recently high-yield formation of soluble, stable, and compositionally discrete AuNR architectures was developed by directed assembly approach in solution. It is accomplished by tailoring the surface chemistry of AuNR anisotropically such that their directed assembly is well controlled by reversible modulation of a solvent quality. Analogous to dimer formation during step-growth polymerization, the initial yield of AuNR side-by-side pairs were achieved, which can be greater than 50% (Figure 11.6). With further processing steps including centrifugation as well as silica encapsulation, the dimer yield $\approx 90\%$ achieved [732].

As illustrated earlier, self and directed assembly of heterostructures are also of strong interest. Here we discuss different design strategy to induce coupling of exciton-plasmon using colloidal nanostructures. For demonstrating some proof-of-concept, hybrid structures were also generated by electron beam lithography (e-beam) approach. E-beam provides routes to complex two-dimension structures, but variability of gap size, surface roughness, and defect content are too high to systematically explore the strongly coupled regime. Additionally, using this approach it is not possible to control precise spacing (< 20 nm) and relative orientation of the molecules to the nanocrystals, which are essential for improving the performance. Since coupling depends very strongly on separation, control of the local structure beyond placing the plasmonic particle in a medium of emitters is crucial to optimize efficiency by deconvoluting the different phenomenon that dominate at small and large separation. These promises have attracted scientists of different disciplines including chemists, materials scientists, physicists and optical engineers. Hence recently the colloidal route to fabricate the hybrid structures has shown significant progress both in terms of materials fabrication and its understanding on the exciton-plasmon interactions. Since the field is rapidly growing, and it is not

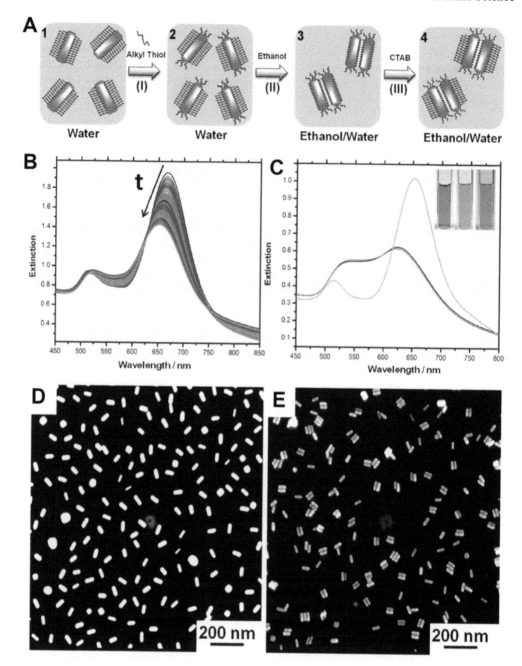

Figure 11.6: A) Schematic diagram showing formation of nanorod pair. Extinction spectra of B) kinetics of side-by-side assembly; C) individually isolated NRs in aqueous solution, and CTAB-quenched products. Inset shows vials containing isolated NR solution, CTAB-quenched solutions. Scanning electron microscopy (SEM) images of individually isolated NRs (D) and NR pairs (E) [732].

possible to integrate all of the literature, we have summarized some of the key findings specially focusing on gold nanostructures based exciton-plasmon coupling. First, we will discuss the synthesis

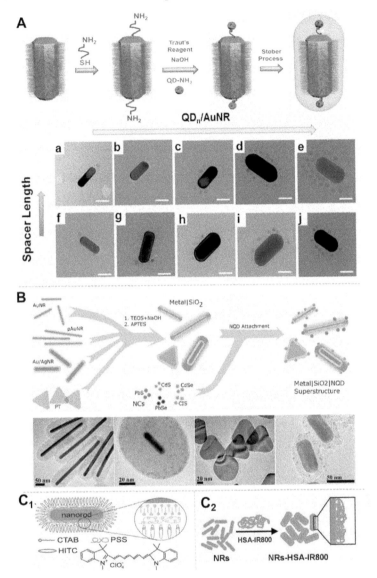

Figure 11.7: Different strategies used for functionalizing anisotropic plasmonic nanostructures (e.g., nanorods) with emitter in solution. A) Organic/polymeric molecule controlled spacing between QD and AuNR the specific chemistry of these organic molecules attached selectively at AuNR ends allows control over position, spacing as well and number of QD attached per AuNR [730]. B) Silica encapsulation using Stober process which enabled controlling spacing between QD and varieties of plasmonic nanostructures [490]. C) Layer-by-layer approach for entrapment of organic dyes close to AuNR [1130, 61].

strategy to fabricate hybrid structure specially focusing on a solution-based approach, and finally showing properties like PL enhancement, quenching, discussing challenges, and the final prospect.

Solution assembly of spherical units via layer-by-layer deposition on particles [61, 1130, 519], direct chemical coupling using molecular/biological linkers [1105, 397] or sequential silica encapsulation [215, 490, 572], have shown impressive precision if the surface functionalization and colloidal stability throughout the assembly process can be optimized (Figure 11.7). Similar assembly with anisotropic NPs, however, still remains a challenge, since location and orientation of the flu-

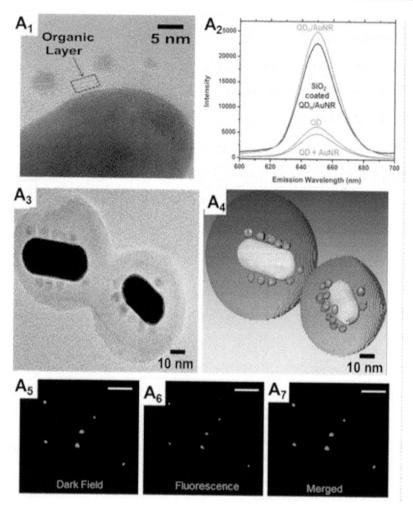

Figure 11.8: Some examples of plasmon enhanced emission obtained in varieties of architectures in solution and film. A) Control over precision on gap and position control of QD on AuNR surface allowed 5 times higher PL emission in solution (A2) than that of control. A1, A3 (TEM), and A4 (STEM-tomography-reconstruction) showing the structure and A5, A6, and A7 are dark field scattering, fluorescence, and merged images. [731] Copyright American Chemical Society 2013.

orophore in addition to gap size must be controlled while avoiding instability of the intermediates during the ligand exchange and assembly. Therefore, additional methods are required to produce anisotropic emitter-plasmon nanostructures so as to forward the understanding of spacing, location and dipole orientation on coupled exciton-plasmon photo-physics [396, 841, 575, 282]. One critical factor limiting the high-yield low-variability assembly is systematic control of the various crystal facets of the initial nanoparticle, such as AuNR. While there has been substantial progress over the last decade in developing recipes for different size of AuNRs [594, 744], only recently has the underlying mechanism of growth been sufficiently understood to provide sufficient control of AuNR shape and surface structure to enable selection of facet reactivity for subsequent assembly [767, 1175].

A key example is demonstrated by Nepal et al. [766] (Figure 11.8 panel A_2) where 5-fold enhancement was observed when colloidal AuNR was selectively functionalized with QD at the AuNR tips. A chemical control directs the assembly of quantum dots (emitter) on gold nanorods (plasmonic units) with precisely controlled spacing. Ratio of quantum dot/nanorod ratio can be controlled based on the reaction conditions. The product had a long-term colloidal stability, which enables the purification and encapsulation of the assembled architecture in a protective silica shell. Overall, such controllability with nanometer precision allows one to synthesize stable, complex architectures at large volume in a rational and controllable manner. Similarly, a notable example of PEF from large NPs is demonstrated using Au nanobowtie [794]. By coating organic dye (TPQDI) which has a low intrinsic quantum yield of 0.07 1300-fold enhancement of the fluorescence was observed. This is the largest fluorescence intensity enhancement factor among those measured so far.

11.3.2   Assemblies on Surfaces

Two- or three-dimensional (2D or 3D) complex architectures based on NP building block assembled on a surface is currently in high demand, as it is closer to the device architectures provided by conventional lithography approach in semiconducting industry. These structures are regarded as the ideal assembly configurations for many technological devices including magnetic memory, switching devices, and sensing devices, due to their unique transport phenomena and the cooperative properties of NPs in assemblies [495]. Consequently, self and directed assembly of NPs on surfaces has greater advantage of improving the throughput, and hence the fidelity of these devices. Despite the strong promise, some challenges exist with the processing in the directed assembly process. This includes high precision and high levels of integration of NPs in desired locations, and ordering and alignment of the assembled structure over large areas.

Taking advantage of existing nanofabrication techniques, with the combination of self-assembly strategies has recently raised tremendous interest. As the synergetic approach, the combination could potentially provide effective and distinctive solutions for fabricating NPs with precise position control and high resolution where the prestructures provide platform for the assembly on the large area surfaces. To take advantage of this approach to achieve spatially selective assembly of NPs at the desired location, various chemical routes are explored. Controlled dewetting process, electrostatically mediated assembly of particles, entropy-driven assembly on confined surfaces are some of the examples. In general, the surface pattern is prepared by various routes including the Langmuir/Blodgett (LB) patterning technique, e-beam lithography (EBL), and nanoimprint lithography (NPL). For successful assembly of NP on the pattern surface, quality of solvent, surface functionality, temperature and concentration are the limiting factors. Where the control over the particle density, particle size, or inter-particle distance in NP assemblies is depends on the geometric parameters of the template structure due to spatial confinement as well as the chemistry of the particles and the surfaces. Complexity on the final pattern can be achieved by incorporating complex design template structures, and good control on surface chemistry of multiplexed NPs is needed such that the structural and functional complexity be achieved (Figure 11.9) [462, 463].

A polymer-based chemical pattern based on e-beam lithography and directed assembly is one of the well-established techniques to create chemical pattern with sub 100 nm feature sizes. Recently, this technique utilized to control surface architecture based AuNR such that control over orientation and spacing between AuNR were achieved on a wafer scale substrate [730]. This is achieved by controlling surface chemistry of AuNR in solution such that it had specific interaction with one of the polymer patterned surface (Figure 11.10). AuNR can be tuned by controlling the Debye length of AuNR in solution and the dimensions of a chemical contrast pattern. Electrostatic and hydrophobic selectivity for AuNR to absorb to patterned regions of poly(2-vinylpyridine) (P2VP) and polystyrene brushes and mats was demonstrated for AuNR functionalized with mercaptopropane sulfonate (MS) and poly(ethylene glycol), respectively. For P2VP patterns of stripes with widths comparable to the length of the AuNR, single- and double-column arrangements of AuNR oriented parallel and perpendicular to the P2VP line were obtained for MS-AuNR (Figure 11.10) [730].

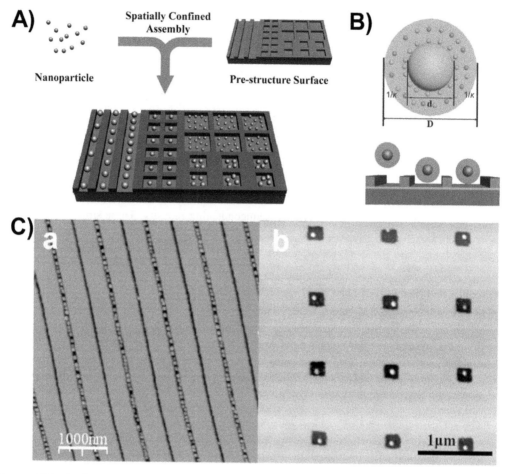

Figure 11.9: A) Schematics showing confined assembly of NPs on pre-patterned grooves [462]. B) Deposition is influenced by effective diameter of a NP (rigid core and an electric double layer) [462]. C) Atomic force microscopy (AFM) images showing successful deposition of NP. Copyright 2009 Wiley-VCH, 2014 American Chemical Society [463].

11.4   Complex Architectures for Metamaterials

In recent years, metamaterials (MM) have shown immense potential to have huge impact on the photonics industry by enabling unique capabilities including super-resolution imaging [892] and optical information processing [104]. Despite the remarkable success shown over the last decade, MMs suffer from drawbacks due to design and fabrication perspectives, limiting their applicability. Their complex architecture requires highly advanced fabrication methods which is predominantly done by lithographic approach, which makes it impractical for application due to the cost and throughput. Thus the current challenge is to develop inexpensive and easy methods to realize three-dimensional MMs at large scale. Self and directed assembly of NPs in solution and surfaces are considered as prime alternatives to create these building blocks. Though there are significant challenges ahead, extensive report on the bottom-up self-assembly approach focusing on negative index properties of MMs of some building blocks has been studied [991]. Consequently, novel properties are being explored either by adding complexity on the architectures or engineering the plasmon with some other property. One of the recent example include active magneto-plasmonic ruler [709],[1211] which is achieved by combining nanoplasmonics and nanomagnetism to conceptualize a magnetoplasmonic

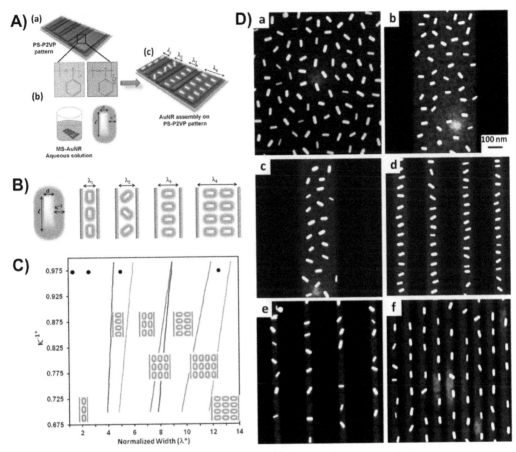

Figure 11.10: A) Schematic of the assembly of gold nanorods (AuNR) on chemically nanopatterned surfaces. Surface selectivity is based on electrostatic interaction. B) Conceptual diagram showing the relationship between size of a particle including Debye length and its packing on a confined surface. C) Preferred packing structure based on surface coverage with regard Debye length (κ^{-1}) and pattern width (λ^*). D) SEM images showing particle ordering within directed assemblies of MS-AuNR on P2VPPS contrast pattern created with conditions supporting long-range particleparticle repulsion [729].

dimer nanoantenna that would be able to report nanoscale distances while optimizing its own spatial orientation.

Recently, by coupling two L-shaped gold nano chiral array structures, optical rotation reversal in the optical response was observed which was demonstrated by inversion in the sign of the CD spectrum (Figure 11.11) [425]. Compare to organic molecule based chiral systems, the plasmonic system exhibits greater control over the geometry and resulting spectrum, and promises applications in switchable nanoscale metamaterials.

11.5   Future Opportunities for Advanced Characterization Methods

11.5.1   Tomography

It is apparent from the complexity in metamaterial architectures that spatial characterization in three dimensions is critical to provide an understanding of the physical mechanisms of structure and plasmonic response. Electron tomography has emerged as a powerful tool for nanoscale character-

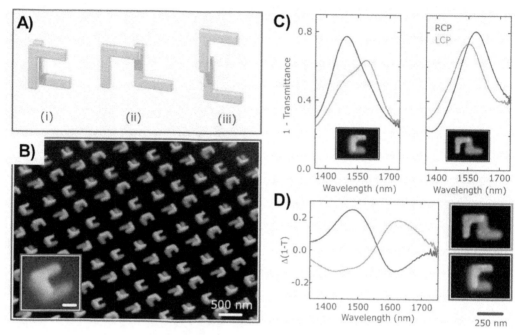

Figure 11.11: A) Schematic of chiral plasmonic system, and (B) SEM image. C) and D) are experimental results showing individual transmission and CD spectra of each of the chiral structure, insets show the corresponding SEM images [425].

ization in 3D, and in the semiconductor industry the technique has been integrated into the R&D process and aspects of the quality control workflow in a semi-automated fashion [730],[1096]. More broadly, sample preparation remains challenging, sample and sample holder geometries often impose a missing wedge of data, and reconstruction of the raw projection data into meaningful 3D data is non-trivial, although often overlooked in practice. In fact, recent advances in reconstruction algorithms are having impact on both sample preparation and geometry limitations. It is less critical to fabricate a perfect sample (exact requirements vary by material, imaging system, and required data) through a time-consuming trial and error fabrication process, if the reconstruction is more robust toward non-optimal data. An example of this is the ubiquitous missing wedge problem in TEM tomography. Although it has been widely accepted that a missing wedge in excess of $(m)70°$ gave significant blurring in the Z direction in reconstructions [142], it has recently been hypothesized that this only holds true for reconstruction methods with Nyquist sampling requirements, such as filtered back projection (FBP) or simultaneous iterative reconstruction technique (SIRT). For other algorithms such as MBIR or other total variation based methods, it has been shown that this empirical relationship between missing wedge and Z-axis blurring does not hold [680], although a new empirical or theoretical relationship has not emerged. For electron tomography to accelerate its growth as a more widespread use there are a few challenges. The robustness of reconstruction algorithms to different types of data is critical. Similarly, non-optimal data (noisy, highly attenuated, diffraction contrast, large missing wedge) is equally important. The resulting reconstructions need to give easily interpreted results with accurate and physically meaningful information. Instead of a 3D volume of arbitrarily scaled grayscale values, we should have calibrated values of atomic number, density, chemical information, or local field intensity.

11.5.2 Scanning Transmission Electron Microscopy/ Electron Energy Loss Spectroscopy (STEM/EELS)

In recent years STEM/EELS has emerged as a powerful characterization technique for detailed spatial mapping of plasmonic materials, especially as electron monochromaters technology has become more widespread. Today's monochromaters routinely reduce the spread of electron energies, or the energy resolution, to less than 300 meV. At these energy resolutions it is possible to resolve plasmonic resonances in the optical frequency regime. With newer instrumentation and improved stability energy resolution has recently been improved to below 10 meV, where vibrational spectroscopy is possible [1096]. While advances in hardware have been critical to achieving these breakthroughs in energy resolution, effective analysis of the large amounts generated requires new methods for data analytics. Many studies have made use of factor analysis, principal component analysis, and non-negative matrix factorization, due to the complex and non-intuitive relationships between data factors, and the large size of the data streams generated. While a typical raster scanned STEM image contains transmitted or scattered intensity information as a function of two-dimensional incident electron beam position, STEM/EELS contains a full EELS spectrum at each position, producing a 3D dataset. Furthermore, when combined with simultaneous acquisition of other spectroscopy techniques such as energy dispersive X-ray spectroscopy and when a tomographic tilt series is acquired for 3D volumetric reconstruction, the data can easily reach 4 or 5 dimensions. Reconstruction of such multimodal tomographic data is a significant challenge and is discussed in detail in the chapter in this book by Venkatakrishnan et al. [1096].

11.5.3 In-Situ Characterization

For a truly accelerated materials designframework to become a reality, real-time high-resolution characterization of growth, assembly, and active device operation will be critical [514]. The old paradigm of trial-and-error, make it and test it, etc., will never yield enough information to make a significant advance how we discover and develop new materials. Algorithms development is critical to in-situ characterization, and it is not just sufficient to speed up existing methods. Some, such as FBP for tomography, are already fast, however the results are valuable for only a small set of problems. It is critical to use advanced algorithms which have been shown to require less raw data to achieve a certain fidelity in information content. Compressed sensing, MBIR, and TV reconstruction algorithms offer this potential, but have not yet found widespread use in in-situ characterization. As data collection gets faster, real-time characterization is enabled, which in turn makes analysis faster because there is less data to analyze. This allows for in-situ feedback of results to the scientist (or algorithm) in the laboratory performing the experiment so that the next experimental course of action can be determined based on the most current and relevant information.

11.5.4 Summary

Novel photophysical properties such as fano-resonances, electromagnetically induced transparency, and negative dielectric susceptibility are currently in high demand for smart applications including SPASER, sensors, ultralight, and flexible electronics. Hierarchical architectures based on plasmonic nanoparticles can exhibit those properties if the structures are well tailored at sub 100 nm length scales. The last years of active research in the materials synthetic community generated complex structures with creative design and hierarchy. Among them, a bottom-up synthetic approach based on colloidal nanoparticles has created a broad array of assembled structures in solution and surfaces. Electron microscopy images and tomography reconstruction of these structures have enabled visualization of these structures in 3D. This type of visualization is crucial for understanding the photophysical properties which subsequently assist in the guidance to the design space. Similarly, advances made in nanoscale spectroscopy combine with electron/scanning probe microscopy have given a new dimension to the characterization of such materials. Morphology visualization along with spectroscopy information on every pixel can generate a large array of datasets. Subsequently, there is a high demand for data analytics in such system, which has a tremendous potential to push the technology further.

Part V

Microstructure

In developing or analyzing materials systems, microstructure is often the common language of all investigations. Early descriptions of microstructure consisted of natural language descriptions of microscopic observations. As the science developed, advances such as grain and particle size quantification, stereological metrics, and some of the early topological descriptions of microstructures set the stage for mathematical representations of microstructure. The development of digital microscopy and mapping techniques, such as Orientation Imaging Microscopy (OIM) Analysis^{TM}, produced a dichotomy in the field, where a microstructure could be described either by pixels in an image or in terms of abstract objects such as interfaces, particles, or grains represented by aggregations of those pixels. The abstract descriptions have been made even more abstract by the development of representations of the statistical properties of these objects. In the broader sense of statistical science, these descriptions would be referred to as *textures*. In materials science, this term is more commonly used to refer to a specific texture description, formed by the distribution of crystallographic orientations in the sample (*crystallographic texture*). The term *microstructure* is commonly used to describe the more general problem of statistical representation of microstructural objects.

Microstructure representation is, by no means, a solved problem and methods are currently being developed for descriptions of microstructures that are amenable to computer methods. We did not attempt to cover the field, but chose several of the emerging methods to highlight here, with the understanding that many new techniques will continue to be developed as further integration of statistical science with materials science occurs.

We start this section with one of two chapters by Niezgoda. Chapter 12 recognizes that the crystallographic texture is a stochastic phenomenon, so that the parameters describing it are not deterministic. Rather, they need to be estimated by statistical methods. The second chapter by the same author, Chapter 13, explicitly recognizes the fundamentally random nature of microstructures and how observations may be classified by methods such as pair correlation functions an other classical metrics. Niezgoda's work heavily borrows from the signal processing literature and has that "flavor." By contrast DeCost and Holm, Chapter 14, use methods of modern computer vision, more commonly associated with the computer science field. In their approach, microstructures form classes and microscope images are classified according to data-driven classifications. Finally, Lazar and Srolovitz, Chapter 15, revisit the classical ideas of topology for describing microstructures, only they quantify the connectivity *between* microstructure objects, instead of the objects themselves. Their "Voronoi topology" then forms a representation of the microstructure.

The above approaches may be recognized as "inverse methods" in which microstructure examples are inverted to give their mathematical descriptions. There is also interest in the forward problem of, given a microstructural description, what would be typical examples of observations consistent with this description? In Chapter 16, Sundaraghavan develops a method for producing these observations by the evolution of *Markov random fields* (MRFs). The MRF is well known in physics as the Ising or Potts models, but is also well known in the modern image processing field as a very common way of smoothing the results of an inversion. This chapter, then, plays somewhat of a unifying role in this book, that of linking these two communities.

Within a given classification scheme, it is always a reasonable question to ask, "how close is this microstructure to the one desired?" In particular, this would allow for the quantitative connection between processing and a target microstructure. It would also allow for quantification of damage evolution, in cases where the microstructural evolution is part of the damage. Callahan, in Chapter 17, develops ways in which statistical metrics may be used in order to quantify these deviations.

Finally, we asked David Furrer (Pratt & Whitney) to round out this section of the book with a chapter giving an industry perspective of the usefulness of techniques of microstructure quantification. Chapter 18, by Furrer et al. forms this final chapter.

Chapter 12

Estimating Crystallographic Texture and Orientation Statistics

Stephen R. Niezgoda
The Ohio State University

The connection between crystallographic preferred orientations (texture) in polycrystalline specimens and anisotropic material response/properties is widely recognized, and research into the quantitative analysis of texture has a long history in materials science and geology [143, 1138, 307, 500, 144, 1136, 1137]. A meaningful overview discussion of the field of texture analysis is well beyond the scope of this chapter. Instead, here we will focus on some recent developments in the estimation and description of microtexture and spatial orientation statistics. For a more complete overview or introduction to the field I recommend the excellent books from Kocks, Tomé, and Wenk [1138] and Engler and Rangle [307].

The relative frequency of crystal orientations within a sample is described mathematically via the orientation distribution function or ODF. Experimental techniques for the estimation of the ODF can be categorized as either macrotexture (or bulk/sample) texture techniques such as X-ray or neutron diffraction [135, 1084], where the relative frequency of orientations is averaged over hundreds of thousands of grains, or meso/micro-texture techniques such as electron backscatter diffraction (EBSD) [6, 263] or high energy X-ray diffraction microscopy (HEDM) [423], which provide spatially resolved 2D or 3D orientation maps of a much smaller number of grains (hundreds to tens of thousands)[1].

The ODF cannot be measured directly by macrotexture diffraction techniques. Instead, multiple lower-order crystallographic axis or pole distribution functions are measured for different families of crystallographic planes, and the ODF is then computed by solving the famous "pole figure inversion" problem [143, 307, 663, 772, 430]. Various methodologies for estimating the ODF from pole figure data have been developed including harmonic polynomial expansions (generalized spherical harmonics) [143], the direct or iterative algorithms such as the WIMV (Williams–Imhof–Matthies–Vinel) algorithm [663] or the arbitrary defined cells (ADC) method [772], and the more recent MTEX algorithm which models the diffraction experiment as a Poisson process parameterized by the unknown ODF [430]. In contrast, microtexture techniques directly measure the lattice orientation at a particular spatial location and the ODF must then be estimated from a set or list of discrete orientations.

---

[1]There is no generally agreed upon definition of meso- vs. microtexture. Some authors use microtexture to describe the orientation distribution of a mesoscale material volume consisting of hundreds to thousands of grains while others prefer mesotexture and reserve microtexture for orientation spread over a single grain or small handful of grains. For the purposes of this chapter the terms meso- and microtexture will be used interchangeably to denote the orientation distribution obtained from a limited set of discrete orientations.

12.1   Orientations and Orientation Distributions

A crystallographic orientation, $\mathbf{g} = e_i \otimes a_j \in SO(3)$, is defined as a proper rotation which maps the macro scale sample or laboratory basis vectors, $\mathbf{e}$, onto the local crystal lattice basis, $\mathbf{a}$. $SO(3)$ denotes the special orthogonal matrix group of proper rotations in 3 dimensions whose members are the 3×3 orthogonal matrices with determinant $+1$. $\mathbf{g}$ accepts multiple equivalent parameterizations including Euler angles, unit quaternions, Rodriguez vectors, and angle-axis pairs (for a review of the different parameterizations, see [649] and the references contained within). Parameterization of $SO(3)$ by unit quaternions will be found particularly convenient for the discussion of analytical probability distributions for orientations. The quaternion and Bunge–Euler representations of orientations is reviewed in the Appendix Sections 12.5 and 12.6, respectively.

$\mathbf{g}^{(2)} \cdot \mathbf{g}^{(1)}$ denotes the composition of rotations and is equivalent to rotation $\mathbf{g}^{(1)}$ followed by rotation $\mathbf{g}^{(2)}$. $\mathbf{g}_I$ is the identity element of $SO(3)$ defined such that

$$\mathbf{g} \cdot \mathbf{g}_I = \mathbf{g}_I \cdot \mathbf{g} = \mathbf{g} \tag{12.1}$$

The inverse rotation is given by

$$\mathbf{g} \cdot \mathbf{g}^{-1} = \mathbf{g}^{-1} \cdot \mathbf{g} = \mathbf{g}_I \tag{12.2}$$

If $\Delta\mathbf{g} \cdot \mathbf{g}^{(1)} = \mathbf{g}^{(2)}$ than

$$\Delta\mathbf{g} = \mathbf{g}^{(2)} \cdot \mathbf{g}^{(1)-1} \tag{12.3}$$

is termed the misorientation between orientations $\mathbf{g}^{(1)}$ and $\mathbf{g}^{(2)}$. The misorientation angle will be represented by $|\Delta\mathbf{g}|$.

12.1.1   Formal Definition of the ODF

Statistics on the relative frequency of orientations in the microstructure are given by the orientation distribution function (ODF). The ODF is usually defined in terms of the volume fractions of orientations in a macroscale sample. Here we will more formally develop the notion of a meso- or microtexture, and the corresponding meso-ODF in a manner that is consistent with traditional interpretation of the ODF and with current interpretation of the microstructure described in Chapter 13 [737]. The probability that the orientation at material point $\mathbf{x}$, lies in some region $\mathfrak{G} \subseteq SO(3)$ is given by

$$\mathscr{P}\{\mathbf{g}(\mathbf{x}) \in \mathfrak{G}\} = \int_{\mathfrak{G}} f(\mathbf{g}; \mathbf{x}) d\mathbf{g} \tag{12.4}$$

$f(\mathbf{g}; \mathbf{x})$ is the first-order density function of the microstructure or equivalently the first-order density of orientations, and will be referred to as the meso-ODF at $\mathbf{x}$. There is a subtle, but important, consideration to be made when interpreting the meaning of the meso-ODF. $f(\mathbf{g}; \mathbf{x})$ is formally defined over an ensemble of independent material samples or microstructure realizations, in the sense that if material point $\mathbf{x}$ were to be examined in many realizations $f(\mathbf{g}; \mathbf{x})$ will give the relative frequency of occurrence of $\mathbf{g}(\mathbf{x}) = g$. While this distinction may, at first glance, appear overly pedantic it is important for the proper definition of orientation statistics in the presence of texture gradients or other texture inhomogeneities. The microstructure is said to be first-order stationary. if the first-order density is invariant to shifts in position,

$$f(\mathbf{g}; \mathbf{x}) = f(\mathbf{g}; \mathbf{x} + \Delta) \tag{12.5}$$

Provided the microstructure is stationary the ODF is no longer dependent on position and $f(\mathbf{g}, \mathbf{x}) = f(\mathbf{g}) \forall \mathbf{x}$. In this case, the concept of ergodicity can be invoked and $f(\mathbf{g})$ measured over a material volume rather than an ensemble of material realizations; the ODF then regains its traditional interpretation as the volume fraction of orientations in a material sample,

$$\int_{\mathfrak{G}} f(\mathbf{g}) d\mathbf{g} = V_{\mathfrak{G}}/V \tag{12.6}$$

When the material possesses texture inhomogeneities, such as gradients or clustering, the microstructure is by definition non-stationary. However, the microstructure may possess stationary increments, or contain neighborhoods or local regions where the microstructure statistics are invariant to small changes in position. In other words, there would exist some neighborhood $\mathscr{X}$ for which Equation 12.5 is satisfied $\forall \Delta \mathbf{x} \subseteq \mathscr{X}$. When discrete data is collected from a stationary increment, ergodicity can also be invoked and the meso-ODF can be interpreted as a distribution over a material volume rather than an ensemble. For the remainder of this chapter, it will be assumed that microstructure contains stationary increments on the scale of the data collection or simulation (hundreds or thousands of grains) and the explicit dependence on $\mathbf{x}$ will be dropped, so that $f(\mathbf{g})$ will be understood to represent a mesoscale ODF.

The Haar measure, or invariant measure, is required to ensure invariant integration over the rotation group such that

$$\int_{SO(3)} f(\mathbf{g})d\mathbf{g} = \int_{SO(3)} f(\mathbf{g} \cdot \mathbf{g}_0)d\mathbf{g} = 1 \; \forall \mathbf{g}_0 \in SO(3) \tag{12.7}$$

$\mathbf{g} \cdot \mathbf{g}_0$ denotes the composition of orientations and is equivalent to a rotation $\mathbf{g}_0$ followed by $\mathbf{g}$. If we denote the uniform random ODF as $f_{UR}(\mathbf{g})$, it is customary to normalize the ODF by choosing a measure $d\mathbf{g}$ such that $\int_{SO(3)} d\mathbf{g} = 1$. This normalization implies that $f_{UR}(\mathbf{g}) = 1 \forall \mathbf{g}$, which is consistent with the custom of expressing $f(\mathbf{g})$ in terms of multiples of the uniform random ODF (M.R.D.). In order to avoid confusion with normalization, $f(\mathbf{g})$ will be used to denote an ODF in units of times random with measure $d\mathbf{g}$ and $p(\mathbf{g})$ will be used to denote a "proper" probability distribution (p.d.f.).

12.1.2 Metrics on the ODF

For comparison purposes one is often interested in a scalar description of the sharpness or peak intensity of a given ODF. Commonly, the texture index

$$J(f(\mathbf{g})) = \int_{SO(3)} f(\mathbf{g})^2 d\mathbf{g} \tag{12.8}$$

is used for this purpose, where J can be loosely understood as the "energy" of the ODF [143].

The author prefers to use the concept of texture information entropy, or texture-entropy, which has a more rigorous connection to information and estimation theory [225]. The entropy of an ODF is defined as

$$H(f(\mathbf{g})) = E\{-\log f(\mathbf{g})\} = -\int_{SO(3)} f(\mathbf{g})\log f(\mathbf{g})d\mathbf{g} \tag{12.9}$$

The information entropy can be understood as a measure of the unpredictability of a random variable, in the same way that thermodynamic entropy is a measure of disorder. If we consider a perfect single crystal, the $f(\mathbf{g})$ is given by a delta function, and we would know exactly what $\mathbf{g}(x)$ would take for any material point in the sample. Our measurement is perfectly predictable and taking additional measurements adds no new information, thus $H(f) = -\infty$. A perfectly uniform random texture represents a measurement with maximum unpredictability, in the sense that $\mathbf{g}(x)$ ranges over the entire orientation space with equal probability, $f_{UR}(\mathbf{g}) = 1 \forall \mathbf{g} \in SO(3)$. Thus the texture-entropy is maximized and $H(f) = 0$.

The information entropy has important theoretical connections to probability theory that are relevant to the discussion of texture estimation. In particular the principle of maximum entropy states that given some set of testable data about a system, the probability distribution that best represents the current state of the system will be the one that maximizes the information entropy. Entropy maximization techniques have been used by Schaeben et al. to solve the pole figure inverse problem [905] and by Böhlke to eliminate "ghosting" in low l_{max} harmonic texture approximation [111].

The principle of maximum entropy is also useful in developing prior distributions for Bayesian inference and in choosing candidate distributions for modeling purposes. The principle of maximum entropy states that if nothing is known about the true distribution that generated some testable information, except that it belongs to a certain class of distributions, the distribution with the largest entropy should be chosen as the prior or candidate distribution. The reasoning is that by maximizing the entropy of the candidate distribution we are maximizing the unpredictability of $\mathbf{g}$, thus minimizing the amount of prior information or equivalently minimizing the assumptions about the data built into the candidate distributions. In the case of our proposed model, the testable information is the set of statistics sufficient to estimate distribution parameters from a measured set of orientations $\mathscr{G}$. This will be an important consideration when choosing distributions for texture modeling in Section 12.3 on modeling ODFs using parameterized probability distributions.

Given an ODF $f(\mathbf{g})$ or a measured set of N orientations $\mathscr{G}$, an important metric is the mean orientation or the expected value of $\mathbf{g}$ or $E\{\mathbf{g}\}$. Unlike probability distributions defined over real numbers where $E\{x\} = \int xf(x)dx$ the appropriate definition of the mean or other moments of distributions defined on matrix groups is not immediately obvious. The modal orientation, defined as $\max_{\mathbf{g}} f(\mathbf{g})$, is often used as a simple approximation of the mean. The quaternion representation of orientation is use for computation of another rapid estimate for the sample mean.

The scatter matrix is formed

$$S = \frac{1}{N} \sum_{i=1}^{N} \mathbf{g}^{(i)} \mathbf{g}^{(i)T} = E\{\mathbf{g}\mathbf{g}^{T}\} \tag{12.10}$$

noting that $\mathbf{g}\mathbf{g}^{T}$ indicates the outer matrix product not quaternion multiplication. An estimate of the mean can be obtained from the 1st eigenvector of the scatter matrix. For the two parametric distributions discussed in Section 12.3, the Von Mises Fisher (VMF) and Bingham distributions, the 1st eigenvector of S has special significance. For the VMF it is the maximum likelihood estimate of the sample mean and for the Bingham distribution it is the modal orientation of the distribution [643, 738, 742, 196].

A formal discussion of the mean of a distribution requires a divergence into the theory of Lie groups, so we will simply present the results without a formal derivation or proof [694]. Given a set of N orientations, $\mathscr{G}$, the distance between any two orientations $\mathbf{g}^{(i)}$ and $\mathbf{g}^{(j)}$ in $\mathscr{G}$ is given by the misorientation angle $|\Delta\mathbf{g}(\mathbf{g}^{(i)}, \mathbf{g}^{(j)})|$. For any arbitrary point in $SO(3)$, $\mathbf{g}^{*}$, the Fréchet variance is defined as

$$\Psi(\mathbf{g}*) = \sum_{i}^{N} |\Delta\mathbf{g}(\mathbf{g}^{(i)}, \mathbf{g}^{*})|^{2} \tag{12.11}$$

The mean orientation is defined as the orientation which minimizes the Fréchet variance. $\bar{\mathbf{g}} = \min_{\mathbf{g}^{*} \in SO(3)} \Psi(\mathbf{g}^{*}) = E\{\mathbf{g}\}$.

12.2   Non-Parametric Estimation of Statistics from Discrete Orientation Data

Techniques for the estimation of orientation statistics from discrete orientation data grew organically out of the existing direct and spherical harmonic techniques for macrotexture analysis, and rely on building up the ODF from the superposition of contributions from the individual orientations. In the direct methods, the orientation space is binned into discrete cells of typically $5°$ to $10°$. The number of points falling into each cell is counted to construct a histogram followed by a smoothing operation to approximate a continuous distribution. This technique was first applied by Perlwitz et al. in 1969 [776] and while there have been numerous advances and variations, the basic technique remains largely unchanged. The most common method for ODF estimation from discrete measurements is based on the generalized spherical harmonics championed by Bunge [143]. In the harmonic method, the Fourier coefficients for the ODF are estimated from the superposition of the Fourier coefficients of the individual orientations convolved with a Gaussian-like smoothing kernel (as the spherical harmonic expansion is a linear operation). Both techniques rely heavily on assumptions made *a priori* on the type and degree of smoothing. For the harmonic case, ad hoc methods are required both for choosing the bandwidth of the spherical harmonics as well as to enforce positivity constraints (to make a proper ODF). The degree of smoothing and bandwidth as well as the number of discrete orientations required for accurate ODF estimation has a long history of debate in the literature [1156, 470, 306, 1114]. A recent advance, on this front, is the development of automatic smoothing kernel optimization available in the MTEX Quantitative Texture Analysis Software [40], which assists the user in determining the appropriate level of smoothing and issues a warning if too low a bandwidth is used.

12.2.1 *Generalized Spherical Harmonics for the Estimation of Mesoscale ODFs*

The most common approach to texture estimation from discrete orientation measurements, is to expand the ODF as a series of generalized spherical harmonic functions $\ddot{T}_l^{\mu\nu}(\mathbf{g})$ following Bunge as [143]

$$f(g;x) = \sum_{l=0}^{\infty} \sum_{\mu=1}^{M(l)} \sum_{\nu=1}^{N(l)} C_l^{\mu\nu}(x) \ddot{T}_l^{\mu\nu}(\mathbf{g}) \qquad (12.12)$$

The dots over the $\ddot{T}_l^{\mu\nu}(\mathbf{g})$ indicate that the harmonic functions have been symmetrized to reflect the crystallographic symmetry of the material and statistical or sample symmetry that arises due to processing (e.g., the statistical two-fold rotation axis about the rolling, transverse, and normal directions of a rolled plate resulting in orthotropic sample symmetry [500]). The generalized spherical harmonics are linear combinations of the classically defined spherical harmonics as

$$\ddot{T}_l^{\mu\nu}(\mathbf{g}) = \sum_{m=-1}^{+l} \sum_{n=-l}^{+l} \dot{A}_l^{m\mu} \dot{A}_l^{n\nu} T_l^{mn}(\mathbf{g}) \qquad (12.13)$$

$\dot{A}_l^{n\nu}$ is the coefficient to account for sample symmetry and $\dot{A}_l^{m\mu}$ asserts the crystallographic symmetry. Note that the coefficients $\dot{A}_l^{m\mu}$ and $\dot{A}_l^{n\nu}$ are not uniquely defined, Bunge removed this ambiguity and chose the coefficients to be real. Some useful mathematical of properties for the manipulation of Bunge's generalized spherical harmonics are presented in the Appendices (Section 12.7).

While the summation in Equation 12.12 goes to ∞, in practice the series is truncated at $l_{max} = 16$ (default TSL EBSD software value [747]) or $l_{max} = 32$ for quantitative texture analysis applications. The Fourier coefficients C_l^{mn} are estimated from a set of N individual orientations $\mathscr{G} = \{\mathbf{g}^{(1)}, \ldots, \mathbf{g}_N\}$ as

$$C_l^{mn} = \frac{1}{N} \sum_{i=1}^{N} K_l(\psi) \ddot{T}_l^{\mu\nu}(\mathbf{g}_i) \qquad (12.14)$$

where $K_l(\psi)$ is the Fourier representation of a smoothing kernel with half-width ψ. In the orientation space, Equation 12.14 can be understood as the convolution of a smoothing function (typically Gaussian-like) with the delta function ODF for a single crystal. In effect, each individual orientation is replaced by a Gaussian of half-width ψ. There has been much debate in the literature of how to choose an appropriate filter and half-width as a function of N and l [1156, 470, 306, 1114]. It should be apparent that the bandwidth and the smoothing will have a large effect on the computed ODF. If the halfwidth of the smoothing kernel is too narrow, the estimated ODF will overfit the data and if the kernel is too broad the important features of the ODF will be smoothed away. The choice of bandwidth is also critical. If too low of a value for l_{max} is chosen, Gibb's oscillations will be observed near sharp peaks and could cause "ghosting" artifacts including non-physical negative ODF values or false peaks. A recent development is the automatic kernel density estimation implemented in the freely available MTEX MATLAB® Toolbox for Quantitative Texture Analysis [40].

A closely related distribution to the ODF is the uncorrelated misorientation distribution function (UMODF), which gives the probability of finding the orientations of two neighboring points, or two neighbor grains, related by the misorientation $\Delta\mathbf{g}$ under the assumption that the grain orientations are spatially uncorrelated and randomly sampled from the ODF. Pospiech describes the UMODF and gives the following harmonic expansion [797]

$$f(\Delta\mathbf{g}) = \sum_{l,\mu,\mu'} \hat{C}_l^{\mu\mu'} \ddot{\hat{T}}_l^{\mu\mu'}(\Delta\mathbf{g}) \tag{12.15}$$

Again assuming that the misorientations are uncorrelated, i.e., grain orientation is independent of the neighbor orientations, the coefficients for the UMODF can be estimated from the texture coefficients as

$$\hat{C}_l^{\mu\mu'} = \frac{1}{2l+1} \sum_{\nu} C_l^{*\mu\nu} C_l^{\mu'\nu} \tag{12.16}$$

12.2.2 Spatial Statistics of Orientations

In the 20 or so years since it was developed, EBSD has become a ubiquitous technique for the study of microtexture as well as the spatial arrangement of orientations, and evolution of orientation during deformation. It is interesting to note that the original motivation of the development of the technique was the characterization of longer-range orientation correlations in polycrystalline samples [7]. The motivating application was the application of Kröner series for the localization of deformation in polycrystalline materials and developing stronger bounds on the homogenization of mechanical response of polycrystals [516, 5]. While EBSD and orientation mapping has been developed into a workhorse technique for the characterization of polycrystals, the description of orientation spatial correlation beyond nearest neighbors via the MODF is not routinely explored.

The following analysis will build off the description of two-point correlations in Chapter 13. The correlation of two functions is properly defined by the integral

$$f(x) \star h(x) = \int_{-\infty}^{+\infty} f^*(y)h(y+x)dy$$

For real functions (such as the microstructure function) the complex conjugate is neglected, however we will find it mathematically convenient to leave the conjugate in place as the GSH coefficients are generally complex.

Consider a macroscale material sample where the microstructure at each material point, x, is defined by a microtexture $f(\mathbf{g};x)$. The second-order spatial statistics of the microstructure are given by the microtexture correlation function

$$f(\mathbf{g}^{(1)},\mathbf{g}^{(2)}|r) = \int_{-\infty}^{+\infty} f^*(\mathbf{g}^{(1)};x)f(\mathbf{g}^{(2)};x+r)dx \tag{12.17}$$

In the limit that $f(g;x) \to \delta(g(x))$ Equation 12.17 becomes the more well-known orientation correlation function. Following Adams et al. the microtexture correlation function adopts the following harmonic expansion

$$f(\mathbf{g}^{(1)},\mathbf{g}^{(2)}|r) = \sum_{l,\mu,\nu} \sum_{\lambda,\alpha,\beta} D_{l\mu\nu}^{\lambda\alpha\beta}(r) \ddot{\hat{T}}_l^{*\mu\nu}(\mathbf{g}^{(1)}) \ddot{\hat{T}}_\lambda^{\alpha\beta}(\mathbf{g}^{(2)}) \tag{12.18}$$

By substituting the harmonic expansion for the micro-ODF into Equation 12.17

$$\tag{12.19}$$

$$
\begin{aligned}
f(\mathbf{g}^{(1)},\mathbf{g}^{(2)}|r) &= \int_{-\infty}^{+\infty} \left[\sum_{l\mu\nu} C_l^{*\mu\nu}(x)\ddot{\hat{T}}_l^{*\mu\nu}(\mathbf{g}^{(1)}) \right] \left[\sum_{\lambda\alpha\beta} C_\lambda^{\alpha\beta}(x+r)\ddot{\hat{T}}_\lambda^{\alpha\beta}(\mathbf{g}^{(2)}) \right] dx \\
&= \sum_{\substack{l,\mu,\nu \\ \lambda,\alpha,\beta}} \left[\int_{-\infty}^{+\infty} C_l^{*\mu\nu}(x)C_\lambda^{\alpha\beta}(x+r)dx \right] \ddot{\hat{T}}_l^{*\mu\nu}(\mathbf{g}^{(1)}) \ddot{\hat{T}}_\lambda^{\alpha\beta}(\mathbf{g}^{(2)})
\end{aligned}
$$

By comparing Equations 12.18 and 12.19 we see that

$$D_{l\mu\nu}^{\lambda\alpha\beta}(r) = \int_{-\infty}^{+\infty} C_l^{*\mu\nu}(x)C_\lambda^{\alpha\beta}(x+r)dx \tag{12.20}$$

which is the spatial correlation of the microtexture coefficients. This correlation can be efficiently computed by making use of the discrete Fourier transform

$$(12.21)$$

$$
\begin{aligned}
D_{l\mu\nu}^{\lambda\alpha\beta}(r) &= \int_{-\infty}^{+\infty} \left[\int_{-\infty}^{+\infty} C_l^{\mu\nu}(-k)e^{2\pi ikx}dk\right]^* \left[\int_{-\infty}^{+\infty} C_\lambda^{\alpha\beta}(k')e^{2\pi ik'x}dk'\right]dx \\
&= \int\int_{-\infty}^{+\infty} C_l^{*\mu\nu}(-k)C_\lambda^{\alpha\beta}(k')e^{2\pi ik'r}\left[\int_{-\infty}^{+\infty}e^{-2\pi ikx}se^{2\pi ik'x}dx\right]dkdk' \\
&= \int\int_{-\infty}^{+\infty} C_l^{*\mu\nu}(-k)C_\lambda^{\alpha\beta}(k')e^{2\pi ik'r}\delta(k-k')dkdk' \\
&= \int_{-\infty}^{+\infty} C_l^{*\mu\nu}(-k)C_\lambda^{\alpha\beta}(k)e^{2\pi ikr}dk \\
&= \mathfrak{F}^{-1}\left[C_l^{*\mu\nu}(-k)C_\lambda^{\alpha\beta}(k)\right](r) \\
&= \mathfrak{F}^{-1}\left[\mathfrak{F}\left[C_l^{*\mu\nu}(x)\right](k)\cdot\mathfrak{F}\left[C_\lambda^{\alpha\beta}(x)\right](k)\right](r)
\end{aligned}
$$

Figure 12.1 demonstrates the application of Equation 12.21. with two examples, the first a synthetic checkerboard structure with only two orientations and the second a highly textured severely deformed Ta plate.

The above analysis can be extended to derive a formulation for a spatially resolved generalized misorientation calculation. Let $\mathbf{g}^{(1)}$ and $\mathbf{g}^{(2)}$ be related to each other by the generalized misorientation

$$\Delta\mathbf{g}\cdot\mathbf{g}^{(1)}\cdot\Delta\mathbf{g}' = \mathbf{g}^{(2)} \tag{12.22}$$

By integrating over $\mathbf{g}^{(1)}$ we can define a generalized misorientation correlation function which gives the probability of finding points at the head and tail of vector r related by the left and right misorientations $\Delta\mathbf{g}$ and $\Delta\mathbf{g}'$ respectively.

$$f(\Delta\mathbf{g},\Delta\mathbf{g}'|r) = \int_{SO(3)} f(g,\Delta\mathbf{g}\cdot g\cdot\Delta\mathbf{g}'|r)dg \tag{12.23}$$

The integration can be carried out as follows

$$f(\Delta\mathbf{g},\Delta\mathbf{g}'|r) = \int_{-\infty}^{+\infty} \sum_{\substack{l,\mu,\nu \\ \lambda,\alpha,\beta}} D_{l\mu\nu}^{\lambda\alpha\beta}(r)\dot{T}_l^{*\mu\nu}(\mathbf{g})\dot{T}_\lambda^{\alpha\beta}(\Delta\mathbf{g}\cdot g\cdot\Delta\mathbf{g}')dg \tag{12.24}$$

By repeated application of the addition theorem and rearranging

$$(12.25)$$

$$
\begin{aligned}
&= \sum_{\substack{l,\mu,\nu \\ \lambda,\alpha,\beta}} D_{l\mu\nu}^{\lambda\alpha\beta}(r)\int_{-\infty}^{+\infty}\sum_{s=-\lambda}^{+\lambda}\sum_{\sigma=-\lambda}^{+\lambda}\dot{T}_\lambda^{\alpha s}(\Delta\mathbf{g})T_\lambda^{s\sigma}(\mathbf{g})\dot{T}_\lambda^{\sigma\beta}(\Delta\mathbf{g}')\dot{T}_l^{*\mu\nu}(\mathbf{g})dg \\
&= \sum_{\substack{l,\mu,\nu \\ \lambda,\alpha,\beta}}\sum_{s,\sigma} D_{l\mu\nu}^{\lambda\alpha\beta}(r)\dot{T}_\lambda^{\alpha s}(\Delta\mathbf{g})\dot{T}_\lambda^{\sigma\beta}(\Delta\mathbf{g}')\int_{-\infty}^{+\infty}T_\lambda^{s\sigma}(\mathbf{g})\dot{T}_l^{*\mu\nu}(\mathbf{g})dg
\end{aligned}
$$

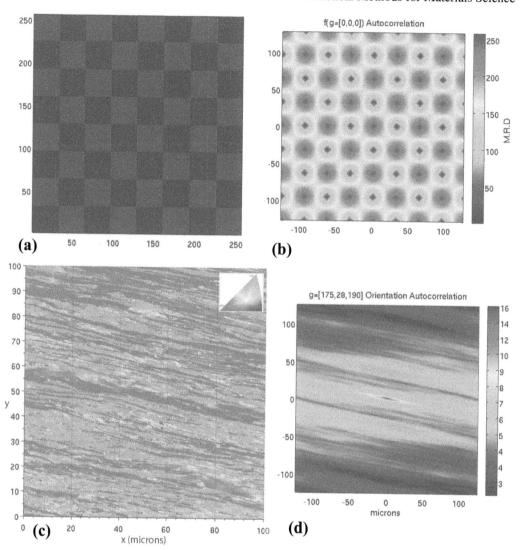

Figure 12.1: (a) Simple synthetic polycrystalline microstructrue. Red grains are perfect cube orientation $g = [0,0,0]$ in Bunge–Euler angles. The blue grains are $g = [0,45,0]$. (b) The cube-cube orientation autocorrelation calculated by Equation 12.21. (c) Inverse pole figure map for severely deformed Ta with strong (001) and (111) texture components (color map shown in the inset). (d) The autocorrelation for orientations with Bunge–Euler angles $g = [172,28,190]$.

Followed by the linear decomposition of $\dot{\ddot{T}}_l^{*\mu\nu}(\mathbf{g})$

$$(12.26)$$

$$
= \sum_{\substack{l,\mu,\nu \\ \lambda,\alpha,\beta}} \sum_{s,\sigma} D_{l\mu\nu}^{\lambda\alpha\beta}(r)\dot{T}_{\lambda}^{\alpha s}(\Delta\mathbf{g})T_{\lambda}^{\sigma\beta}(\Delta\mathbf{g}') \sum_{m=-l}^{+l}\sum_{n=-l}^{+l} \dot{A}_l^{m\mu}\dot{A}_l^{n\nu} \int_{-\infty}^{+\infty} T_{\lambda}^{s\sigma}(\mathbf{g})T_l^{*\mu\nu}(\mathbf{g})dg
$$

$$
= \sum_{\substack{l,\mu,\nu \\ \lambda,\alpha,\beta \\ m,n}} \sum_{s,\sigma} D_{l\mu\nu}^{\lambda\alpha\beta}(r)\dot{T}_{\lambda}^{\alpha s}(\Delta\mathbf{g})T_{\lambda}^{\sigma\beta}(\Delta\mathbf{g}')\dot{A}_l^{m\mu}\dot{A}_l^{n\nu}\frac{1}{2l+1}\delta_{\lambda l}\delta_{ms}\delta_{n\sigma}
$$

$$
= \sum_{\substack{l,\mu,\nu \\ \alpha,\beta}} \sum_{m,n} \frac{1}{2l+1}D_{l\mu\nu}^{l\alpha\beta}(r)\dot{T}_l^{\alpha m}(\Delta\mathbf{g})T_l^{n\beta}(\Delta\mathbf{g}')\dot{A}_l^{m\mu}\dot{A}_l^{n\nu}
$$

By summing over m and n

$$= \sum_{\substack{l,\mu,\nu \\ \alpha,\beta}} \sum_n \frac{1}{2l+1} D_{l\mu\nu}^{l\alpha\beta}(r) \ddot{T}_l^{\mu\alpha}(\Delta\mathbf{g}) \dot{T}_l^{n\beta}(\Delta\mathbf{g'}) \dot{A}_l^{n\nu} \tag{12.27}$$

$$= \sum_{\substack{l,\mu,\nu \\ \alpha,\beta}} \frac{1}{2l+1} D_{l\mu\nu}^{l\alpha\beta}(r) \ddot{T}_l^{\mu\alpha}(\Delta\mathbf{g}) \ddot{T}_l^{\nu\beta}(\Delta\mathbf{g'})$$

By setting $\Delta\mathbf{g'} = \mathbf{g}_I$ (identity rotation), we recover a more traditional misorientation correlation function (left multiplication/composition only) $f(\Delta\mathbf{g}|r)$ which gives the probability of finding the head and tail of vector r related by the crystallographic misorientation $\Delta\mathbf{g}$. If we substitute in $\ddot{T}_l^{\nu\beta}(I) = \sum_{m=-l}^{+l} \dot{A}_l^{m\beta} \dot{A}_l^{m\nu} = \delta_{\beta\nu}$ (see properties above)

$$f(\Delta\mathbf{g}|r) = \sum_{\substack{l,\mu,\nu \\ \alpha,\beta}} \frac{1}{2l+1} D_{l\mu\nu}^{l\alpha\beta}(r) \ddot{T}_l^{\mu\alpha}(\Delta\mathbf{g}) \delta_{\beta\nu} \tag{12.28}$$

$$= \sum_{l,\mu,\alpha} \overline{D}_l^{\mu\alpha}(r) \ddot{T}_l^{\mu\alpha}(\Delta\mathbf{g})$$

where

$$\overline{D}_l^{\mu\alpha}(r) = \frac{1}{2l+1} \sum_{\nu=-l}^{+l} D_{l\mu\nu}^{l\alpha\nu}(r) \tag{12.29}$$

Notice that the form of Equation 12.28 matches the form of the expansion of the misorientation distribution function given by Pospiech (Equation 12.15), and the form of Equation 12.29 can be understood as a spatially resolved version of Equation 12.16.

12.3   Parametric Estimation of Orientation Distributions

Despite the popularity of the generalized spherical harmonic approaches for microtexture estimation from EBSD or other discrete orientation data, the reliance on ad hoc smoothing parameter and bandwidth choices remains problematic. There are no information-theoretic guarantees as to the optimality (in some sense) of a fit ODF in describing a given dataset. Further there is no framework for determining the uncertainty in the calculated ODF or in other words a methodology for putting error bars on texture. These limitations motivate interest in exploring the modeling of microtexture using analytical probability distributions with known statistical properties. In this section we will specifically look at two distributions, the Von Mises Fisher and Bingham distributions, to model orientation spreads and unimodal ODFs.

Historically speaking, the main advantage to using spherical harmonics was computational simplicity and speed coupled with widely available subroutines and libraries. The Bingham distribution was first described in 1974 [98], and efficient computational schemes for Bingham parameter estimation were not available until recently. In contrast, symmetrized spherical harmonics have been available to the texture community for almost 50 years. The explicit consideration of crystallographic and sample symmetry also posed a challenge, as methodology for parameter estimation with respect to the equivalence classes of crystallographic and sample symmetry were only recently developed [196, 742].

With the current generation of EBSD cameras, capable of taking and indexing > 100 diffraction patterns per second in ideal materials, the generation of detailed orientation maps that contain several thousand grains is now fairly routine. It is a fair question to ask why a radically different approach to ODF estimation is even required; as when the number of orientations are increased, the ODFs produced from EBSD via harmonic analysis converge to the macrotexture ODFs produced from diffraction experiments. The counterargument is that by measuring datasets from a large area, we are using EBSD to mimic a macrotexture measurement, and for that case the direct methods or harmonic analysis will yield acceptable results. However, there are many instances where we are interested in a true microtexture measurement, such as when looking at highly localized phenomena such as duplex microstructures and macrozone formation in α/β Ti alloys [883, 134], ridging and roping in rolled Al [525], orientation clustering due to texture memory in HDDR produced magnets [403], localized plastic deformation at crack tips [486], etc. The description of texture gradients [944, 508] and other non-homogeneous materials poses a challenge particularly if the gradients are steep and transitory regions are of interest. There is also a critical need for quantitative analysis of texture on mesoscale volumes resulting from crystal plasticity simulations. The comparison of macrotexture ODFs with those resulting from crystal plasticity (CP) based micromechanical simulations is often a primary method of model validation and verification. Typically these simulations are performed on statistical volume elements that contain a few hundred to a thousand grains for full field simulations, such as CP-FEM [809], and a few thousand grains for homogenized or mean-field models, such as self-consistent models [550]. Quantitative comparison of the simulated textures with experimentally measured ODFs is difficult, as the intensity values are strongly dependent on the degree of smoothing and bandwidth selection as well as the resolution on which the ODFs are computed.

12.3.1 Bingham and Von Mises Fisher Distributions

The Bingham and Von Mises Fisher (VMF) distribution are antipodally symmetric probability distributions defined on the unit hypersphere $\mathbb{S}^d \subset \mathbb{R}^{d+1}$ [98]. Points on $\mathbb{S}^3$ are represented by the set of unit quaternions U, which is isomorphic to special unitary group of degree 2, $SU(2)$, and is a double covering group of the rotation group SO(3). The VMF is max entropy isotropic distribution on the unit quaternions and is suited to modeling orientation spreads about an orientation, such as the degree of orientation spread in a single grain after deformation [196]. The Bingham distribution on $\mathbb{S}^3$, or the quaternion Bingham, is a convenient probability distribution to model texture and can represent common texture components including fibers, sheets, uniform, or anisotropic spreads around individual orientations [904, 524, 932, 374, 40]. The Bingham distribution is the maximum entropy anisotropic distribution in $\mathbb{S}^d$.

The probability density function (p.d.f) for the quaternion Bingham is given by

$$p(\mathbf{g}; \boldsymbol{\Lambda}, \mathbf{V}) = \frac{1}{F(\boldsymbol{\Lambda})} \exp \sum_{i=1}^{4} \lambda_i (\mathbf{v}_i \cdot \mathbf{g})^2 \qquad (12.30)$$

where $\mathbf{g}$ is a unit quaternion representing an orientation, $\boldsymbol{\Lambda}$ is a 4-vector of concentration parameters λ_i, and F is a normalization constant. $\mathbf{V}$ is a 4×4 matrix the columns of which, $\mathbf{v}_i$, are orthogonal unit quaternions representing the principal directions of the distribution. The dot operator $(\cdot)$ denotes the quaternion inner product. The concentration parameters, $\boldsymbol{\Lambda}$, are unique only up to an additive constant. For this work we choose the convention $\lambda_1 \leq \lambda_2 \leq \lambda_3 \leq \lambda_4 = 0$ to resolve the ambiguity. The concentration parameters determine the sharpness of the distribution along the associated principal direction with $\boldsymbol{\Lambda} = [0,0,0,0]$ corresponding with the uniform distribution. The primary difficulty in working with the Bingham distribution is the computation of the normalization constant $F(\boldsymbol{\Lambda})$ which is a generalized hypergeometric function with matrix argument. For fast processing the authors have precomputed $F(\boldsymbol{\Lambda})$ to a lookup table for a discrete grid of $\boldsymbol{\Lambda}$ values. Interpolation is then used to quickly estimate normalization constants on the fly for arbitrary $\boldsymbol{\Lambda}$ values [739, 367].

The maximum likelihood estimators (MLE) for the quaternion Bingham parameters, denoted $\hat{\boldsymbol{\Lambda}}$ and $\hat{\mathbf{V}}$, are straightforward to calculate. Given a set of N discrete orientations, $\mathscr{G} = \{\mathbf{g}^{(1)}, \ldots, \mathbf{g}^{(N)}\}$, the scatter matrix

$$\mathbf{S} = \frac{1}{N} \sum_{i=1}^{n} \mathbf{g}^{(i)} \mathbf{g}^{(i)T} = E\{\mathbf{g}\mathbf{g}^{T}\} \qquad (12.31)$$

is a sufficient statistic to calculate both $\hat{\boldsymbol{\Lambda}}$ and $\hat{\mathbf{V}}$. $\hat{\mathbf{V}}$ is found by performing an eigenvalue decomposition of $\mathbf{S}$. The MLE mode of the distribution is the eigenvector of $\mathbf{S}$ with the largest eigenvalue and the columns of $\hat{\mathbf{V}}$ are the eigenvectors corresponding to the 2nd, 3rd, and 4th eigenvalues. $\hat{\boldsymbol{\Lambda}}$ is found by setting the partial derivatives of the log-likelihood function, $\log p(\mathscr{G}; \boldsymbol{\Lambda}, \mathbf{V}) = \sum_{i=1}^{N} \log p(\mathbf{g}^{(i)}; \boldsymbol{\Lambda}, \mathbf{V})$ with respect to the components of $\boldsymbol{\Lambda}$ to zero yielding

$$\frac{1}{F(\boldsymbol{\Lambda})} \frac{\partial F(\boldsymbol{\Lambda})}{\partial \lambda_j} = \mathbf{v}_j^T \mathbf{S} \mathbf{v}_j. \tag{12.32}$$

The values of the derivatives are precomputed and stored as a lookup table. Since the tables for F and ∇F are indexed by $\boldsymbol{\Lambda}$, a kD-tree is used to efficiently find the nearest neighbors for a computed $\nabla F / F$ to compute $\hat{\boldsymbol{\Lambda}}$ by interpolation.

The VMF distribution, also defined on $\mathbb{S}^d$, is a comparatively simpler distribution with only 2 parameters: the mean orientation $\boldsymbol{\mu}$ and a single concentration parameter $\kappa \geq 0$ [643]. The p.d.f. is given by

$$p(\mathbf{g}; \boldsymbol{\mu}, \kappa) = c_p(\kappa) \exp(\kappa \boldsymbol{\mu} \cdot \mathbf{g}) \tag{12.33}$$

The normalization constant, c_p is given as

$$c_p(\kappa) = \frac{\kappa^{(d+1)/2-1}}{(2 * \pi)^{(d+1)/2} I_{(d+1)/2-1}(\kappa)} \tag{12.34}$$

where $I_d(\cdot)$ is a modified Bessel function of the first kind.

The ML estimator of the parameters are also straightforward to calculate. Given a set of N discrete orientations $\mathscr{G}$,

$$\hat{\boldsymbol{\mu}} = \frac{\boldsymbol{\gamma}}{||\boldsymbol{\gamma}||}, \quad \boldsymbol{\gamma} = \sum_{i=1}^{N} \mathbf{g}^{(i)} \tag{12.35}$$

and

$$\hat{\kappa} = A_p^{-1}\left(\frac{||\boldsymbol{\gamma}||}{N}\right), \quad A_p(u) = \frac{I_{(d+1)/2}(u)}{I_{(d+1)/2-1}(u)} \tag{12.36}$$

12.3.2 Symmetrized Probability Distributions

In order to practically apply the above distributions to ODF representation we need to account for the underlying crystallographic and sample symmetries of the material and processing operations. Let $\mathscr{Q}^c = \{\mathbf{q}_1^c, \dots \mathbf{q}_M^c\}$ denote a group whose elements transform one orientation to a crystallography equivalent one. If $\mathbf{q}^c \in \mathscr{Q}^c$ then any function where $f(\mathbf{q}^c \cdot \mathbf{g}) = f(\mathbf{g})$ is said to be invariant under $\mathscr{Q}^c$. A material sample can also contain statistical symmetry due to processing. The classic example is a statistical twofold rotation axis about the rolling, transverse, and normal directions of a rolled plate resulting in orthotropic sample symmetry [500]. The sample symmetry group $\mathscr{Q}^s = \{\mathbf{q}_1^s, \dots, \mathbf{q}_P^s\}$ is defined in an identical way to the crystal symmetry group except the group operation is right multiplication. For materials with both symmetries any function of $\mathbf{g}$ must be invariant under both symmetries as [143],

$$f(\mathbf{q}^c \cdot \mathbf{g} \cdot \mathbf{q}^s) = f(\mathbf{g}). \tag{12.37}$$

Chen et al. recently derived the form that all probability density functions must have in order to be invariant under spherical symmetry groups [196]. Here we trivially extend their result to include sample symmetry and state that the density function $p : SO(3) \to \mathbb{R}$ is jointly invariant under $\mathscr{Q}^c$ and $\mathscr{Q}^s$ if and only if

$$p(\mathbf{g}; \boldsymbol{\Theta}) = \frac{1}{M} \frac{1}{P} \sum_{i=1}^{M} \sum_{j=1}^{P} p(\mathbf{q}_i^c \cdot \mathbf{g} \cdot \mathbf{q}_j^s; \boldsymbol{\Theta}) \tag{12.38}$$

where $\boldsymbol{\Theta}$ are the parameters of the probability density. Equation 12.38 states that any probability density $p(\mathbf{g})$ over the orientations which is invariant to crystallographic and sample symmetry can be represented as a finite mixture with equal weights of the rotated density under the combined crystallographic and sample symmetry groups actions.

Theorem 1. *Let $\mathbf{g}$ be a random orientation defined on $SO(3)$. The probability density function $p : SO(3) \to \mathbb{R}$ is invariant under $\mathscr{Q}^c$ and $\mathscr{Q}^s$ if and only if*

$$p(\mathbf{g};\boldsymbol{\Theta}) = \frac{1}{M}\frac{1}{P}\sum_{i=1}^{M}\sum_{j=1}^{P} p(\mathbf{q}_i^c \cdot \mathbf{g} \cdot \mathbf{q}_j^s;\boldsymbol{\Theta}). \tag{12.39}$$

Proof. Assuming 12.38 then

$$p(\mathbf{q}^c \cdot \mathbf{g} \cdot \mathbf{q}^s;\boldsymbol{\Theta}) = \frac{1}{M}\frac{1}{P}\sum_{i=1}^{M}\sum_{j=1}^{P} p(\mathbf{q}_i^c \cdot \mathbf{q}^c \cdot \mathbf{g} \cdot \mathbf{q}^s \cdot \mathbf{q}_j^s;\boldsymbol{\Theta}) \tag{12.40}$$

By definition $\mathscr{Q}^c$ and $\mathscr{Q}^s$ must be closed under left and right multiplication, respectively. Therefore

$$\frac{1}{M}\frac{1}{P}\sum_{i=1}^{M}\sum_{j=1}^{P} p(\mathbf{q}_i^c \cdot \mathbf{q}^c \cdot \mathbf{g} \cdot \mathbf{q}^s \cdot \mathbf{q}_j^s;\boldsymbol{\Theta}) = \frac{1}{M}\frac{1}{P}\sum_{r=1}^{M}\sum_{s=1}^{P} p(\mathbf{q}_r^c \cdot \mathbf{g} \cdot \mathbf{q}_s^s;\boldsymbol{\Theta}). \tag{12.41}$$

From the definition of invariance under groups

$$p(\mathbf{q}_r^c \cdot \mathbf{g} \cdot \mathbf{q}_s^s;\boldsymbol{\Theta}) = p(\mathbf{g};\boldsymbol{\Theta}) \tag{12.42}$$

On the other hand, assuming 12.42

$$
\begin{aligned}
p(\mathbf{g};\boldsymbol{\Theta}) &= \frac{1}{M}\frac{1}{P}\sum_{i=1}^{M}\sum_{j=1}^{P} p(\mathbf{q}_i^c \cdot \mathbf{g} \cdot \mathbf{q}_j^s;\boldsymbol{\Theta}) \tag{12.43}\\
&= \frac{1}{M}\frac{1}{P}\sum_{i=1}^{M}\sum_{j=1}^{P} p(\mathbf{g};\boldsymbol{\Theta})\\
&= p(\mathbf{g};\boldsymbol{\Theta})
\end{aligned}
$$

$\square$ $\square$

Applying Equation 12.38 allows us to directly write the p.d.f for the symmetrized Bingham as

$$p(\mathbf{g};\mathscr{Q},\boldsymbol{\Lambda},\mathbf{V}) = \frac{1}{MP}\sum_{i=1}^{M}\sum_{j=1}^{P} p(\mathbf{q}_i^c \cdot \mathbf{g} \cdot \mathbf{q}^s;\boldsymbol{\Lambda},\mathbf{V}) \tag{12.44}$$

$$= \frac{1}{MP}\sum_{i=1}^{M}\sum_{j=1}^{P} p(\mathbf{g};\boldsymbol{\Lambda},\mathbf{Q}_i^c\mathbf{V}\mathbf{Q}_j^s) = \frac{1}{MP}\sum_{i=1}^{M}\sum_{j=1}^{P} p(\mathbf{g};\boldsymbol{\Lambda},\mathbf{V}_{ij}) \tag{12.45}$$

$$= \frac{1}{MP}\frac{1}{F(\boldsymbol{\Lambda})}\sum_{i=1}^{M}\sum_{j=1}^{P}\left[\exp\sum_{k=1}^{4}\lambda_i\left([\mathbf{V}_{ij}]_k \cdot \mathbf{g}\right)^2\right] \tag{12.46}$$

where $\mathscr{Q}$ denotes the symmetry groups, $\mathbf{Q}_r$ denotes the quaternionic matrix where the product $\mathbf{Q}_r\mathbf{V}$ is equivalent to applying rotation $\mathbf{q}_r$ to each column of $\mathbf{V}$, and $[\mathbf{V}_{ij}]_k$ denotes the kth column of V_{ij}. Going from Equation 12.44 to Equation 12.45 requires the application of the inner quaternion product $\mathbf{v}_k \cdot \mathbf{g}$ in Equation 12.30 and the observation that the inverse of symmetry elements must also be elements of the symmetry group. Equation 12.46 states that the symmetrized Bingham distribution is a finite mixture of the standard quaternion Bingham distributions, with each component a) having equal weight, b) having principal directions rotated by $\mathbf{Q}_i^c\mathbf{V}\mathbf{Q}_j^s = \mathbf{V}_{ij}$ and c) having the same concentration parameters $\boldsymbol{\Lambda}$. A similar weighted mixture was defined in [374] and termed the Pseudo-Bingham distribution. The symmetrized group-invariant VMF is completely analogous [196].

12.3.3 EM-ML Algorithm for Parameter Estimation

S is a sufficient statistic for parameter estimation for the standard Bingham distribution. Equation 12.45 shows that the symmetrized Bingham is a finite mixture of rotated standard Bingham distributions. If **S** could be calculated for the symmetrized case the standard ML estimates $\hat{\mathbf{\Lambda}}$ and $\hat{\mathbf{V}}$ could be trivially computed. Consider a set of N discrete orientations, $\mathscr{G} = \{\mathbf{g}^{(1)}, \ldots, \mathbf{g}^{(N)}\}$ which was generated by sampling a symmetrized Bingham distribution. In order to calculate **S** we are missing information. The complete dataset would also contain a label which identifies which of the MP rotated components of the mixture generated each sample [739, 196, 701]. The EM-ML algorithm seeks the Bingham ODF which maximizes the probability of measuring the orientation data by a) estimating the labels given an estimate of the Bingham parameters (E-step), then b) using this new estimate of the labels to update the Bingham parameters (M-step). The algorithm is described more formally below.

For compactness let $\Theta = \{\mathscr{Q}, \mathbf{\Lambda}, \mathbf{V}\}$ denote the complete set of parameters necessary to specify the symmetrized Bingham $p(\mathbf{g}; \Theta)$. Further let $\theta_{ij} = \{\mathbf{\Lambda}, \mathbf{V}_{ij} = \mathbf{Q}_i^c \mathbf{V} \mathbf{Q}_j^s\}$ denote the set of parameters necessary to specify an individual rotated component of the mixture, $p(\mathbf{g}; \theta_{ij})$. Define a set of binary label vectors $\mathscr{Z} = \left[\mathbf{z}^{(1)}, \ldots, \mathbf{z}^{(N)}\right]$, where the elements of each vector $z_{ij}^{(n)}$ take the value 1 if $\mathbf{g}^{(n)}$ was generated by $p(\mathbf{g}; \theta_{ij})$ and 0 otherwise. If $\mathscr{Z}$ could be measured, then computing the scatter matrix and finding the ML estimate of the parameters $\hat{\Theta}$ would be trivial.

The complete data log-likelihood is given by

$$\log p(\mathscr{G}, \mathscr{Z}; \Theta) = \frac{1}{MP} \sum_{n=1}^{N} \sum_{i=1}^{M} \sum_{j=1}^{P} z_{ij}^{(n)} \log p(g^{(n)}; \theta_{ij}). \tag{12.47}$$

An ideal algorithm would maximize Equation 12.47 directly. However, optimization of functions of binary variables (i.e., $\mathscr{Z}$) is problematic [491]. Instead we define the conditional expectation, $\mathscr{W} = E\left[\mathscr{Z} | \mathscr{G}, \hat{\Theta}\right]$. $w_{ij}^{(n)}$ gives the probability that $z_{ij}^{(n)} = 1$ or equivalently the probability that orientation $\mathbf{g}^{(n)}$ was generated by the rotated Bingham $p(\mathbf{g}; \theta_{ij})$. By substituting $\mathscr{W}$ into Equation 12.47 we can use Bayes' rule to update the probabilities of mixture assignments. In the EM literature this is termed the Q function. If $\hat{\Theta}$ represents a current estimate of the parameters then

$$\begin{aligned} Q(\Theta | \hat{\Theta}) &= E\left[\log p(\mathscr{G}, \mathscr{Z} | \Theta) | \mathscr{G}, \hat{\Theta}\right] \\ &= \log p(\mathscr{G}, \mathscr{W} | \Theta) \end{aligned} \tag{12.48}$$

The E-step consists of using Bayes' rule to update $\mathscr{W}$ as

$$\begin{aligned} w_{ij}^{(n)} &= E[z_{ij}^{(n)} | \mathscr{G}, \hat{\Theta}] = \mathscr{P}[z_{ij}^{(n)} = 1 | \mathbf{g}^{(n)}, \hat{\Theta}] \\ &= \frac{(MP)^{-1} p(\mathbf{g}^{(n)}; \hat{\theta}_{ij})}{\sum_{s,t} (MP)^{-1} p(\mathbf{g}^{(n)}; \hat{\theta}_{st})}. \end{aligned} \tag{12.49}$$

Then in the M-step the parameter estimates $\hat{\Theta}$ are updated to maximize Equation 12.48. The ML estimate for $Q(\Theta | \hat{\Theta})$ can then be derived, by setting the derivatives with respect to the parameters to zero and solving. The ML estimates take exactly the same form as those given for the standard Bingham in Section 12.3.1 if the scatter matrix is replaced by $\mathbf{S}(\mathscr{Q})$ the symmetrized scatter matrix

$$\mathbf{S}(\mathscr{Q}) = \frac{1}{N} \sum_{n=1}^{N} \sum_{i=1}^{M} \sum_{j=1}^{P} w_{ij}^{(n)} \mathbf{g}_{ij}^{(n)} \mathbf{g}_{ij}^{(n)T} \tag{12.50}$$

where $\mathbf{g}_{ij}^{(n)} = (\mathbf{q}_i^s)^{-1} \cdot \mathbf{g}^{(n)} \cdot (\mathbf{q}_j^c)^{-1}$. The EM-ML algorithm alternates between Equation 12.49 and Equation 12.50 until convergence is reached. For this work, convergence was defined as the change in Equation 12.48 between iterations was less than some small value , $\Delta Q(\Theta | \hat{\Theta})/n \leq \delta$.

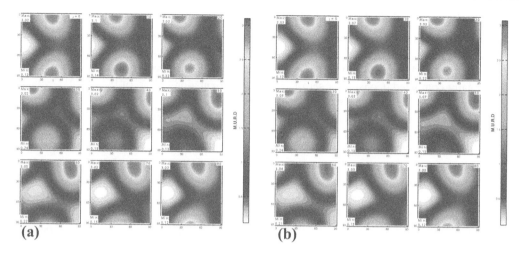

Figure 12.2: Comparison of the ground truth ODF (a) against the fit ODF for a material with cubic crystal symmetry and orthotropic sample symmetry. For convenience the ODFs are plotted as ϕ_2 sections of the Bunge–Euler angles [143], as is routinely done in the quantitative texture analysis literature [742].

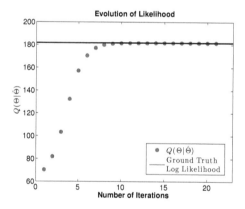

Figure 12.3: Comparison of ground truth versus fit symmetrized Bingham ODF [742].

EM-ML for the VMF again proceeds 100% analogously with γ replacing $\mathbf{S}$. For the VMF the M-step becomes

$$\gamma(\mathscr{Q}) = \sum_{n=1}^{N}\sum_{i=1}^{M}\sum_{j=1}^{P} w_{ij}^{(n)}\mathbf{g}_{ij}^{(n)} \tag{12.51}$$

Figure 12.2 shows the results for fitting the symmetrized Bingham distribution for a fiber texture. One thousand samples were drawn from an anisotropic Bingham distribution with $\Lambda = [-25, -20, -15]$ and V chosen as a random orthogonal matrix. The progression of the algorithm, shown by the evolution of $Q(\Theta|\hat{\Theta})$ with each iteration, is highlighted in Figure 12.3.

12.4   Brief Discussion and Conclusions

A large portion of the above content was focused on describing some basic features and ideas behind non-parametric and parametric approaches to texture analysis, with only limited guidance for the practitioner as to when one approach may be better for analyzing their data. In some ways that guidance is impossible to give, as modeling texture via parametric distributions is relatively new and even the techniques described above are in their nascent stages of development. As of the writing of this chapter there are no commercially or readily available open-source software packages that support either the symmetrized VMF of Bingham distributions.[2] For the "average" user it is largely an academic discussion, as they largely rely on the EBSD analysis software from the camera vendor due to ease of use and familiarity. It has been the author's experience that most EBSD users rarely plot the ODF much less use it as a comparative tool; instead they rely on qualitative or semi-quantitative interpretation of pole figures or inverse-pole figure colored maps. This is a testament to the robustness of the commercial EBSD packages that they can largely be operated as "black boxes" and return informative results. For the general practitioner, the author hopes that the above discussion encourages further exploration of the effect common analysis parameter such as bandwidth, smoothing, and plotting resolution has on the presentation and interpretation of the data.

Non-parametric analysis, both direct and spherical harmonic approaches, are computationally efficient and have a clear advantage when the number of orientations is "large." The magnitude of the effect of the smoothing kernel on the resulting ODF decreases rapidly as the number of measured grains increases. Provided the number of independent measurement points, or grains, is large enough so that the meso-ODF measured by non-parametric methods will converge to the macro-ODF measured by diffraction techniques. However for quantitative or semi-quantitative analysis it is strongly recommended that effect of changing bandwidth on the resultant ODF be explored, particularly checking for convergence with respect to the magnitude of modal intensities and minor peaks.

The author believes that in the near future the non-parametric approaches described above will mature and a host of new tools and techniques will be made available. This is likely to be driven by the continued development of integrated computational materials engineering (ICME) and related activities. ICME is predicated on the replacement of expensive experiment and testing paradigms with computational modeling and simulation. This necessitates the incorporation of uncertainty quantification and quantitative comparison between simulations and across simulation and experiment. It is telling that despite the ubiquity of texture analysis there is no standard way of reporting error or uncertainty in a texture measurement or plot. We don't have a framework for reporting error bars on a pole figure, for example. The adoption of parametric approaches will allow materials to leverage the recent developments in the area of UQ being developed by the statistics, applied mathematics, design, and machine learning communities.

Even in their early developmental stages parametric approaches have definite advantages when analyzing small datasets or for characterizing highly heterogeneous systems, where the number of orientations is too small for accurate ODF estimation using direct or harmonic techniques. The advantage of parametric techniques is that the estimate is "optimal" in some sense, for example either a maximum likelihood estimate (MLE) or maximum a posteriori (MAP) of distribution parameters. Also bounds on the estimates can be formulated making a direct link to uncertainty quantification and error analysis.

12.5   Appendix A: Quaternion Representation of Orientations

Working with the quaternion parameterization offers significant advantages over other orientation representations, such as Euler angles, including a simple and computationally efficient multiplication rule for the composition of rotations, a continuous group space with non-degenerate representation of all rotations, and a simple and intuitive physical interpretation [649, 650]. A unit quaternion can be thought of as representing a point on the surface of the unit sphere in four dimensions, $\mathbb{S}^3 \subset \mathbb{R}^4$. Visualizing quaternions and equivalently orientations as points on $\mathbb{S}^3$ will be particularly convenient in this work, as it allows for a straightforward extension of directional statistics on

---

[2]The MTEX toolbox for MATLAB has some support for the Bingham distribution; however, key features like fitting a complex multimodal ODF's and accounting crystallographic and sample symmetries are not yet available.

the unit circle and sphere to the more abstract notion of statistics on rotations. The mapping from quaternions to orientations is two to one in that each orientation has an equivalent representation by antipodal quaternions. This antipodal symmetry allows any hemisphere of $\mathbb{S}^3$ to be considered as a fundamental zone or fundamental region of orientations. However, when considering directional statistics on the hypersphere it is convenient to consider the whole space and work with a distribution that is also antipodally symmetric, such as the Bingham, to avoid introducing jumps or discontinuities at the equator.

An arbitrary quaternion is represented as a vector in a four-dimensional space over the field of real numbers

$$\mathbf{q} = q_0 + q_1\hat{\mathbf{i}} + q_2\hat{\mathbf{j}} + q_3\hat{\mathbf{k}} = q_0 + \vec{\mathbf{q}} \tag{12.52}$$

where q_0 is referred to as the scalar part of $\mathbf{q}$, and $\vec{\mathbf{q}}$ as the vector part. When discussing quaternions in the context of rotations, we specifically mean the unit quaternions, i.e., $q_0^2 + q_1^2 + q_2^2 + q_3^2 = 1$, which describe a point on the surface of the unit sphere in four dimensions, $\mathbb{S}^3 \subset \mathbb{R}^4$. The quaternion representation can be related to the angle-axis pair by

$$
\begin{aligned}
q_0 &= \cos\left(\tfrac{\omega}{2}\right) \\
\vec{\mathbf{q}} &= \hat{\mathbf{u}}\sin\left(\tfrac{\omega}{2}\right)
\end{aligned}
\tag{12.53}
$$

where ω is the angle of rotation and $\hat{\mathbf{u}}$ is a unit vector representing the rotation axis. The inverse rotation or inverse quaternion $\mathbf{q}^{-1}$ is formed by changing the sign of the $\vec{\mathbf{q}}$ in complete analogy with the angle-axis representation. Adding 2π to the rotation angle, ω, results in a rotation which is physically indistinguishable from the original, however as can be seen in Equation 12.53, results in the antipodal quaternion, $-\mathbf{q}$, with the signs of both the vector and scalar part reversed. Thus both $\mathbf{q}$ and $-\mathbf{q}$ correspond to rotations resulting in the same orientation g. This is a direct consequence of the group structures of the quaternions and $SO(3)$. Quaternions are isomorphic to the Lie group $SU(2)$ which is related by a 2-to-1 homomorphism to $SO(3)$, the well-known double cover of the three-dimensional rotations by quaternions. While slightly abstract, the structure of this mapping from quaternions to orientations and the properties of $SO(3)$ and $SU(2)$ as Lie groups manifests itself in the lack of singular points and discontinuities for the quaternion representation, such as the singularity near the identity operation in Euler angles, or near rotations by π with the angle-axis or Rodrigues–Frank representation [650].

In analogy with describing a point on the surface of the unit sphere by the spherical angles θ and ϕ, any quaternion can be described by three angles, the hyperspherical angle $0 \le \alpha \le \pi$ and the spherical angles $0 \le \theta \le \pi$ and $0 \le \phi \le 2\pi$. These angles relate to the quaternion $\mathbf{q}$ via

$$
\begin{aligned}
q_0 &= \cos\alpha \\
q_1 &= \sin\alpha\sin\theta\cos\phi \\
q_2 &= \sin\alpha\sin\theta\sin\phi \\
q_3 &= \sin\alpha\cos\theta
\end{aligned}
\tag{12.54}
$$

where the spherical angles describe the direction of the rotation axis $\hat{\mathbf{u}}$ and $\alpha = \tfrac{\omega}{2}$, see Equation 12.53. When parameterized by quaternions, the invariant or Haar measure, $d\mathbf{g}$, is given as [650]

$$d\mathbf{g} = \frac{1}{C}(\sin\alpha)^2\,d\alpha\sin\theta d\theta d\phi \tag{12.55}$$

The value of C depends on the symmetry of the material and any applied sample symmetry. For integration over $SO(3)$ or triclinic crystal symmetry $C = 8\pi^2$. For material with cubic and hexagonal crystal symmetries $C_c = \pi^2/3$ and $C_h = 2\pi^2/3$, respectively.

12.6 Appendix B: Bunge–Euler Angle Representation of Orientations

Bunge–Euler angles are probably the most well-known representation of orientations due to being the default representation used in EBSD software. The Bunge–Euler representation is perhaps the

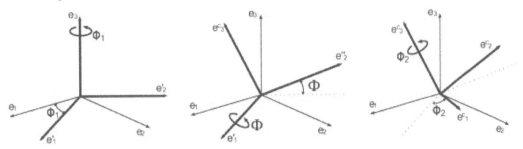

Figure 12.4: Schematic description of the Bunge's Euler angles used to establish the relationship between two arbitrarily defined Cartesian reference frames as a sequence of three rotations [341].

most cumbersome representation for manipulating discrete orientation data, as there is no simple way to compose rotations or straightforward multiplication rules. Therefore when computing with orientations it is common to utilize quaternion representations "under the hood" and use the familiar Euler angles when plotting ODFs or describing common orientations or texture components. The ubiquity of the Bunge–Euler angles is due to the compact Fourier series representation in terms of the generalized spherical harmonic basis functions [143].

Physically, Bunge–Euler angles represent a set of three rotations (ϕ_1, Φ, ϕ_2) that brings the sample frame into coincidence with the local crystal lattice. Figure 12.4 schematically shows the rotations and order. We can create the equivalent rotation matrix representation of the Bunge–Euler angles as

$$(12.56)$$

$$
g = \begin{bmatrix} \cos\phi_2 & \sin\phi_2 & 0 \\ -\sin\phi_2 & \cos\phi_2 & 0 \\ 0 & 0 & 1 \end{bmatrix} \begin{bmatrix} 1 & 0 & 0 \\ 0 & \cos\Phi & \sin\Phi \\ 0 & -\sin\Phi & \cos\Phi \end{bmatrix} \begin{bmatrix} \cos\phi_1 & \sin\phi_1 & 0 \\ -\sin\phi_1 & \cos\phi_1 & 0 \\ 0 & 0 & 1 \end{bmatrix}
$$

$$
= \begin{bmatrix} \cos\phi_1\cos\phi_2 - \sin\phi_1\cos\Phi\sin\phi_2 & \sin\phi_1\cos\phi_2 + \cos\phi_1\cos\Phi\sin\phi_2 & \sin\Phi\sin\phi_2 \\ -\cos\phi_1\sin\phi_2 - \sin\phi_1\cos\Phi\cos\phi_2 & -\sin\phi_1\sin\phi_2 + \cos\phi_1\cos\Phi\cos\phi_2 & \sin\Phi\cos\phi_2 \\ \sin\phi_1\sin\Phi & -\cos\phi_1\sin\Phi & \cos\Phi \end{bmatrix}
$$

A closer look at the definition of Bunge–Euler angles reveals a redundancy in the definition. There are indeed two possible choices for the selection of $\hat{e}'_1$ while satisfying the requirement that this direction be perpendicular to both $\hat{e}_3$ and $\hat{e}^c_3$. To address this redundancy, the following limits are set for the three Bunge–Euler angles

$$0 \leq \phi_1 < 2\pi, \quad 0 \leq \Phi < \pi, \quad 0 \leq \phi_2 < 2\pi \qquad (12.57)$$

Even with the restricted range specified above for the Bunge–Euler angles, there is an additional redundancy. This occurs in situations when Φ, for which we can specify only $(\phi_1 + \phi_2)$, and not ϕ_1 and ϕ_2 individually. Except for the situation noted above, it is possible to select a unique set of Bunge–Euler angles to describe the relationship between any two arbitrarily selected Cartesian reference frames in Euclidean 3-space.

The invariant measure is given as [143]

$$d\mathbf{g} = \frac{1}{C}\sin\Phi d\phi_1 d\Phi d\phi_2 \qquad (12.58)$$

12.7   Appendix C: Useful Properties of Generalized Spherical Harmonics

The following properties of Bunge's generalized spherical harmonics are useful for manipulating and developing mathematical expressions involving orientation relationships. They are all carefully developed and proved by Bunge in [143]; however, as that book is out of print and difficult to find we reprint them here (omiting proofs).

- Orthogonality of basis functions and symmetry coefficients

$$\int_{SO(3)} \dot{T}_l^{\mu\nu}(\mathbf{g})\dot{T}_{l'}^{*\mu'\nu'}(\mathbf{g}) = \frac{1}{2l+1}\delta_{ll'}\delta_{\mu\mu'}\delta_{\nu\nu'} \tag{12.59}$$

$$\sum_{n=-l}^{+l} \dot{A}_l^{n\nu}\dot{A}_l^{n\nu'} = \delta_{\nu\nu'} \tag{12.60}$$

- Addition theorem or composition of rotations

$$T_l^{mn}(\mathbf{g}^{(1)}\cdot\mathbf{g}^{(2)}) = \sum_{s=-l}^{+l} T_l^{ms}(\mathbf{g}^{(1)})T_l^{sn}(\mathbf{g}^{(2)}) \tag{12.61}$$

$$\dot{T}_l^{mv}(\mathbf{g}^{(1)}\cdot\mathbf{g}^{(2)}) = \sum_{s=-l}^{+l} T_l^{ms}(\mathbf{g}^{(1)})\dot{T}_l^{sv}(\mathbf{g}^{(2)}) \tag{12.62}$$

$$\dot{T}_l^{\mu n}(\mathbf{g}^{(1)}\cdot\mathbf{g}^{(2)}) = \sum_{s=-l}^{+l} \dot{T}_l^{\mu s}(\mathbf{g}^{(1)})T_l^{sn}(\mathbf{g}^{(2)}) \tag{12.63}$$

$$\ddot{T}_l^{\mu\nu}(\mathbf{g}^{(1)}\cdot\mathbf{g}^{(2)}) = \sum_{s=-l}^{+l} \dot{T}_l^{\mu s}(\mathbf{g}^{(1)})\dot{T}_l^{sv}(\mathbf{g}^{(2)}) \tag{12.64}$$

- Symmetries of Fourier coefficients and generalized spherical harmonic basis functions

$$C_l^{-m-n} = (-1)^{m+n}C_l^{*mn} \tag{12.65}$$

$$\dot{T}_l^{*\mu\nu}(\mathbf{g}) = \dot{T}_l^{\nu\mu}(\mathbf{g}^{-1}) \tag{12.66}$$

- Identity relationship

$$\ddot{T}_l^{\mu\nu}(I) = \sum_{m=-1}^{+l} \dot{A}_l^{m\mu}\dot{A}_l^{m\nu} \tag{12.67}$$

Chapter 13

Representation of Stochastic Microstructures

Stephen R. Niezgoda
The Ohio State University

13.1 Overview

The concept of microstructure is fundamental to the field of materials science and engineering. The origins of the field as a physical science, rather than a craft discipline, has its origin with the development of microscopy and metallographic techniques. Since the mid 1800s, developments in the field have been linked with the understanding that materials are not homogenous, but instead possess an internal structure or microstructure that spans several disparate length scales, from the atomistic to the macroscale. This structure while having a definite order at each scale is stochastic, or random, in the sense that the structure, for all length scales, is not identical from sample to sample or even spatially within the same sample. The near simultaneous development of metallography and stereology (or geometric probability as it was then called) to obtain point estimates of statistical three-dimensional (3D) microstructural descriptors (e.g., grain size or volume fractions) from two-dimensional (2D) sections highlights how significant the concept of randomness is to the field and how deeply it permeates our understanding of materials [254, 227, 861, 452]. Despite the central role that structural randomness plays in materials performance and processing, as a field, we have only just begun to leverage the knowledge-base and literature from the statistics, signal processing, and the machine learning and data analytics communities to develop microstructure descriptions which facilitate the quantification and visualization of structural variance in materials.

The slow progress in this area is largely due to the visual nature of materials structural data. In this regard materials science is transitioning from a *data limited* to a *data driven* field. The recent revolution in characterization techniques and sensors (most notably inexpensive CCD cameras) has enabled the rapid collection of cast amounts of digital 2D and 3D structural information. As the ability to collect data increases the development of analysis tools to synthesis raw data into materials knowledge must also progress in parallel. It is essential that metrics on the quality or sufficiency of microstructure data are be developed. For example critical questions such as "How much new information can be gained by characterizing additional samples, or is additional characterization redundant?" or "Given the current level of characterization what types of bounds can be placed on the probability of observing extreme property values or outliers of performance?" are currently open. In the opinion of the author the major obstacle is the difficulty in formulating appropriate mathematical descriptions of 2D and 3D visual materials data which can then be coupled with modern data analytics.

Recently, significant advances have been made in the quantitative description of microstructures via statistical measures and descriptors. The most common measures being the n-point correlation functions [475, 1189, 341, 1055, 1010], nearest neighbor distributions [1051, 843], the lineal path function and chord length distributions [341, 1055, 593, 475, 939, 593], topological descriptors [740, 357, 472], and information content (entropic) descriptors [786, 48]. In this chapter we will engage the mathematical formalism of stochastic processes to describe the inherent randomness in microstructure and look at what these various descriptors tell us about the underlying process. As a point we will look at the application of this work to the identification of representative volume element sets for use as input to computational materials simulations.

The reader will note that perhaps the most well-known statistical microstructure descriptors for polycrystalline materials, the orientation distribution function or other related crystallographic texture measures, are conspicuously absent from this chapter. Given the importance of texture to property prediction and processing of structural metals, and the richness of the field of quantitative texture analysis, these topics have been merged into Chapter 12. Also due to space limitations, the subject of the chapter was narrowly kept to the description of structure to keep the focus on materials characterization. Unfortunately that means that connections to properties, processing, and performance are not covered in this chapter as a meaningful discussion would be a book in itself. Interested readers can start with *Microstructure Sensitive Design for Performance Optimization* by Adams, Kalidindi, and Fullwood [9] and various reviews and current research papers [668, 400, 708, 667, 136] to get an overview of the field.

13.2   Interpreting Microstructure as a Stochastic Process

13.2.1   Brief Review of the Terminology and Notation of Stochastic Processes

In this section we will briefly review the terminology and notation of stochastic processes. (For a more complete review see [763].)

Consider a probability space defined by the ordered triplet $(\Omega, \mathscr{F}, \mathscr{P})$. Ω denotes the sample space and is the set of all possible individual outcomes, ω. $\mathscr{F}$ is a subset on Ω and is the set of observed events and is formally defined as a Borel σ algebra [763]. Each event is a set of zero or more experimental individual outcomes or observations. All the events in $\mathscr{F}$ that contain the specific outcome ω are said to have occurred. $\mathscr{P}$ denotes the probability measure on $\mathscr{F}$. A random variable $\mathbf{x}$ is then defined as a function, with domain Ω, that maps to each experimental outcome ω a number $\mathbf{x}(\omega)$ such that the standard axioms of probability are satisfied. The probability distribution function (pdf), $f(\mathbf{x})$, associates each possible experimental outcome with a probability such that $\mathscr{P}\{x_1 \leq \mathbf{x} \leq x_2\} = \int_{x_1}^{x_2} f_\mathbf{x}(x)dx$ and $\int_{-\infty}^{+\infty} f_\mathbf{x}(x)dx = 1$. Important statistical parameters of a random variable include the expected value or mean given by $\mu = E\{\mathbf{x}\} = \int_{-\infty}^{+\infty} x f_\mathbf{x}(x)dx$ and the variance $\sigma^2 = E\{(\mathbf{x} - \mu)^2\}$ which represents the average square deviation of a random variable from its mean. The concept of a random vector or field is a straightforward extension of a random variable. A random vector $\mathbf{X} = [\mathbf{x}_1, \ldots, \mathbf{x}_n]$ whose components are random variables. For real random vectors the correlation, R_{ij}, and covariance, C_{ij}, matrices are defined as

$$R_{ij} = E\{X_i X_j\} \tag{13.1}$$
$$C_{ij} = E\{(X_i - \mu_i)(X_j - \mu_j)\} = R_{ij} - \mu_i \mu_j \tag{13.2}$$

A stochastic process, $\mathbf{x}(t)$, is by extension a set of rules which assigns a function $x(t, \omega)$ to every experimental outcome ω of the experiment Ω. These rules take the form of a set of associated probability distributions. To completely determine the statistical properties of a stochastic process the nth order probability distribution function $f(x_1, \ldots, x_n; t_1, \ldots, t_n)$ must be known for all t_i, x_i, and n. For virtually all processes this in impossible. Instead the following first- and second-order statistical descriptors are used to partially quantify the process. The mean, $\mu(t)$, of the process $\mathbf{x}(t)$ is the expected value

$$\mu(t) = E\{\mathbf{x}(t)\} = \int_{-\infty}^{\infty} x f(x, t)dx \tag{13.3}$$

and the autocorrelation is defined as

$$R(t_1, t_2) = E\{\mathbf{x}(t_1)\mathbf{x}(t_2)\} = \int_{-\infty}^{\infty} x_1 x_2 f(x_1, x_2; t_1, t_2)dx_1 dx_2 \tag{13.4}$$

The covariance of a process is defined analogously to Equation 13.1. A process is considered wide sense stationary (WSS) if it has a constant mean and its autocorrelation depends only on $\tau = t_1 - t_2$ as

$$E\{\mathbf{x}(t)\} = \mu \quad \forall t, \tag{13.5}$$

$$E\{\mathbf{x}(t + \tau)\mathbf{x}(t)\} = R(t + \tau, t) = R(\tau) \tag{13.6}$$

Often when dealing with collected or measured data only a limited number of realizations of the process are available. In these cases it is convenient to invoke the concept of ergodicity and replace expectations calculated over a large ensemble of independent realizations of the process with expectations or integrals over a long time from a single process. The development of rigorous conditions under which ergodicity holds is beyond the scope of this overview, In general though such conditions depend on statistics and moments of higher-order than the autocorrelation. For the purpose of this overview all processes and microstructures will be considered WSS and ergodic up to their second-order (two-point) statistics unless specifically stated.

13.2.2   The Microstructure Process

The microstructure of virtually all materials exhibit rich details that span several hierarchical length scales. A microstructure constituent that can be assigned a distinct local structure or combination of properties can be identified as a distinct local state, For example Figure 13.1 shows the structure of a "typical" polycrystalline Ni-based superalloy for turbine disk applications at various length scales along with the relevant local states at each length scale. The local state can be denoted h and is an element of the local state space H that identifies the complete set of distinct local states that could be theoretically encountered [341]. For example at the scale of individual grains in a polycrystal the local state may include thermodynamic phase ρ, local gradient in chemistry or other thermodynamic potential μ, the lattice orientation g, the state of dislocation α, etc. In this way the local state can be considered as a random vector $\mathbf{h}(x) = [\rho, \mu, g, \alpha, \dots]$. For WSS microstructures (no microstructural gradients on the length scale of interest), the local state distribution $f_\mathbf{h}(h)$ over the local state space is understood as the volume density of material points in the microstructure associated with local state h. Note that the individual local state descriptors are not independent (components of the local state vector are not statistically independent); instead strong correlations exist between local states and local state descriptors at lower length scales may be conditional on higher length scales. Borrowing from the common notation of orientation distributions used in the texture community (see Chapter 12), the local state distribution is formally defined as

$$f_\mathbf{h}(h)dh = \frac{V_{h(m)dh/2}}{V} \tag{13.7}$$

where $V_{h(m)dh/2}$ denotes the volume fraction of the microstructure lying within an invariant (Haar) measure dh of local state h.

In principal the local state descriptors can be associated with length scales that span several orders of magnitude; thus individual material points are tied to the smallest relevant local state lengths scale. For computational modeling the scale of the material is not necessarily the smallest length scale relevant to the physics, but instead to a spatial discretization which is computationally practical. In order to effectively integrate the local state distribution across multiple length scale Adams et al. introduced the microstructure function $m(x,h)$ which was interpreted as a spatially resolved local state distribution [341, 8]. In terms of the probability space and language of stochastic processes defined in Section 13.2.1 the set of all possible events $\mathscr{F}$ is the set of all possible spatial arrangements of local states $h \in H$. The sample space Ω consists of an ensemble of microstructure realizations (characterized sample, or computational volume) where each realization is considered as an experimental outcome. The microstructure function $\mathbf{m}(x,h)$ is interpreted as a stochastic process that assigns a local state field $m(x,h,\omega)$ to every realization. Analogously with the large stochastic process literature $\mathbf{m}(x,h)$ can take on different interpretations depending on the context:

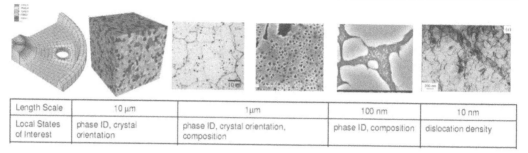

Length Scale	10 µm	1µm	100 nm	10 nm
Local States of Interest	phase ID, crystal orientation	phase ID, crystal orientation, composition	phase ID, composition	dislocation density

Figure 13.1: Example of hierarchical length scales and local states in a polycrystalline Ni-based superalloy.

1. An ensemble of microstructure realization. In this case x,h, and ω are all variables and $\mathbf{m}(x,h)$ specifically refers to the stochastic process or set of probability rules that gave rise to the ensemble. Note that the ensemble does not have to actually exist (either physically or on a computer); instead the ensemble may only be a mathematical construct.

2. A specific realization or experimental outcome of the microstructure process $\mathbf{m}(x,h)$. In this case x and h are variables but ω is fixed. The local state found at each spatial position x is the random variable or vector $\mathbf{h}(x)$ with associated probability distribution $f_{\mathbf{h}}(h(x))$. This interpretation is consistent with the work of Adams et al. [341, 8]. It is assumed that every infinitesimal material point can be associated with a distinct local state but for any finite region there will exist a distribution of local states.

3. If x is fixed but h and ω are variable then $\mathbf{m}(x,h)$ may also indicate the expected state of the local state at position x over the ensemble Ω. This is equivalent to asking what is the expected local state at point x or $E\{\mathbf{h}(x)\} = \int_{h \in H} h f_{\mathbf{h}}(h(x))dh$. For microstructures with gradients or other non-stationary process this allows for a description of how the expected structure varies with spatial position.

4. If x,h, and ω are fixed then $\mathbf{m}(x,h)$ is simply a number that gives the volume fraction of local state h at position x in realization ω.

It is important to remember that the microstructure process cannot be observed directly. Instead we can only see the output of the process. The process is best understood as a series of associated high-order probability distributions which describe how the local states are spatially placed in a material relative to each other. The statistics of the process must be estimated from the realizations of the process, specifically the local state fields, $m(x,h,\omega)$, associated with each sample.

13.2.3   Statistics of the Microstructure Function

Let's consider a material system where that local state space is defined by a combination of k local state descriptors of interest. For any material position x the local state is described by the random vector $\mathbf{h} = [\beta_1, \beta_2, \ldots, \beta_k]$. The probability that the local state $\mathbf{h}$ at material point x is in a region $\mathscr{H}$ of the local state space H is given by

$$\mathscr{P}\{\mathbf{h}(x) \in \mathscr{H}\} \int_{\mathscr{H}} f(\beta_1, \beta_2, \ldots, \beta_k; x)d\beta_1 d\beta_2 \ldots d\beta_k \qquad (13.8)$$

and from the axioms of probability $\int_H f(\mathbf{h}(x))dh = 1 \quad \forall x$. $f(h,x)$ is referred to as the first-order density of the microstructure, and can be interpreted as a spatially dependent fraction of local state $\mathbf{h}$. The marginal distribution, obtained by integrating over x, yields the sample average volume or area fraction of local state h. It is also important to consider the pdf defined by Equation 13.8 considers the expected value obtained from the ensemble of microstructure realizations Ω.

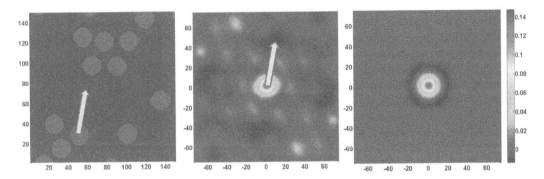

Figure 13.2: Toy example of the two-point covariance function (see Equation 13.12). The material is a simple-two-phase two-dimensional composite composed of circles embedded in a second phase (analogous to a fiber reinforced composite sectioned orthogonal to the fiber direction). One realization of the process is shown on the left. The center image shows the estimate of the two-point correlation function (Equation 13.10) obtained from the microstructure realization on the left. The yellow arrow in both images highlights the same vector r. The image on the right shows the 2-point correlation computed over an ensemble of 50 realizations. The correlation shows that the circles are uniformly distributed and maintain an minimum separation.

By extension the nth order pdf of the microstructure is given by the nth order joint density $f(h_1, h_2, \ldots, h_n; x_1, x_2, \ldots, x_n)$. Except for a few special cases, e.g., Gaussian process [763] where it can be shown that the process can be completely specified by lower-order distributions, knowledge of the nth order statistics is necessary to completely specify microstructure. Unfortunately, calculation of higher-order (> 2) pdfs is at best impractical and more often than not impossible given the level of feasible material characterization. Instead we must rely on the more convenient framework of hierarchical spatial statistics, the n-point correlations, to describe the stochastic geometry of heterogeneous materials (see [1010, 1055, 1189, 475, 465, 5, 736] and the references therein).

Provided the microstructure is WSS, the first-order density becomes independent of position $f(h, x) = f(h)$. Distributions on local-state are called one-point statistics as they reflect the probability of finding a specific local state at a randomly selected point in the material. Expanding on this basic idea, the two-point correlation $f_2(h_1, h_2 | x_1, x_2)$ is the joint density of occurrence of local state h_1 and h_2 at points x_1 and x_2, respectively.

$$f_2(h_1, h_2 | x_2, x_2) = E\{\mathbf{m}(x_1, h_1)\mathbf{m}(x_2, h_2)\} \tag{13.9}$$

Assuming a stationary microstructure, the correlation does not depend on the absolute endpoints but only on their vector separation $r = x_2 - x_1$ and the correlation becomes

$$f_2(h_1, h_2 | r) = E\{\mathbf{m}(x, h_1)\mathbf{m}(x + r, h_2)\} \tag{13.10}$$

The term auto-correlation is often used when $h_1 = h_2$ and cross-correlation when $h_1 \neq h_2$. Three-point and higher-order correlations can be defined in an analogous manner.

Equation 13.10 implies that the correlation functions are computed over an ensemble of realizations. If we consider a sample space, Ω, consisting of a set of P volumetric regions, denoted $(\omega_1, \omega_2, \ldots, \omega_P)$, each described by the local state field $m(x, h, \omega_i)$. The two-point correlation for the microstructure can be estimated as

$$f_2(h_1, h_2 | r) \approx \widehat{f}_2(h_1, h_2 | r) - \langle f_2(h_1, h_2 | r, \omega_i)\rangle \tag{13.11}$$

$$= \left\langle \frac{1}{\text{Vol}(\omega | r)} \int_{x \in \omega_i | r} m(x, h_1, \omega_i) m(x + r, h_2, \omega_i) dx \right\rangle$$

where $\omega_i|r = \{x|x \in \omega_i \cap x + r \in \omega_i\}$ and $\langle \cdot \rangle$ denotes the ensemble average. $f_2(h_1, h_2|r, \omega_i)$ is the estimate of the microstructure statistics obtained from a single realization or volume. If the individual realization is "large enough" then one can envoke the concept of ergodicity and replace the ensemble average with the volume average over a single sample.

Several authors have found it convenient to introduce a normalization of the 2-point correlation function sometimes misleadingly called the spatial covariance as [1055, 1010]

$$\tilde{f}(h_1, h_2|r) = \frac{f(h_1, h_2|r) - f(h_1) \cdot f(h_2)}{f(h_1) \cdot f(h_2)} \tag{13.12}$$

This normalization is especially useful in analyzing the 2-point correlation of polycrystalline materials where the local state includes orientation information [5]. In Equation 13.12 the magnitude of the correlation is normalized by the volume fractions of each local state. When one of the local states (h_1, h_2) is a minor constituent of the microstructure (e.g., a minor texture component or secondary phase particles with a small volume fraction) crucial spatial correlation information may be lost without renormalization.

The definition of the 2-point correlation implies the symmetry relation $f(h_1, h_2|r) = f(h_1, h_2| - r)$. From the basic properties of distribution functions

$$\int_{h_2} f(h_1, h_2|r)dh_2 = f(h), \quad \int_{h_2} \int_{h_1} f(h_1, h_2|r)dh_2dh_1 = 1 \tag{13.13}$$

In other words the 1-point distribution is the marginal distribution for the higher-order microstructure correlation functions.

Equation 13.11 is readily identified as a convolution-like operation which can be readily computed using Fourier transform techniques [341]. In practice spatial statistics are typically calculated on a digital image (i.e., two- or three-dimensional array or pixels/voxels). Equation 13.11 can be recast in a discrete form as

$$F_t^{h_1 h_2} = \frac{1}{S} \sum_{s=0}^{S-1} m_s^{h_1} m_{s+t}^{h_2} \tag{13.14}$$

where the indices s and t indicate the discretized spatial position and separation vector respectively [341]. Making use of the convolution properties of the discrete Fourier transform (DFT) Equation 13.14 can be recast as

$$\Im\left(f_t^{h_1 h_2}\right) = F_k^{h_1 h_2} = \frac{1}{S} M_k^{h_1 *} M_k^{h_2} \tag{13.15}$$

where the $*$ indicates complex conjugation. In practice Equation 13.15 is the most efficient method to calculate 2-point correlations. In using the DFT one must be cognizant of the implicit periodic boundary conditions imposed on the structure and if the periodicity is problematic implement appropriate padding or other correction [133]. Three-point and higher-order correlations are also efficiently computed using spectral approaches [341].

13.3 Microstructure Descriptors

13.3.1 Metrics from the Two-Point Correlations and Characteristic Length Scales

The spatial correlation functions contain a great deal of information about the microstructure; however, it is not always easily interpreted. Conversely, scalar measures are easy to understand and often carry a simple physical meaning but are limited in information content. The connections between the hierarchical spatial statistics and common descriptors such as mean particle size are not always straightforward. There are however some important scalar metrics closely connected with the correlation functions, several of which can often be estimated from the correlations easier than from the original micrograph.

Geometrically we also expect the following relationships to hold for very short and very long r

$$\lim_{||r|| \to 0} f(h_1, h_2 | r) = \begin{cases} f(h_1) & \text{for } h_1 = h_2 \\ 0 & \text{otherwise} \end{cases} \tag{13.16}$$

$$\lim_{||r|| \to \infty} f(h_1, h_2 | r) = f(h_1) f(h_2) \tag{13.17}$$

Equation 13.17 motivates the definition of a critical length scale of interest, the coherence length. The coherence length, r_c, will be defined to be the separation at which local states in a material become spatially uncorrelated and will be defined as

$$f(h_1, h_2 | r) - \langle f(h_1) \rangle \cdot \langle f(h_2) \rangle \le \varepsilon \quad \forall ||r|| \ge r_c \tag{13.18}$$

For example the microstructure shown in Figure 13.2 has an exceptionally short correlation length as the second phase circles were uniformly distributed and non-interacting.

There is a strong connection between microstructure statistics and the field of stereology. The connection between local state autocorrelations and stereology was actually partially developed in the 1950s by Guinier, Fournet, and Debye in their work on radiation scattering in isotropic porous three-dimensional materials [247, 248, 398]. This work was extended by Berryman who showed that it holds for anisotropic solids as well, provided the 2-point statistics were integrated over the angular portion or the vector r [85]. The 2-point autocorrelation can be written as a power series of the scalar distance $s = ||r||$ with the form

$$f(h, h | s) = f(h) - \frac{S_v^h}{4} s + \mathcal{O} s^2 + \dots \tag{13.19}$$

where S_v^h is the specific surface area or interface area per unit volume associated with local state h. The derivative of the autocorrelation as $r \to 0$ is proportional to the specific surface area for three-dimensional materials. In general for a d-dimensional material [1055, 247]

$$\left. \frac{d\left(f(h, h | r)\right)}{dr} \right|_{r=0} = \begin{cases} -S_v^h / 2, & d = 1 \\ -S_v^h / \pi, & d = 2 \\ -S_v^h / 4, & d = 3 \end{cases} \tag{13.20}$$

Readers familiar with stereology will recognize the well-known relations $S_v = \frac{4}{\pi} L_a = 2 P_l$ (L_a is the interface length per unit area on a 2-D cut and P_l is the number of interface points per unit length on a 1-D cut) [1069]. This result allows for the calculation of surface area from microstructures data collected on two-dimensional sections as well as three-dimensional volumes. The above results can be extended to show that the mean chord length of local state h, l_c^h is also related to the slope of the angular averaged two-point correlation at the origin via [1054]

$$l_c^h = \begin{cases} 2 f(h) / S_v^h, & d = 1 \\ \pi f(h) / S_v^h, & d = 2 \\ 4 f(h) / S_v^h, & d = 3 \end{cases} \tag{13.21}$$

The coherence length, r_c is the first microstructural length scale of interest related to the two-point correlation functions (Equation 13.18). Another important length scale is the average feature length (average particle or grain size, etc.), which Tewari has defined as the short range length, r_0. By examination Tewari has suggested that r_0 can be determined from the autocorrelation of the local state of interest as

$$\left. \frac{d(f(h, h | r)}{dr} \right|_{r_0} = \frac{1}{10} \left. \frac{d(f(h, h | r)}{dr} \right|_{r=0} \tag{13.22}$$

Torquato has proposed the decomposition of the two-point correlation into a connected part and a disconnected part [1053]

$$f(h,h|r) = C(h|r) + E(h,h|r) \qquad (13.23)$$

where $C(h|r)$ is termed the two-point cluster function that gives the probability of finding the end-points of vector r in the same particle, and $E(h|r)$ is the two-point blocking function or the probability of finding both ends of the vector in local state h but in different particles or disconnected regions. Decomposing the autocorrelation is a non-trivial problem. The support of $C(h|r)$ is specified by the largest particle in the system. Given the symmetry of the autocorrelation with respect to $(m)r\, C(h|r)$ can only be non-zero over a region exactly $2\times$ the largest particle. McSleyene et al. proposed approaches to estimate the cluster and blocking functions in anisotropic materials (they called these the particle autocorrelation and the particle cross-correlation, respectively, and observed that $E(h|r)$ is strongly dependent on the individual particle shapes [614, 615]. Instead they proposed an approximation by replacing each particle with a point mass and performing a statistical analysis to estimate the blocking function from the relative particle positions and the two-point cluster function.

Tewari et al. observed that beyond the correlation length the normalized 2-point correlation (Equation 13.12) oscillates around zero with the amplitude of the oscillations decaying with increasing r. The characteristic wavelength of those oscillations, λ_∞, is related to the equilibrium spacing between clusters of particles and is another important length scale of interest.

13.3.2 Other Microstructure Descriptors

Numerous other higher-order statistical and topological descriptors have been developed and utilized for characterization of complex materials and microstructures. Often these parameters and descriptors are developed for a specific material/property combination and are not broadly applicable. Here a few of the more general descriptors with relevance to a wider range of material systems are reviewed. For a more detailed description of many other descriptors the reader is urged to start with the more comprehensive (almost a classic by now) *Random Heterogeneous Materials* by Salvatore Torquato [1055]. Note that discussion of entropic or other information theory based descriptors of microstructure are omitted, purely for space limitations. Interested readers should review the work of Piasecki and collaborators for more information on this interesting and underutilized (in the opinion of the author) class of microstructure analysis tools [785, 786, 333].

13.3.2.1 Surface Correlation Function

The surface correlation functions can be thought of as a generalization of the n-point correlations where the interface between two phases or regions is considered a distinct local state. The spatially resolved specific surface area, $S_v^{h_1}(x)$ is analogous to the 1-point correlation function and gives the interfacial area connecting local states h_1 and $h \neq h_1$ in a neighborhood about x. For simplicity consider a material system consisting of only two local states. Torquato introduces higher-order surface correlations through the use of indicator functions $\chi^{h_1}(x)$ and $\aleph^{h_1}(x)$. $\chi^{h_1}(x)$ takes the value 1 if $\mathbf{h}(x) = h_1$ and 0 otherwise. $\aleph^{h_1}(x)$ is the spatial derivative of $\chi^{h_1}(x)$ and only takes value at the interface and is zero everywhere else. Using $\nabla h_1(x)$ to indicate the surface, the surface-h_1 and the surface-surface correlations are then defined as

$$f(\nabla h_1, h_1 | x_1, x_2) = \langle \aleph^{h_1}(x_1)\chi^{h_1}(x_2) \rangle \qquad (13.24)$$
$$f(\nabla h_1, \nabla h_1 | x_1, x_2) = \langle \aleph^{h_1}(x_1)\aleph^{h_1}(x_2) \rangle \qquad (13.25)$$

Higher-order surface correlations can be defined analogously and provided the microstructure is stationary the dependence on absolute position can be replaced by relative separation r.

13.3.2.2 Chord Length Distributions and Lineal Path Functions

The chord length distribution and lineal path function are two important measures of particle or grain size and the connectedness of microstructure features along a straight path. Chords are defined as

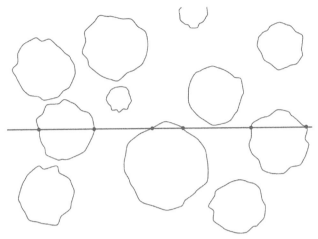

Figure 13.3: Schematic of measuring the chord-length distribution function in a two-phase random heterogenous material. Chords are defined by the intersection of a line with the random phase-interfaces. After Torquato and Liu [1054].

line segments between intersections of an infinitely long line with random interfaces in the material. Figure 13.3 schematically shows the identification of chords in a two-phase material. The chord length distribution can be denoted as $p(h|z)$ and $p(h|z)dz$ gives the probability of finding a chord of length z and $z+dz$ connecting the interfaces of local state h. The chord length distribution plays an important role in stereology as first moment of the distribution is the mean lineal intercept length l_c; used to determine average grain size in ASTM E112 and other similar intercept counting methods.The chord length distribution provides information about the "discrete free path" of a species important in transport problems such as Knudsen diffusion or radiative transport [865, 1054]. Another related statistical measure is the lineal-path function, $L(h|z)$, which gives the probability that length z lies entirely in local state h [593, 1055]. $L(h,z)$ is monotonically decreasing with z with the asymptotic bounds $L(h|0) = f(h)$ and $L(h,|\infty) = 0$. For a system of simple convex particles the lineal path function of the particle phase is equivalent to the 2-point cluster function $C(h,r)$ [1055]. Again the lineal path function has important connections with stereology specifically with calculating the projected area fractions or projected length fractions from three-dimensional or two-dimensional structures respectively. Lu and Torquato have demonstrated that the lineal path function is identically the area fraction of phase h measured from the projection of a slab of thickness z onto a plane in the limit of a very large ensemble of microstructure realization. This projection is shown schematically in Figure 13.4.

Lu and Torquato used a simple probability argument to shown that the chord length distribution can be obtained directly from the lineal path function in the following manner

$$p(h|z) = \frac{l_c(h)}{f(h)} \frac{d^2 L(h|z)}{dz^2} \tag{13.26}$$

The above relation is very useful in a theoretical sense for modeling stochastic materials, however the difficulty associated with calculating the second derivative from measured intercept data makes it difficult to apply in practice.

13.3.2.3 Topological Invariants

The chord length distribution and lineal path function are two common measures that provide information on the distribution on the individual particle or grain shape that is independent of the spatial arrangement of the population of particles. However these measures are not invariant to scaling of

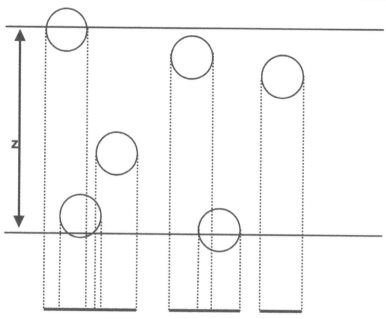

Figure 13.4: Schematic of measuring the lineal path function in a two-phase random heterogenous material. Chords are defined by the intersection of a line with the random phase-interfaces. After Torquato and Liu [1054].

the particles or rigid body rotation. Recently MacSleyne et al. have applied the concept of shape moment invariants to characterize the absolute shape of microstructural features independent of scale, rotation, and transformation [614, 615]. The technique is discussed in detail in Chapter 17.

Many properties such as conductivity are dependent on the connectivity of microstructural features, in addition to the volume fractions and shape. For high-contrast composites and porous solids it is natural to assume that the directionally dependent connectivity of the stronger phase will have a strong effect on the mechanical response. Percolation refers to the existence of a connected path through the material along which transport can occur (fluid flow, electrical conduction, short circuit diffusion). Classical percolation theory uses statistical arguments to show that for random materials there is a critical percolation threshold $f(h) = \rho_c$ where the probability of finding a percolating path jumps from near zero to near one [390]. For two-phase "random" structures (where the probability of finding h_1 at each material point is given by p and h_2 is $1 - p$, and the local state assignment of each point is independent of its neighbors) the percolation threshold can be calculated; however, in materials where spatial correlation exists between local states the percolation threshold can be significantly higher or lower [919]. Unfortunately the identification of percolating paths in three dimensions is not trivial, and simply determining if such a path exists does not provide much information about the local connectivity in a material (e.g., branching and looping of paths), the number of paths, or other microstructure details which might have relevance. Porous solids and foams (with infinite contrast between the material and void/air phases) must percolate otherwise they would simply break apart, and it is obvious that the number and nature of connections is critical to properties. For materials *above and near* the percolation threshold scaling relationships can often be developed which link properties with volume fraction, $f(h)$ [919, 357]. However when the volume fraction is well above the percolation threshold simple scaling relationships break down. In this case a more general measure of connectivity is required, such as topological metrics.

Topological invariants refer to proper properties of an object that are preserved under homeomorphisms (the continuous stretching or bending of an object into a new shape). Here we will limit the discussion to metrics related to homology or the mathematical classification of objects based on the number of "holes" and connected/disconnected components in the structure. The strength of this approach to characterizing stochastic materials is that it provides a description of the long-range connectivity in a microstructure. The drawback of homology as a metric is the lack of scale in the topological relationships. It is intuitive to expect that the size and shape of clusters or percolating paths are important [390], however here we are only consider metrics that are invariant to stretching and bending. Size and anisotropy are neglected from a topological point of view.

Structures can be classified with respect to their homology, connectedness of local state h, via the Betti numbers $\beta_n(h)$. The nth Betti number gives the number of cuts which must be made to an n-dimensional surface before separating it into two disconnected surfaces [714]. The number β_0 gives the number of disconnected regions of local state h, β_1 gives the number of independent one-dimensional circular loops in the structure, β_2 gives the number of two-dimensional voids or holes, etc. For example a hollow torus has Betti numbers $[\beta_0 = 1, \beta_1 = 2, \beta_3 = 1]$. The torus has 1 connected component. two circular holes (the center of the donut and the middle of the hollow), and one two-dimensional void. Figure 13.5 shows a simple toy two-dimensional composite microstructure with three disconnected components and two independent loops giving Betti numbers $[\beta_0 = 3, \beta_1 = 2]$. In general for an m-dimensional structure $\beta_{n \geq m} = 0$.

Let $H_n(h)$ denote a homology group in n dimensions of local state h in a microstructure realization. Formally $H_n(h)$ is an algebraic group that contains information about the connectivity of local state h in the realization and can be decomposed into the product of a free group and a torsion group. The nth Betti number is the rank of $H_n(h)$ (a simplification provided the structure is torsion free [714]) A more complete discussion of the mathematics is beyond the scope of this text, but it can be shown that the Betti numbers completely define the homology thus the connectivity of the structure.

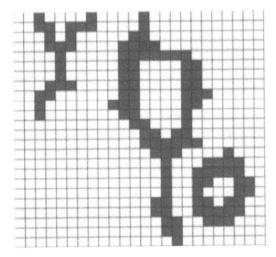

Figure 13.5: A simple two-dimensional structure with 3 disconnected regions, and two independent loops. The resulting Betti numbers are $[\beta_0 = 3, \beta_1 = 2]$. Figure from Gerrard et al. [357].

In order to connect the homology directly with percolation, Gerrard et al. introduced the concept of relative homology [357]. Percolation theory relates to connectedness between the boundaries of a given domain. If the boundaries are considered as elements in the structure, a relative homology can be used to determine the connectedness of the boundaries through the structure (e.g., will it percolate). Let A denote the top boundary of Figure 13.5 and B be the bottom interface. Ignoring the formality of defining the relative homology groups, let the relative 1st Betti number $\beta_1(h, A)$ give the number of independent loops or paths *that begin and end in A*. For the example in Figure 13.5, $\beta_1(h, A) = 2$ as the addition of the top boundary completes the loop in the upper left corner. The number of percolating paths, n_p, between A and B can be given by

$$n_p = \beta_1(X+A+B, A+B) + \beta_1(X) - \beta_1(X+A) - \beta_1(X+B) \quad (13.27)$$

If the number of paths is not important, a binary determination of whether percolation has occurred can be obtained from

$$\beta_1(X+A+B, A+B) - \beta_1(X+A+B) = \begin{cases} 1 & \text{if percolation has occurred} \\ 0 & \text{otherwise} \end{cases} \quad (13.28)$$

The Euler characteristic is another well-known invariant which has been applied to the analysis of percolation [917, 672, 674]. The Euler characteristic, Ψ, was originally defined for the classification of complex polyhedra and relates the number of faces F, edges E, and vertices C as $\Psi = V - E + F$. For all convex polyhedra in n-dimensions $\Psi = 2$. Ψ can be generalized for all topological objects as the alternating sum

$$\Psi = \beta_0 - \beta_1 + \beta_2 - \beta_3 \ldots \quad (13.29)$$

Recently, Scholz et al. observed the following relationship for the permeability of porous materials, k, where the pores are modeled as interpenetrating spherical or elliptical particles

$$k = cl_c^2 \left(\frac{1-\Psi}{N}\right)^\alpha \quad (13.30)$$

c and α are constants, l_c is a characteristic hydrodynamic distance and N is the total number of spheres of ellipsoids used to model the pore structure.

13.4 Reduced Order Descriptions and Relational Statistics

The above sections provided an overview of a small subset of the available higher-order microstructure descriptors. In general the discussion above was also limited to the simplest of local state spaces, generally only composites of two binary local states at one relevant length scale. The application to more realistic materials necessitates a much more complex local state space and sets of microstructural descriptors that describe the structure at each length scale of relevance. Even for the example material systems described above, generally combinations of multiple descriptors provide significantly more detail about the microstructure than any single one. For example it has been observed that using the 2-point correlations in conjunction with either the lineal path function or entropic descriptors provided to computationally generate microstructure realizations results in structures that are "more representative" than using the correlation functions alone and are generally much more efficient [786, 1189, 1055].

The difficulty in using multiple higher-order descriptors is the curse of dimensionality. For example, consider a realization simple two-local state composite, a micrograph of 256×256 pixels. The autocorrelation of one phase is also an array of 256×256 elements, which for convenience can be reshaped into a vector of length 65,536. Granted the symmetry of the correlation reduces the independent component by a factor of 2, however it is clear that while such metrics contain much more information they are much more difficult to work with than simple scalar measures like volume fraction. This problem grows exponentially as more descriptors are added for more interesting materials. In order to effectively manipulate and visualize this data in some sort of analysis or design framework some sort of dimensionality reduction is required.

While there are numerous approaches to dimensionality reduction, the needs of the materials data analytics places strong constraints on the type and complexity of the dimensionality reduction scheme. In order to determine an appropriate low-dimensional representation the following factors should be considered:

1. The approach must be computationally robust and be insensitive to the type or structure of the underlying data.

2. The representation should be able to be updated in (near) real time as new datasets or realizations are added.

3. The representation should be invertible, meaning that given the low-dimensional representation the original high-dimensional data should be recoverable.

4. In order to facilitate the description of comparative and relational statistics the resulting low-dimensional variables should be independent or at least uncorrelated, suggesting an orthogonal decomposition of some sort.

Based on these requirements, dimensionality reduction via principal component analysis (PCA) has been the primary tool over non-linear approaches which may yield a more efficient representation [741, 743, 988, 986].

As a quick reminder, consider an ensemble of N microstructure realizations $\Omega = \omega_1, \ldots, \omega_N$. Let $\theta(H, \omega_n)$ be a feature vector that contains all the microstructure descriptors of interest for a single realization and let $\Theta(H)$ be a matrix whose columns are the feature vectors of the complete ensemble. PCA seeks to find a decomposition

$$\theta(H, \omega_n) = \sum_{j-1}^{N-1} \alpha_j^p \phi_j + \widehat{\theta} \tag{13.31}$$

where ϕ_j are the orthogonal principal component vectors, α_j^p is the weight for the jth principal component vector for the pth microstructure realization, and $\widehat{\theta}$ is the average feature vector over the ensemble. PCA is most easily interpreted as a projection of the high-dimensional data onto a low-dimensional orthogonal coordinate frame where the axes are defined by the directions of highest variance. The weights form a new set of independent variables on which further analyses, relational statistics, or visualization can be performed.

In order to effectively utilize the low-dimensional representation, a measure of distance in the space must be introduced. For comparing different realizations of the same microstructure the Euclidean distance is sufficient. However, when comparing two different ensembles of potentially different materials, for example when trying to quantify the microstructural differences that result from two different heat treatments or deformation processing routes in a metal alloy, the Euclidean distance is no longer appropriate. In this case we are asking how close are the clouds of points representing the two microstructural ensembles? Or how close is a new data point from the cloud of points from each ensemble? For this sort of analysis the Mahalanobis distance is more appropriate, which is commonly used in multivariate statistics to gauge the similarity between an unknown sample set to a classified set [969].

Consider a multivariate vector $X = (x_1, \ldots, x_n)^T$; the Mahalanobis distance between X and a distribution of vectors $f(Y)$ with mean μ_Y and covariance matrix C_Y is defined as

$$D_m(X, f(Y)) = \sqrt{(X - \mu_Y)^T C_Y^{-1} (X - \mu_Y)} \tag{13.32}$$

The key feature of the Mahalanobis distance is that it takes the shape of the known probability distribution into account, in that all X that lie on the same iso-probability surface of $F(Y)$ will have the same Mahalanobis distance regardless of the shape of $f(Y)$, provided that the known distribution is "well behaved" and unimodal. The Mahalanobis distribution can be generalized to a similarity measure between multivariate vectors X and Y where Y comes from a distribution $f(Y)$ as

$$d(X, Y) = \sqrt{(X - Y)^T C_Y^{-1} (X - Y)} \tag{13.33}$$

When computing distances in the PCA space the individual components of X and Y, x_i and y_i, respectively, are independent (orthogonal) variables and the covariance matrix is by definition diagonal. In this case $d(X, Y)$ reduces to a normalized Euclidean distance $d(X, Y) = \sqrt{\Sigma_i (x_i - y - 1)^2 / \sigma_i^2}$.

In answering the first hypothetical question "Are two microstructures produced by different processing routes equivalent?", we really wish to answer two distinct but equally important questions. The first is, do the ensembles share the same mean or average representative structure. The second is, do the two ensembles possess the same degree of microstructure variability or scatter about the mean. For quality control and process certification it is critical to have quantitative measures that describe any drift in the process or to be able to compare material produced on two different processing lines. It is natural to ask whether two different lots of material are indeed the same, and if not, how different are they. If they are confirmed to possess the same structure, then analysis and comparison of the variability of the ensembles may be carried out to determine the likely effects on properties and performance.

The standard tool for hypothesis testing on the means of multivariate data is the 1-way multivariate analysis of variance (1-way MANOVA). The 1-way MANOVA tests the null-hypothesis that the means of each group lie in the same d-dimensional subspace. Choosing $d = 0$ tests that the means are the same or that the ensembles have the same average structure. The major assumption in performing a 1-way MANOVA analysis is that the different classes or ensembles have the same covariance matrix; however, the analysis has been demonstrated to be robust to violations of this assumption provided the individual distributions are "well behaved" and unimodal [969]. The basic notion of the MANOVA analysis is to transform the original n-dimensional dataset into a new variable on a d-dimensional subspace, termed the canonical variable, that maximizes the separation between the classes, then to perform a statistical test on the ratio of within group scatter of the canonical variable to the total scatter across all groups. Note that the transformation seeks to maximize separation between ensembles in contrast to PCA which seeks the direction of maximum variability across all ensembles.

If we consider a dataset of K material volume ensembles where $\Omega_k = \{\omega_1^k, \ldots, \omega_p^k\}$ denotes the p members of the kth ensemble, after performing PCA each ω_p^k is represented by the J weights $^k\alpha_j^p$. For compactness, let A_{pk} denote the J dimensional column vector of PCA weights for the pth member of the kth ensemble. Further let $\widehat{A}_k$ represent the mean vector of PCA weights over the P_k members of ensemble k. The within group scatter is characterized by the intra-class sum of squares and cross product (SSCP) matrix W defined

$$W = \sum_{k=1}^{K} \sum_{p=1}^{P} (A_{pk} - \widehat{A}_k)(A_{pk} - \widehat{A}_k)^T \qquad (13.34)$$

The between group scatter is characterized by the inter-class SSCP matrix defined as

$$B = \sum_{k=1}^{K} P^k (\widehat{A}_k - \widehat{A})(\widehat{A}_k - \widehat{A})^T = \sum_{k=1}^{K} P^k \widehat{A}_k \widehat{A}_k^T \qquad (13.35)$$

where $\widehat{A}$ is the mean weight across all K ensembles which is by definition the zero vector. The total scatter is then $T = W + B$. The test statistic for the 1-way MANOVA can be chosen as Wilks's lambda $\Lambda = |W|/|T|$ which is the multivariate extension of the F-test, or Pillai's trace $\text{TR}(WT^{-1})$ which tends to be more robust to violations of the assumptions.

The representation of the data in terms of the canonical variables can be found by performing an eigenvalue decomposition of $W^{-1}B$ then projecting the PCA weights onto the eigenvectors. If L is a matrix of eigenvectors then the canonical representation can be found $c_{pk} = LA_{pk}^T$.

The canonical representation is useful for visualizing clustering and other relationships between groups. We will briefly show an example described in detail by Niezgoda, Kanjarla, and Kalidindi in [741]. For this example, ensembles of 50 realizations each were constructed for 8 different classes of porous composites. The microstructure metric of interest was the 2-point correlation function, $\Theta = f(h, h|r)$. Representative examples of each class are shown in Figure 13.6 and the reduced order representation in the PCA space and in terms of the canonical variables are shown in Figure 13.7. Figure 13.7 shows that the microstructure classes can be separated by projection into the first

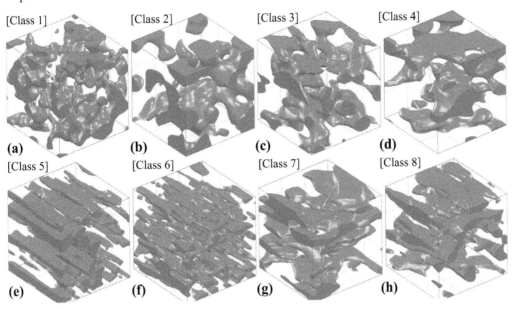

[Class 1] (a) [Class 2] (b) [Class 3] (c) [Class 4] (d) [Class 5] (e) [Class 6] (f) [Class 7] (g) [Class 8] (h)

Figure 13.6: Representative examples of the 8 classes of porous solids (material classes) used in this study. The pores are shown in red, while the isotropic matrix is transparent [741].

two canonical variables. In other words, there exists a plane in the 50 dimensional PCA space, that when the microstructure data is projected onto it there is virtually no overlap between the classes. Projection onto the first three canonical variables yields perfect separation. In contrast the projection onto the first three principal components shows that many of the classes are strongly overlapped.

That perfect separation can be achieved by only three canonical variables does not imply that the data lies on a hyper-plane in the 50 dimensional PCA space. A 1-way MANOVA analysis indicates that at 5% significance we can reject the null hypothesis that the means of the 8 microstructure classes lie in a 6 or lower dimensional space. Performing the test pairwise between the groups indicates that we can state with statistical certainty that the means of the 8 classes are all different or that the microstructures of the 8 ensembles are all distinct.

This notion of Mahalanobis distance can be used to further explore the relationships between the various groups. A key question to ask is what is the expected distance between a randomly selected member of class a to class b. By computing this expectation for all combinations of classes the entire dataset (8 material classes) can be broken up into groups and subgroups of classes that share similar features by a hierarchical cluster analysis. Hierarchical cluster analysis is a technique to build a multi-level hierarchy of groups and subgroups that are easily represented by a tree-like graph. The analysis is performed by first computing the distance between the different groups, in case the distance measure is the expected Mahalanobis distance between a randomly selected member of class a and the cloud of points corresponding to class b, $\overline{D}_M(a,b) = E\{D_M(x \in a, f(y \in b)\}$. Then the two closest groups are combined into a larger group. The procedure is then repeated until all the smaller groups have been combined into a single large class.

The results of the hierarchical cluster analysis are shown in Figure 13.8 as a cluster tree. The cluster tree is interpreted by moving up on the y-axis, which shows the expected Mahalanobis distance; the height of the inverted U connecting two groups is the expected distance between those groups. Note that $\overline{D}_M(a,b) \neq \overline{D}_M(b,a)$ due to the differences in covariance between the groups; for plotting the cluster tree the minimum of the two expectations is taken. Groups that are linked lower down are more similar than groups that are linked higher up. For example, classes 5 and 6 are more similar than 7 and 8, and the expected distance between a group formed from classes 1-4 and the

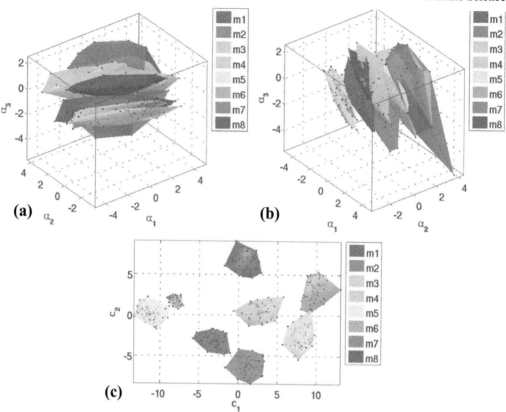

Figure 13.7: (a) and (b) Projection of the members of the 8 microstructure classes into the inter-class PCA space. The points corresponding to each class are enclosed in different colored convex hulls to help visualize the space spanned by each material. Two different views are presented for clarity. (c) Projection of the microstructure data onto the first two canonical variables, c_1 and c_2. This is the two-dimensional view that offers the best linear separation of the dataset into microstructure classes. There is a slight overlap between microstructure classes 2 and 4. Projection onto the top three canonical variables yields a perfect separation.

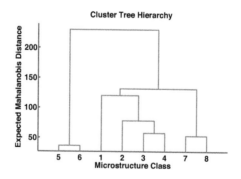

Figure 13.8: Cluster tree graph showing the relationship between the different microstructure classes. The height of the inverted U joining two groups gives the expected distance between groups.

group consisting of classes 7 and 8 is approximately 135. While the cluster tree graph is a useful visualization of the relationships between the microstructure classes, it is up to the end user to decide on the number of clusters or groups in the dataset or where to cut the tree to define clusters. For this example a likely pruning scheme is to break the data into three groups: the first consisting of classes 5 and 6, the second containing classes 1-4, and the third classes 7 and 8. It is interesting to note that this grouping naturally follows the degree of anisotropy in the pore shape as seen in Figure 13.6. Classes 1-4 all have a low degree of anisotropy with a covariance ratio of 2 or less between the highest and lowest direction, while the other two groups are more highly anisotropic with 5 and 6 having highly elongated pores along 1 primary axis, while classes 7 and 8 have pores elongated along two directions.

13.5 Closing Thoughts

This chapter provided a very brief overview of some of the current research developments regarding the statistical modeling and characterization of microstructure. It is expected that research interest in these areas will continue to grow, driven by the need within the materials community for effective and efficient materials databases and design frameworks as envisioned by the Materials Genome Initiative and the continued growth of integrated computational materials engineering [18, 760]. In order to transition from ideas to applications there are still several key open technological questions:

1. What measures of the material microstructure best capture its salient features?

2. How do we describe the inherent variability in a material and how do we link this variability with scatter in properties or performance?

3. How do we quantitatively compare materials from different classes or material systems?

4. How do we place metrics on the quality or sufficiency of the microstructure data collected, or how do we place measures on how well we have characterized a particular material?

5. What objective data-driven reduced-order measures of the material microstructure are best suited for establishing the invertible structure-property relationships needed for materials design?

While a concerted effort from the larger materials community is required if workable microstructure data analytics tools are to ever come to fruition, the author feels that the overview presented in this chapter represents a critical first step in their development.

Chapter 14

Computer Vision for Microstructural Image Representation: Methods and Applications

Brian L. DeCost
Carnegie Mellon University

Elizabeth A. Holm
Carnegie Mellon University

14.1 Introduction

Analysis and interpretation of complex mesoscale materials structure – *microstructure* – has long been considered a distinctly human task requiring deep physical intuition, especially where the materials community has not yet succeeded in formulating reductionist models for mesoscale physics [947, 857]. The rise of materials data science along with the Materials Genome Initiative of 2011 [665] has been accompanied by increasing interest in applying to microstructure data the concepts and techniques from the fields of computer vision and machine learning.

Computer vision is a broad field organized around the deceptively difficult task of describing and interpreting image data, with roots extending to the early years of artificial intelligence research in the early 1970s [996]. Many contemporary tasks in computer vision are heavily motivated by robotics applications, such as object recognition, localization, and tracking; interpretation of human gestures, facial expressions, and actions; and the inverse problem of 3D scene reconstruction from multi-viewpoint images. Other focus areas include image- and video-based information retrieval; image restoration, completion, and synthesis; object segmentation; image texture analysis; and scene interpretation. These last two tasks are perhaps the most apt analogy for the task of microstructure interpretation.

The concept of texture in this context takes the same meaning as in the visual arts, as opposed to the crystallographic texture that the materials scientist will be familiar with. Image texture describes the spatial distribution of pixel-level intensity values within a region, without regard to the global shape of the region. Common image texture tasks include identifying macroscopic material surfaces [177], extracting geospatial information from satellite imagery [178], medical screening and diagnosis using 2D and 3D imaging techniques [221, 384], and various process monitoring systems in the minerals industry and surface inspection [90, 1162, 17]. Microstructure is similar in the relative unimportance of global geometry, but often the features of interest span a greater breadth of length scales than in traditional image textures. For example, a lamellar structure such as pearlite represents an image texture with a particular pixel intensity distribution; a lath martensite would evince a different distribution, and thus a different texture. However, a representative hypereutectoid steel microstructure might have multiple pearlite colonies interspersed with primary cementite particles, with potentially differing morphology depending on processing conditions.

Many microstructure analysis and characterization tasks have historically been (and in many cases still are) heavy manual workflows, but are now ripe for application of modern computer vision techniques. Image registration algorithms might use better keypoint matching techniques to reconstruct microstructures from large composite micrographs and serial sections. Automated microscopy systems could use computer vision techniques to search large samples for interesting microstructure constituents or rare features. For example, high-level image representations might be used to screen a sample using secondary electron imaging before applying more time-intensive electron backscatter diffraction (EBSD) imaging to areas of interest. Automated identification, segmentation, and quantification of complex hierarchical microstructure features, such as prior austenite grain size in martensitic steels, could be realized through semantic segmentation techniques, which jointly perform object or texture recognition with segmentation. These approaches will enable expert human metallurgists and materials scientists to more efficiently collect statistical amounts of microstructure data in a highly automated fashion, and extend current segmentation-based microstructure analysis workflows to more complex microstructure features at higher levels of abstraction.

Microstructure descriptors that permit quantitative comparison of visual similarity can enable a variety of materials science applications. For example, image-based information retrieval techniques may help the materials science community to search and leverage historical archives of microstructure data more efficiently. The ability to submit a microstructural image as a search query could facilitate connections between disparate sub-fields of microstructure science, and increase the discoverability of the materials science literature. Similarly, computer-vision-based microstructure descriptors coupled with data visualization algorithms enable novel ways of interacting and exploring microstructure databases. These techniques will only become more important as the materials science community continues to digitize current and past research data.

The ability to quantitatively compare general microstructural images can support a wide variety of materials engineering tasks, including process control and material qualification. For example, a microstructure might be designated as "acceptable" or "not acceptable" based on similarity to a baseline standard, and similarity to known high-performing microstructures might become materials or process design goals. It may also be possible to parameterize physical microstructure evolution models (e.g., phase field simulations) by searching for simulation parameters that yield microstructures that are statistically equivalent to experimentally captured microstructures. High-level image features also enable automated identification and localization of defects and rare features of interest.

Another area of interest is the generation of synthetic microstructures. The microstructure community is only just beginning to apply data-driven image texture synthesis methods to generate statistically representative microstructures to serve as input for physics-based modeling of mesoscale materials phenomena, as discussed in Section 14.3.3. Modern image synthesis and texture transfer algorithms could enable computational materials scientists to generate much more realistic simulation inputs for more complex microstructure systems.

Finally, it remains an open question whether microstructure representations drawn from modern computer vision techniques will directly support quantitative microstructure $\rightarrow$ properties and processing $\rightarrow$ structure mappings for complex hierarchical microstructure features. The success of local feature descriptors and deep convolutional neural networks in object classification, scene recognition, and image retrieval suggest that microstructure science applications will be very fruitful, but quantitative microstructure characterization supporting properties and processing relationships is a much higher bar than discriminative classification. To support progress in this area, the materials science community should establish open microstructure datasets annotated with processing and properties metadata to serve as benchmark tasks for evaluating the performance of novel microstructure characterization approaches.

Computer vision and machine learning are both fast-paced areas of active research with rich histories. The present discussion is intended to give a broad introduction to some of the important computer vision techniques for the materials scientist. These are broken down into three coarse (and roughly chronological) groups: low-level texture features, local/mid-level descriptor methods, and deep representations. Though the latter two groups are typically associated with higher-level recognition tasks, all three of these groups of techniques have been successfully applied to image texture and scene characterization, particularly addressing more complex textures and scenes in the case of deep representations. For more comprehensive and detailed coverage of particular topics, the inquisitive reader might begin with several review articles on low-level texture features [837, 1193], local image features [1192, 569], and deep learning [551, 399, 915].

14.2   A Brief Tour of Computer Vision

The choice of image representation is the keystone in current computer vision applications [190]. As the field has developed, the state-of-the art image representations have evolved from shallow features to deep (i.e., hierarchical and compositional) features. Since the early 2000s, the once-distinct fields of object, scene, and texture recognition have begun to converge, first with local feature representations [1192] and support vector machine (SVM) [222] classification, and more recently with the advent of scalable training for deep convolutional neural networks [515] and their application to texture recognition tasks [206]. Choosing among the many recognition approaches entails tradeoffs in training data requirements, computational complexity, simplicity of implementation, and model size (number of parameters) and capacity (i.e., the ability to fit the data). This discussion attempts to outline the conceptual basis for and advantages of some of the important image representations that have been applied to texture recognition, roughly broken down into texture-specific features, local pattern descriptors, and deep learning.

14.2.1   Texture Features

The classic approach to image texture recognition focuses on descriptive statistics of low-level image features [1062, 837, 1193]. These texture representations summarize a distribution describing some aspect of the local structure of the image while managing a tradeoff between spatial and frequency resolution. The simplest statistical representations are first-order point statistics, such as grayscale and/or color intensity histograms [992]. To improve the spatial locality of these statistics, they are sometimes computed within a small neighborhood window at each pixel. Local higher-order statistics and pattern descriptors are necessary for texture recognition and segmentation because first-order point statistics do not capture local structural relationships between adjacent pixels. Some of the popular approaches include statistical methods such as autocorrelation features, signal-processing approaches, and structural models [837].

One of the earliest successful statistical texture representations are the gray-level co-occurrence matrix (GLCM) features, which are often called the Haralick texture features [406]. GLCM features estimate the local second-order co-occurrence statistics of an image by tabulating and summarizing the conditional joint probabilities of discrete image intensity values over a set of displacements within a small neighborhood [210]. GLCM features tend to perform best for structured textures with a neighborhood radius of 4 pixels or less, and the choice of image quantization (i.e., number of intensity values) and GLCM statistics can have significant impact on classification performance and the variance of the texture representation [210]. After symmetrizing and normalizing the co-occurrence matrix, several texture features are calculated, including intuitive measures of image texture like contrast, homogeneity, and gray-level correlation, as well as more abstract statistical measures such as variance and entropy. GLCM features are still an important texture characterization method, though they are often used in conjunction with other feature sets.

Another dominant texture recognition paradigm is the filtering or signal processing approach [814]. These methods process the image with a set of filter banks, and characterize the distribution of filter responses to derive both pixel-level and image-level texture features. The most widely known example is the family of Gabor filterbanks [1064, 458], whose use was inspired by research on human perception of visual patterns [469]. Gabor filters are oriented sinusoidal plane waves weighted by two-dimensional Gaussian functions, approximating a Gaussian-windowed Fourier transform [557]. Pixel-level filter responses are used for texture segmentation tasks, and simple filter response statistics (e.g., the mean, variance, and/or entropy of the distribution over each channel) are typically used for image-level texture classification.

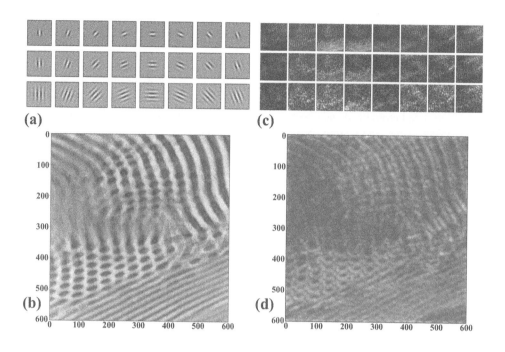

(a)

(c)

(b)

(d)

Figure 14.1: (a) A set of Gabor filters spanning 3 different scales and eight orientations. (b) A transmission electron micrograph of a block copolymer material exhibiting both lamellar and hexagonal phases. (c) The output of each individual Gabor filter channel applied to (b). (d) The overall Gabor feature map reduced by PCA.

Figure 14.1(a) shows a Gabor filterbank spanning three different scales and eight orientations. Application of each of these 24 Gabor filters to the block copolymer transmission electron micrograph in Figure 14.1(b) yields the 24 pixel-level filter response images in Figure 14.1(d); brighter regions indicate stronger local filter activations. Individual filters primarily respond to the lamellar structures with corresponding orientations, and to a lesser extent respond to the matrix region in the hexagonal phase. Figure 14.1(c) shows the overall distribution of filterbank responses: the 24-channel Gabor feature image is reduced to 3 channels via PCA and normalized to be displayed as a valid RGB image. The lower lamellar region is most clearly resolved here as a pink region in this visualization, while the upper lamellar region corresponds to less well-defined green and blue edges. The more crisp hexagonal region at the bottom left of the micrograph shows up as a combination of the blue, green, and pink edges, as the matrix region of the hexagonal phase yields similar oriented edge responses to the two lamellar regions.

Jain et al. [458] show that appropriate Gabor filterbank features can be used to discriminate between two textures with identical second-order or third-order point statistics. Gabor filterbanks are closely related to the discrete wavelet transforms used in image processing for applications such as compression; statistics based on wavelet transforms have also been applied to texture recognition [1081, 32].

The *texton* models [562] were developed as an extension to filtering methods to provide 3D surface texture representations that are consistent under lighting and viewpoint changes [814]. These models characterize image appearance by modeling the joint distribution of filterbank responses, in contrast to the first-order filter response statistics. The filterbanks used with these methods are conceptually similar to Gabor filters, though in practice they typically perform better [814]. Important filter banks include edge, bar, and spot filters (oriented Gaussian derivatives) with various scales and orientations [562, 230]; isotropic "Gabor-like" harmonics [914]; and a set of maximum-response filters similar to those from [562] where only the maximally activated rotational variant for each type of filter is registered [1092]. The *texton* primitives are learned from a training set by clustering in the filter response space; textured images are represented by mapping the joint filter responses for each pixel to the nearest texton and constructing a normalized histogram over these texton labels.

14.2.2   Midlevel: Local Features

The local features approach, which dominated the field in the 2000s, fundamentally represents the contents of an image as a distribution of quantized local pattern descriptors [1192, 569]. The canonical local features recognition pipeline consists of 1) extracting local image features, 2) encoding these local features, 3) aggregating or pooling the encoded features into a vector representation of the image, and 4) applying some machine learning algorithm, typically SVM classification. Many variants of the local features recognition pipeline have been evaluated, combining different feature identification techniques, pattern descriptors, encoding methods, and means of pooling features within spatial regions and image scale ranges [569, 189].

The baseline among this class of methods is the Bag of Visual Words (BoW) [940, 229], which is often referred to as the Bag of Features or Bag of Keypoints method. The BoW model encodes 1) continuous local pattern descriptors by 2) mapping them to a discrete set of "visual words," so that a global image representation can be obtained by 3) counting occurrences of each visual word, yielding a fixed-length vector. Figure 14.2 schematically illustrates this process for the same block copolymer micrograph shown in Figure 14.1(b). The difference of Gaussians (DoG) [591, 592] selects multiscale blob-like interest points by constructing a multi-resolution image representation [228, 777]: first the original image is convolved with a series of Gaussian filters of increasing scale parameter (Figure 14.2(a)), which is used to approximate the image Laplacian at each scale by subtracting adjacent scale-space images (Figure 14.2(b)). The interest points are any local extrema in the DoG image pyramid that exceed some threshold. Figure 14.2(c) shows the DoG keypoints for the block copolymer micrograph in yellow; the radius of each circle indicates the characteristic scale of the keypoint, and the tick mark indicates the orientation of the keypoint, determined by computing the average gradient orientation within the keypoint neighborhood.

The region surrounding each interest point is normalized with respect to its characteristic scale and shape, and is characterized using SIFT (scale invariant feature transform) [591, 592] descriptors, illustrated by the blue frame in Figure 14.2(d). SIFT computes Gaussian image gradient orientations and magnitudes at each point in the keypoint neighborhood, and constructs a 3D histogram of the gradient magnitudes in a 4×4 grid of spatial bins, each with eight gradient orientation bins. SIFT weights the image gradient values according to their distance from the bin edges and the keypoint to avoid introducing boundary artifacts. The result is a 128-dimensional vector describing the appearance of the keypoint in a manner that is consistent under image rotation, scaling, noise, and changes in image intensity [592].

These local features are encoded by quantizing the SIFT vectors into a dictionary (or "codebook") of "visual words," most commonly via k-means clustering [589]. This dictionary learning step is usually performed on a representative set of training images, and its size is generally selected to optimize the empirical performance of the overall recognition pipeline. The final BoW representation of an image is the normalized occurrence frequency histogram of each of the SIFT descriptors mapped to the nearest cluster center in the dictionary, as illustrated in Figure 14.2(e).

The BoW model was an important improvement in object recognition, as relying on local feature descriptors allowed the method to effectively detect even partially occluded objects, irrespective of object geometry and image rotation and without relying on image segmentation algorithms.

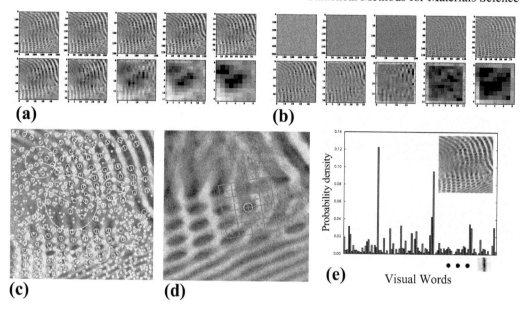

Figure 14.2: (a) Gaussian scale space, (b) Difference of Gaussians output, (c) Keypoints obtained, (d) SIFT descriptor, and (e) Bag of visual words histogram for the block copolymer micrograph from Figure 14.1(b).

Improvements to this general approach focus on feature localization, design of the local pattern descriptor, and pooling strategies to incorporate spatial relationships between keypoint features. Relative recognition performance of these various approaches can depend highly on the dataset and task, and on differences in viewpoint, illumination, and scale; typically researchers evaluate methods by comparing them with several approaches over several different datasets [684, 1011].

14.2.2.1   Feature Localization Techniques

Because the BoW method was originally developed in the context of object detection, early variants used sparse sets of keypoints, in contrast to the densely extracted features traditionally used in texture recognition. Keypoint localization algorithms have been highly engineered to yield consistent keypoint selection over a wide range of viewpoints and image scales commonly encountered in object and scene recognition tasks. The original model of [229] represents an image as a set of distinctive interest points obtained via the Harris affine detector [682], which uses the local second moments of the image to locate corner-like points and an elliptical neighborhood shape, and determines a characteristic keypoint scale using the image Laplacian.

In addition to the affine region detectors [683] based on the Harris corner detector [411], several other region detectors have been applied. The Difference of Gaussians (DoG) [591, 592] and related detectors locate blob-like interest points by searching for local extrema in various approximations of the multiscale image Laplacian. Blob-based interest point detectors often find complementary sets of interest points to corner-based detectors, so they are sometimes used in tandem [1192]. The maximally stable extremal regions (MSER) [652] detects distinctive regions by (effectively) adaptively thresholding the input image and searching for local regions that have a stable segmentation over a range of threshold values.

14.2.2.2 Alternate Pattern Descriptors

Once interest points have been selected, many other local pattern descriptors have been designed for and applied to object and texture recognition [569, 189]. In addition to SIFT, Gabor filters [1064, 458] and texton filter banks [562, 230, 914, 1092] have been applied at sparse sets of interest points. Mindru et al. used first- and second-order moment invariants [444, 1087, 689] computed on each of the RGB color bands as local affine-invariant feature descriptors, characterizing the shape and intensity distributions over the pixels within regions of interest.

Local binary pattern (LBP) descriptors [748] are one of the earliest local descriptors used for scale and rotation invariant texture classification. The classic LBP descriptor characterizes the (interpolated) image intensity at a set of points on the perimeter of a circle, subjected to a binary threshold at the image intensity value of the center point. Scale and rotation invariance are achieved by extracting LBP descriptors at multiple resolutions and rotating the descriptors into a canonical reference frame. LBP features are simple and efficient, but are not particularly stable in relatively uniform image regions.

Many local descriptors are inspired by SIFT, and attempt to more effectively or more efficiently characterize the local image gradient orientation distribution. Some of these are compared in Table 14.1. Among these diverse approaches, SIFT and GLOH are generally regarded as the best-performing local feature descriptors for matching and recognition.

Table 14.1: Comparison of several SIFT-inspired midlevel local feature descriptors.

Descriptor	Concept	Advantages/Disadvantages	Ref.
Histogram of Oriented Gradients (HOG)	SIFT-like, extracted from a dense grid of overlapping windows	Captures fine scale gradient, orientation, and shape information; fewer false positives in some domain-specific tasks	[234]
PCA-SIFT	Performs PCA on SIFT's normalized neighborhood gradient histograms	More compact feature representation; more expensive initial computation	[482]
Speeded Up Robust Features (SURF)	Similar to dense SIFT, but employs more efficient numerical methods	More efficient; omits spatial weighting used by SIFT to mitigate binning boundary artifacts	[69]
Gradient Location-Orientation Histogram (GLOH)	Like SIFT, with a polar grid of spatial bins; uses finer gradient orientation quantization; applies PCA to resulting gradient histogram	More expressive than SIFT with the same dimensionality; more expensive initial computation	[684]
RootSIFT	Implicitly applies the Hellinger distance to SIFT by applying element-wise square-root to SIFT features and normalizing	Hellinger kernel without increased storage or computational requirements	[29]
Augmented SIFT	Concatenates keypoint scale and/or position to SIFT features	Incorporates some measure of scale and geometry	[888, 190]

Some more recent recognition approaches have returned to using dense local features with improved descriptors and encoding and pooling strategies. One major advantage of densely sampling local features on multiple scales on a regular grid is that the sheer number of samples yields better statistics, which can lead to better recognition performance [471]. Dense features are also more amenable to efficient numerical implementation. One potential drawback is that the image patch representations obtained by dense descriptors may have weaker rotation invariance than the oriented keypoints obtained via sparse sampling.

14.2.2.3 Alternate Image Encoding Methods

Alternate image encoding methods have played perhaps the most influential role in improving the performance of the BoW approach [189, 507]. Important developments in this regard include the use of specialized kernel functions for computing image descriptor similarity, directly optimizing the dictionary to yield good empirical performance [1148, 544], using improved visual word assignment strategies to mitigate quantization effects, and strategies for pooling (aggregating) local descriptors to yield higher-level image representations [569, 189].

The kernel methods are the most straightforward of these extensions, replacing the Euclidean distance used to compare image representations with metrics that are more appropriate for comparison of distributions, such as the Hellinger and χ^2 kernels [1192]. The Hellinger distance D_h, also referred to as the Bhattachayya distance, compares two normalized m-dimensional histograms X and Y by

$$D_h(X,Y) = \sum_{i=1}^{m} \sqrt{x_i y_i} \tag{14.1}$$

The χ^2 distance D_{χ^2} is

$$D_{\chi^2}(X,Y) = \frac{1}{2} \sum_{i=1}^{m} \frac{(x_i - y_i)^2}{x_i + y_i} \tag{14.2}$$

There is also an explicit approximation for the χ^2 kernel that can be efficiently computed for large datasets [1093]. Often, the exponential version of the χ^2 or Hellinger kernels is used with SVM classification:

$$K(X,Y) = \exp\left(-\frac{1}{A} D(X,Y)\right) \tag{14.3}$$

The exponential scaling parameter A empirically yields good classification results when set to the mean distance between all of the training images [1192], though it is sometimes chosen to minimize prediction error during a cross-validation procedure.

Potential quantization errors in assigning local features to visual words can be mitigated by various soft-assignment strategies [122, 507], where individual local feature descriptors are mapped to multiple visual words. The boundaries between visual words in the baseline BoW method are generally arbitrary, as the clustering step is a simple vector quantization procedure rather than an earnest attempt to identify cluster structure in the training set of local feature descriptors [1086]. This violates the bag of visual words assumption that each cluster of local feature descriptors corresponds to some prototypical image texture feature, and introduces uncertainty in the mapping of local feature descriptors to visual words. One simple strategy to account for this is to assign each local feature descriptor to a weighted combination of several of the nearest visual words [783]. Alternatively, Gaussian kernel density estimation can be used to distribute the probability mass contributed by each local feature descriptor onto nearby visual words according to the probability density of the local feature training set [1085, 1086]. These methods tend to be most effective for very large dictionary sizes when a large training set is available.

Pooling sets of local feature descriptors can significantly improve performance by yielding smoother estimates of the distribution of local features, and by capturing spatial variations in the image [123]. Table 14.2 summarizes several important pooling strategies. Like the other components of the local features recognition pipeline, the choice of pooling strategy will generally be application-driven.

Table 14.2: Comparison of several higher-level feature encoding methods.

Pooling Method	Concept	Advantages/Disadvantages	Ref.
Spatial Pyramid Pooling	Divides image into a pyramid of smaller regions and concatenates bow representations for each region.	Captures global visual content along with some coarse-grained spatial context	[545]
Fisher Encoding / Fisher Vector (FV)	Similar to dense bag of SIFT with a smaller, soft dictionary explicitly modeled via a Gaussian mixture model (GMM). Derives probabilistic image representation by computing the gradient of the log-likelihood of local features w.r.t. the GMM distribution.	Captures first- and second-order occurrence statistics of visual words; efficient classification with linear SVM; small vocabularies lower computational costs; large feature vector sizes; in practice fails to significantly improve over non-linear classification with bow features	[778]
Improved Fisher Vector (IFV)	Introduces normalization techniques to reduce the sparsity of the FV representation, and employs spatial pyramid pooling	Yields excellent classification results with efficient linear classification (SVM) techniques; dense feature representation compresses poorly leading to higher storage costs	[779]
Vector of Locally Aggregated Descriptors (VLAD)	A simplified FV representation using a hard dictionary. Calculates the distance between each local feature and the nearest visual word. For each visual word, sums these distances, and concatenates the results into a single image descriptor.	Simpler implementation than FV/IFV; adaptable to new or incrementally updated (online) dictionaries; often yields somewhat lower performance compared to IFV	[459]

Performance for methods based on local features can depend heavily on the particular dataset, task, and implementation details [189], such as proper normalization for linear SVM classification [1093]. Generally, larger vocabularies yield better (but diminishing marginal) accuracy for classification; vocabularies as large as 10^6 visual words may be required for fine-grained classification tasks [378]. Large image signatures may also necessitate explicit computation of the kernel matrix for use with the dual formulation SVM, particularly for methods like IFV when the dataset size is large [189].

14.2.3 Deep Learning

Since Krizhevsky's 2012 breakthrough in training deep convolutional neural networks for large-scale image classification [515], deep learning has dominated the field of object recognition, and increasingly scene and texture recognition and segmentation as well [551]. Deep learning refers to models that are "composed of multiple processing layers to learn representations of data with multiple levels of abstraction" [551]. Enabled by very large datasets and advances in computational power, deep learning models have recently even surpassed human-level performance for some tasks [915]. While deep learning encompasses many different types of models [551, 915, 399], the current discussion focuses on deep convolutional neural networks (CNN), one of the most popular deep learning approaches in computer vision.

Convolutional neural networks were first applied to handwritten digit recognition by LeCun in 1989 [552], though the basic idea of neural networks is much older. At their core, multilayer neural networks are compositions of simple non-linear functions, in which the functions in each layer operate on the output of the functions in the previous layer, as schematically indicated in Figure 14.3(a), where nodes represent variables and edges represent weights (i.e., function parameters). This schematic neural network operates on a 9-element vector input layer (green nodes), with two sparsely connected hidden layers (blue nodes) and a single fully-connected output node (red node). Modern neural networks often employ linear functions for each connection: For example, the value of the first node in the first hidden layer is based on a weighted sum of the first two input nodes, plus a bias term: $h_{1,1} = x_1 w_{011} + x_2 w_{021} + b_{11}$. This weighted sum is then passed through a non-linear function such as tanh, the sigmoid function, or the more modern "rectified linear unit" (ReLU) function [718]: $ReLU(z) = \max(0, z)$.

In convolutional neural networks used for image recognition, each node in the input layer corresponds to a pixel in the input image, and the weights for each hidden layer are arranged into a set of convolution filters that are learned from the training data. The same set of convolution filters is applied to each node in a given layer, and their output yields a higher-level image representation, sometimes referred to as the "feature map" for that layer. The filters for a given layer are applied to the local neighborhoods in the input for that layer to yield a higher-level image representation, sometimes referred to as the feature maps for that layer. Convolutional layers are interleaved with spatial pooling layers [553] that down-sample the feature maps, merging similar features and improving the spatial invariance of the feature detectors [82]. The highest layers in many CNNs are traditional fully-connected layers, which fix the allowable size of the input and capture some of the spatial aspects of the input.

Figure 14.3(b) shows a schematic diagram of the feature maps for the VGG16 deep neural network [938] with five groups of convolution layers and three fully-connected layers; the annotations for each layer indicate the number of 3×3 convolution filters contained in that layer. The VGG16 network takes as input a $224 \times 224 \times 3$ color image; the first convolution layer yields a 64-channel 224×224 feature map that captures low-level image features. Before being fed into the second convolution layer, the first-layer feature map is downsampled by max-pooling: within each sub-sampling region, the maximum activation for each convolution feature is kept, resulting in a first-layer representation of size $112 \times 112 \times 64$. Each subsequent convolution layer proceeds similarly, applying an increasing number of convolution filters to an increasingly coarse image structure representation, reducing the spatial resolution by a factor with each application of pooling. After five layers of convolution and spatial pooling, a $7 \times 7 \times 512$ set of feature maps is fed into the sixth, seventh, and eighth fully-connected layers to arrive at a global image representation for classification among the 1000 different object classes in the ImageNet dataset.

Figure 14.3(c) is a PCA visualization of the feature maps after each convolutional layer in the VGG network, using the same method as for Figure 14.1(c). The input image (top left) is a transmission electron micrograph of the same block copolymer microstructure as in Section 14.2.1; omission of the fully-connected layers of the VGG network allows for a larger input image. The colors in each feature map visualization show the variation in local response to the set of convolution filters in the current layer; similar colors indicate similar filterbank responses. (However, note that color similarity between different layer visualizations is incidental, because PCA is performed for each convolution layer separately.) The first- and second-layer feature maps respond to similar features as the Gabor filterbank in Figure 14.1(a), highlighting low-level oriented edge features of individual lamellae, but the remaining convolution layers respond to higher-level microstructure features

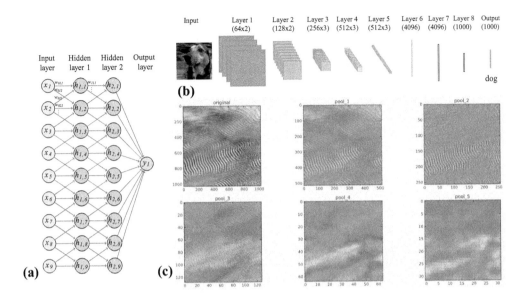

Figure 14.3: (a) Schematic neural network with 9 input nodes, two hidden layers, and a single output node. (b) A block diagram showing the internal feature maps in each layer of the VGG convolutional neural network [938]. (c) Output of the VGG ConvNet after pooling for each group of layers.

corresponding to the hexagonal and lamellar block copolymer phases. The third-layer feature maps begin to resolve the finer lamellar regions (particularly in the bottom left of the layer 3 visual-ization) as a cohesive image feature, though coarser lamellae are still resolved and appear similar to the hexagonal regions. The fourth-layer feature maps clearly reflect the difference between the hexagonal regions (bright pink) and lamellar regions of varying orientation and lamellar thickness (blue-green). Finally, the fifth convolution layer seems to respond most strongly to the disparity in crispness between different lamellar regions.

Deep learning is attractive because it offers a way to learn hierarchical data representations without the intensive feature engineering and parameter tuning that was responsible for much of the success of the local feature representations outlined in Section 14.2.2. Interestingly, the lower-level CNN filters trained on image data often resemble Gabor filter banks, while the higher-level filters become increasingly abstract and specific to the training dataset [1186, 205]. A recent correlational study suggests that hierarchical internal image representations of a deep CNN trained on an object recognition task are similar to image representation in the visual centers of primate brains [205]. However, even though parameters are learned from data, architectural design of deep models re-mains somewhat of an art form, and theoretical understanding of these models has been elusive [551]. Contemporary deep learning approaches require a large amount of data and computational resources to achieve state-of-the-art performance: typical modern deep neural networks have hun-dreds of millions of model parameters learned from hundreds of millions of training examples, and can require weeks of training time, though large multi-GPU systems can cut this down to a matter of hours [551]. This data requirement likely places direct application of deep learning squarely outside the reach of most microstructure informatics tasks for the foreseeable future.

Transfer learning [1187, 267, 1186, 206] may provide a way forward, in which the feature maps from deep models trained on one dataset are adapted for use with a different dataset, as in the feature map visualizations of Figure 14.3(c). This approach is already being explored successfully in the medical imaging community [384]. Either the final classification layer of the CNN can be retrained on the new task, or the highest-level feature map can be used (with careful normalization, and optional dimensionality reduction [190]) as input to another classifier, such as a linear SVM. The output features of CNNs are typically of much lower dimensionality than in state-of-the-art local features methods like IFV, and can be further reduced in dimensionality without sacrificing performance [190]. Furthermore, CNN feature maps actually take less time (50× faster) to compute [190] than IFV representations.

However, high-level CNN features are specific to the type of task for which the CNN was trained [1206, 206]: naturally, a CNN trained on an object recognition dataset will not perform to a scene recognition task as well as CNN trained on a scene dataset [1206]. However, additional training can potentially mitigate the feature map bias and significantly improve performance [364, 399], even when the available training set is small (e.g., $\sim$ 5000 images total for the 20 class VOC dataset) [190].

Deep learning researchers have also begun to address the dataset bias problem in transfer learning by exploring some of the spatial and multiscale pooling methods that enjoyed great success with local feature methods. Data augmentation and/or multiscale pooling can mitigate this deficiency when transfering feature maps learned from object datasets to scene recognition [371, 832, 419, 206].

14.3   Materials Applications

Multiscale image texture is a natural way to think of microstructure, where the texture encompasses: the spatial distribution of constituents, crystallographic orientations, chemical composition, or similar materials structure or property values. Notably, microstructure classification was among the earliest applications of image texture analysis, along with analysis of aerial and satellite imagery for land-use categorization. In 1973, Haralick et al. classified photomicrographs of five different types of sandstone when introducing their GLCM texture features [406]. In intervening years, microstructural images have not been a prominent application in image texture research, likely due to the expensive nature of micrograph acquisition. However, since the 1990s various image texture methods have been widely applied in surface analysis in support of industrial process quality monitoring and surface defect screening [90, 1162, 17], mostly focusing on the texture models outlined in Section 14.2.1.

Applications of image analysis in microstructure science have traditionally focused heavily on segmentation of microstructural constituents for subsequent quantification. Two microstructure quantification techniques that are related to image texture analysis have been thoroughly developed in recent years: applications of higher-order point statistics [475] and moment-invariant shape descriptors [614, 615]. As these approaches have been extensively discussed elsewhere, the present discussion focuses on the remaining microstructure science applications of image texture characterization, with emphasis on the microstructure representation rather than the particular classification algorithms used.

14.3.1   *Microstructure Characterization*

14.3.1.1   *GLCM and Wavelet Features*

Many applications of image texture to microstructure and surface characterization seem to revolve around GLCM and wavelet features [1162]. Some typical applications of GLCM features for surface characterization include online monitoring of flotation froths in minerals processing [64, 17] and classification of different types of corrosion damage [201, 789], and assessing the surface quality of rolled steels [90]. In the context of microstructure, GLCM features have sometimes been used for microstructure constituent classification, including a set of copper alloys [1] and a set of steel microconstituents including pearlite, ferrite, martensite, and cementite [488]. Other interesting applications include classification of weld defects in radiograms using both Haralick features and Gabor filters [676]. These methods are also sometimes applied for image-texture-based segmentation of microstructural features, such as differently-oriented groups columnar grains in polycrystals [1135] and morphologically different graphitic inclusions in gray, malleable, and ductile cast irons [761]. Such studies often evaluate multiple image texture methods and select the one with the best performance on the application at hand.

14.3.1.2 EM/MPM Texture-Based Segmentation

Markov random field (MRF) models are commonly used in textured image segmentation for microstructure analysis [935]. For example, the expectation maximization/maximizing posterior marginal (EM/MPM) method [219] has been used to segment textured micrographs of primary γ' precipitates in Ni-based superalloys and α lath structures in Ti-6V-4Al [935], as well as carbon foam structures [204, 648].

14.3.1.3 Characterizing Two-Phase Microstructures

Often in microstructure science, two-phase structures such as those in precipitation-hardened materials are characterized by the size and morphological distributions of the precipitate particles. The method is often specific to the microstructure system of interest, requiring expert evaluation of segmentation algorithms, parameter tuning, and selection of morphological descriptors. Some recent approaches are similar to generic image texture characterization and often employ *synthetic* microstructure data created with physical or statistical microstructure models. For examples, Simmons et al. [936] used image patch features to model the morphology and neighbor correlation of the distinctive γ' particles in Ni-based superalloys. Beginning with 41×41 pixel image patches centered on each particle, they first linearly reduce the dimensionality of the patches to 5 dimensions using principal component analysis (PCA) [467], then cluster these reduced patch features to obtain sets of particles with similar coordination patterns.

Similarly, Nguyen et al. [733] used an image texture representation based on moment invariants to discriminate between austenite, martensite, pre-martensitic, and strain glass states in ferroelastic materials. After segmenting microstructures generated by a phase field model, they obtained pixel-level texture features by computing windowed second- and fourth-order moment invariants on the binarized phase map. They analyzed how the distribution over these features varies for a range of chemical compositions in order to quantitatively identify and classify the different microstructure variants, and the transitions between them.

Xu et al. [1164] applied a machine learning approach to characterize segmented polymer nanocomposite microstructures using a library of 56 microstructure descriptors, such as volume fractions, particle volume, surface area, and compactness. They winnowed the full set of descriptors to a low-order physics-based microstructure representation via a series of unsupervised and supervised feature selection techniques. First, they filtered out redundant microstructure descriptors using rank correlation analysis. Then they computed microstructure representations based on two-point, surface, lineal path, and radial distribution functions, and finally used the RreliefF [847] method to identify a subset of microstructure descriptors that explains the variance in the correlation function representations. Finally, they applied RreliefF a second time to learn a model relating the reduced descriptor set to materials properties computed with an appropriate physical model.

14.3.1.4 Midlevel Features

DeCost and Holm [251] evaluated the bag of sparse SIFT representation for microstructure characterization using a small, diverse microstructure dataset. They demonstrated that using BoW fea-

Input image   Closest match   Second match   Third match

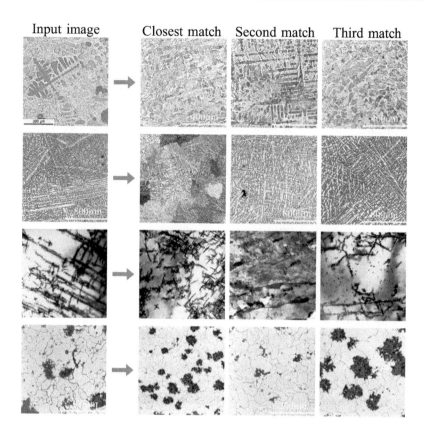

Figure 14.4: Results of the visual microstructure search of the DoITPoMS micrograph library [57] for four query images (left column). The right three columns show the top three matches for each query image.

tures, χ^2 kernel SVM can achieve reasonable (~80%) accuracy on a 7-way microstructure classification task, placing images into categories such as "twinned," "superalloy," and "ductile cast iron." They also found that this representation provides promising results on a microstructure-based image search task. Correct matching of similar microstructures over widely different image scales indicates that mid-level image texture features could serve as a useful tool for exploratory microstructure dataset analysis.

Figure 14.4 shows the top three microstructure search results for four different microstructure queries using the sparse bag of SIFT representation from [251] on the DoITPoMS microstructure library [57], which is a diverse dataset containing over 750 micrographs.

It remains an open question as to what extent the image texture descriptors described in this review can be effectively used as quantitative microstructure characterization tools. DeCost and Holm have begun exploring this issue by testing the limits of the bag of sparse SIFT representation on synthetic [252] and experimental [253] powder micrographs, with potential applications in material qualification for powder metallurgy and powderbed-fusion additive manufacturing. Bag of words and VLAD encoded SIFT features can discriminate well between synthetic powder micrographs differing primarily in particle size distribution [252], and between SEM powder micrographs with similar particle size distributions and different morphological and textural characteristics [253]. Further study is needed to evaluate the utility of these microstructure representations for determining quantitative structure → properties mappings, and for identifying, localizing, and quantifying defects such as agglomerated and/or remelted powder particles.

14.3.2 Deep Features

Within the last year at the time of this writing, modern deep learning methods have begun to be applied to microstructure data. Chowdhury et al. [203] provide a broad comparison of low-level texture features, midlevel patch representations, and pretrained deep neural network features, applied to the task of classifying experimentally-collected dendritic microstructures. Lubbers et al. [596] apply bilinear CNN representations [351, 582, 581] to synthetically generated lamellar microstructures, using dimensionality reduction techniques to relate the structure of the CNN-based microstructure representations to microstructure parameters such as lamellar spacing and orientation. Finally, [249] visualize processing–structure relationships in ultrahigh carbon steel microstructures [250], comparing midlevel BoW and VLAD encoded SIFT features with VLAD encoded pretrained convolution features, following the approach to texture segmentation presented in [206]. These pretrained CNN features enable classification accuracy of up to 98% on a microconstituent classification task, and over 80% on a 13-way annealing condition classification task. Overall, deep microstructure representations show great promise for the development of quantitative processing–microstructure–properties regression models, contingent upon carefully collected training datasets.

14.3.3 Microstructure Generation

Although synthetic microstructures are usually generated using physics-based materials models, some researchers have begun investigating image texture generation algorithms for this purpose. Liu et al. performed statistical reconstruction of 2D and 3D microstructures [588] by applying an MRF texture synthesis algorithm [1129], as well as the image quilting method of Efros and Freeman [294]. The MRF algorithm generates a new textured image by sampling one pixel at a time from a source image, conditioned on the partial local neighborhood of the new pixel in the synthesized image. The image-quilting algorithm approximates this process by combining patch samples selected from the source image so that the sampled image patches overlap in the quilted image. Each successive patch is selected from the source image such that the difference between overlapping pixels is minimized, and the overlapping patches are joined along a ragged boundary determined by finding the minimum cost path through the overlapped pixels. Bostanabad et al. [117] also use a Gaussian MRF model for texture synthesis, along with a decision tree for supervised classification.

These methods apply equally to continuous (grayscale) images and to the binary phase map microstructures in [588], though the image-quilting heuristic [294] is necessary for reasonable computational efficiency when many discrete states are allowed. The patch sample size needs to be carefully chosen so that the relevant structure of the textured image can be recreated without excessive verbatim copying. Figure 14.5 shows the results of the Efros and Freeman algorithm [294] for several complex microstructures from the DoITPoMS micrograph library [57], including (a) γ' particles in a Ni-based superalloy, (b) a slow-quenched carbon steel, (c) pearlite with graphitic inclusions in a malleable cast iron, and (d) dendritic Ni-hard iron.

These image-quilting syntheses reproduce low-level texture details of the input microstructures with reasonable fidelity, but fail to reproduce longer range structure, particularly for the dendritic iron in Figure 14.5(d). Recently, Lubbers et al. [596] have demonstrated impressive microstructure syntheses with more realistic higher-level structures using texture transfer methods based on CNNs.

14.4 Outlook and Call for Standardization around Datasets

Modern image texture recognition techniques show promise for microstructure characterization, but the microstructure community has only recently begun to explore them. These models tend to have very high capacity; prudence in model selection and fitting are necessary to avoid overfitting

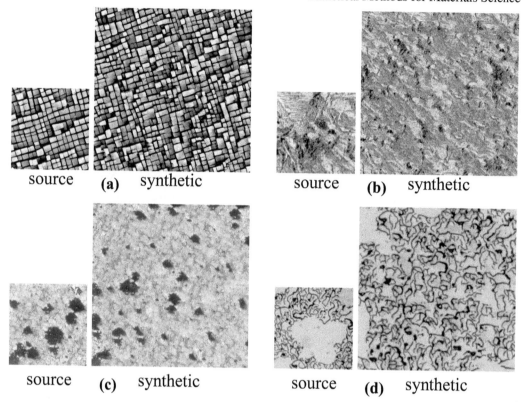

source   **(a)**   synthetic   source   **(b)**   synthetic

source   **(c)**   synthetic   source   **(d)**   synthetic

Figure 14.5: Synthetic micrographs generated by the Efros and Freeman image quilting algorithm [294] using 16×16 patches with 3px overlap for (a) Ni-based superalloy, (b) ferrite in a carbon steel, (c) pearlitic malleable cast iron, and (d) dendritic Ni-hard iron.

and to achieve models that will prove generally applicable outside of a controlled academic setting. There are many open questions regarding materials science applications of computer vision techniques, particularly regarding quantitative microstructure representations. Can existing image texture recognition approaches support quantitative microstructure, processing, and properties research? If not, what modifications will be necessary, and can these be feasibly implemented within the constraints of current computing resources? How can modern image recognition methods best be integrated with contemporary microstructure analysis and data collection capabilities? The microstructure characterization community needs to establish standard datasets (c.f. [250]) for microstructure recognition and characterization tasks, along with a culture of sharing code, so that various approaches can be evaluated reproducibly.

The extreme expense of collecting microstructure datasets, relative to natural scene, object, and even medical datasets, means that a direct application of current deep learning techniques is infeasible for many microstructure characterization tasks. Transfer learning techniques may enable applications of deep spatial representations for some microstructure data, and indeed has begun to be investigated [203, 596, 249]. Microstructure scientists should pursue close collaboration with the computer scientists who are working on reducing the data requirements of next-generation computer vision and machine learning systems.

14.5   Acknowledgments

We gratefully acknowledge funding for this work through the National Science Foundation grant numbers DMR-1307138 and DMR-1507830, and through the John and Claire Bertucci Founda-

Computer Vision for Microstructure Representation

257

tion. Block-copolymer micrographs in Figures 14.1, 14.2, and 14.3 are courtesy of Bongjoon Lee and Prof. Michael Bockstaller of Carnegie Mellon University. Micrographs from the image query example in Figure 14.4 and source microstructures in Figure 14.5 are from the DoITPoMS microstructure library [57]. We are also grateful for helpful and inspirational discussions with Prof. Abhinav Gupta and Xinlei Chen of the Carnegie Mellon University School of Computer Science.

Chapter 15

Topological Analysis of
Local Structure in Atomic Systems

Emanuel A. Lazar
University of Pennsylvania

David J. Srolovitz
University of Pennsylvania

15.1 Introduction

Increasingly powerful experimental and computational resources have made possible large-scale studies of an enormously broad range of physical systems on the atomic scale. A central challenge in analyzing atomic systems is the automated characterization, visualization, and analysis of structure, so that meaningful results can be extracted from massive sets of raw data. Given only atomic coordinates, what can be said about the underlying structure of the sample? In what crystalline phases, if any, are atoms arranged? What types of defects appear in the sample and where are they located?

Such questions are easy to ask and surprisingly difficult to answer. The last several decades have witnessed the development of numerous automated approaches for characterizing, visualizing, and analyzing structure in atomic systems [983, 566]. Despite much progress in this area, conventional methods tend to be ineffective in analyzing high-temperature systems without quenching or time-averaging atomic coordinates. This chapter describes a topological approach for automated structure analysis in atomic systems which is substantially more robust than conventional approaches, especially for high-temperature and imperfect systems.

This chapter begins by detailing the related problems of structure characterization and defect identification, considers several conventional approaches and their limitations, and then explains how Voronoi topology can be used to analyze structure in atomic systems. We provide several examples of applications that benefit from this approach. We conclude with a brief discussion of *VoroTop*, a new set of open-source tools built to automate this analysis.

15.1.1 Local Structure in Atomic Systems

Molecular dynamics simulation techniques are now widely applied to investigate material properties in systems with billions of atoms, often over the course of many millions of time-steps. Molecular statics and Monte Carlo methods are also routinely employed in studies of large systems with atomic resolution. More recently, experimental methods such as atom probe tomography have been introduced that provide three-dimensional atomic coordinate datasets directly from physical samples [688]. To a large degree, these enormous datasets must be substantially simplified to allow for the extraction of interesting conclusions from a vast sea of mostly uninteresting data. Countless terabytes of data might be ultimately distilled into a short statement such as: "The migration of point defects is inhibited by grain boundaries." The task of the scientist – as distinct from that of the programmer – is to understand how to properly reduce and interpret the massive data generated

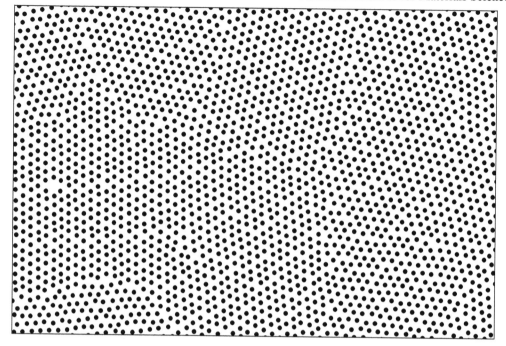

Figure 15.1: A typical two-dimensional atomic system [541].

by experiment or simulation. Without proper tools to help with this task, many atomic coordinate datasets look almost identical.

Consider the two-dimensional system illustrated in Figure 15.1. Although the sample is primarily crystalline, it contains defects of various types, including grain boundaries and vacancies. Although many of these defects can be seen with our eyes, defining them precisely enough to make them amenable to automated analysis can be difficult. A simple exercise helps illustrate some of the underlying complexities. Look again at Figure 15.1 and choose an arbitrary atom. Now think about the following two questions. First, does the chosen atom belong to a crystal or to a defect? Second, how can the answer to the first question be made precise? Even brief reflection on these questions will persuade most readers that answers are neither simple nor unambiguous.

Although Figure 15.1 may be analyzed with only an intuitive idea of what it means for an atom to "belong to a crystal," making this intuition precise is necessary for automating the analysis of larger systems, especially three-dimensional ones. For example, while studying phase transitions containing billions of atoms, determining which atoms belong to which phases will depend on precise definitions of what it means for an atom to belong to one phase or another. Without precise definitions, quantitative analysis of such systems is not possible.

15.1.2 *Conventional Characterization Approaches*

The characterization considered in this chapter is a form of *local* structure analysis, as its objective is to describe the local environment around each atom by considering how its neighbors are arranged relative to it. Only after local structure is characterized can larger-scale structure be determined.

Conventional approaches to local structure characterization can be classified as either *physical* or *aphysical*. Physical approaches take into account physical properties of a system, such as energy and stress, in addition to atomic coordinates. As atoms belonging to defects often have higher energies than those belonging to bulk crystals, or different stress patterns, such quantities can be helpful in identifying structural defects.

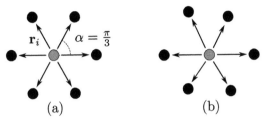

Figure 15.2: A central particle and neighbors in (a) an unperturbed triangular-lattice crystal, and (b) a perturbed triangular-lattice crystal.

In physical approaches, a threshold value is typically chosen, and atoms with property values above or below that threshold are identified as defects. In high-temperature systems, choosing a threshold can be difficult because the variance in energy, stress, or other physical quantities associated with thermal fluctuations is typically of the same order of magnitude as that associated with structural defects. In practice, physical quantities are rarely useful in directly studying high-temperature atomic systems [566].

Aphysical approaches, in contrast, consider only atomic coordinate data, and not the underlying energetics. In coordination number analysis, perhaps the simplest example of an aphysical method, each atom is assigned a number counting atoms within a specified radius. The expected number of neighbors in many crystalline systems is known *a priori*, and atoms with different coordination numbers are associated with defects. For example, for a properly chosen threshold, most atoms in Figure 15.1 will have a coordination number of 6, while those belonging to defects will have coordination numbers other than 6.

More sophisticated aphysical approaches consider not only the number of neighbors of an atom, but also the manner in which those neighbors are arranged. In an ideal triangular lattice, for example, every atom is surrounded by six equally-spaced neighbors; adjacent pairs of neighbors make identical angles $\alpha = \pi/3$; see Figure 15.2. Centroysymmetry analysis [483] and bond-angle analysis [4] are but two examples of widely-used methods that measure deviation from this ideal structure. Centrosymmetry measures the extent to which neighbors of an atom can be matched in equally-spaced but oppositely-directed pairs, so that their displacement vectors cancel. Bond-angle analysis measures the extent to which angles between "bonds" from a central atom to neighboring atoms remain close to an idealized value, $\pi/3$ in the example of Figure 15.2.

Several other aphysical approaches, including bond-order analysis [964] and neighbor distance analysis [983], have been developed and applied, sometimes at greater computational cost. As with physical approaches, most aphysical approaches tend to be ineffective in analyzing high-temperature systems, in which deviations from ideal values are affected by thermal fluctuation as much as by structural defects. In a sense that can be made mathematically precise, this problem is a necessary limitation of all continuous structure-characterization methods [534]. To circumvent this problem, systems must be quenched or time-averaged before analysis [566]. It is unknown whether this pre-conditioning of the data results in changes to its meaning. Consider, for example, the effect of quenching on a material that has different high- and low-temperature polytypes and a displacive transition between them.

15.1.3 *Voronoi Analysis*

Voronoi analysis traces its history to the mathematical study of crystal lattices [1111], and in recent decades has been used to analyze many physical and biological systems [749, 974]. In this section we introduce the central ideas of Voronoi analysis and consider how they can be used to characterize local structure in atomic systems. We do this first in the context of two-dimensional systems, and

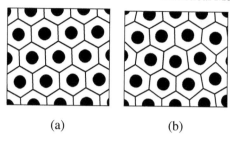

(a)   (b)

Figure 15.3: (a) atoms in a zero-temperature crystal with regular six-sided Voronoi cells; (b) atoms in finite-temperature crystal with slightly irregular six-sided Voronoi cells.

defer discussion of the conceptually similar, but practically more complex, three-dimensional case to Section 15.2.

The *Voronoi cell* of an atom is the region of space closer to that atom than to any other; Voronoi cells which meet along a boundary are called *neighbors*. Voronoi cells of atoms are always convex polygons (in two dimensions) or polyhedra (in three dimensions), though not necessarily regular. Figure 15.3(a) shows several atoms in an unperturbed triangular lattice along with their Voronoi cells. Figure 15.3(b) shows a finite-temperature version of the same system, in which atoms are slightly perturbed from their perfect lattice sites. Although cell perimeters and areas have slightly changed, the topology of each Voronoi cell has not; in particular, all cells remain six-sided. This stability in the presence of small perturbations is a powerful feature of Voronoi analysis, and is the source of its robustness in analyzing high temperature systems.

The Voronoi cell of an atom can be characterized both geometrically and topologically. Its area, for example, can be used to calculate the "free volume" of an atom or to estimate local atomic density. These geometric characterizations of Voronoi cells tend to be sensitive to small distortions of atomic coordinates. In contrast, topological descriptions of the Voronoi cell tend to be robust under such distortions. In particular, the number of sides of a Voronoi cell, equivalently its number of neighbors, is generally invariant under small perturbations of atomic coordinates. For reasons that will be explained in Section 15.2.2, we use the word topology to refer to the number of sides of a two-dimensional Voronoi cell.

In defect-free crystals with triangular lattice structure, the Voronoi cell of every atom has six sides [561]. Atoms whose Voronoi cells have more or fewer than six sides can be considered as belonging to defects. Local structure near an atom in a two-dimensional system can thus be characterized and analyzed according to its number of Voronoi neighbors. Figure 15.4 illustrates the same polycrystal shown in Figure 15.1, this time with each atom colored according to the number of sides of its Voronoi cell. Most atoms have six sides and are colored yellow; other Voronoi cells have five or seven sides and are colored blue and red, respectively.

Characterizing local structure through Voronoi cell topology facilitates the visualization and analysis of defect structure in polycrystalline materials. In Figure 15.4, for example, atoms whose Voronoi cells have six sides can be classified as belonging to crystals, whereas those with more or fewer sides can be classified as belonging to defects. After we have identified defects at the single-atom level, we can identify larger-scale defects as contiguous regions of defect atoms. Notice, for example, a ring of 3 red and 3 blue atoms on the left side of Figure 15.4. This ring identifies the presence of a vacancy. Similarly, a chain of alternating red and blue atoms indicates the presence of a high-angle grain boundary. This analysis requires no choice of threshold or preconditioning of the data, and is robust against thermal vibrations and elastic strains.

In the following section, we present a similar analysis for studying three-dimensional systems. The topology of Voronoi cells in three dimensions is more complicated, yet can be approached in much the same way.

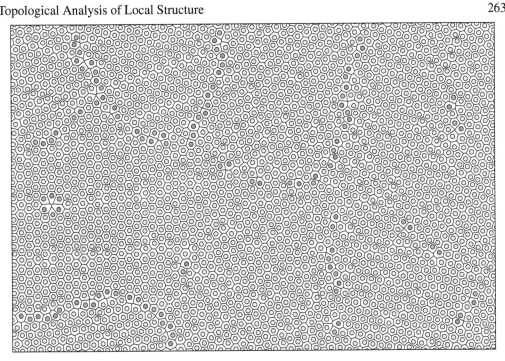

Figure 15.4: The two-dimensional polycrystal from Figure 15.1 with Voronoi cells shown and atoms colored by the number of sides of their Voronoi cells.

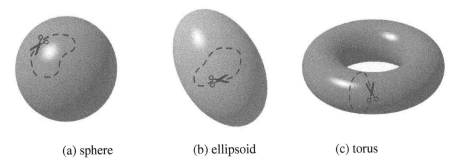

(a) sphere   (b) ellipsoid   (c) torus

Figure 15.5: The sphere and ellipsoid can be deformed into one another without cutting or gluing, while the torus cannot. Cutting a loop in the first two shapes necessarily divides them into two pieces; a loop cut in the third shape can leave it connected.

15.2 Voronoi Topology Structure Analysis

15.2.1 Topology Basics

Topology is the mathematical study of properties of objects that do not change under continuous deformations; these properties are often related to the manner in which objects are connected to themselves and to other objects. To illustrate this idea, consider the shapes shown in Figure 15.5. While the sphere can be continuously deformed into the ellipsoid without cutting or gluing, the torus cannot. In the language of topology, the sphere and ellipsoid are *isomorphic* with one another, whereas the torus is not isomorphic to either.

Many properties of these shapes do not change when they are continuously deformed. For example, imagine taking a pair of scissors and cutting a closed loop in the three shapes, as illustrated

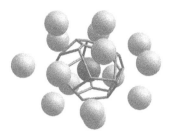

Figure 15.6: The Voronoi cell of a central atom (blue), surrounded by neighboring atoms (gold).

in Figure 15.5. This procedure inevitably divides each of the sphere and ellipsoid into two disconnected pieces. In contrast, the same procedure can leave the torus connected as one piece. This type of "connectedness" is a topological property of a shape, and if two shapes are isomorphic then they either both have this property or both do not.

This brief exercise illustrates one property that is preserved under continuous deformations; the field of topology develops powerful tools to study shapes and other topological invariants [715]. An overview of one recent application of topology to the study of materials can be found in 13, in which Betti numbers are used to characterize random microstructures.

As topology focuses on studying the connectivity of shapes such as spheres and tori, it is not immediately clear that it would have much relevance to studying sets of discrete points, such as those encountered in studying atomic systems. In what meaningful way can points in space be considered connected? Over the last decade or so, however, powerful tools such as discrete Morse theory and persistent homology [288, 1134] have been developed to analyze data of diverse kinds [360]. Voronoi topology continues in this spirit. In what follows we show how considering the topology of a Voronoi cell can provide keen insight into the manner in which a set of points is arranged in space. In this sense, Voronoi topology forms a bridge between the discrete and continuous, and enables the application of ideas from topology to the study of atomic systems.

15.2.2   Voronoi Topology

In atomic systems, small perturbations resulting from thermal noise or small strains are often unimportant for understanding crystal structure and defects. Whereas geometrical characterizations of local structure are generally sensitive to such perturbations, topological ones are generally not. Figure 15.6 illustrates the three-dimensional Voronoi cell of a central atom surrounded by neighbors. When atomic positions are perturbed, geometrical properties of this Voronoi cell, including edge lengths and face areas, almost always change. In contrast, the manner in which edges and faces are connected does not. Such topological properties are thus naturally suited for studying atomic systems when we wish to ignore small perturbations.

In the remainder of this chapter, we use the term *topology* of a Voronoi cell to refer to the manner in which neighboring Voronoi cells are connected to a central Voronoi cell and to each other. In two dimensions, neighbors of a central Voronoi cell can be connected in only one manner – cyclically, with each neighbor connected to two others – and so a count of sides of a Voronoi cell completely describes its topology. In three dimensions, however, completely describing the arrangement of all neighboring cells is more complicated. Consider, for example, that while a count of faces of a Voronoi cell indicates its number of neighbors, it says nothing about how those neighbors are arranged.

Additional topological information about a Voronoi cell is provided through consideration of the number of sides of its faces. This additional data provides a more refined description of the manner

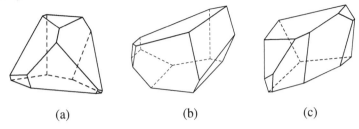

(a)   (b)   (c)

Figure 15.7: Three topologically distinct Voronoi cells, each with 4 four-sided, 4 five-sided, and 2 six-sided faces.

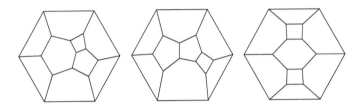

Figure 15.8: Planar edge graphs of the three Voronoi cells shown in Figure 15.7.

in which neighbors are connected to a central cell and to each other, as the number of sides of a face indicates the number of neighbors shared in common between two Voronoi cells.

Even this additional information, however, does not completely describe the manner in which neighbors of a Voronoi cell are connected, as seen through the examples in Figure 15.7. While all three Voronoi cells have 4 four-sided faces, 4 five-sided faces, and 2 six-sided faces, these faces are arranged differently in the three examples. For example, a pair of six-sided faces are adjacent in (b), but not in (a) or (c). Differences in the arrangements of faces indicate differences in the manner in which neighboring Voronoi cells are connected to one another and, consequently, the manner in which neighboring atoms are arranged. Our desire to completely describe the arrangement of neighbors motivates the development of a method to record all information about the topology of a Voronoi cell.

15.2.3   Recording Voronoi Topology

The language and tools of graph theory can be used to record complete topological information of a Voronoi cell by looking at it as a planar graph. Briefly, a graph is a set of points called vertices, and a set of connections between those vertices are called edges. A planar graph is one whose vertices and edges can be drawn in the plane without any edges crossing. Two graphs are *isomorphic* if there is a correspondence between their vertices so that two vertices are connected by an edge in one graph if and only if corresponding vertices are connected by an edge in the other graph [1059]. Mathematical theorems from the early twentieth century [965, 1142] guarantee that every Voronoi cell can be uniquely represented as a planar graph, thus allowing us to make precise statements about Voronoi cells using the language of graph theory. Figure 15.8 illustrates planar graphs corresponding to the three Voronoi cells of Figure 15.7.

We employ an algorithm introduced by Weinberg [1133, 1132] to calculate a unique "code" for each planar graph that completely describes the manner in which its edges and faces are connected. Equivalently, this code captures complete information about the manner in which neighbors of a Voronoi cell are arranged relative to a central cell and to each other. Determining whether two atoms have the same local structure then reduces to comparing codes of two Voronoi cell graphs [540].

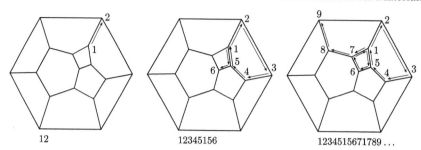

Figure 15.9: Vertices are labeled as they are initially encountered while traversing the graph following the rules described in the text. The code lists all vertices in the order in which they have been visited.

The algorithm of Weinberg is as follows: (a) An initial vertex is chosen and assigned the label 1. (b) An edge incident with that vertex is chosen and travel begins along it. (c) If an unlabeled vertex is reached, it is labeled with the next unused integer and we "turn right" and continue. (d) If a labeled vertex is reached after traveling along an untraversed edge, we return to the last vertex along the same edge but in the opposite direction. (e) If a labeled vertex is reached after traveling along an edge previously traversed in the opposite direction, we "turn right" and continue; if that right-turn edge has also been traversed in that direction, we instead continue along the next right-turn edge available; if all outgoing edges have been traversed, we stop.

Throughout this procedure, the list of the vertices in the order in which they are visited, including multiplicities, is recorded; we call this ordered list a *code*. Figure 15.9 illustrates the process of constructing a code. Codes are then produced for each choice of initial vertex and edge, and for each of two possible orientations of the graph; the lexicographically smallest code is used as the canonical code for the graph.

Constructing canonical codes allows us to capture complete information about the topology of a Voronoi cell, and hence about the manner in which neighboring atoms are arranged. After determining the Voronoi topologies that are associated with a particular crystal structure we can identify which atoms belong to the bulk, and which ones are associated with defects.

15.2.4 Topological Instability and Families of Topologies

Although many Voronoi cells are topologically stable under small perturbations, many are not. Consider, for example, the Voronoi cells in a two-dimensional square lattice, illustrated in Figure 15.10(a). When unperturbed, the Voronoi cell of each atom has exactly four edges, and exactly four Voronoi cells meet at every corner. These corners are unstable in the sense that small perturbations of the atomic coordinates transform them into pairs of corners at which only three cells meet; examples of these transformations can be seen in Figures 15.10(b) and (c).

A consequence of these instabilities is that Voronoi cells in finite-temperature square-lattice crystals can have between 4 and 8 edges. Voronoi cells with fewer than 4 or more than 8 edges cannot be obtained through small perturbations of the atomic coordinates and should thus be considered defects. We use the term *family* to refer to a set of Voronoi cell topologies that can be obtained from a perfect structure through infinitesimal perturbations of atomic coordinates; topologies that cannot be obtained in this manner are classified as defects.

This approach is directly applicable to analyzing three-dimensional systems. Consider, for example, the Voronoi cell of an atom in an unperturbed FCC crystal, illustrated in Figure 15.11. Many of its corners are topologically unstable, and small perturbations of the atomic coordinates result in topological changes [543]; Voronoi cells of several atoms in a finite-temperature FCC crystal are illustrated in Figure 15.12. We refer to the set of Voronoi cell topologies that can be obtained in this manner as the family of FCC topologies; Voronoi cells that cannot be obtained in this manner are classified as defects in FCC crystals. We can likewise associate to any structure a family of topologies which can be obtained through infinitesimal perturbations. In the following section we outline a method by which these families can be determined.

15.2.5   Determination of Families of Topologies

Determination of families of topologies associated with particular structures enables us to anticipate which topologies will appear in finite-temperature systems, and consequently identify and characterize defects in those systems. We briefly consider both analytic and Monte Carlo approaches.

The unique Voronoi cells of atoms in perfect crystalline systems are easily determined and well known; Figure 15.11 shows several such Voronoi cells. In three dimensions, Voronoi cell corners that are adjacent to more than four Voronoi cells are topologically unstable, and random perturbations of nearby atomic coordinates will change the manner in which neighboring Voronoi cells are connected. For example, Voronoi cells of atoms in perfect FCC crystals have six corners which are each adjacent to six different cells; these can be seen in Figure 15.11 as corners at which four edges meet. When atomic positions are perturbed, each of these topologically-unstable vertices can resolve in one of seven ways, illustrated in Figure 15.13. Unstable corners that appear in other systems can resolve in different manners, and each system must be analyzed independently.

To enumerate all topologies associated with FCC, we must consider all permutations of these resolutions over all unstable corners. Since each of the six unstable vertices can resolve in one of seven ways, or remain unstable, we must check a total of $8^6 = 262,144$ possible permutations. We compute canonical codes, as described in Section 15.2.3, for each of these, and find 6250 unique topologies in the family associated with FCC; due to symmetries, many topologies appear multiple times in the initial analysis. Similar analysis for other crystalline systems proceeds in a similar fashion.

The analytic approach described here is not practical for analyzing all systems. For example, the unique Voronoi cell associated with the diamond cubic crystal has 12 topologically-unstable corners, each of which can resolve in one of 7 possible manners, or remain unstable. Enumerating all possible topologies associated with diamond cubic crystal would thus require consideration of $8^{12} = 68,719,476,736$ topologies, the determination and storage of which can be computationally restrictive. In these cases, Monte Carlo sampling allows us to determine Voronoi topologies that occur with significant probability within finite-temperature versions of a particular structure. In particular, atoms in a perfect structure are randomly displaced to simulate the effects of temperature, and the resulting Voronoi topologies are recorded. This is repeated for a large number of atoms in order to sample the set of expected topologies. Care must be taken in choosing appropriate perturbations and sample size.

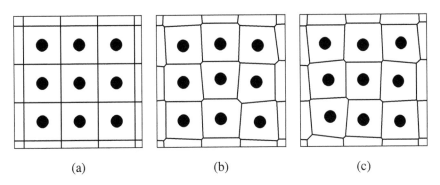

                (a)   (b)   (c)

Figure 15.10: (a) Atoms in an unperturbed square lattice; (b-c) atoms in finite-temperature square-lattice crystals.

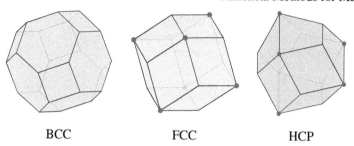

BCC   FCC   HCP

Figure 15.11: Voronoi cells of atoms in unperturbed BCC, FCC, and HCP crystals [543]; red circles indicate unstable corners.

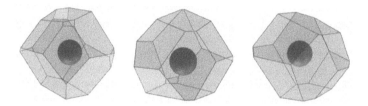

Figure 15.12: Voronoi cells of atoms in a finite-temperature FCC crystal; each unstable corner has transformed into one of the stable configurations illustrated in Figure 15.13.

15.2.6   Ambiguous Topologies and Their Disambiguation

One complication of the topological classification approach results from individual Voronoi topologies belonging to multiple families. Consider for example a two-dimensional Voronoi cell with six sides. We have previously observed that such Voronoi cells can appear both in triangular and perturbed square-lattice crystals. Thus, knowledge that the Voronoi cell of an atom has six sides is insufficient to unambiguously identify its local structure. Similar ambiguities arise in three-dimensional structures, as many Voronoi topologies that belong to the FCC family also belong to the HCP family [543].

Such ambiguities can be resolved by reconsidering topological instabilities under small perturbations as follows. We first consider the case of the ambiguous two-dimensional six-sided Voronoi cell; similar analysis for three-dimensional systems will be described below. We have noted before that Voronoi cells in triangular-lattice crystals are topologically stable under small perturbations;

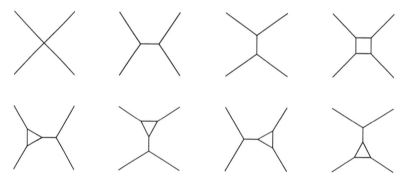

Figure 15.13: Unstable corner in an FCC Voronoi cell and seven possible ways in which it can transform under small perturbations of atomic positions.

Figure 15.14: Voronoi cells of atoms in binary alloys, with faces colored according to neighboring element.

in contrast, Voronoi cells in square-lattice crystals are not. Therefore, if we slightly perturb atoms near a six-sided Voronoi cell, its topology will unlikely change if it belongs to a triangular lattice. In contrast, if it belongs to a square-lattice crystal, then the topology will be more likely to change. Notice, for example, that the six-sided Voronoi cell in the center of Figure 15.10(b) changes upon the small perturbation illustrated in Figure 15.10(c). Observing whether the topology of a Voronoi cell changes when atomic positions are perturbed can thus help disambiguate ambiguous topologies.

Similar analysis is both necessary and effective in studying three-dimensional systems. For example, many Voronoi topologies belong both to the families of FCC and HCP crystals. When neighboring atoms are slightly perturbed, topologies of such Voronoi cells will often change. If small perturbations of neighboring atoms result in unambiguous FCC topologies then we can identify these atoms as belonging to FCC crystals; similarly, if small perturbations of neighboring atoms result in unambiguous HCP topologies then we can identify these atoms as belonging to HCP crystals. Understanding how small perturbations of atomic coordinates change the topology of a Voronoi cell thus helps resolve ambiguous topologies. An example in which this method of resolving ambiguous topologies improves the identification of defect structure is described in Section 15.3.1.

15.2.7 Alloys

The topological approach described so far does not distinguish between atoms of different element types, but can be generalized to do so. Each code described in Section 15.2.3 records an implicit ordering of the faces. In particular, since each code "travels" along each edge of every face exactly once in a particular direction, faces can be ordered by the point along the path at which all of their directed edges have been traversed. Then, in addition to the code which describes the Voronoi topology, we can also record the element types associated with each face in that order. Figure 15.14 shows three different colorings of a Voronoi cell with BCC structure. If red faces indicate neighbors with element types R and blue faces indicate neighbors with element types B, then lists of the corresponding element types are: RRRRRRRRRRRRRR, BRBRBRBRBRBRRR, and RBR-BRRRBRRRRBR. This additional information about element types distinguishes between different arrangements of neighbors in alloys, even when the geometry and topology of the Voronoi cells are identical.

15.3 Applications

Voronoi topology enables the analysis of complex systems in ways not possible using convention methods. In this section we consider two primary applications of this approach. First we consider its utility in identifying defects in high-temperature systems, something that is difficult to do using conventional approaches. We then consider how Voronoi topology can be used to characterize complex grain boundary structures, and subsequently be used to analyze their evolution.

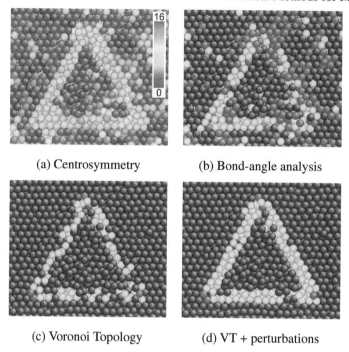

(a) Centrosymmetry   (b) Bond-angle analysis

(c) Voronoi Topology   (d) VT + perturbations

Figure 15.15: Cross-section of a stacking-fault tetrahedron in copper at 85% of its melting temperature, colored using (a) centrosymmetry, (b) bond-angle analysis, (c) Voronoi topology, and (d)
Voronoi topology with perturbation disambiguation. In (a), colors represent the centrosymmetry parameter as per the color bar; in (b-d) dark blue, yellow, and red indicate atoms with FCC, HCP, and
other local structures, respectively.

15.3.1 Defect Identification in High-Temperature Crystals

The first step in studying defect structure evolution in high-temperature systems is the accurate
identification of defect structure. We consider here how the topological approach, and its extension described in Section 15.2.6, can be used to precisely identify a stacking-fault tetrahedron in a
high-temperature FCC copper crystal; we also contrast it with two popular conventional methods.
A stacking-fault tetrahedron (SFT) is a three-dimensional defect consisting of four stacking faults
that form faces of a tetrahedron. The interior and exterior of an SFT are FCC crystals, bounded by
stacking-fault planes with a local structure resembling HCP, and edges which are stair-rod dislocations [433]. Figure 15.15 illustrates a cross-section through the center of an SFT and parallel to one
of its faces; the intersection of the SFT with the viewing plane is an equilateral triangle. This SFT
was constructed in an FCC copper crystal and then thermalized at 85% of its melting temperature.

Figure 15.15 illustrates the SFT visualized using two conventional approaches, and the topological methods described in this chapter. Figure 15.15(a) shows atoms colored using the centrosymmetry order parameter [483]. In this coloring, atoms belonging to faces of the SFT have higher
centrosymmetry values than those in the FCC crystal, as expected. However, many atoms inside
and outside the SFT also have high centrosymmetry values, making the automated location of the
SFT difficult at the simulation temperature.

Figure 15.15(b) shows atoms colored using bond-angle analysis [4]. Many atoms belonging
to the SFT faces are classified as having HCP local structure, as expected. However, many atoms
away from the SFT are also erroneously classified as structural defects, despite the absence of other
defects in the crystal. Moreover, bond-angle analysis incorrectly identifies many atoms in the bulk

as having HCP local structure. Although the general shape of the SFT can be discerned, its structural details are ambiguous, restricting automated analysis.

These figures should be contrasted with the pictures produced using the approaches discussed above. Figure 15.15(c) illustrates the same sample colored using Voronoi topology. Atoms whose Voronoi cell topologies belong to FCC are colored dark blue. Atoms whose Voronoi topologies are unambiguously HCP are colored yellow; all remaining atoms are colored red. Every atom characterized as have an HCP local structure is on an SFT face. Moreover, all atoms not at the surface of the SFT are correctly identified as belonging to an FCC crystal. Finally, atoms lying at the corners of the triangular cross-section through the SFT triangle are identified as having a local structure that is distinct from the bulk crystal and the stacking-fault faces; these are the stair-rod dislocation cores. The sole weakness of this visualization procedure results from Voronoi topologies which belong to both FCC and HCP families and whose local structure identified as FCC rather than HCP.

Figure 15.15(d) shows the clearest picture of the SFT using the extension of the Voronoi analysis method described in Section 15.2.6. To resolve the ambiguous topologies, we perturbed the system 50 times using a random Gaussian perturbation with standard deviation 1% of the average interatomic distance. Atoms which transform into exclusively HCP topologies are characterized as having HCP local structure, while those that occasionally transform into unambiguously FCC topologies, or which always remained ambiguous, are characterized as FCC. Here many more atoms belonging to the stacking fault faces are correctly identified as having HCP-like local structure. In [543], we further demonstrated the successful application of the Voronoi topology method to the identification of grain boundaries, twins, stacking faults, vacancies, and dislocations in a high temperature, plastically deformed nanocrystalline metal.

The correct identification of atoms as belonging to bulk or defect structure allows for a wide range of possible applications in studying many physical mechanisms. We briefly consider the analysis of heterogeneous melting in superheated metals.

15.3.2   Melting

The melting of crystals is one of the most ubiquitous and also least-understood phase transformations. Due to its intrinsic high-temperature nature, and because one of the two phases involved in the process is amorphous, melting has been notoriously difficult to study at the atomic level, even through simulation. We provide here a brief description of how Voronoi topology analysis can be used to identify liquid phases in a superheated crystal matrix.

Even at temperatures just below the bulk melting point, the Voronoi topologies of over 98% of atoms in a single copper crystal belong to the FCC family [543]. This suggests identifying liquid phases as contiguous regions of atoms with Voronoi topologies that do not belong to the FCC family. To study melting, a liquid nucleus was constructed inside a large copper crystal. The size of the liquid nucleus was large enough so that it would grow when the copper crystal was thermalized at 120% of its bulk melting temperature. Figure 15.16 shows a single (111) plane inside this system at several times. Every atom in the system is first identified as FCC or not, based on whether its Voronoi topology is in the FCC family. Next, atoms with non-FCC topologies are clustered so that any pair of neighboring non-FCC Voronoi cells belong to the same "cluster." Figure 15.16 shows all atoms with Voronoi topologies belonging to the FCC family in dark blue, and atoms in the largest cluster of non-FCC topologies in gold. Other atoms with non-FCC topologies, belonging to smaller liquid nuclei, are not shown.

Although the system is heated to well above its melting point, the crystalline and liquid phases in the system are correctly identified without quenching or time-averaging, as validated by visual inspection. This ability to precisely identify the two phases allows for quantitative analysis of the rate at which the supercritical liquid nucleus grows. Figure 15.17 shows the volume of the largest liquid nucleus, measured in number of atoms, as a function of time. While a detailed analysis of this phase transformation is beyond the scope of this chapter, this example shows the potential of the Voronoi topology approach for studying questions inaccessible to conventional analysis methods.

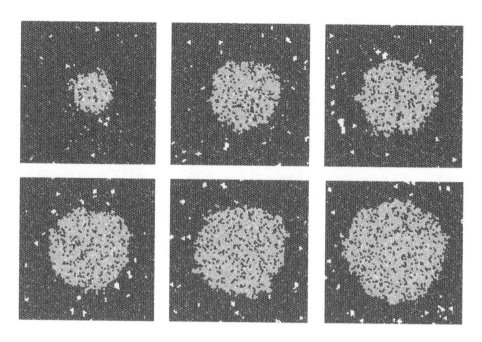

Figure 15.16: A single (111) plane in an FCC single crystal, heated to 120% of the bulk melting temperature, and annealed for 200 ps at this temperature. Blue atoms are those with local FCC structure; gold atoms belong to a single liquid nucleus; all other atoms are not shown.

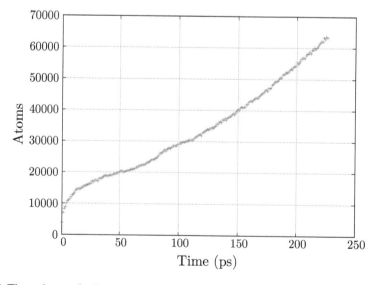

Figure 15.17: The volume of a liquid nucleus as a function of time in a crystal heated to 120% of its bulk melting temperature.

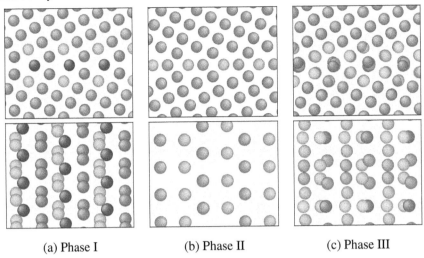

(a) Phase I   (b) Phase II   (c) Phase III

Figure 15.18: Three meta-stable phases of the $\Sigma5$ [001] (310) symmetric tilt boundary in BCC tungsten. Each Voronoi topology is assigned a distinct color. In the profile view, atoms with BCC topologies are shown in gray; in the planar view, these atoms are not shown.

15.3.3 Grain Boundary Characterization

In addition to accurately identifying known defects, Voronoi topology can also be used to characterize structure in complex systems and automate their analysis.

At 0 K, a $\Sigma5$ [001] (310) symmetric tilt boundary in BCC tungsten has three metastable states. The atomic structure of each state can be characterized as a pattern of several Voronoi cell topologies. This can be seen in Figure 15.18, where each atom is assigned a color according to its Voronoi topology. Phase I consists of three distinct Voronoi topologies, and its atoms are colored in three shades of blue. Phase II consists of two Voronoi topologies, and its atoms are colored in two shades of green. Phase III consists of six Voronoi topologies, and its atoms are colored in shades of red, orange, and yellow. These Voronoi topologies are stable under small perturbations of the atomic coordinates, suggesting Voronoi topology as a robust method for characterizing, and subsequently identifying, grain boundary structure.

In one study, Voronoi topology was used to study the evolution of grain boundaries under irradiation conditions [543]. In particular, a $\Sigma5$ [001] (310) symmetric tilt boundary was constructed in the Phase I state of body-centered cubic tungsten and equilibrated at 1500 K, or roughly 40% of its melting temperature. Self-interstitial atoms were inserted in random locations in the grain boundary at a constant rate to mimic the effects of radiation damage. The insertion of these atoms transformed the grain boundary from a Phase I state to a mixture of Phases I, II, and III states. By characterizing atoms using their Voronoi topologies, domains of different grain boundary phases can be readily observed in the irradiated grain boundary, illustrated in Figure 15.19. This automated identification of complex structure allows for further automated analysis of the manner in which this grain boundary changes over time [543]. Although a complete analysis of grain boundary evolution under irradiation is beyond the scope of this chapter, this example highlights the potential of Voronoi topology to automate analysis of complex defect structures.

15.4 Automation through Software

The analysis described in this chapter can be automated through publicly-available computer software. *VoroTop* is an open-source software package that characterizes local structure in atomistic datasets based on Voronoi cell topology [542]. *VoroTop* was developed and is maintained by the

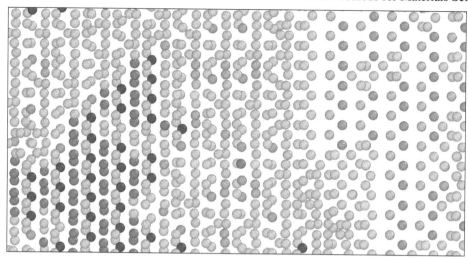

Figure 15.19: Grain boundary, initially of Phase I state, transformed under irradiation conditions; atoms are colored by Voronoi topology.

first author with the generous support of the NSF under grant DMR-1507013; both the software and source code are available at http://www.vorotop.org.

VoroTop reads atomistic data in several standard formats, including those of LAMMPS [792] and AtomEye [570], and then uses the Voro++ software library [876] to compute the Voronoi cell of each atom. Using algorithms described in Section 15.2.3, *VoroTop* computes the topology of each Voronoi cell and then compares it against a precomputed family of topologies associated with a user-chosen structure, such as BCC, FCC, or HCP crystals. The program also enables users to create new families of topologies, with techniques described in Section 15.2.5, and then use those user-generated families to analyze other datasets.

After characterizing atoms through their Voronoi cell topology, *VoroTop* outputs data in standard formats, including those of LAMMPS and AtomEye; these and other programs can then be used for visualization and further analysis. *VoroTop* can also output the topology of each atom (through codes described in Section 15.2.3), and calculate distributions of Voronoi topologies; these capabilities allow for further independent analysis. Much of the analysis described in this chapter has also been integrated into the popular OVITO visualization and analysis software package [982] for convenient use. Additional documentation about *VoroTop*, and information about other functionality, are available on the website.

The use of Voronoi topology for structure identification is computationally efficient, and its runtime scales linearly with the number of atoms in a system. In preliminary tests, the Voronoi topology of one million atoms could be calculated on a single core of a desktop computer in under one minute. Although this is slower than most conventional methods, Voronoi topology obviates the need for quenching, a step otherwise necessary for analysis of high-temperature systems. Furthermore, Voronoi analysis also provides more accurate characterization of structure even in quenched systems; see Supporting Information of [543].

Extensions of the basic topological approach, such as those described in Sections 15.2.6 and 15.2.7, and the clustering method described in Section 15.3.2 are currently being developed for future implementation.

Chapter 16

Markov Random Fields for Microstructure Simulation

Veera Sundararaghavan

University of Michigan

16.1 Introduction

Microstructures are stochastic in nature and a single snapshot of the microstructure does not give its complete variability. However, we know that different windows taken from a polycrystalline microstructure generally "look alike." In mathematical terms, this amounts to the presence of an underlying probability distribution from which various microstructural snapshots are sampled. There are various ways of modeling this probability distribution indirectly. Feature-based algorithms have long been used that categorize various microstructural snapshots based on a common set of underlying features, and generate new synthetic images with similar features [421, 114, 937]. These features could include marginal histograms [421], multiresolution filter outputs (Gaussian [114] and wavelet [937] filters), and point probability functions (e.g., autocorrelation function) [1052]. These methods are good at capturing the global features of the image, however local information in the form of per-pixel data is lost. Thus, features such as grain boundaries are smeared out when reconstructing polycrystalline structures [937].

Alternatively, one could start with sampling the conditional probability distribution for the state of a pixel given the known states of its neighboring pixels using reference 2D or 3D experimental images. If only the nearest neighbors are chosen, this amounts to sampling from an Ising–type model [453]. For general microstructures, the correlation lengths can span several pixels [1052] and a larger neighbor window may be needed. In this chapter, we employ generalized Ising models called Markov random fields (MRFs) to model the probability distribution. While in Ising models, a lattice is constructed with pixels (with binary states) interacting with their nearest neighbors, in MRFs, pixels take up integer or vector states and interact with multiple neighbors over a window. The sampling of conditional probability of a pixel given the states of its known neighbors is based on Claude Shannon's generalized Markov chain [927]. In the one-dimensional problem, a set of consecutive pixels is used as a template to determine the probability distribution function (PDF) of the next pixel.

Efros and Leung [292] developed a non-parameteric sampling approach for extending the sampling technique to 2D microstructures. In this approach, microstructures are grown layer-by-layer from a small seed image (3x3 pixels) taken randomly from the experimental micrograph. Here, the algorithm first finds all windows in an experimental micrograph that are similar to an unknown pixel's neighborhood window. One of these matching windows is chosen and its center pixel is taken to be the newly synthesized pixel. This technique is popular in the field of "texture synthesis" [795, 1159, 756, 292], in geological material reconstruction literature where such sampling methods are termed "multiple-point statistics" [644], and more recently, has been applied for modeling polycrystalline microstructures [520].

An alternate methodology based on optimization has become popular in recent years. The nonparametric sampling method of Efros [292] is posed in the form of an expectation-maximization algorithm [527, 509, 985]. The approach involves minimizing a neighborhood cost function defined using the difference between various x-, y- or z- slices of a synthetic 3D microstructure and the corresponding best matching windows in the 2D experimental images provided by the user. This reconstruction problem generates 3D anisotropic microstructures that have similar high-order statistics [985], which is in contrast to other such works that use assumptions of microstructural isotropy [986] or methods that use lower-order statistics such as two-point correlation functions to synthesize 3D microstructures [1176, 636]. The sampling approach and the optimization approach can also be applied in tandem for various applications involving MRFs. Two other applications are presented in this chapter. One is to embed 3D microstructures into a CAD model geometry and the other is to extend MRFs to synthesize time evolving microstructures.

16.2   Mathematical Modeling of Microstructures as Markov Random Fields

Some of the early attempts at microstructure modeling were based on Ising models [453]. In the Ising model, a $N \times N$ lattice (L) is constructed with values X_i assigned for each particle i on the lattice, $i \in [1,..,N^2]$. In an Ising model, X_i is a binary variable equal to either $+1$ or -1 (e.g., magnetic moment [453]). In general, the values X_i may contain any one of G color levels in the range $\{0,1,..,G-1\}$ (following the integer range extension of the Ising model by Besag [87]). A *coloring* of L denoted by X maps each particle in the lattice L to a particular value in the set $\{0,1,..,G-1\}$. Ising models fall under the umbrella of *undirected graph models* in probability theory. In order to rewrite the Ising model as a graph, we assign neighbors to particles and link pairs of neighbors using a bond as shown in Figure 16.1(a). The rule to assign neighbors is based on a *pairwise Markov property*. A particle j is said to be a neighbor of particle i only if the conditional probability of the value X_i given all other particles (except (i,j), i.e., $p(X_i|X_1,X_2,..,X_{i-1},X_{i+1},..,X_{j-1},X_{j+1},..,X_{N^2})$) depends on the value X_j.

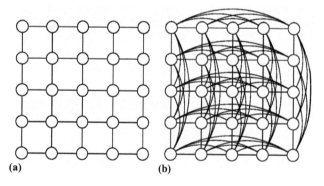

Figure 16.1: Markov random field as an undirected graph model, circles are pixels in the image and bonds are used to connect neighbors: (a) Ising model with nearest neighbor interactions; (b) microstructure modeled by including higher-order interactions in the Ising model.

Note that the above definition does not warrant the neighbor particles to be close in distance, although this is widely employed for physical reasons. For example, in the classical Ising model, each particle is bonded to the nearest neighbors, as shown in Figure 16.1(a). For modeling of microstructures, a higher-order Ising model (Figure 16.1(b)) is used. The particles of the lattice correspond to pixels of the 2D microstructure image. The neighborhood of a pixel is modeled using a square window around that pixel and bonding the center pixel to every other pixel within the window [388]. Using this graph structure, a *Markov random field* can be defined as the joint probability distribution $P(X)$ on the set of all possible colorings X, subject to a *local Markov property*. The *local Markov property* states that the probability of value X_i, given its neighbors, is conditionally independent of the values at all other particles. In other words, $P(X_i|\text{all particles except } i) = p(X_i|\text{neighbors of particle } i)$. Next, we describe a method based on [292] to sample from the conditional probability distribution $p(X_i|\text{neighbors of voxel } i)$. In this chapter, we describe methods to sample from the conditional probability distribution for various applications, including generation of synthetic 2D and 3D microstructures and modeling temporal evolution of 2D microstructures using experimental data.

16.2.1 Spatial Sampling

In the following discussion, the color (X_i) of a pixel i is represented using G color levels in the range $\{0, 1, .., G-1\}$ each of which maps to an RGB triplet. The number of color levels is chosen based on the microstructure to be reconstructed, e.g., for binary images $G = 2$. Let E and S denote the experimental and synthesized microstructure, respectively. Let v be a pixel in S whose color needs to be inferred using the sampling procedure. Let S_v denote the colors in a neighborhood window around pixel v. Let E^w denote the colors of pixels in a window of the same size in the input 2D micrograph.

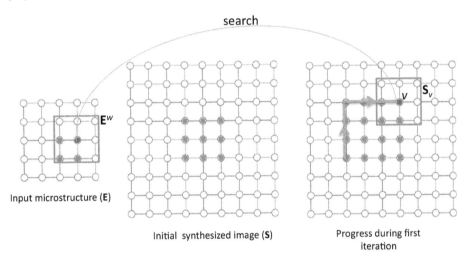

Figure 16.2: The Markov random field approach [292, 1128]: The image is grown from a 3x3 seed image (center). As the algorithm progresses along the path shown (right), the unknown output pixel (shown in blue) is computed by searching for a pixel with a similar neighborhood in the input image (left).

In order to find the coloring of pixel v, one needs to compute the conditional probability distribution $p(X_v|S_v)$. Explicit construction of such a probability distribution is often computationally intractable. Instead, the most likely value of v is identified by first finding a window E_v in the input 2D micrograph that is most similar to S_v (see Figure 16.2). This is done by solving the following problem (where $S_{v,u}$ denotes the color of pixel u in S_v and E_u^w denotes the color of pixel u in E^w):

$$E_v = \arg \min_{E^w} \sum_u \omega_{v,u}(S_{v,u} - E_u^w)^2, \tag{16.1}$$

where, $D = \sum_u \omega_{v,u}(S_{v,u} - E_u^w)^2$ is a distance measure defined as the normalized sum of weighted squared differences of pixel colors. In order to preserve the short range correlations of the microstructure as much as possible, the weight for a nearby pixel is taken to be greater than for pixels farther away (a Gaussian weighting function ω is used). If the pixel u is located at position (x, y) (in lattice units) with respect to the center pixel v (located at $(0,0)$), $\omega_{v,u}$ is given as:

$$\omega_{v,u} = \frac{\exp\left(-\frac{(x^2+y^2)}{2\sigma^2}\right)}{\sum_i \sum_j \exp\left(-\frac{(i^2+j^2)}{2\sigma^2}\right)}. \tag{16.2}$$

Here, the summation in the denominator is taken over all the known pixels in S_v. The weights $\omega_{v,u}$ for the unknown pixels in S_v are taken to be zero. This ensures that the distance measure is computed only using the known values and is normalized by the total number of known pixels. The standard deviation (σ) is taken to be $0.16w$.

The problem in Equation 16.1 is solved using an exhaustive search by comparing all the windows in the input 2D micrograph to the corresponding neighborhood of pixel v. In our approach, a measure of stochasticity is introduced by storing all matches with a distance measure that is within 1.3 times that of the best matching window [292]. The center pixel colors of all these matches produce a histogram for the color of the unknown pixel (X_v), which is then sampled using a uniform random number.

The microstructure is grown layer-by-layer starting from a small seed image (3x3 pixels) taken randomly from the experimental micrograph (Figure 16.2). In this way, for any pixel the values of only some of its neighborhood pixels will be known. Each iteration in the algorithm involves coloring the unfilled pixels along the boundary of filled pixels in the synthesized image as shown in Figure 16.2. An upper limit of 0.1 is enforced for the distance measure initially. If the matching window for a unfilled pixel has a larger distance measure, then the pixel is temporarily skipped while the other pixels on the boundary are filled. If none of the pixels on the boundary can be filled during an iteration, then the threshold is increased by 10% for the next iteration.

The window size is the only adjustable parameter for different microstructures. Window size plays an important role in the MRF model. At window sizes much smaller than the correlation lengths, false matches lead to high noise in the reconstructions. At very high window sizes, not enough matching windows can be identified. Hence, there is an ideal window size that needs to be found through numerical trial. In Figure 16.3, we demonstrate the effect of window size on the quality of the synthesized image. Note that only odd values are allowed for the window size such that the window is symmetric around the center pixel. This example is of a two-phase W-Ag composite from [1068].

The fundamental approximation in this numerical implementation is that the probability distribution of an unfilled pixel is assumed to be independent of that of its unfilled neighbors. In other words, the probability distribution used for filling in a pixel may not stay valid as the rest of its neighbors are also filled in. One problem with this approach is the tendency to sample outlier windows in the experimental image and grow large grains, for example. Feature mapping [558] and k-coherence techniques [34] have been developed recently that largely address this issue. Simple color differences often fail to recognize meaningful structures in an image (e.g., grain boundary segregated phases between grains). Wu and Yu [1158] introduced the notion of a feature mask to help guide the synthesis process that can be used within an MRF pixel-based scheme. The idea is to include structures such as grain boundaries as an additional image channel that is compared. The weight w given to this "feature" channel can be varied [558]. The tradeoff is that a larger weight w downplays color differences, and could result in a noisy reconstruction.

When colors are copied from input to output during the synthesis process, it is very unlikely that they will land on random output locations. Instead, colors that are together in the input (e.g., each grain in a polycrystal) ought to have a tendency to be also together in the output. This concept is called coherence [34]. Similar ideas have also appeared in other methods such as jump maps [1188] and k-coherence [1050]. During initialization, a similarity-set is built for each input pixel which contains a list of other pixels with similar neighborhoods. During synthesis, the algorithm copies a pixel from the input to the output, but in addition to colors, the source pixel location is also copied. By forcing the matching windows to cluster together for neighboring pixels in the synthesized microstructure, the algorithm gives significant improvement in reproduction of color

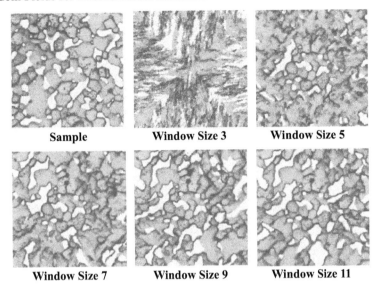

Figure 16.3: Effect of window size: The sample image of a W-Ag composite is shown and its reconstruction using our algorithm is shown for various window sizes. The reconstructed images and the input image intuitively "look alike" at higher window sizes. In the later examples, we quantify this notion by comparing the features.

information (in the form of crystal orientations or phases) [1128]. The sampling algorithm can also be extended towards generating 3D synthetic microstructures from 3D experimental images.

16.2.2 Temporal Sampling

The following problem can also be addressed: *Can one reconstruct the temporal evolution of large-scale synthetic microstructures given experimental measurements of microstructure evolution over a small spatial domain?* Here, a movie of microstructure evolution is measured over a small window in an experimental sample. Using this data, a Markov random field (MRF) algorithm is proposed to estimate the evolution of microstructures over a larger region, or perhaps the entire experimental sample. Such an algorithm would decrease the cost of full-scale microstructure measurements by coupling mathematical estimation with targeted small-scale spatiotemporal measurements.

Consider an input microstructure movie which is a collection of F frames. In the temporal reconstruction algorithm, the first frame is reconstructed using the algorithm in Section 16.2.1. The reference location (in the experimental frame) of each pixel in the synthesized microstructure is stored. The Nth synthesized frame is computed by updating the $(N-1)^{th}$ synthesized frame using pixels at the previously stored locations in the Nth experimental frame. As the microstructure evolves, the reference locations may change. To update the stored reference locations, an optimizer is used on the Nth synthesized frame. The optimization is posed as a minimization of an energy function:

$$S^* = \arg \min_{S} \sum_v \sum_u \omega_{v,u}(S_{v,u} - E_{v,u})^2, \tag{16.3}$$

where S^* is the optimum synthetic microstructure. The optimization is carried out in a two-step iterative process. In the first step, the energy is minimized with respect to E_v. This step is identical to the sampling algorithm (Equation 16.1) and finds the best matching neighborhood of each pixel v by solving the following problem:

$$E_v = \arg \min_{E^w} \sum_u \omega_{v,u}(S_{v,u} - E_u^w)^2. \tag{16.4}$$

The neighborhood is taken over a small user-assigned window around pixel v. This is an exhaustive search that finds a matching experimental image neighborhood for each pixel v in the synthesized image. Because the matching neighborhoods contain other pixels too, multiple values of coloring are obtained for each pixel depending on how many windows overlap over that pixel. The optimal color of pixel v is computed by setting the derivative of the energy function (Equation 16.3) with respect to X_v to zero. This leads to a simple weighted average expression for the color of pixel v:

$$X_v = \left(\sum_u \omega_{u,v} E_{u,v}\right) / \left(\sum_u \omega_{u,v}\right). \tag{16.5}$$

Note that the subscripts u and v are switched in the above expression as compared to Equation 16.3. This implies that the optimal color of pixel v is the weighted average of the colors at locations corresponding to pixel v in the best matching windows (E_u) of pixels (u) in the synthesized microstructure. Since X_v changes after this step, the set of closest input neighborhoods E_v will also change. Hence, these two steps are repeated until convergence, i.e., until the set E_v stops changing. To get optimal reconstruction speed, the optimizer is only used once every k frames, where k is specified by the user.

16.2.3 3D Optimization

In the following discussion, let E^x, E^y, and E^z denote the set of orthogonal (x, y, and z, respectively) slices of the experimental microstructure. Let V denote the solid (3D) microstructure. The color of voxel v in the 3D microstructure is denoted by X_v. In addition to the color (e.g., RGB triplet), the vector X_v may also contain other values including grain orientation and phase index. In this work, the color is represented using G color levels in the range $\{0, 1, .., G-1\}$ each of which maps to an RGB triplet. The number of color levels is chosen based on the microstructure to be reconstructed, e.g., for binary images $G = 2$.

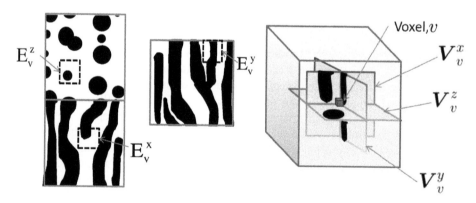

Figure 16.4: The neighborhoods of v in the slices orthogonal to the x, y, and z axis, respectively, are shown. The windows in the input 2D micrograph shown in dotted lines are denoted by E_v^i ($i = x, y, z$). These windows closely resemble the neighborhoods of v.

The vectors denoting the spatial neighborhood of voxel v in the slices orthogonal to the x, y, and z axis, respectively, are denoted as V_v^x, V_v^y, and V_v^z (see Figure 16.4). The neighborhood is taken over a small user-assigned window around the voxel v. Let $E^{x,w}$, $E^{y,w}$, and $E^{z,w}$ denote a window of the same size in the input 2D micrographs. In order to find the coloring of voxel v based on the neighbor voxels in the x–plane, one needs to compute the conditional probability distribution $p(X_v|$color of x–plane neighbors of $v)$. Explicit construction of such a probability distribution is often computationally intractable. Instead, the most likely value of v is identified by first finding a window $E^{x,w}$ that is most similar to V_v^x in the input 2D micrograph. This window is denoted by E_v^x (see Figure 16.4). Similarly, matching windows to the y– and z–plane neighborhoods of voxel

v in the corresponding 2D sectional image denoted as E_v^y, E_v^z are found. Each of these matching windows E_v^x, E_v^y, E_v^z may have different coloring of the center pixel. Thus, we need an optimization methodology to effectively merge these disparate values and identify a unique coloring for voxel v. The optimization approach is described next.

Let the value $V_{v,u}^x$ denote the color of voxel u in the neighborhood V_v^x. Similarly, the value $E_{v,u}^x$ and $E_u^{x,w}$, respectively, denote the color of pixel u in the window E_v^x and $E^{x,w}$. The 3D microstructure is synthesized by posing the problem as an L^2 minimization of the energy [527]:

$$V^* = \arg \min_V \sum_{i\in\{x,y,z\}} \sum_v \sum_u \omega_{v,u}^i (V_{v,u}^i - E_{v,u}^i)^2, \tag{16.6}$$

where V^* is the optimum synthetic microstructure.

The optimization is carried out in two steps as in the 2D case. In the first step, the energy is minimized with respect to E_v^i. In this step, we assume that the most likely sample from the conditional probability distribution of the center pixel in the 3D image (e.g., $p(X_v|\text{colors of } x\text{–plane neighbors of } v)$) is the center pixel of a best matching window in an experimentally obtained 2D slice on the corresponding plane. The best matching neighborhood of voxel v along the x–plane is selected by solving the following problem:

$$E_v^x = \arg \min_{E^{x,w}} \sum_u \omega_{v,u}^x (V_{v,u}^x - E_u^{x,w})^2. \tag{16.7}$$

This is an exhaustive search that compares all the windows in the input 2D micrograph to the corresponding x–slice neighborhood of voxel v and identifies a window that leads to a minimum weighted squared distance. In this process, for 2D images of size 64×64 with a 16×16 neighborhood window, a matrix of size $16^2 \times (64-16)^2$ is built containing all possible neighborhoods of pixels that have a complete 16^2 window around it. The column in this matrix that has a minimum distance to the 3D slice V_v^x is then found through a k–nearest neighbor algorithm [20]. Note that, we are only given a limited (in this work, a single) 2D experimental sample along each cross-section, which means that the best match may not be an exact match for V_v^x.

Thus, for each voxel v, a set of three best matching neighborhoods is obtained, possibly with different colors corresponding to the center pixel. A unique value of v thus needs to be found by weighting colors pertaining to location v in not only the matching windows of voxel v but also its neighbors. This is exactly done in the second step of the optimization procedure, where the optimal color of voxel v is computed by setting the derivative of the energy function with respect to X_v to zero. This leads to a simple weighted average expression for the color of voxel v:

$$X_v = \left(\sum_{i\in\{x,y,z\}} \sum_u \omega_{u,v}^i E_{u,v}^i \right) / \left(\sum_{i\in\{x,y,z\}} \sum_u \omega_{u,v}^i \right). \tag{16.8}$$

As in the 2D case, the subscripts u and v are switched in the above expression as compared to Equation 16.6. This implies that the optimal color of the voxel v is the weighted average of the colors at locations corresponding to voxel v in the best matching windows (E_u^i) of voxels (u) in the solid microstructure. Since X_v changes after this step, the set of closest input neighborhoods E_v^i will also change. Hence, these two steps are repeated until convergence, i.e., until the set E_v^i stops changing.

As a starting condition, a random color from the input 2D images is assigned to each voxel v. The process is carried out in a multiresolution (or multigrid) fashion [509]: starting with a coarse voxel mesh and interpolating the results to a finer mesh once the coarser 3D image has converged to a local minimum. Three resolution levels (16^3, 32^3, and 64^3) were used. Recent literature points to the use of improved weighting schemes for matching global features, such as color histograms [509] and positional histograms [1063], in addition to matching the local Markovian neighborhood. Kopf et al. [509] used a reweighting scheme to match the global color histograms. In this approach, the weights of colors ($E_{u,v}^i$) involved in Equation 16.8 are reduced when these colors contribute to an increase in the difference between the color histogram of the 3D synthesized image (H^s) and the color histogram of the 2D microstructure image (H^e).

$$\omega_{u,v}^{i'} = \omega_{u,v}^i / (1 + \max(0, H^s(E_{u,v}^i) - H^e(E_{u,v}^i))), \tag{16.9}$$

where $H^s(E_{u,v}^i)$ denotes the frequency of occurrence of color $E_{u,v}^i$ in the synthesized 3D image. For better convergence, histograms are dynamically updated during computation of the color of each pixel, rather than once every iteration. In addition, the voxels are updated in a random order to avoid bias. A key challenge is to reduce the number of neighborhood matching iterations to keep computations fast. In the approach of Dong et al. [269], this is achieved by precomputing small 3D neighborhoods by interleaving well-chosen 2D neighborhoods. Among the large number of neighborhoods at each pixel, the best ones were chosen based on color consistency and k-coherence across pixel lines in the three orthogonal neighborhoods.

16.3 Examples

In order to test the physical features and properties of the synthesized microstructure, we are interested in the following criteria:

- The synthetic microstructure must "look like" the seed image. The similarity measures in this work come from the field of metallography/crystallography including lower-order statistics (e.g., grain size/shape distribution, orientation distribution function) and finer statistical features (e.g., higher-order correlation functions, statistics of stress distribution).

- Physical properties of the synthetic microstructure, such as elastic moduli and yield strength, must be within reasonable bounds to the properties of the original microstructure. The properties were tested using finite element models and compared to experiments.

16.3.1 Example 1: 2D Synthesis of an Aluminum Alloy AA3002 Representing the Rolling Plane

Voronoi construction has been used in several studies in modeling polycrystalline microstructures [1195, 549, 1123]. A variety of microstructures can be generated by altering the Voronoi cell generator points, altering aspect ratios [127] and orientations. However, Voronoi constructions are largely an idealization and do not account for the complexity of real microstructures (e.g., nonconvex grain shapes). Physics-based simulations based on Monte Carlo and phase field methods are computationally intensive and several parameters (e.g., nucleation models, free energy models) need to be carefully calibrated from experiments. In this section, we explore the use of MRFs to generate polycrystalline microstructures. A microstructure of the aluminum alloy AA3002 measured using polarized light microscopy ([1150]) was used. The microstructure represents the rolling plane and reveals a fully recrystallized grain structure with randomly distributed intermetallic phases (dark spots in the image). The microstructure is colored based on the occurrence of near-cube and non-cube orientations. This analysis is based on observed contrast effects when the object is rotated relative to the polarized light directions. Purple regions are cube or near-cube orientations, whereas the yellow/red regions are non-cube. The microstructure was reconstructed using a 150×170 pixel input image shown in Figure 16.5(a) using a window size of $w = 7 \times 7$ pixels. The Markov random field reconstruction is shown in Figure 16.5(b). Only a small part (in 16.5(a)) of the larger experimental image (in 16.5(c)) was used for the reconstruction and the reconstructed image of larger size (in 16.5(b)) is to be compared with the larger experimental image.

The fraction of cube versus non-cube orientations and distribution of intermetallic phases was studied using color clouds. The color cloud used here is an attempt at showing the pixels in "color space" rather than Euclidean space [1077]. Color densities are converted into scattered random dots around the spatial position assigned to the color, with the extent of the spatial position determined by the frequency with which that RGB triplet appears in the image. The results shown in Figure 16.6

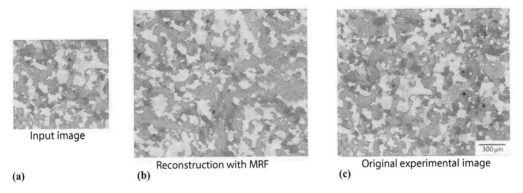

Input image

(a)

Reconstruction with MRF

(b)

Original experimental image

300 µm

(c)

Figure 16.5: Reconstruction of an experimentally measured AA3002 aluminum alloy microstructure [1150] using a Markov random field algorithm. (a) Input micrograph, (b) MRF reconstruction; purple regions are cube/near cube grain orientations, yellow/red regions are non-cube orientations. The fine dark spots are the intermetallic phases. (c) The larger microstructure from which the input image is taken is also shown for comparison.

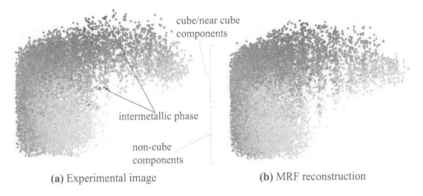

cube/near cube
components

intermetallic phase

non-cube
components

(a) Experimental image   (b) MRF reconstruction

Figure 16.6: Color clouds are used to compare the distribution of cube/near cube regions, and intermetallic phases in the input and synthesized images.

show good visual correlation of the color clouds of the reconstructed image with the experimental input image. The texture components (cube versus non-cube orientations) are well reproduced in the larger synthesized image.

To measure the similarity of properties between sample and synthesized image, we assigned a unique orientation to each pixel based on its color. In Figure 16.7, we have shown the orientation distribution function (ODF) for sample and two synthesized images. We performed finite element calculations on synthesized and sample image in Figure 16.5 and compared the Young's modulus with angle of rotation for sample and synthesized image. The calculations were based on a crystal plasticity model for aluminum (from [987]) and were performed at a constant strain rate of 6.667×10^{-4} s^{-1} and a temperature of 300 K. In Figure 16.8(a) we have plotted the equivalent stress-strain response for sample and synthesized image for a shear test. In Figure 16.8(b), we have plotted the variation of Young's modulus (E) with sample rotation angle for the original and reconstructed images. We also compared the stress histograms for the original and reconstructed image. The histogram plots the number of pixels in the microstructure within a given stress range. The color histograms of stresses shown in Figure 16.9 reveal that the global response of the synthesized image is very similar to that of sample microstructure.

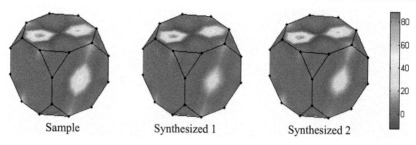

Figure 16.7: ODFs of sample micrograph is compared with the two synthesized images.

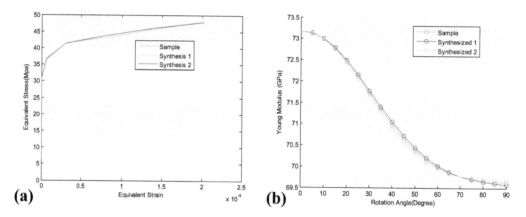

Figure 16.8: (a) Comparison of the equivalent stress-strain curve predicted through homogenization for sample and two synthesized image; (b) variation of Young's modulus with angle of rotation for sample and synthesized images.

16.3.2   3D Reconstruction: A Polycrystal and a Lamellar Composite

The 3D reconstruction approach in Section 16.2.3 has been tested for two test cases with 2D images corresponding to a polycrystalline microstructure and an anisotropic case with solid circles in the z-slice but an interconnected lamellar structure in the x- and y- slices. In the case of the polycrystal, all three slices (x-, y- and z-) were assigned to the same 2D image depicted in Figure 16.10(a). The resulting 3D microstructure is shown in Figure 16.10(b) and its internal structure revealed through x–axis slices in Figure 16.10(c). The results show that the grains built by the algorithm are also equiaxed with a variety of 3D shapes identified by the algorithm.

The lamellar microstructure introduces anisotropy in the x- and y- planes. Three 2D images corresponding to x–, y– and z– slices (as shown in Figure 16.11(a)) were used in the reconstruction. The z–plane image allowed merging of the solid circles to allow for a more complex microstructure. In the algorithm, we match the 2D images with all three orthogonal slices through every voxel. The resulting anisotropic 3D microstructure shown in Figure 16.11(b) is quite complex. The internal structure of the darker phase as shown in Figure 16.11 reveals an intricate internal structure that includes merging of lamellae while still maintaining statistical similarity to the experimental 2D sections.

16.3.3   3D Reconstruction of a Two-Phase Composite

The properties of the reconstructed image were studied based on an experimental dataset in [1068]. The author provides a high-resolution planar microstructure image (Figure 16.12(a)) of a silver-

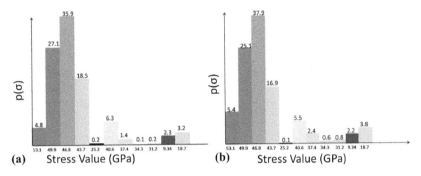

Figure 16.9: Comparison of the distribution of the equivalent stress in a finite element simulation using a color histogram: (a) Experimental image; (b) synthesized image 1.

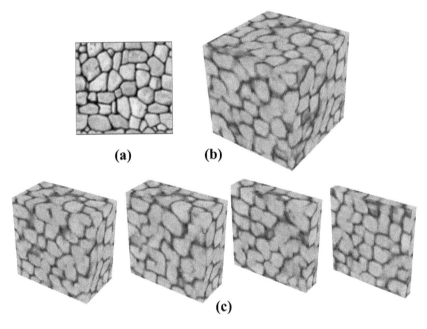

Figure 16.10: (a) An experimental 2D polycrystalline microstructure; (b) 3D reconstruction; (c) 3D sectional images of the reconstructed microstructure.

tungsten composite with porous tungsten matrix and molten silver (volume fraction of silver phase $p = 20\%$). The microstructure has been employed for several reconstruction studies [845, 846]. A 657×657 pixel region of the microstructure corresponding to $204 \times 204 \ \mu m^2$ square area was converted to a black and white image for distinguishing the two phases. This was done by selecting a threshold color below which phases were set to white (the silver phase) and the rest of the image was set to black (the tungsten phase). The final black and white image is shown in Figure 16.12(a)(inset). A $64 \times 64 \ \mu m^2$ square cell within this image was chosen to reconstruct the 3D image.

An instance of the reconstructed microstructure is shown in Figure 16.12(b,c) with the distribution of each phase shown separately. The two-point correlation measure, $S_{(2)}^i(r)$, of this image can be obtained by randomly placing line segments of length r within the microstructure and counting the fraction of times the end points fall in phase i. The auto-correlation function for the silver phase $\gamma(r) = (S_{(2)}^1(r) - p^2)/(p - p^2)$ of the reconstructed 3D microstructure and the experimental image are compared in Figure 16.13(a). The decay in the two-point correlation function is almost identical with the reconstructed image up until 3 μm, showing excellent reproduction of the short range cor-

Experimental

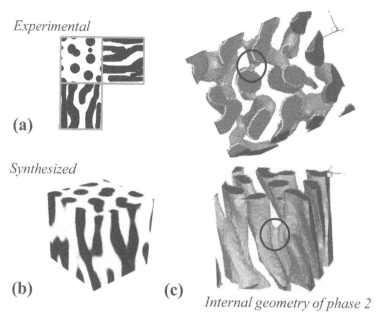

(a)

Synthesized

(b)   **(c)**

Internal geometry of phase 2

Figure 16.11: (a) An anisotropic case with solid circles in the z-slice (similar to case (i)) but an interconnected lamellar structure in the *x*- and *y*- slices; (b) 3D reconstruction; (c) 3D internal structure of the lamellae in the reconstructed microstructure.

relation. Although the longer range correlations match qualitatively, there is a drift as the distance between pixels increases. Both the excellent match in short range correlation and the small drift in the long range correlation can be explained based on the reconstruction algorithm, which models a stronger interaction of a center pixel to pixels in its immediate local neighborhood than to pixels farther away. In effect, the algorithm gives a stronger weighting towards matching the short range correlations in the microstructure. Short range correlations carry the greatest weightage in determining mechanical properties, such as elastic modulus (e.g., [1052]), although long range correlations have been found to be important for phenomena such as surface roughening during plastic deformation [556]. To test if the elastic properties are well captured in the reconstructed 3D microstructure, we compared against the experimental data from [1068] of the elastic modulus as a function of temperature. The elastic properties of individual components at different temperatures are available from [845]. The data was used within a finite element simulation to compute the elastic modulus of the reconstructed microstructure using the method described in [347]. The computed properties of the reconstructed 3D microstructure closely follow the experimentally measured Young's modulus from [1068], as shown in Figure 16.13(b), with a typical error with respect to the experimental data of about 5%.

16.3.4   Spatio-Temporal Sampling (2D + time) of Grain Growth

The temporal Markov random field (MRF) algorithm described in Section 16.2.2 is used to synthesize the evolution of microstructure over a larger region given a small input movie. The movie is obtained from a phase field simulation of grain growth [356]. The image size of the original grayscale movie is 71×71 mesh. The image sizes of the synthesized movies are larger (100×100 pixels) but cover the same time steps as the original microstructure movie. The snapshots from the original phase field simulation and the synthesized movie for initial, an intermediate, and the final time are compared in Figure 16.14. The three synthesized windows correspond to different window sizes used in the MRF model. The window size of 5×5 does not produce a good quality reconstruction

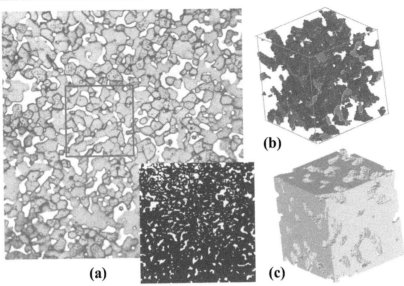

Figure 16.12: (a) Experimental tungsten-silver composite image ($204 \times 236 \mu m$) from Umekawa et al. [1068]. The black and white image corresponds to a thresholded image with white representing the silver phase and black representing tungsten. A 64 μm square cell shown in inset was used to reconstruct the 3D image. (b) A 64 μm length cell of reconstructed 3D microstructure of the experimental image showing silver distribution. (c) The tungsten phase of the reconstructed microstructure.

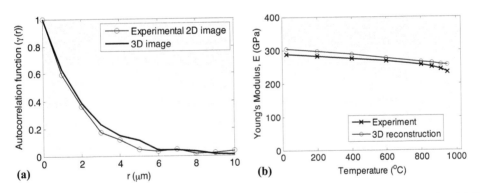

Figure 16.13: Comparison of properties of 3D reconstruction of silver-tungsten composite. (a) The autocorrelation function for the silver phase. (b) Experimental Young's modulus is shown along with the FEM results for the reconstructed 3D microstructure.

as seen in the clusters of small grains that persist at longer times. The window sizes of 7×7 and 9×9 visually look similar to the phase field simulation. To quantitatively compare the grain size and shapes of the input and synthesized images, two global feature vectors were extracted from the input microstructure:

1. Heyn's intercept histogram [1109, 1108], that contains histograms of the intercept length distribution (mean intercept length versus number of test lines possessing the mean intercept length).

2. Rose of intersections [887] is used as the feature vector for assessing grain shapes. To obtain the rose of intersections, a network of parallel equidistant lines is placed over the microstructure image at several angles and the number of grain boundary intersections with each test line is

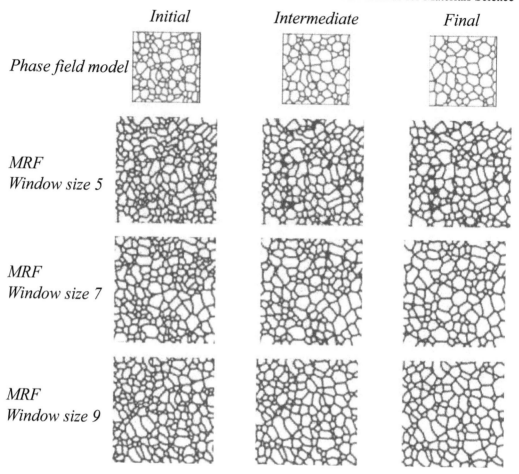

Figure 16.14: Snapshots of the grayscale movie are compared to the synthesized movie snapshots at initial, intermediate, and final times. The synthesized microstructures are obtained using three different window sizes.

measured. The histogram of intersections with the angle of orientation of the lines is called the rose of intersection.

The mean intercept length of the original and synthesized images at the final time are compared in Figure 16.15(a). Figure 16.15(b,c) compares the rose of intersections of the phase field and MRF snapshots at initial and final times, respectively. The features are compared for reconstructions with three different window sizes. A window size of 5×5 results in the worst match, while window sizes of 7×7 and 9×9 compare favorably with the input image. The shape histogram depicts the decrease in overall grain size with time (shrinking of the contour) while the overall grain shape as indicated by the contour shape remains mostly equiaxial. Window sizes of 7×7 and 9×9 are able to track the global statistics of the original microstructure reasonably well.

16.3.5 Microstructure Embedding in CAD Models

The 3D reconstruction methodology shown in this chapter can be extended to any geometry. In principle, one could use sampling and optimization methodologies to embed microstructures over an engineering (computer aided design (CAD)) model. To demonstrate this, we have embedded a microstructure (stainless steel) into a CAD geometry in Figure 16.16. The microstructure was ob-

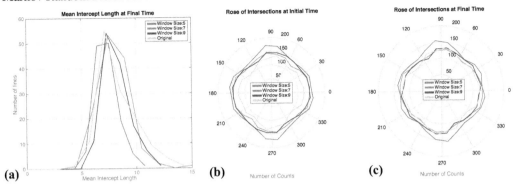

Figure 16.15: (a) Mean intercept length at final time; (b,c) rose of intersections at initial time and final time

tained using serial sectioning and diffraction techniques. The microstructure is in the form of voxels colored by grain numbers [1012]. The 3D microstructure is then sampled onto the CAD geometry using an extension of the sampling method described in Section 16.2.1. Such methodologies will be useful for developing 3D computational models of microstructures at a component level using limited experimentally sampled volumes.

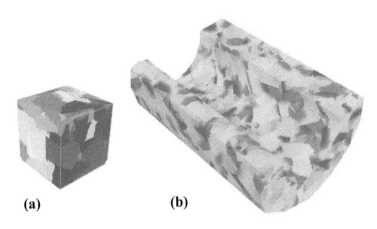

Figure 16.16: (a) AL6XN microstructure from serial sectioning, Al6XN RVE. (Courtesy of Alexis Lewis, NRL.) (b) Synthesized geometry.

16.4   Conclusion

It is human intuition that different windows taken from a polycrystalline microstructure generally "look alike." In this chapter, we explored the notion that these microstructural windows are indeed samples from a common underlying high-dimensional probability distribution. While directly quantifying such a high-dimensional distribution is computationally intractable, we looked at methods to indirectly sample from this distribution. We used the Markovian assumption that the probability of the state of a pixel depends only on the states of its neighboring pixels. In the 2D implementation, the microstructure is grown layer-by-layer from a small seed image (3×3 pixels) taken randomly from the sample. To synthesize a pixel, the algorithm first finds all windows in the sample image that are similar to the unknown pixel's neighborhood window. One of these matching windows is chosen and its center pixel is taken to be the newly synthesized pixel. Previous methods for reconstructing polycrystals using algorithms based on a common set of underlying features such as marginal histograms and point probability functions have often failed to capture the local information, such as sharp grain boundaries. We find that the MRF approach is an attractive solution in this regard. We show that not only are the global features such as grain size/shape distribution captured, but also the texture distribution and mechanical (elastic) properties of the synthesized microstructures as computed using finite element method were found to closely reproduce the experimental values. We also presented an extension of this MRF algorithm for synthesizing microstructure evolution. A promising method for reconstructing 3D microstructures from two-dimensional microstructures imaged on orthogonal planes was also presented. The algorithm reconstructs 3D images through matching of 3D slices at different voxels to the representative 2D micrographs. This is posed as an iterative optimization problem where the first step involves searching of patches in the 2D micrographs that look similar to the 3D voxel neighborhood, followed by a second step involving optimization of an energy function that ensures various patches from the 2D micrographs mesh together seamlessly in the 3D image. The method is particularly promising for anisotropic cases where the $x-$, $y-$ and $z-$ slices look different. With these developments, we believe that Markov random fields present an exciting avenue for computational synthesis of microstructures.

16.5   Acknowledgments

The author would like to acknowledge the Air Force Office of Scientific Research, MURI contract FA9550-12-1-0458, for financial support. The author would like to thank former PhD student Abhishek Kumar and current PhD student Pinar Acar for help in developing the examples in this chapter.

Chapter 17

Distance Measures for Quantifying the Differences in Microstructures

Patrick G. Callahan

University of California Santa Barbara

17.1 Introduction

There are various quantitative descriptors for characterizing microstructural features, such as grain size, shape, number of neighbors, or misorientation. With the development of advanced experimental tools and methods such as serial sectioning with focused ion beam scanning electron microscopes [1065], the Tri-Beam system [287], the RoboMet.3D [960], and high energy x-ray diffraction mapping microscopy (HEDM) [577], a large number of microstructural features can be determined in an automated fashion. Once observed, it is straightforward to generate distributions of descriptors from these microstructural ensembles. But how can these distributions be compared? That is the topic of this chapter. Several distance measures for comparing distributions will be discussed, and also a method for developing quantitative criteria for comparing distributions will be described. For demonstrative purposes, object volume and object shape or morphology will be the primary descriptors used to generate the distributions that are compared. Be aware that these methods can be applied to a distribution generated by many descriptors, whether the descriptor represents size, morphology, number of neighbors, or some other feature of interest.

Since this chapter will partially focus on using object shapes to characterize and compare microstructures, we will start with a brief introduction to moment invariants and the shape quotient, both of which are effective at quantifying shape. The microstructures considered here will be compared in 3-D, so the 3-D Cartesian moments and 3-D version of the shape quotient will be described. The methods described can be applied to 2-D data, and there are moment invariants and a shape quotient that can be calculated for 2-D microstructural features as well. To compare microstructures, or quantify differences between them, a descriptor, volume V for example, is calculated for each feature of interest, in this case each grain. V for every observed grain is assembled to yield a distribution for each microstructure. These distributions represent the microstructures and can be compared using distance measures that quantify similarity between the distributions. There are many distances that can be used to determine the similarity between distributions. Here the discrete representation of distributions, histograms, will be compared with a variety of distance metrics. Both bin-by-bin and cross-bin distances will be introduced and used to quantify the difference or similarity between microstructures. After the introduction of shape descriptors and distance measures, an experimentally observed polycrystalline IN100 microstructure and three synthetic microstructures generated from the experimental microstructures statistics will be briefly described. These four microstructures are used as examples to demonstrate the distance measures for quantifying the distance between microstructures.

17.2 Moment Invariants and the Shape Quotient

The moment invariants of an object can be used to quantify its shape. Moment invariants are combinations of moments that are invariant with respect to some transformation, for example, similarity or affine transformations. The Cartesian moments of an object in 3-D can be defined by the equation:

$$\mu_{pqr} = \iiint d\mathbf{r}\, x^p y^q z^r D(\mathbf{r}), \tag{17.1}$$

where $\mathbf{r} = (x, y, z)$ and $D(\mathbf{r})$ is the shape function of an object, given by:

$$D(\mathbf{r}) = \begin{cases} 1 & \text{for } \mathbf{r} \text{ inside object;} \\ 0 & \text{for } \mathbf{r} \text{ outside object.} \end{cases} \tag{17.2}$$

As such, they are essentially the coordinates of an object raised to some power; the order of the moment, n, is given by the sum of these exponents:

$$n = p + q + r. \tag{17.3}$$

If the moments are computed with respect to the object's center-of-mass coordinates (x_c, y_c, z_c), then they are invariant to translation. These are central moments, given by:

$$\mu_{pqr} = \iiint_D dx\,dy\,dz\, (x - x_c)^p (y - y_c)^q (z - z_c)^r. \tag{17.4}$$

The three second-order moment invariants in 3-D are

$$\mathcal{O}_1 = \mu_{200} + \mu_{020} + \mu_{002}, \tag{17.5a}$$

$$\mathcal{O}_2 = \mu_{200}\mu_{020} + \mu_{200}\mu_{002} + \mu_{020}\mu_{002} - \mu_{110}^2 - \mu_{101}^2 - \mu_{011}^2, \tag{17.5b}$$

$$\mathcal{O}_3 = \mu_{200}\mu_{020}\mu_{002} + 2\mu_{110}\mu_{101}\mu_{011} - \mu_{200}\mu_{011}^2 - \mu_{020}\mu_{101}^2 - \mu_{002}\mu_{110}^2; \tag{17.5c}$$

for a detailed derivation see [631] or [615]. $\mathcal{O}_1$ and $\mathcal{O}_2$ are similarity invariants when normalized by volume, while $\mathcal{O}_3$ is an affine invariant. These moment invariants are well suited for shape characterization as they are independent of position, size (when normalized by volume), and orientation; here orientation means the orientation of an object in space, not crystallographic orientation. The moment invariants $\mathcal{O}_i$ have dimensions of $(\text{length})^{5i}$ so they are normalized by the appropriate power of volume to yield dimensionless parameters:

$$\Omega_1 \equiv \frac{3V^{5/3}}{\mathcal{O}_1}, \quad \Omega_2 \equiv \frac{3V^{10/3}}{\mathcal{O}_2}, \quad \Omega_3 \equiv \frac{V^5}{\mathcal{O}_3}. \tag{17.6}$$

The sphere has the minimum surface area for a given volume, so it is the most energetically efficient shape [431]. Therefore, the $\mathcal{O}_i$ parameters take their minimum values for the sphere, and likewise, the dimensionless Ω_i's are maximal for the sphere. They have the following values [615]:

$$\Omega_1^S = \left(\frac{2000\pi^2}{9}\right)^{1/3}, \quad \Omega_2^S = \left(\frac{2000\pi^2}{9}\right)^{2/3}, \quad \Omega_3^S = \frac{2000\pi^2}{9}. \tag{17.7}$$

It is convenient to limit the range of the dimensionless moment invariants from 0 to 1 by normalizing them by the value of Ω_i^S to give $\bar{\Omega}_i$:

$$\bar{\Omega}_1 \equiv \frac{\Omega_1}{\Omega_1^S}, \quad \bar{\Omega}_2 \equiv \frac{\Omega_2}{\Omega_2^S}, \quad \bar{\Omega}_3 \equiv \frac{\Omega_3}{\Omega_3^S}. \tag{17.8}$$

Hence, the maximum value of $\bar{\Omega}_i$ equals 1, and that value corresponds to a sphere, or in the case of $\bar{\Omega}_3$, the affine invariant, an ellipsoid. $\bar{\Omega}_1$ and $\bar{\Omega}_2$ are similarity invariants, so they contain information about the aspect ratio of the object.

The shape quotient Q, also called the isoperimetric quotient, is defined in 3-D as:

$$Q = \frac{36\pi V^2}{S^3} \tag{17.9}$$

where V and S are the volume and surface area of the object, respectively. It is based on the spatial isoperimetric inequality [810], which relates V and S for objects defined by a closed surface as

$$V^2 \leq \frac{S^3}{36\pi}. \tag{17.10}$$

Q is always between 0 and 1, the latter being the value for a sphere. As surface area increases for a given volume, Q decreases. In other words, as the complexity of an object increases, Q decreases. Readers may be familiar with the 2-D analog of Q, sometimes called the circularity or shape factor of an object, defined by:

$$Q_{2D} = \frac{4\pi A}{L^2}. \tag{17.11}$$

To generate a distribution from a dataset, in this case a microstructure, we calculate the descriptor for each feature, or grain, in that microstructure. These descriptors for each feature are assembled in a histogram, and the histogram is normalized yielding a discrete representation of a distribution. These are the distributions representing our datasets that will be compared using various distance metrics.

17.3 Distances

Several distance metrics for comparing distributions will be discussed. Among these are two classes, those that compare each equivalent bin in two distributions, and those that compare multiple bins between distributions. These have been referred to as bin-by-bin measures and cross-bin measures, respectively [869]. The bin-by-bin distance measures discussed below are the Hellinger distance, the histogram intersection distance, the χ^2 distance, and the Jeffrey divergence; the cross-bin distance measures are the Kolmogorov–Smirnov distance, the Earth Mover's distance, and the quadratic form distance. Bin-by-bin distances tend to be more sensitive to bin size and quantization effects than cross-bin distances [773]. Four conditions must hold for the distance $d(P,Q)$ between two distributions P and Q to be a true metric [948]:

1. Identity, $d(P,P) = 0$
2. Non-negativity, $d(P,Q) \geq 0$
3. Commutativity, $d(P,Q) = d(Q,P)$
4. Triangle inequality, $d(P,R) \leq d(P,Q) + d(Q,R)$.

In general, smaller distances between distributions indicate that they are more similar. Later, when we compare multiple synthetic microstructures to an experimentally observed microstructure, the synthetic microstructure distribution that is situated the smallest distance from experimental microstructure's distribution is the most similar to it.

Five normalized distributions are shown in Figure 17.1. Four of the distributions are Gaussian, $G(\mu,\sigma)$. They are given by $P_1 = G(0.8, 0.04)$, $P_2 = G(0.8, 0.08)$, $P_3 = G(0.5, 0.04)$, and $P_4 = G(0.3, 0.02)$; the fifth distribution is given by $P_5 = P_1 + P_3$, the sum of two Gaussian distributions, which is subsequently normalized. The distances between P_1 and the other four distributions, $d(P_1, P_i)$, $(i = 2...5)$, will be used to demonstrate some of the different properties of the distance measures that are introduced in Sections 17.3.1-17.3.7. These distances are tabulated in Table 17.1.

294

Statistical Methods for Materials Science

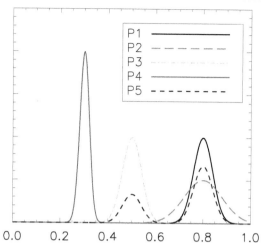

Figure 17.1: The distributions of the normalized affine moment invariant $\bar{\Omega}_3$ for each of the microstructures in S_{IN100}.

Table 17.1: The different distance measures comparing the example distributions discussed in the text.

	Hellinger	Hist Int	χ^2	Jeffrey	K-S	EMD	Quad Form
$d(P_1,P_2)$	0.320	0.319	0.157	0.181	0.164	0.031	0.007
$d(P_1,P_3)$	1.000	1.000	1.000	1.385	1.000	0.300	0.515
$d(P_1,P_4)$	1.000	1.000	1.000	1.386	1.000	0.500	0.942
$d(P_1,P_5)$	0.428	0.333	0.200	0.264	0.333	0.100	0.0572

17.3.1 Hellinger Distance

The Hellinger distance, or modified Bhattacharyya coefficient, is defined as [217]

$$d_H(P,Q) = \sqrt{1 - \beta(P,Q)};$$ (17.12)

β is the regular Bhattacharyya coefficient, which is given by

$$\beta(P,Q) = \sum_{i=1}^{N} \sqrt{P(i)Q(i)}, \left(\text{with} \sum_{i=1}^{N} P(i) = \sum_{i=1}^{N} Q(i) = 1 \right),$$ (17.13)

P and Q are the normalized histograms representing the distributions being compared, and N is the number of bins in each histogram [91]. The Hellinger distance is a true metric, whereas the Bhattacharyya coefficient is not, as it does not satisfy the triangle inequality. The Hellinger distance is convenient as it is straightforward to compute and interpret. Since it is a bin-by-bin similarity measure, we expect that $d(P_1,P_3)$, and $d(P_1,P_4)$ should be 1.0, indicating that they are completely different distributions and do not overlap. This is indeed the case as indicated in Table 17.1. This also illustrates one limitation of bin-by-bin measures; P_1 and P_3 are identical except for a lateral shift, while P_4 is not identical to P_1. Even so, the Hellinger distance, and any other bin-by-bin distances, do not give any indication that P_3 is more like P_1 than P_4.

17.3.2 Histogram Intersection Distance

The histogram intersection, defined by

$$d_{HI}(P,Q) = 1 - \frac{\sum_{i=1}^{N} \min(P(i), Q(i))}{\sum_{i=1}^{N} P(i)}, \qquad (17.14)$$

is a metric when P and Q are normalized and is appealing because it performs well when only part of the histograms are similar [993]. If P is considered the ground truth distribution, the histogram intersection indicates how many objects in Q are similar to objects in P. Referring to Table 17.1, the histogram intersection distance between P_1 and P_2 is approximately equal to the Hellinger distance. Contrast that with $d(P_1, P_5)$, where the histogram intersection distance is smaller than the Hellinger distance because the histogram intersection indicates when parts of one distribution are similar to another.

17.3.3 χ^2 Distance

The χ^2 distance is defined as

$$d_{\chi^2}(P,Q) = 0.5 \sum_{i=1}^{N} \frac{(P(i) - Q(i))^2}{P(i) + Q(i)}. \qquad (17.15)$$

Because there is a sum in the denominator, d_{χ^2} is less sensitive to bins with large values and more sensitive to bins with smaller values [773]. It is based on the familiar χ^2 test statistic, and is also symmetric. $d_{\chi^2}(P_1, P_2)$ is less than $d_H(P_1, P_2)$ because the difference in peak height between P_1 and P_2 has a less pronounced effect on d_{χ^2}. This is particularly useful if distributions have peaks of different magnitudes in the same locations.

17.3.4 Jeffrey Divergence

The Jeffrey divergence is given by

$$d_J(P,Q) = \sum_{i=1}^{N} \left(P(i) \log \frac{2P(i)}{P(i) + Q(i)} + Q(i) \log \frac{2Q(i)}{P(i) + Q(i)} \right) \qquad (17.16)$$

It is a symmetric modification of the Kullback–Leibler divergence and is less affected by noise. Like the bin-by-bin distances, it is easy to implement and fast to calculate. There are various other divergences available as well, including the Kullback–Leibler, Jensen–Shannon, and Cauchy–Schwarz divergences. The Jeffrey divergence does not satisfy the triangle inequality, so it is not a true metric; hence it is sometimes referred to as a pseudo-distance measure [1067].

17.3.5 Kolmogorov–Smirnov Distance

The cross-bin Kolmogorov–Smirnov distance is given by

$$d_{KS}(P,Q) = \max_i |C_P(i) - C_Q(i)| \qquad (17.17)$$

where C_P and C_Q are the cumulative distributions for the histograms for P and Q. d_{KS} is a computationally very cheap cross-bin measure [869].

17.3.6 Earth Mover's Distance

The Earth Mover's Distance (EMD) is a cross bin metric for comparing histograms. It has seen significant use recently in the computer vision field for grayscale [774] and color image retrieval

[868, 869]. It has been used to compare probability distributions or signatures; signatures are unnormalized distributions or histograms that can have different weights for different bins. When comparing normalized probability distributions, the EMD is equivalent to Mallow's distance from statistics [563]. It is called the EMD because it measures the amount of work that must be done to move mass (or earth) around from the bins of one histogram to transform it to another. It is essentially a linear programming method and is based on a solution to the transportation of goods problem.

The EMD is greatly simplified when calculated between two 1-D distributions P and Q [868], and is given by:

$$d_{EM} = \sum_{i=i}^{n} |C_P(i) - C_Q(i)| \qquad (17.18)$$

where C_P and C_Q are the cumulative distributions for the histograms for P and Q. While it is computationally costly for 2-D distributions, the simplification for 1-D distributions make the EMD computationally cheap [869], and there are fast implementations of the EMD available [125].

The capability of cross-bin measures to find similar distributions at different locations is readily apparent in Table 17.1. Since P_3 is just a shifted P_1, $d_{EMD}(P_1, P_3) = 0.300$, whereas the bin-by-bin distances d_H, d_{HI}, and d_{χ^2} indicate that the distributions are completely different because of the lack of overlap between them. This can be particularly useful if there is a systematic error in measurements between two datasets, or to determine how a distribution may be evolving.

17.3.7 Quadratic-Form Distance

The quadratic form distance is given by

$$d_Q(P, Q) = \sqrt{(P - Q)^T A (P - Q)} \qquad (17.19)$$

where A is a similarity matrix that makes d_Q cross-bin in nature [773]. For A, Niblack [734] used $A_{i,j} = 1 - d_{ij}/d_{max}$ where d_{ij} is the Euclidean distance between elements. d_Q shows similar properties to d_{EM}, and it can be intuitive to choose a similarity matrix that indicates the "penalty" of being in different bins. With the Euclidean distance similarity matrix used here, d_Q tends to increase more quickly than d_{EM}. One can choose other distances to generate the similarity matrix, for example the taxicab distance, or the Chebyshev distance.

17.4 IN100 Experimental and Synthetic Microstructures

The IN100 experimental dataset was collected via serial sectioning with a focused ion beam-scanning electron microscope equipped with an electron backscatter detector (EBSD); for details on the collection procedure and the serial sectioning process see [392] or [1065]. The volume of data collected is approximately $96 \times 36 \times 46$ μm^3 with voxels that are 0.25 μm on each side. The orientation maps collected using electron backscattered diffraction (EBSD) were used in reconstructing the volume. Statistics from this reconstructed volume were used to generate synthetic microstructures with the software environment *DREAM.3D* version 6.1.77 [393]. For a detailed description of the microstructure generation process, see [391], but a brief description follows. The volume, shape (via $\bar{\Omega}_3$), number of neighbors, and various other statistics are collected for each grain in the experimental microstructure. Synthetic grains are generated from these statistics and placed in a volume. The statistics of the synthetic microstructure are checked against the observed statistics, and grains are added or removed accordingly. The grains are generated from one of several shape classes that can be described using $\bar{\Omega}_3$. Here, we will consider three synthetic microstructures; one uses the shape class of ellipsoids, the second one the shape class of superellipsoids, and the third one the shapes generated by truncating a cube with an octahedron (details in [157, 160]). These microstructures are basically sets of packed grains, so they will be represented by the symbols M_{EX}, M_{EL}, M_{SE}, and M_{CO} for the experimental, ellipsoidal, superellipsoidal, and truncated cube microstructures, respectively. These four microstructures are members of a set representing experimental and

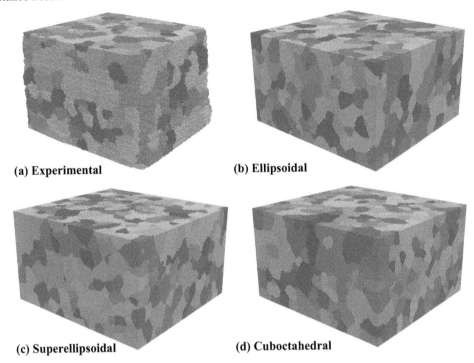

(a) Experimental

(b) Ellipsoidal

(c) Superellipsoidal

(d) Cuboctahedral

Figure 17.2: Visualizations of the (a) experimental, (b) ellipsoidal, (c) superellipsoidal, and (d) cuboctahedral datasets.

synthetic IN100 microstructures, $S_{\text{IN100}} = \{M_{EX}, M_{EL}, M_{SE}, M_{CO}\}$. The microstructures are shown in Figure 17.2. Note that while only surface-intersecting grains are shown in Figure 17.2, they were removed from each dataset before characterization. There are a few differences to note between the different synthetic microstructures. M_{EL} has rather smooth and gradually changing grain boundaries as compared with the others; the grains tend to have more simple shapes than in M_{EX}. This is not surprising, as no grains in M_{EX} have $\bar{\Omega}_3 = 1$, while all ellipsoids have $\bar{\Omega}_3 = 1$. M_{SE} and M_{CO} have more abrupt, less smooth grain boundaries, which is more in line with M_{EX}. M_{SE} appears to have particularly jagged and rough grain boundaries, while M_{CO} appear to have more compact grain shapes.

17.4.1 Volume

The size distributions of microstructures are often informative and also relatively easy to determine. Some of the basic statistics for the volume distributions from S_{IN100} are given in Table 17.2. M_{EX} has the largest grain of all the microstructures, the largest mean, and also the smallest median, indicating that there are some large outliers that are the cause of the increased mean. Hence, it is unsurprising that M_{EX} also has the largest standard deviation of volume. By simply comparing the values in Table 17.2, it appears that the volume distribution for M_{EL} is the best match for that of M_{EX} as the smallest, largest, and median grain size of M_{EL} correspond well with those of M_{EX}. M_{SE} has the mean that is nearest to the mean of M_{EX}, but because M_{SE}'s largest grain is only half the size of M_{EX}'s largest grain, it is difficult to discern whether this nearness is due to a simple increase in the average grain size, or due to outliers. M_{CO} has the smallest mean grain size, but it also has the smallest large grain of the four microstructures, so it is unclear from Table 17.2 if the small mean is due to not having large outliers or just smaller grains on average. The volume distributions are shown as histograms $P_{EX,V}$, $P_{EL,V}$, $P_{SE,V}$, and $P_{CO,V}$ in Figure 17.3, and the seven distance measures introduced in Section 17.3 have been calculated between M_{EX} and the three synthetic microstructures; they are tabulated in Table 17.3.

Table 17.2: The minimum, maximum, median, mean, and standard deviation of volume in μm^3 for M_{EX}, M_{EL}, M_{SE}, and M_{CO}.

Statistic	Experimental	Ellipsoidal	Superellipsoidal	Cuboctahedral
Min	2.1	1.7	0.5	2.3
Max	891.9	827.4	411.8	352.4
Median	22.3	24.9	25.6	26.3
Mean	50.1	44.0	44.7	42.9
Std Dev	79.0	63.1	54.9	45.2

Table 17.3: The different distance measures comparing the volume distributions between the M_{EX} and the three synthetic microstructures.

	Hellinger	Hist Int	χ^2	Jeffrey	K-S	EMD	Quad Form
$d(P_{EX,V}, P_{EL,V})$	0.176	0.139	0.041	0.050	0.076	0.010	0.0007
$d(P_{EX,V}, P_{SE,V})$	0.172	0.133	0.040	0.048	0.077	0.010	0.0006
$d(P_{EX,V}, P_{CO,V})$	0.197	0.181	0.058	0.066	0.110	0.015	0.0012

In Figure 17.3, the volume distribution for M_{EX} and M_{EL} tend to have more large grains while M_{CO} and M_{SE} have fewer large grains; all have the majority of their weight between 0 and 0.2. We expect the cross-bin distance measures to be smaller than the bin-by-bin measures in this case, as they will be less affected by the large grained bins in M_{EX}. This is indeed the case; the EMD and Quadratic-Form Distance are smaller than all the bin-by-bin distance measures. In all cases, the $d(P_{EX,V}, P_{CO,V})$ is larger than $d(P_{EX,V}, P_{EL,V})$ and $d(P_{EX,V}, P_{SE,V})$, so M_{EL} and M_{SE} have volume distributions that are more similar to the experimentally determined microstructure.

17.4.2   Grain Morphology

Both grain size and shape largely influence the mechanical properties of polycrystalline materials [1143]. Including size in theoretical models is relatively easy and not uncommon; however, including morphology is much more difficult. This is because of the rather complex and difficult to describe shapes that exist in polycrystalline materials. Determining and quantifying the differences between the morphology of grains in a polycrystalline material is the main goal of this section. Two measures of shape morphology, the affine moment invariant $\bar{\Omega}_3$ and the shape quotient Q (see Section 17.2) will be used to generate distributions representing the grain shape in the set of microstructures S_{IN100}.

17.4.2.1   Moment Invariant $\bar{\Omega}_3$

The basic statistics describing the distribution of the affine moment invariant $\bar{\Omega}_3$ are shown in Table 17.4. The mean and median $\bar{\Omega}_3$ for M_{EL} and M_{CO} are both larger than the mean and median $\bar{\Omega}_3$ for M_{EX}, while for M_{SE} they are both slightly smaller. This provides a rough indication that the grains in M_{EL} and M_{CO} have less complex shapes than the grains in M_{EX}, while M_{SE} potentially has more complex grain shapes. A standard statistical test for comparing datasets is Student's t-test.

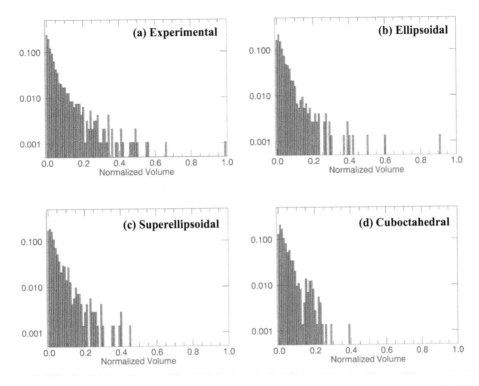

Figure 17.3: The grain volume histograms for each microstructure. Volume was normalized by the largest grain in S_{IN100}. Note that the vertical axis has a logarithmic scale to more clearly show the large grains.

The t-test statistics were calculated to determine if the mean $\bar{\Omega}_3$ were the same between M_{EX} and the synthetic datasets. The t-test statistics between all datasets were all very small, less than 10^{-9}. These small values are not particularly meaningful since they merely indicate that the synthetic microstructures are morphologically different from M_{EX}; we are interested, however, in determining how similar they are, which we will accomplish by using the distance measures introduced in Section 17.3.

Table 17.4: The minimum, maximum, median, mean, and standard deviation of the second-order moment invariant $\bar{\Omega}_3$ for the experimental, ellipsoidal, superellipsoidal, and cuboctahedral datasets.

Statistic	Experimental	Ellipsoidal	Superellipsoidal	Cuboctahedral
Min	0.14	0.52	0.01	0.46
Max	0.87	0.91	0.86	0.87
Median	0.70	0.80	0.67	0.74
Mean	0.68	0.79	0.65	0.73
Std Dev	0.11	0.06	0.11	0.06

The $\bar{\Omega}_3$ distributions for the microstructures in S_{IN100}, $(P_{EX,\bar{\Omega}_3}, P_{EL,\bar{\Omega}_3}, P_{SE,\bar{\Omega}_3}, P_{CO,\bar{\Omega}_3})$, are shown in Figure 17.4. It is immediately apparent that $P_{EL,\bar{\Omega}_3}$ and $P_{CO,\bar{\Omega}_3}$ are strongly peaked and have little mass in their tails, whereas $P_{EX,\bar{\Omega}_3}$ and $P_{SE,\bar{\Omega}_3}$ on average have lower values, and have more mass in the tail to the left of the peak. As $\bar{\Omega}_3$ decreases as a shape becomes more complex, this indicates that M_{EX} and M_{SE} have more complex grains than M_{EL} and M_{CO}. $P_{EL,\bar{\Omega}_3}$ is shifted to the right of the other distributions; this is not surprising as the starting shape for each grain in M_{EL} is an ellipsoid, which has $\bar{\Omega}_3 = 1$. The distances between the $\bar{\Omega}_3$ distributions of the experimental microstructure and the synthetic microstructures using the seven measures introduced in Section 17.3 are shown in Table 17.5. The minimum distance between distributions of $\bar{\Omega}_3$ are between M_{EX} and M_{SE} for all seven distance measures considered. This matches what is apparent in Figure 17.4, where the $P_{SE,\bar{\Omega}_3}$ overlaps quite well with $P_{EX,\bar{\Omega}_3}$. It also matches what qualitatively appears to be the case from a cursory look at the grain shapes in Figure 17.3 where the complex shapes of the grains in M_{SE} appear to look most similar to those in M_{EX}. The grains in M_{CO} also look similar to the grains in M_{EX} but the shapes are more compact. This difference leads to the better match of $\bar{\Omega}_3$ for M_{SE}. It should be noted that $\bar{\Omega}_3$ is relatively robust to noise at the surface of a shape, especially when computed using voxelized data.

Table 17.5: The different distance measures comparing the $\bar{\Omega}_3$ distributions between the M_{EX} and the three synthetic microstructures.

	Hellinger	Hist Int	χ^2	Jeffrey	K-S	EMD	Quad Form
$d(P_{EX,\bar{\Omega}_3}, P_{EL,\bar{\Omega}_3})$	0.502	0.525	0.356	0.425	0.520	0.111	0.077
$d(P_{EX,\bar{\Omega}_3}, P_{SE,\bar{\Omega}_3})$	0.207	0.177	0.070	0.077	0.146	0.032	0.006
$d(P_{EX,\bar{\Omega}_3}, P_{CO,\bar{\Omega}_3})$	0.286	0.258	0.119	0.137	0.244	0.049	0.016

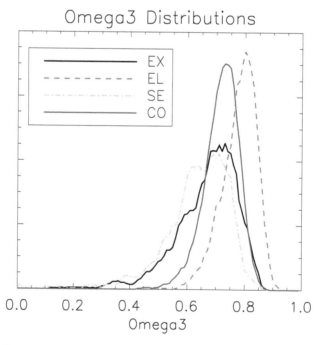

Figure 17.4: The distributions of the normalized affine moment invariant $\bar{\Omega}_3$ for each of the microstructures in S_{IN100}.

17.4.2.2 Shape Quotient

As the shape quotient Q is another measure of an object's morphology, we expect to see results similar to those for $\bar{\Omega}_3$ shown in Section 17.4.2.1. However, Q is more sensitive than $\bar{\Omega}_3$ to variations at the object's surface, particularly for voxelized data. This is intuitively clear since if one voxel protrudes from an object's surface, the surface area exposed is $5\times$ the area of the voxel's face although the volume only increases by one voxel. The basic statistics describing the distributions of Q are provided in Table 17.6.

Table 17.6: The minimum, maximum, median, mean, and standard deviation of the shape quotient Q for the experimental, ellipsoidal, superellipsoidal, and cuboctahedral datasets.

Statistic	Experimental	Ellipsoidal	Superellipsoidal	Cuboctahedral
Min	0.11	0.51	0.18	0.38
Max	0.84	0.89	0.91	0.85
Median	0.53	0.69	0.52	0.59
Mean	0.52	0.69	0.51	0.60
Std Dev	0.11	0.06	0.11	0.07

As was the case with $\bar{\Omega}_3$, M_{SE}'s mean, median, and standard deviation best match those of M_{EX} for the parameter Q. The Q distributions for the microstructures in S_{IN100}, ($P_{EX,Q}$, $P_{EL,Q}$, $P_{SE,Q}$, $P_{CO,Q}$), are shown in Figure 17.5. $P_{EL,Q}$ and $P_{CO,Q}$ have well-defined peaks that are higher than and shifted to the right of the peak in $P_{EX,Q}$. The peak in $P_{EL,Q}$ is shifted further to the right than $P_{CO,Q}$. This indicates that the grains in M_{EL} tend to be more spherical than the grains in the other microstructures, as $Q = 1$ corresponds to a sphere. As surface area for a given volume increases, Q decreases and the shape becomes more complex, so clearly M_{EX} and M_{SE} are the more complex microstructures as measured by Q. $P_{EX,Q}$ and $P_{SE,Q}$ match up quite well, $P_{SE,Q}$ is only slightly shifted to the left from $P_{EX,Q}$, so we expect that the distance measures will be less between $P_{EX,Q}$ and $P_{SE,Q}$ than between the other two synthetic microstructures.

The distances between the Q distributions of the experimental microstructure and the synthetic microstructures using the seven measures introduced in Section 17.3 are shown in Table 17.7. The distance is minimum between $P_{EX,Q}$ and $P_{SE,Q}$ using all distances, as expected by qualitatively considering the distributions in Figure 17.5. Unsurprisingly, M_{EL} is composed of the least complex grains; an ellipsoidal starting grain shape is just too simple for matching the morphology of a real microstructure. In all the bin-by-bin comparisons, the distance between $d(P_{EX,Q},P_{CO,Q})$ is much larger than $d(P_{EX,Q},P_{SE,Q})$; the difference is not quite as pronounced for the χ^2 distance because of its property that the difference in height of the bins has less of an effect than in the other distances. While all of the distributions compared here were normalized because there were different numbers of grains in each microstructure, in situations where the distribution is not normalized, this is a particularly useful property.

17.4.2.3 Volume and Morphology

Comparing the results from Sections 17.4.1 and 17.4.2, it is clear that more than size should be considered when characterizing a microstructure. Although all synthetic microstructures did a reasonably good job of matching the volume distribution for M_{EX}, there were clear differences in how well they matched the shape distributions of $\bar{\Omega}_3$ and Q.

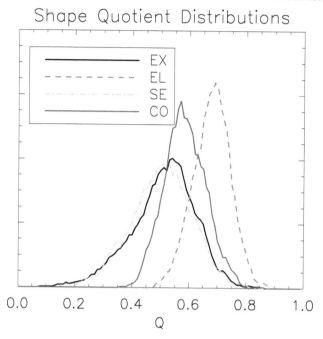

Figure 17.5: The distributions of the shape quotient Q for each of the microstructures in S_{IN100}.

17.4.3 Developing a Criterion for Microstructure Comparison

When one is interested in comparing several distributions, distance measures that determine the similarity between distributions are quite effective in determining which distributions, or datasets, are most alike with regard to the descriptor used to generate the distributions, as demonstrated in Sections 17.4.1 and 17.4.2. This was done by comparing an experimental microstructure with three classes of synthetic microstructures. We saw that when comparing volume distributions, M_{EL} and M_{SE} were nearest to M_{EX}, and when comparing grain shape distributions, M_{SE} was the most like M_{EX}, or had the most realistic grain shapes. Additionally, it would be useful to have a quantitative criterion for determining if one dataset is indistinguishable from another distribution. That is the aim of this section, to determine a threshold distance between distributions below which they are indistinguishable. This can only be accomplished if the ground truth distribution is known. In this situation, we will assume that the experimental microstructure is the ground truth, i.e., the true microstructure. As we are comparing normalized distributions from datasets with the same resolution, we will use the bin-by-bin Hellinger distance measure since we do not need the properties of cross-bin distances that compare nearby bins as well; in fact, since there is no experimental error from data collection for the synthetic datasets, a bin-by-bin measure is preferable.

Table 17.7: The different distance measures comparing the Q distributions between the M_{EX} and the three synthetic microstructures.

	Hellinger	Hist Int	χ^2	Jeffrey	K-S	EMD	Quad Form
$d(P_{EX,Q}, P_{EL,Q})$	0.655	0.680	0.549	0.686	0.679	0.168	0.159
$d(P_{EX,Q}, P_{SE,Q})$	0.157	0.135	0.036	0.041	0.065	0.011	0.001
$d(P_{EX,Q}, P_{CO,Q})$	0.367	0.336	0.177	0.215	0.321	0.075	0.032

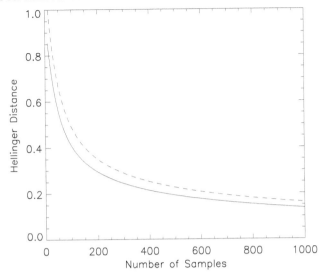

Figure 17.6: The mean Hellinger distance as a function of the number of grains sampled from the experimental microstructure is shown as a solid line and the threshold distance, $t_{d_H} = \mu_{d_H} + 2.5\sigma_{d_H}$, is shown as a dotted line.

To determine the threshold distance below which two distributions are indistinguishable, N_s random grains will be chosen from the ground truth microstructure M_{EX} to obtain a distribution P_{EX}. Then N_s random grains will be chosen from M_{EX} again to obtain a second distribution Q_{EX}. The Hellinger distance between these distributions $d_H(P_{EX}, Q_{EX})$ is determined, and this process is repeated many times (100,000 times, in this case). The mean and standard deviation of the Hellinger distance μ_{d_H} and σ_{d_H} are determined, and from these the threshold distance is calculated as $t_{d_H} = \mu_{d_H} + 2\sigma_{d_H}$. The threshold distance for the distribution of $\bar{\Omega}_3$ as a function of the number of grains sampled from the ground truth microstructure, $t_{d_H, \bar{\Omega}_3} = t_{d_H, \bar{\Omega}_3}(N_s)$ is shown in Figure 17.6. The function $t_{d_H, \bar{\Omega}_3}$ was fit with a power-law of the form $f(x) = P_1 x^{P_2} + P_3$ where P_i are constants. The resulting fit is given by

$$t_{d_H, \bar{\Omega}_3}(N_s) = 4.284 N_s^{-0.456} - 0.182. \tag{17.20}$$

There are approximately 750 complete grains in each of the synthetic microstructures; that is, grains that do not intersect the surface of the volume. Plugging 750 into Equation (17.20) yields $t_{d_H, \bar{\Omega}_3} = 0.192$. Comparing this to the first column in Table 17.5, all Hellinger distances between the experimental and synthetic microstructures are larger than this value. Thus, we say that all of the synthetic microstructures are distinguishable from M_{EX} with regard to grain shape. Using this criterion, 2,423 randomly chosen distributions out of a set of 100,000, or 2.4%, from M_{EX} would have been incorrectly classified as being distinguishable from the ground truth dataset.

Using the same technique just described, we also determine a threshold for volume, namely $t_{d_H, V}(N_s = 750) = 0.174$. Comparing this with the Hellinger distances in Table 17.3, much more concrete conclusions can be made than in Section 17.4.1. We now see that with this criterion the volume distribution of M_{SE}, while very close to the threshold, is below it, so we would conclude that the microstructures are indistinguishable from each other when volume is the descriptor of interest. M_{EL} and M_{CO} both have quantitatively different volume distributions than M_{EX}. Using this criterion, 2,444 randomly chosen distributions out of 100,000 would be incorrectly classified. This is 2.4%, as was the case with $\bar{\Omega}_3$. This makes sense considering that with a normal distribution, about 95% of the data lies within 2 standard deviations of the mean [838]. The criterion is only concerned with one side of the distribution; if d_H is less than the μ_{d_H} we simply conclude that the distributions are indistinguishable. Since it's one side of the distribution, we can expect that the incorrect conclusion will be made approximately 2.5% of the time. The number of standard deviations away from μ_{d_H} can be adjusted accordingly to control how often the incorrect conclusion may be made. For example, since 99.7% of data lies within 3σ if the criterion was adjusted to $t_{d_H} = \mu_{d_H} + 3\sigma_{d_H}$, the wrong conclusion would be made about 0.15% of the time.

17.5 Summary

There are many different choices for measuring the distance between two distributions, and only a small selection of them have been discussed here. The researcher interested in comparing microstructures or distributions should be aware of the strengths and limitations of the different distance measures and choose accordingly. If the microstructures being compared are of the same resolution, have been collected or generated with the same techniques, and the descriptor distributions are normalized, then a bin-by-bin measure like the Hellinger distance is a good choice for data comparison. If the microstructures were collected with different techniques, say one with serial sectioning in an electron microscope, and one with an x-ray technique such as High Energy X-ray Diffraction Microscopy (HEDM), then cross-bin measures like the Earth Mover's Distance or Quadratic Form Distance are likely a better choice as there will be different experimental errors and biases in each dataset, which will likely manifest themselves as shifted peaks and different spread in the data. If the researcher knew the distribution of grain size from a part that undergoes a dual microstructure heat treatment, and so has a small grain region and a large grain region, and wanted to verify the distribution in the small grain region of a newer part, the histogram intersection distance would be a reasonable choice as it is a good indicator for when only parts of distributions are similar. For working with unnormalized distributions, the χ^2 distance discriminates well because it is less affected by differences in peak amplitude. These are just a few examples that illustrate situations where certain distance measures are more suitable than others for microstructure comparison.

There is also the issue of determining if two microstructures are distinguishable from each other. This was discussed in Section 17.4.3, where a method for determining a criterion to distinguish between microstructures was described. The approach relies on assuming one microstructure is the correct, or ground truth, microstructure to which the others are compared. This method yields a threshold distance that indicates if two microstructures are distinguishable from each other. In the examples used to illustrate this method, synthetic microstructures were compared to a real microstructure M_{EX}, and it was shown that one synthetic microstructure, M_{SE} was indistinguishable from M_{EX} when volume distributions were considered. When the shape of the grains in the microstructures were considered using the morphological descriptor $\bar{\Omega}_3$ all synthetic microstructures were distinguishable, or different, from M_{EX}. While the Hellinger distance was used in Section 17.4.3, other measures can be used instead. Just as when measuring the distance between two distributions, the limitations of the different distances should be considered before choosing which distance to use.

Acknowledgments

The author would like to thank the AFRL for support through the STEP and SCEP programs, and the ONR for support through contract number N00014-16-1-2982. The author would like to thank Jean-Charles Stinville, Will C. Lenthe, McLean P. Echlin, and Tresa M. Pollock for helpful discussions.

Chapter 18

Industrial Applications of Microstructural Characterization — Current and Potential Future Issues and Applications

David Furrer
Pratt & Whitney

Ryan Noraas
Pratt & Whitney

David B. Brough
Georgia Institute of Technology

18.1 Introduction

Control of microstructure is a major goal and requirement for industrial materials and processes. Microstructure to a large extent controls the properties and behavior in materials. Engineering materials are developed to produce a controlled range of properties. This control often is accompanied by a need to control the variability of the microstructure. The issue that industry and research engineers developing new materials and processes have to contend with is the ability to actually, objectively and robustly characterize microstructure. Microstructures are very complicated and subtle changes in microstructure at various length scales can produce a marked change in properties. These subtle variations in microstructural features and potential synergistic interactions of microstructures make it very difficult for engineers to quantitatively compare and rate microstructures for many engineering materials. Methods to simplify the quantification of microstructure using limited parameters with simplified definitions, such as average grain size, have been established for many materials, but these simplified representations no longer support all needs for microstructural characterization. Advances in characterization methods and definitions of microstructures are continuing with the application of statistical methods for data sampling and analysis, along with new quantitative imaging techniques that provide new information and higher fidelity on legacy microstructural parameters.

18.2 Industrial Application of Microstructural Characterization

Characterization of microstructures has been a critical part of materials science and engineering from the early days of microscopy [946] to today's development and application of advanced imaging technologies such as the electron backscatter diffraction (EBSD) method [262]. Characterizing the microstructure of materials can provide the necessary insight to the relationship between structure and properties, including mechanical, electrical, magnetic, etc. Physical metallurgy or today's materials science strives to provide understanding of the linkages between microstructure and properties, but this has traditionally been largely empirical and largely based on the ability to measure and quantify microstructural features.

Quantifying microstructure has been a challenge for many reasons. Preparation methods to measure microstructural features often require multiple steps that can result in variability in observations, including sample preparation and imaging techniques [1091]. Quantifying observations is a primary requirement relative to enabling establishing quantitative, mathematical relationships.

In the search for quantitative relationships, theoretical analysis and hypotheses have led to searching for new methods to image and measure microstructure. As an example, theories regarding dislocation generation and interactions were proposed before these microstructural features were imaged [448]. To prove this and countless other theories and associated models, microstructural characterization methods have been developed and further advanced. Modeling and experimentation have continued to support and drive each scientific and engineering discipline further forward. This synergy between modeling and experimentation can be seen in nearly all current advances in materials science and engineering.

The development and application of quantitative microstructural methods are for much more than the foundational purpose of theoretically linking quantitative structure and properties. Quantitative characterization of materials is required for materials definitions. This is a seemingly obvious purpose, but it is still a major opportunity for materials science and engineering.

Materials definitions in the form of specifications provide a means of communicating requirements for a material to ensure equivalency to a desired capability. Materials science and engineering within industry is largely focused on development and assurance of material equivalence either in a global component sense or on a location-specific basis for advanced processing and component applications such as dual-microstructure/dual-property disks for aircraft turbine engines [352, 653].

Materials definitions that include microstructural elements can range greatly in description method and format. Many definitions in the form of written descriptions are often open to interpretation and have a limited capability to ensure quantitative equivalency. Numerical values for microstructural features deemed critical to a material definition are also applied, such as grain size, but how to measure and analyze raw data to establish a simplified quantitative value is not without challenge.

Similar to materials definitions, quality control methods related to microstructural features is another major industrial activity related to microstructural characterization. The most important attributes relative to industrial quality control assessment are the standardization of robust methods and ease of analysis process. These two attributes are often opposed to the ultimate goal of accurate quantitative characterization methods. Standards require concurrence and acceptance of a single approach that may require specialty preparation, imaging, image analysis, and process control methods and associated equipment. Though this may seem straightforward, standards are a challenge to establish and evolve.

As part of microstructural quality control, ease of characterization is vital. Pragmatic, "go/no-go gages" have been established and are strongly engrained in the materials and manufacturing communities. Rapid, low-cost characterization tools can lead to reduced microstructural feature quantification and fidelity of characterization results. This may or may not be acceptable for the material definition and component application.

An inadequate method of defining materials can actually limit the overall usefulness and applicability of a material. If a critical microstructural feature is not defined and controlled, then the resultant properties will have a commensurately large range of variation. It is important to note that the worst expected property levels for a controlled material are used for design purposes. This fact means that the greater the microstructural quantification and control of any given material can lead to enhanced property and application capabilities. Obviously this may bring cost implications, but it is well known throughout industry that novel material definitions can be effectively developed and deployed. Limitations are often in the definition and quantification of specific features that can be readily manipulated by manufacturing processes. Enhanced materials definitions are starting to incorporate enhanced microstructural definitions with greater quantification of critical features.

18.3 Industrial Microstructural Characterization Examples

As noted previously, microstructure measurement and characterization for industrial applications are for the purposes of material definitions, development, enhancement, and quality control. One of the most common microstructural characterization type for crystalline materials is grain size. Characterization of this feature, at first glance, may seem straightforward and able to be standardized. Grain size, however, is very complicated. ASTM standards define how average (ASTM E112) and largest grain size (ASTM E930) have been established and employed for a range of research, development and quality control applications [326].

ASTM definitions of grain size have been successfully utilized to characterize and differentiate materials. The use of these standards require imaging microstructures to enable identification of grain boundaries. Various methods for calculating ASTM grain size have been developed and qualified for industrial use, including line intercept methods by a range of measurement techniques [1091] and comparison methods. The former methods provide a means of calculating an average grain size for a planar section of microstructure. The latter requires subjective comparison between photographic standards or standard image overlays for subjective comparison with the structure being characterized.

There are several challenges and deficiencies with this type of characterization of grain size, including:

1. Grain size values are defined in terms of average values, maximum values or for special cases average values of bi-model populations, and

2. 2D grain size values may not be representative of 3D values.

The first issue of limited grain size information can limit or even eliminate the use of grain size as a clear differentiator between materials. Average grain size can provide trends relative to material properties, such as tensile properties. The Hall–Petch relationship (Figure 18.1) provides a relationship between grain size and tensile properties, where tensile strength varies linearly with inverse of the square-root of the average grain size. It can be seen for crystalline materials that deform by dislocation motion that as grain size decreases the strength increases. Average grain size may not fully describe how the material actually behaves as it is not common to have a material with a unimodal grain size distribution. Grain sizes exist as a distribution that can vary to a wide range. Variation in grain size distribution can result in differences in properties. Bi-modal grain structures where fine recrystallized grains surround or necklace unrecrystallized grains can have a significantly different creep capability for example.

Special boundaries are also important features in defining materials, and assessing capabilities and equivalency to specifications or standards. $\Sigma3$ (twin) boundaries are important features in nickel-base superalloys and stainless steels for mechanical property and corrosion capabilities reasons. The Hall–Petch relationship relates grain size to tensile properties as a result of dislocation intersection with grain boundaries and an accompanied increase in critical resolved shear stress to continue dislocation motion into the adjacent grain. Based on this fact, twin boundaries must also be included in the Hall–Petch relationship, so the grain size of austenitic materials must take the primary grain boundaries and twin boundaries into account, which is often not the case for typical average grain size determination and definitions.

The ability to distinguish between recrystallized and unrecrystallized grains can also be crucial for metallic material systems. Percentage of recrystallized grains can provide an important metric that can enable prediction of material performance capabilities. Figure 18.2 shows an example of the nickel-base superalloy Waspaloy that has been processed to produce a 100% warm-worked condition where there is substantial grain-level strain and recovery, but no recrystallization [344]. This has significant implications for high temperature tensile and creep strength [971]. The ability to characterize and determine which grains are recrystallized and unrecrystallized enables the ability to define and track required threshold values of such features.

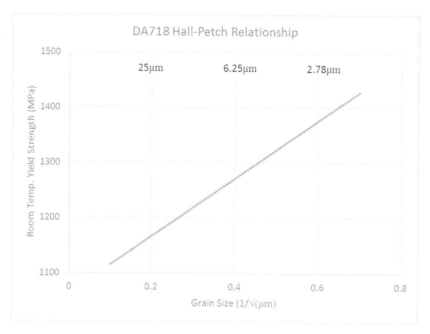

Figure 18.1: An example Hall–Petch relationship for nickel-base superalloy DA718 [831].

In addition to the lack of grain size distribution information, average grain definitions alone also do not provide information about grain size orientation distribution. Many materials can recrystallize to fine uniform grain sizes, but neighboring grains or clusters of grains can exhibit very low angles of crystallographic misorientation. These regions of near-common crystallographic orientation can effectively act as a single, larger grain assemblage. This has been seen in nickel-base superalloys [11] and titanium alloys [874]. When nickel-base superalloys are statically recrystallized in special cases, sub-grains within the prior larger grains can result with the sub-grains having very low angle boundaries. This condition has been previously termed "ghost grains" where many of the properties such as ultrasonic inspectability (or attenuation) and tensile strength are controlled by the larger prior grain structure. In titanium materials, local colonies of aligned alpha grains or often termed micro-texture regions (MTRs) can result and can cause a similar impact to material and component properties [1152]. It can be seen from these examples that average grain size of single or multi-phase materials cannot completely describe capability and equivalency of these materials and associated components.

Single crystal materials, such as silicon for electronic applications and nickel-base superalloys for high temperature, high stress applications can also have subtle characteristics relative to definition of grains and grain boundaries. For electronic purposes, single crystal silicon material must have a very high degree of lattice order. The presence of low angle boundaries can impact electrical performance. Within industrial nickel-base single crystals, there are low angle boundaries that can be accepted and still be considered within the definition of a single crystal. Means of measuring and quantifying these features is required in both examples to ensure components meet final material and component design intent.

Precipitation and precipitate phase characterization are critical features in many materials. The size, shape, and spacing, and the distributions of these parameters provide important information about the properties of many engineered, precipitation-strengthened materials. Precipitate size can be measured in a similar manner to that of grain size, but definitions of precipitate morphology can also be additionally critical. Size distribution can be strongly linked to mechanical properties (Figure 18.3) [343].

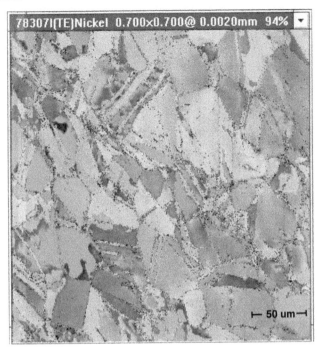

78307I(TE)Nickel 0.700×0.700@ 0.0020mm 94%

⊢ 50 um ⊣

Figure 18.2: Photomicrograph of a Waspaloy sample that has been processed to develop a completely warm-worked structure without the formation of recrystallized grains.

Distribution of precipitate size and spatial placement can also be significant elements controlling mechanical properties. Denuded regions in alloys result in local property variations. Similarly, preferential precipitation at grain boundaries, dislocation networks or slip bands can directly control the performance of materials. Preferential precipitation of γ' on superalloy grain boundaries can result in wavy or serrated grain boundaries (Figure 18.4). If these precipitation features are not included in the microstructure definition within a material definition, then control of material and component capabilities will be challenging.

There are nearly endless examples of microstructural features that need to be characterized and controlled within materials to produce specific unique properties and capabilities. Coatings of various types have special features such as columnar grain size and spacing for electron beam physical vapor deposition (EB-PVD) thermal barrier coatings, to porosity size and distribution in thermal sprayed coatings for thermal barrier coating and abradable coating systems.

Industrial microstructural quantification standards and application methods: As noted previously, current industrial characterization methods have been established into standards that are readily accomplished by a large extent of the industrial and research communities. These standards are very pragmatic and are often simplified to enable ease of application from an expertise, equipment, and cost standpoint. The trade between cost and pragmatism, and technical correctness and completeness may be questionable for some applications. Some standards are not capable of effectively differentiating subtle, but critical microstructural features.

Characterization methods and standards are often established with a "fit-for-purpose" mindset. Not all methods to quantify microstructures are equal nor do they provide the same results. Some standards, especially optical comparison-based standards, provide the ability to categorize microstructures, but do little to quantify the specific constituent features. This may be fully acceptable for some specific applications, but may be completely unacceptable for use in property predictions. Cleanliness characterization, relative to size, volume, and distribution of inclusion phases or particles, has been previously established as a pragmatic subjective photographic comparison

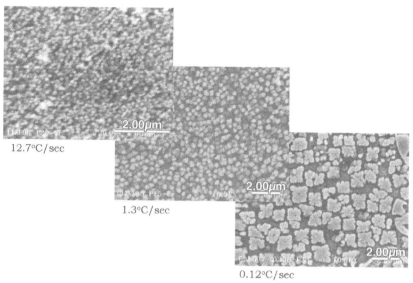

12.7°C/sec

1.3°C/sec

0.12°C/sec

Figure 18.3: Photomicrographs showing the size, distribution, and morphology change in alloy U720 as a function of cooling rate.

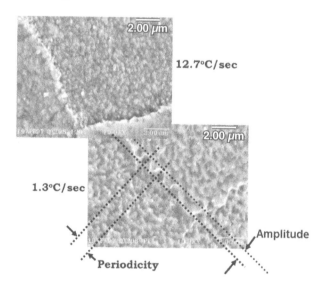

12.7°C/sec

1.3°C/sec

Amplitude

Periodicity

Figure 18.4: Photomicrographs of U720 that show the change in grain boundary serration as a function of cooling rate.

method. Figure 18.5 shows an example of a standard established to categorize the quantity and spatial distribution of inclusions. With this method and standard, prepared samples are viewed and compared to the standard template and each evaluated sample is given a rating for quantity/severity and distribution.

Secondary phase and cleanliness ratings can be quantitatively characterized by means of image analysis [327]. Methods of segmenting a field of view and manually counting have been and continue to be successfully utilized. Digital image analysis is a common approach where images are processed through a threshold of brightness, color, or other feature to enable differentiating regions of different phases. Each separated phase can then be assessed by wide ranges of algorithms such

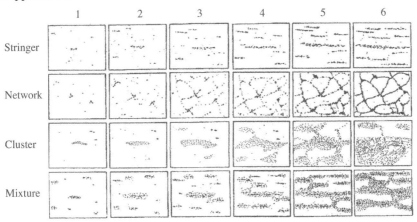

Figure 18.5: Comparative photomicrographic standard for cleanliness that rates distribution of inclusions from stringers, networks, clusters, and mixtures; and severity or relative quantity of secondary particles. (Reprinted, with permission, from ASTM STP 672 MiCon 78: Optimization of Processing, Properties, and Service Performance Through Microstructural Control, copyright ASTM International, 100 Barr Harbor Drive, West Conshohocken, PA 19428 [258].)

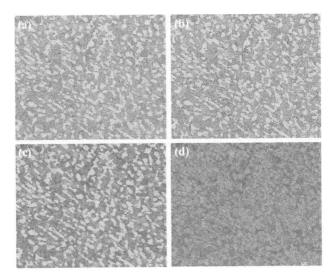

Figure 18.6: Photomicrographs of a Ti64 sample, where (a) is the originally captured image and (b) and (c) are processed digital images, and (d) shows the detection of primary α phase particles that can be used for digital analysis of average grain size and size distribution, along with many other possible parameters.

as area percentage, particle size, average particle size, particle shape, particle aspect ratio, particle orientation, etc. Image analysis methods have moved characterization from subjective human impression of structures to objective capture of the essence of the microstructure assemblage.

Titanium alloys can produce a wide range of microstructures. Image analysis has utilized many features within this family of alloys [235]. Image analysis has been successfully applied to the characterization of percentage of primary α phase in $\alpha - \beta$ titanium alloys. Figure 18.6 shows an example of the image processing steps for a typical $\alpha - \beta$ processed Ti64 sample.

Figure 18.7: Microstructure of a Ti64 sample after digital image processing for use in subsequent image analysis and microstructural feature characterization.

Very complicated β-processed titanium microstructures can also be analyzed by image analysis methods. Figure 18.7 shows an example of Ti64 photomicrographs that have been processed to reveal secondary α laths and their associated orientations.

Microstructure feature characterization has evolved from very subjective methods to very quantitative, objective approaches. Repeatability of characterization methods is very important for any type of characterization. Sampling size and sampling location need to be also considered relative to the local and global microstructure features to be quantified. Current industrial standards do define the sample size required for various microstructural features, but the sampling of structure over a wide spatial distribution of industrial components can lead to large variation and ambiguity in the overall quantified microstructure. Industrial characterization of component microstructure often requires specification of test locations. If these locations are arbitrarily defined, then they could reside in regions of large microstructure gradient that can lead to large microstructure quantification variability with small shifts in characterization area or volume.

18.4   Future State in Microstructural Characterization

One of the major issues in characterizing materials for industrial applications is the current need to a priori define the feature that must be measured and controlled. This is both a challenge and an opportunity. The need to understand the one-to-one type relationships between microstructure and properties can be very difficult, as there are few one-to-one relationships. Even the ubiquitous Hall–Petch relationship has limitations as there are other microstructural features that also synergistically influence tensile properties. Targeting one-to-one (or several-to-one) relationships does have an advantage in that they provide some physical meaning between the structure property relationships. Knowing the specific features that influence properties can guide researchers and engineers to specific material and process changes that directly impact these key features. Future methods for microstructure characterization and quantification must allow for non-a priori understanding of critical features, but must be able to link established characterization quantification metrics back to tangible, physical-based parameters that can be addressed by material and processing engineers.

The future-state of microstructural characterization is a type of "big data" problem. Characterization tools are being developed that can readily capture millions of statistically relevant data points for nearly endless features. Electron microscopy and diffraction that can differentiate crystal

Microstructure   **2-point auto correlation**

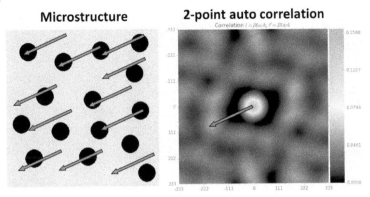

Figure 18.8: Example of schematic microstructure and resultant 2-point correlation statistical description using PyMKS software.

orientation, lattice parameters, local chemistry and boundary sharpness and character are being developed and employed to many traditional problems and are leading to new understanding relative to microstructure and property relationships. Analytical tools are being developed that can capture data at unprecedented rates and resolution. The quantity of data is also unprecedented in the world of industrial material characterization.

As microstructure characterization methods are evolving to generate larger and larger amounts of data, utilization of data and informatics tools is becoming a requirement. Statistical treatment of the data and data mining methods are being developed along with emerging sample characterization tools and methods. The larger quantity and variation in microstructural data types would be difficult to manage and interpret by traditional methods, but introduction of statistical and data analytics methods is making possible the use of large quantities and types of data. This is leading to much more complete understanding of complex microstructure interactions and associated properties. The linkage of microstructural characterization and materials informatics is leading to and will continue to drive significant changes in microstructure characterization and microstructure definitions.

Emerging tools, such as Dream3D [393] and PyMKS [1140], are supporting the capture and analysis of large quantities of microstructural data. Automated routines are allowing very complex analysis methods to be readily accomplished with commercially available personal computers. The intersection of the advances in sample characterization, data capture, data analytics, and computational capabilities is giving rise to tools that can conceivably be applied by industry for material development, definition, and quality control applications.

Advanced microstructural analysis tools that capture, analyze microstructure in a statistical manner are evolving into user-friendly tools that are making their way into industrial application, such as PyMKS. N-point correlation function methods have been developed and applied to many microstructural characterization problems [1140, 474]. These approaches are often in the form of 2-pt correlation function tools that systematically sample microstructural images with vector placement and assessment (Figure 18.8). This advanced approach to objectively characterize and describe microstructure does not require an a priori list of microstructural features that need to be assessed. The large datasets generated by these statistical characterization approaches can be distilled to a manageable size using dimensionality reduction techniques such as principal component analysis (PCA) which results in the ability to statistically represent microstructure in n-component vector space while still capturing the import differences in microstructure features.

The method is a statistical-based data generation and mining process. Cluster analysis can be utilized to determine if examples of the microstructure are related in "families" and separated from other structures within the established vector space. This is an objective and statistically relevant method that can "recognize" subtle single or multiple feature differences from sample to sample.

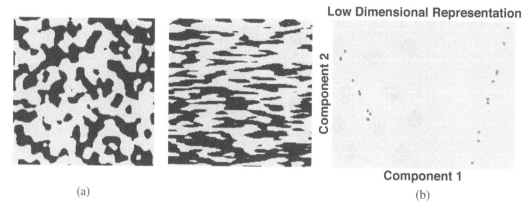

(a)   (b)

Figure 18.9: An example of cluster analysis, which shows 2-point statistics from individual microstructures that group into distinct, statistically relevant families. The microstructures in (a) are statistically separated in the PCA graph in (b).

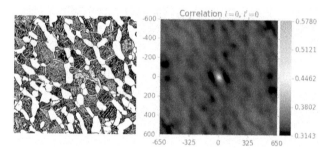

Figure 18.10: Example of an $\alpha - \beta$ titanium microstructure and the associated 2-point correlation function description.

Figure 18.9 shows an example of a cluster analysis for the microstructure of a series of material samples. The material samples are seen to group or cluster with distinct characteristics. This can be used to guide further analysis relative to the specifics behind this clustering and how this can be used to define, optimize, and control this material.

Two-point correlation function descriptions of microstructure are extremely useful for industrial applications. This method can be effectively used to assess microstructural equivalence of different material samples. Figure 18.10 shows an example of microstructures that have been characterized with a 2-pt correlation function.

Using statistically relevant descriptions of microstructures enables quantitative comparison of different materials. Mathematical operations can be conducted with these statistical quantitative definitions. Subtracting one statistical description from another can provide information relative to similarity or differences. Figure 18.11 shows a comparison of 2-pt statistical representations of two microstructures. These statistical descriptions can be subtracted from each other or even averaged to produce a resultant statistical description of the similarities or differences. Figure 18.12 shows examples of a subtraction and an averaging operation on these microstructure descriptions. It can be seen that these different microstructures are statistically equivalent. Thresholds or other criteria can also be introduced for establishing the extent of equivalency.

This statistical description of microstructure has enabled the ability to create artificial microstructures that have equivalent statistical descriptions. These microstructures can be used with models to enable the critical exploration and assessment of potential material capability space. Efforts are being conducted to develop statistical representations of microstructures for material where

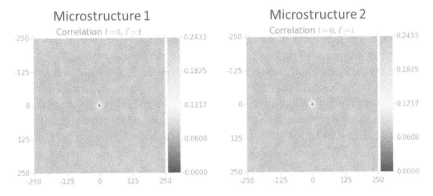

Figure 18.11: 2-pt statistical descriptions of two different example microstructures.

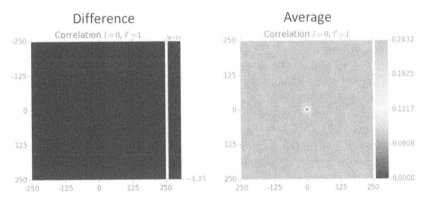

Figure 18.12: Mathematical difference and average of the statistical descriptions of the two microstructures represented in Figure 18.11.

specific properties are known. Variation in microstructure and associated properties allows development of models for structure-property prediction. This approach greatly enhances the capability for statistically relevant characterization of microstructures that can be more accurately linked to properties. The ability to characterize real and artificial microstructures enables the capability to assess the sensitivity of microstructure definitions relative to resultant properties. This analysis approach can be used to bound the microstructure space for a given material and microstructure definition.

These same microstructure descriptors can be used to quantify microstructure evolution during manufacturing processes and can provide insight as to which process control parameters have the largest impact on the variation in microstructure. This insight can be translated into appropriate SOPs to control variation in material properties. With the combination of structure evolution and material property data, processes-structure-property linkages can be generated to provide guidance on optimizing desired material properties by moving microstructures into a desirable location in microstructure space.

In these statistical methods, however, it is difficult to relate the statistical vector space values to physical features within the microstructure, since contributions of many features are compiled into the complex statistical description. In situations where there is an inadequate amount of structure evolution data, it is vital that the vector-space descriptions can be de-convoluted into specific microstructural features that engineers and researchers can understand. Understanding the relationship between microstructural features and the 2-pt statistical description will enable assessment of what processing methods can be altered to alter the microstructural features in a desired manner. As an example, if percent primary α phase in a titanium microstructure is a critical feature as defined by the 2-correlation function and PCA analysis, then process engineers can readily assess methods to increase or decrease the quantity of this feature through changes in processing temperature relative to measured or thermodynamically predicted phase boundaries.

Since the 2-pt correlation function analysis process can be used to analyze measured or artificially created microstructures, it is possible for engineers to create artificial microstructures using CAD-type tools where each microstructure feature is an element of a working parametric solid model. Systematically varying features within the artificial microstructures can provide direct correlation between the vector space statistical description and specific microstructural features. This approach can readily lead to sensitivity analysis of complex microstructural features and feature combinations relative to resultant material properties.

Linking the ability to statistically quantify microstructure and the ability to create realistic artificial microstructures will enable rapid material development and optimization. Using such approaches may also lead to discovery of unique microstructure assemblages that have unique properties, and which have not been observed or realized previously.

Convolutional neural networks have recently enjoyed great success in large-scale image and video recognition tasks, owing to their ability to learn discriminative features and hierarchical semantic information along their deep architecture, eliminating the need for manual feature engineering. The prevalence of large, labeled datasets and advancements in GPU computing have enabled such models to achieve state-of-the-art performance in image modeling [938, 451]. These models have also been applied to tasks like image de-noising, synthesis, and artistic style transfer, showcasing the models' capability to learn robust, transferrable spatial statistics [350, 351].

Similar to 2-pt correlation statistics, high-performance convolutional neural networks can be used to extract statistically equivalent, but lower-dimensional representations of image data, including material microstructures. These texture vectors have been shown to be effective as input into machine learning classification algorithms to predict material pedigree variations due to heat treatment, thermal exposure temperature, processing conditions, etc. By running the images through a pre-trained convolutional neural network like VGG19 [938] in Figure 18.13, spatial summary statistics can be calculated by computing the correlations between layer outputs for any given set of layers in the network [938]. It is a combination, or often subset, of these network layer activations correlations that contributes to a statistical representation of an image at differing length scales. Dimensionality reduction techniques like PCA, as mentioned before, can further compress the data to lower dimensions, and be used as feature vectors for inputs into other machine learning algorithms like classification or regression. If sufficient microstructure image data are available, convolutional neural networks can be fine-tuned.

The question regarding where to analyze microstructure and what area/volume of material must be characterized to establish an appropriate microstructure description is still a major open issue that is also being addressed by various efforts. The location of where samples of material need to be analyzed is actually an issue of product definition and whether the product has a single microstructure definition or if it is zoned for specific purposes. For components that have a material definition that requires a single microstructure definition, the location that must be analyzed for microstructure equivalence must be the location with the greatest potential of failing the requirement. Manufacturing process models are being developed that can provide prediction of local or location-specific microstructure. The fidelity of these models varies greatly, but in general they are outstanding at providing gradients and trends in microstructure spatially throughout material volumes within components. Using these models to guide placement of microstructure analysis samples can provide enhanced capabilities in defining limiting locations and limiting microstructures. Performing quantitative microstructural analysis linked with microstructural modeling on an ongoing basis can enable generation of data that can be used for Bayesian updating of the microstructural prediction models. This approach is being applied to many modeling tools [842, 485] to establish and reduce the uncertainty of the predictions from the established models. Linking advanced microstructure characterization, microstructure modeling/prediction, and industrial quality control can

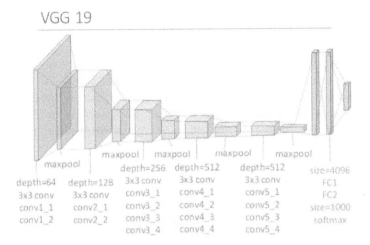

Figure 18.13: Schematic diagram of VGG19 convolutional neural network, showing iterative combinations of convolutional and max pooling layers, tied to a fully-connected softmax output layer for class prediction [938].

Figure 18.14: An EBSD image of a titanium sample that shows regions of local texture. It length-scale of the texture regions and spatial distribution drives the size of the required microstructure sampling to achieve a statistically representative microstructure description.

lead to rapid advancement of tools that can be used for design and optimization of future component designs produced from the same material.

The size of the area or volume that must be characterized to establish a statistically relevant microstructure description is related to two major criteria: (1) local variation and gradients within a component, and (2) the length scale that defines the physics-based mechanisms that control the specific property of interest (e.g., tensile, dwell-fatigue, etc.). Local variation and gradients in microstructure within a component may drive a requirement to enlarge the size of field or volume of interest to capture all variation into a single definition, or it might be of interest to reduce the step size within the area or volume that is interrogated to allow defining the specifics of the variation or gradient in the microstructure. This question may be aided through an assessment of what properties are being controlled via the microstructure characterization and associated control. For properties that have limited sensitivity to the variation in the microstructure, such as tensile properties and variation in microtexture in titanium, the critical volume to establish a statistically relevant description may be relatively small. Conversely, dwell-fatigue properties in titanium materials are sensitive to microtexture, so the size, orientation, and spatial distributions must be captured, which may require a much larger area or volume to be characterized. Figure 18.14 shows an example of a titanium microstructure that shows regions of local texture. This means that there is a mechanism-based statistical volume element (SVE) for each microstructure. The type of characterization that will be required will be based on the purpose for the characterization and what property is the goal for control based on the microstructure.

18.5   Conclusion

Microstructural characterization is a very important element of materials science and engineering. Traditional subjective methods are making way for new characterization tools and objective, statistically relevant methods. It is believed that industry will rapidly accept and adopt these new approaches for assessing and comparing microstructures as materials and components are being designed to produce maximum capability, efficiency, and durability.

Part VI

Anomalies

It is often said that the difference between physics and materials science is that physics speaks of the ideal state and materials science of the defect state. This would imply that a very important set of tools would be those that highlight defects, or deviations from the idealized structure. A standard approach to this would be to proceed with physics-based models and identify areas of high energy and declare them to be defects. Statistical science has a different take: find where the statistical properties are so different that it is *unlikely* that would have resulted from mere fluctuations. This is the realm of *anomaly testing*.

Anomaly testing is not, at this point, widely used in microstructure characterization. With that in mind, we asked Theiler to prepare a chapter on the method of anomaly testing with some examples from his own work, which were not in the materials science realm. In order to make the applications more concrete, we asked Przybyla et al. to prepare a chapter on a specific study in anomaly testing in SiC/SiC ceramic matrix composites.

Chapter 19

Anomaly Testing

James Theiler
Los Alamos National Laboratory

19.1 Introduction

An anomaly is something that in some unspecified way stands out from its background, and the goal of anomaly testing is to determine which samples in a population stand out the most. This chapter presents a conceptual discussion of anomalousness, along with a survey of anomaly detection algorithms and approaches, primarily in the context of hyperspectral imaging. In this context, the problem can be roughly described as target detection with unknown targets. Because the targets, however tangible they may be, are unknown, the main technical challenge in anomaly testing is to characterize the background. Further, because anomalies are rare, it is the characterization not of a full probability distribution but of its periphery that most matters. One seeks not a generative but a discriminative model, an envelope of unremarkability, the outer limits of what is normal. Beyond those outer limits, *hic sunt anomalias*.

Traditionally, anomalies are defined in a negative way, not by what they are but by what they are not: they are data samples that are not like the data samples in the rest of the data. "There is not an unambiguous way to define an anomaly," one review notes, and then goes on to ambiguously define it as "an observation that deviates in some way from the background clutter" [657]. Anomalies are defined "without reference to target signatures or target subspaces" and "with reference to a model of the background" [963]. Indeed, as Ashton [36] remarks, "the basis of an anomaly detection system is accurate background characterization."

This exposition will concentrate on anomaly detection in the context of imagery, with particular emphasis on hyperspectral imagery (in which each pixel encodes not the usual red, green, and blue of visible images, but a spectrum of radiances over a range of wavelengths that often includes upwards of a hundred spectral channels). Testing for anomalies is an exercise that has application in a variety of scenarios, however. Since anomalies are deviations from what is normal, particularly in situations where the nature of that deviation is not predictable or well characterized, anomaly detection has been used in a variety of fault detection contexts [1141, 1153, 633, 1041, 155].

What we are calling anomaly testing here is essentially the same as what the machine learning community calls "novelty detection" [916, 646, 647] or "one-class classification" [1005, 632].

19.1.1 Anomaly Testing as Triage

Although we may have difficulty *defining* anomalies, the reason we seek them is that (being rare and unlike most of the data) they are potentially interesting and possibly meaningful. We have to acknowledge that "interesting" is even harder to define than "anomalous" but in this exposition, we will make this distinction, partly because it separates the mystical "by definition undefined" [1021] aspect of the interesting-data detection problem into two components, which correspond to the two boxes in Figure 19.1.

We can indeed define "anomalous," but will leave "interesting" and "meaningful" to be domain-specific concepts. From this point of view, anomaly detection is a kind of triage. If anomalies can

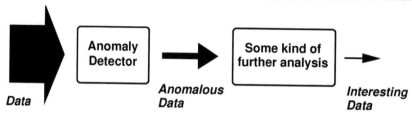

Figure 19.1: Anomaly testing as triage. In this picture, anomaly detection provides a way to reduce the raw quantity of data that needs to be considered by further analysis. That further analysis might be expensive domain-specific automated processing, or it may involve human inspection, with trained analysts making judgments about whether the *potentially* interesting anomaly is indeed interesting or meaningful or important.

be defined and detected in a relatively generic way, then experts in the specific area of application can decide which anomalies are interesting or meaningful.

Here the goal of anomaly detection is to reduce the quantity of incoming data to a level that can be handled by the more expensive downstream analysis. It is this later analysis that judges which of the anomalous items are in fact meaningful for the application at hand. This judgment can be very complicated and domain-specific, and can involve human acumen and intuition. What makes anomaly detection useful as a concept is that the anomaly detection module has more generic goals, and is consequently more amenable to formal mathematical analysis.

19.1.2   Anomalies Drawn from a Uniform Distribution

Anomalies are rare, and where we expect to find them is in the far tails of the background distribution $p_b(\mathbf{x})$. We can express "anomalousness" as varying inversely with this density function, and can derive this expression in two distinct ways, each providing its own insight.

In the first and most direct approach, we make an explicit generative model for anomalies, and say that they are samples drawn from a uniform distribution. This is a simple statement, but it is in some ways revolutionary; in contrast to conventional wisdom [657, 963, 36], we are defining anomalies directly, *without respect* to the background distribution. To distinguish these anomalies from the background, we treat the detection as a hypothesis testing problem. The null hypothesis is that the measurement $\mathbf{x}$ is drawn from the background distribution ($\mathbf{x} = \mathbf{z}$ with $\mathbf{z} \sim p_b$), and the alternative hypothesis is that $\mathbf{x}$ is drawn from the anomalous distribution ($\mathbf{x} = \mathbf{t}$ with $\mathbf{t} \sim u$). This leads to the likelihood ratio

$$\mathscr{L}(\mathbf{x}) = \frac{P(\mathbf{x} = \mathbf{t})}{P(\mathbf{x} = \mathbf{z})} = \frac{u(\mathbf{x})}{p_b(\mathbf{x})} = \frac{c}{p_b(\mathbf{x})} \qquad (19.1)$$

where $u(\mathbf{x}) = c$ is the uniform distribution from which anomalies are drawn.[1] The expression in Equation (19.1) is a candidate for "anomalousness" because a large value of $\mathscr{L}(\mathbf{x})$ indicates a higher likelihood that $\mathbf{x}$ is drawn from the anomalous distribution.

In the second approach, we avoid making an explicit model of what an anomaly is, but we assume that the effect of the anomaly on the scene is additive. In particular, we say that a pixel with an anomaly in it is of the form $\mathbf{x} = \mathbf{z} + \mathbf{t}$ where $\mathbf{t}$ is the unknown target, and $\mathbf{z}$ is the background. We employ the *generalized* likelihood ratio test to write

$$\mathscr{L}(\mathbf{x}) = \frac{P(\mathbf{x} = \mathbf{z} + \mathbf{t})}{P(\mathbf{x} = \mathbf{z})} = \frac{\max_{\mathbf{t}} p_b(\mathbf{x} - \mathbf{t})}{p_b(\mathbf{x})} = \frac{\max_{\mathbf{z}} p_b(\mathbf{z})}{p_b(\mathbf{x})} = \frac{c'}{p_b(\mathbf{x})} \qquad (19.2)$$

where the (again, irrelevant) constant c' is the maximum value of p_b and does not depend on $\mathbf{x}$. This second approach produces the same result as the first, but the argument used to get that result has

---

[1]The constant c is irrelevant to our purposes, and in fact it is possible to make this argument in a more formal way that treats $u(\mathbf{x})$ not as a proper probability distribution, but more generally as a measure [966].

two problems: one, the additive assumption is very restrictive and may not apply to the scenario of interest; and two, it uses a generalized likelihood ratio test (GLRT). Although the GLRT is a popular and often practical tool, it has ambiguous properties, and can produce detectors that are not only sub-optimal, but in some cases *inadmissible* [559, 1017]. This is not to say that the detector in Equation (19.2) is inadmissible; indeed, it is identical to the detector in Equation (19.1). The objection is not to the detector but to this second derivation of the detector *via* the GLRT. A supposed advantage of this second derivation is that it makes "no assumptions" about the nature of the anomaly, but this is a spurious argument. In fact, it is making implicit assumptions about the distribution of an anomaly, but it is not providing the algorithm designer with access to alter those implicit assumptions.

Although there are many reasons to favor the first approach, it is the second argument that is most widely invoked in the hyperspectral anomaly detection literature.

We remark that the first approach can also accommodate additive anomalies. Here, the numerator of the likelihood ratio is a Bayes factor, but it still evaluates to a constant independent of $\mathbf{x}$:

$$\mathscr{L}(\mathbf{x}) = \frac{P(\mathbf{x} = \mathbf{z} + \mathbf{t})}{P(\mathbf{x} = \mathbf{z})} = \frac{\int p_b(\mathbf{x} - \mathbf{t})u(\mathbf{t})d\mathbf{t}}{p_b(\mathbf{x})} = \frac{c \int p_b(\mathbf{z})d\mathbf{z}}{p_b(\mathbf{x})} = \frac{c''}{p_b(\mathbf{x})} \tag{19.3}$$

We again obtain the result that anomalousness varies inversely with $p_b(\mathbf{x})$, the probability density function of the background. Contours of anomalousness will be level curves of the background density functions. For a Gaussian distribution, these contours are ellipsoids of constant Mahalanobis distance [617], with larger distances corresponding to smaller densities and greater anomalousness; we can therefore use Mahalanobis distance as a measure of anomalousness

$$\mathscr{A}(\mathbf{x}) = (\mathbf{x} - \mu)^T R^{-1}(\mathbf{x} - \mu) \tag{19.4}$$

where μ is a vector-valued mean and R is a covariance matrix.

The Mahalanobis distance is the basis of the Reed–Xiaoli (RX) detector [836, 972, 907]. Although RX, as originally introduced [836], refers specifically to multispectral imagery, and in fact is a local anomaly detector, the term "RX" is often used as a shorthand for Mahalanobis distance based anomaly detection.

19.1.2.1   *Nonuniform Distributions of Anomalousness*

Since the assumption of a uniform distribution for anomalies is explicit in the derivation of Equation (19.1) – and by extension, Equation (19.4) – that allows us to consider other, nonuniform, distributions if we are looking for other kinds of anomalies. Examples include anomalous change (to be further discussed in Section 19.6), and anomalous "color" in multispectral imagery. Here, in place of a uniform distribution of anomalies, a distribution of anomalously colored pixels are generated from the product of marginal distributions associated with each individual spectral band. To sample at random from this distribution, one creates a vector-valued pixel where each component is independently sampled from the corresponding component of the multispectral image [1025].

Another example is given by the blind gas detection algorithm. In the traditional gas detection problem, one is looking for plumes that contain a gas-phase chemical of interest. When the absorption spectrum of that chemical of interest is known, one can derive a matched filter that can detect impressively low concentrations of the chemical [418]. In the blind gas detection problem, one does not have a single chemical (or even a short list of chemicals) of interest; one wants to detect chemical plumes without knowing the chemical species in the plume. But one does know that gas-phase chemicals usually have very sparse spectral signatures – i.e., there is zero or nearly-zero absorption at all but a few wavelengths. A sparse RX (or "spaRX") algorithm was developed for detecting spectrally sparse additive anomalies $\mathbf{t}$ based on the RX derived in Equation (19.2), but with $\mathbf{t}$ constrained to a limited number of nonzero components [1036].

19.1.3   *Anomalies as Pixels in Spectral Imagery*

Traditional statistical analysis treats data as a set of discrete samples that are drawn from a common distribution. Because each pixel in a hyperspectral image contains so much information (a many-channel spectrum of reflectances or radiances), one can often quite profitably treat the pixels as independent and identically distributed. It is as if the image were a "bag of pixels." But however spectrally informative individual pixels are, they comprise an image, and the spatial structure in an image provides further leverage for characterizing the background and discovering anomalies.

Hyperspectral imagery provides a rich and irregular dataset, with complex spatial and spectral structure. And the more accurately we can model this cluttered background, the better our detection performance. Simple models can be very effective, but the mismatch between simple models and the complicated nature of real data has driven research toward the development of more complex models [660].

19.1.3.1 Global and Local Anomaly Detectors

A pixel is anomalous in the context of a background. In *global* anomaly detection that background is the full image, but in *local* anomaly detection that background is restricted to the immediate neighborhood, often defined in terms of an annulus that surrounds the pixel. Local anomaly detection is one of the most straightforward (and, in practice, more effective) ways to exploit the spatial structure of imagery.

Indeed, the initial formulation of the RX algorithm [836] computed anomalousness at each pixel in terms of a local mean and a local covariance matrix, each computed from the pixels in an annulus surrounding that pixel. The tradeoff in choosing the size of the annulus is that a larger annulus will have better "statistics" – since it has more pixels, it will better average out the fluctuations in the pixels values; but a smaller annulus will be less affected by spatial nonstationarity [214]. Matteoli et al. [657] observed that one could use a smaller annulus for the local mean and a larger annulus for the local covariance. This is very sensible, since the need for "good statistics" is greater for the covariance matrix than for the mean vector. As the covariance annulus approaches the size of the image, this approach can be simplified by using a local mean and a global covariance, but with the global covariance based on subtraction of the spatially varying local mean. A broad survey of approaches used to improve estimates of local covariance, including regularization, segmentation, and robustification, is provided by Matteoli et al. [656].

The importance of regularization derives from the fact that RX requires the *inverse* of the covariance matrix. If the covariance matrix is singular, then the inverse does not exist, and regularization is required. But even for a well-conditioned covariance matrix, the best estimator of the inverse is not the same as the inverse of the best estimator, and some amount of regularization is still beneficial. The most common and straightforward regularization is by shrinkage. Here, we estimate the covariance matrix with a linear combination of the sample covariance R_s (which tends to overfit the data) and a very simple estimator R_o that tends to underfit the data

$$\widehat{R} = (1 - \alpha)R_s + \alpha R_o \tag{19.5}$$

where typically $\alpha \ll 1$. In the simplest case, R_o is just a multiple of the identify matrix [334, 719] (choosing the multiple so that R_o has trace equal to R_s ensures that α is dimensionless). An argument can be made for shrinkage against the diagonal matrix [435, 1018], an approach that is generalized in the sparse matrix transform [176, 1026]. Caefer et al. [147] recommended a quasi-local estimator that combines local and global covariance estimators by using local eigenvalues with global eigenvectors.

The idea of segmentation is to replace the moving window with a static segment of similar pixels that surround the pixel of interest in a more irregular way. Here the image is partitioned into distinct segments of (usually contiguous) pixels, and a pixel's anomalousness is based on the mean and covariance of the pixels in the segment to which the pixel belongs [179, 147]. This sometimes leads to extra false alarms on the boundaries of the segments, and one way to deal with this problem is with overlapping segments [1033].

The estimation of covariance in the local annulus can be corrupted by one or a few outliers[2] and robust estimates of covariance [862, 658, 659] exclude or suppress the outliers in the computation. Excluding outliers is advisable in global estimates of covariance as well [67, 394].

19.1.3.2 Regression Framework

By estimating what the target-free radiance *should* be at a pixel, we have a point of comparison with what the measured value of that pixel actually happens to be. It is common to make this estimate using the mean of pixels in an annulus around the pixel of interest. But there is more information in the annulus than this mean value [1024, 148], and that suggests using more general estimators than just the mean. The derivation in [1022] uses multivariate regression of the central pixel against the pixels in the surrounding annulus. This can be done on a band-by-band basis, or with multiple bands simultaneously. Other variants use median instead of mean [413], and a patch-based nearest neighbors regressor was also developed [1037].

It is worth noting that the problem of estimating a central pixel, using the specific pixels that immediately surround it, along with the statistical context of the rest of the image, is a special case of the "inpainting" problem. In this case, only a single pixel is being inpainted at a time, but this inpainting is done for every pixel in the image.

19.2 Evaluation

A conceptual problem with anomaly detection is the ambiguous nature of "interestingness." But a more technical problem is that, since anomalies are by definition rare, it is difficult to find enough of them to do statistical comparisons of algorithms, and it is easy to be misdirected by anecdotal results.

Section 19.1.2 described how to resolve the ambiguity issue by treating anomalies as samples from a specific and well-defined, yet very broad and flat (e.g., uniform), distribution. This resolution led, for instance in Equation (19.1), to likelihood-based algorithms for anomaly detection.

An arguably more important advantage to treating anomalies as samples from a distribution is that it leads to a direct quantitative way to measure the performance of an anomaly detection algorithm, without relying on anecdotally identified anomalies in a given scene. The most straight-forward way of exploiting this model is to use it to create artificial targets that can be "implanted" into the scene [67, 214]. This is usually performed at a judiciously chosen subset of locations, to avoid contaminating the background estimation with an unrealistically large number of targets. But in some cases one can more efficiently place a target at effectively *every* location in the image, producing matched pairs of with-target and without-target pixels from which to learn a target detector [1019].

An advantage of explicit implanting is that it can be made as realistic as the simulation will allow. A potential disadvantage is that this explicit simulation can be expensive, may require *ad hoc* choices. Particularly for anomaly detection, where physical properties of the target are generally not well known, or at least not well specified, more generic approaches may be desirable.

With anomalies drawn from a uniform distribution, the volume inside a contour of anomalousness corresponds to undetected anomalies. Thus, a small volume is a proxy for a low missed detection rate; for a given false alarm rate, smaller volumes imply better anomaly detectors (Figure 19.2). In an early example of this principle, Tax and Duin [1006] used a uniform distribution of points to estimate volumes produced by different choices of kernel parameters, thereby providing a way to choose parameters without ground truth. Steinwart et al. [966] formalize the classification framework for anomaly detection and recommend a two-class classifier (a support vector machine,

---

[2]Outliers and anomalies are essentially the same thing, and we make no formal distinction between them. But informally we think of anomalies as rare nuggets deserving of further analysis, while outliers are nuisance samples that contaminate the data of interest.

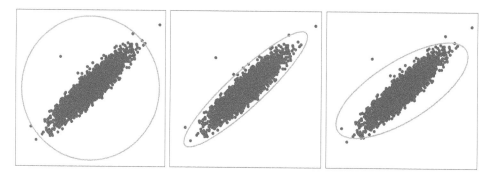

Figure 19.2: Three anomaly detectors with the same false alarm rate ($P_{\text{fa}} = 0.001$). The contour marks the boundary between what is normal (inside) and what is anomalous (outside). If anomalies are presumed to be uniformly distributed, then the anomaly detector with smallest volume will have the fewest missed detections.

specifically) that distinguishes measured data from artificially generated uniformly distributed random samples. (See also Hastie et al. [414].)

Plotting volume (or log-volume, which is often more convenient, especially in high dimensions) against false alarm rate provides a ROC-like curve that characterizes the anomaly detector's performance [1029, 1016].

For a covariance matrix, the volume is proportional to the determinant of the matrix. For local anomaly detectors described in Section 19.1.3.1, one can still use a global covariance based on the difference between measured and estimated (e.g., by local mean) values at each pixel, and the smaller that covariance, the better the estimator. A natural choice, from a signal processing perspective is the total variance of that difference,

$$\sum_n (\mathbf{x}_n - \widehat{\mathbf{x}}_n)^T (\mathbf{x}_n - \widehat{\mathbf{x}}_n) \tag{19.6}$$

which corresponds to the trace of the covariance matrix. Smaller values of this variance imply that $\widehat{\mathbf{x}}$ is closer to $\mathbf{x}$, but Hasson et al. [413] point out that, in terms of target and anomaly detection performance, closer is not necessarily better.

When, instead of a global covariance estimator, we use a separate covariance for each pixel, based on the local neighborhood of that pixel, then it is more complicated. It is clear that the volumes of the individual covariances should be small, but it is not obvious how best to combine them. In Bachega et al. [39], it is argued, more on practical than theoretical grounds, that an average of the log volume is a good choice.

19.3 Periphery

For anomaly detection, low false alarm rates are imperative. So the challenge is to characterize the background density in regions where the data are sparse; that is, on the periphery (or "tail") of the distribution. Unfortunately, traditional density estimation methods, especially parametric estimators (e.g., Gaussian), are dominated by the high-density core. And it bears pointing out that "robust" estimation methods (e.g., [163, 862, 863]) achieve their robustness by paying even less attention to the periphery.

Robustness to outliers can be achieved by essentially removing the outliers from the dataset. This direct approach is taken by the MCD (minimum covariance determinant) of Rousseeuw et al. [862, 863]. For a dataset of N samples, the idea is to take a core subset $\mathscr{H}$ of $h < N$ samples and to compute the mean and covariance from just the samples in $\mathscr{H}$, ignoring the rest. The formal aim

	All data samples	Subset of h samples
Sample covariance matrix	Mahalanobis/RX	MCD
Minimum volume matrix	MVEE	MVEE-h

More robust
against outliers
↓

More attention →
to periphery

Figure 19.3: Four algorithms for estimating ellipsoidal contours. All four algorithms seek ellipsoidal contours for the data, and can all four be expressed with an equation of the form $\mathscr{A}(\mathbf{x}) = (\mathbf{x} - \mu)^T R^{-1}(\mathbf{x} - \mu)$. The top two algorithms use sample mean and sample covariance to estimate μ and R, respectively; the bottom two seek a minimum volume ellipsoid that strictly encloses the data. The left two algorithms use all of the data in the training set; the right two algorithms use a subset $\mathscr{H}$ that includes *almost* all of the data. Note that the MVEE-h algorithm [394] is both robust to outliers and sensitive to data on the periphery of the distribution.

is to choose the subset so as to minimize the volume of the ellipsoid corresponding to the sample covariance. Specifically,

$$\min_{\mathscr{H}} \det(R) \quad \text{where } R = (1/h) \sum_{\mathbf{x}_n \in \mathscr{H}} (\mathbf{x}_n - \mu)(\mathbf{x}_n - \mu)^T$$
$$\text{and} \quad \mu = (1/h) \sum_{\mathbf{x}_n \in \mathscr{H}} \mathbf{x}_n \quad (19.7)$$
$$\text{and} \quad \#\{\mathscr{H}\} \geq h$$

As stated, this is an NP-hard problem, but an iterative approach can be employed to find an approximate optimum. Given an initial set of core samples $\mathscr{H}$, we can compute μ and R as sample mean and covariance of the core set. With this μ and R, we can use Equation (19.4) to compute $\mathscr{A}(\mathbf{x})$ for *all* of the samples. Taking the h samples with smallest $\mathscr{A}(\mathbf{x})$ values yields a new core set $\mathscr{H}'$. This process can be iterated, and is guaranteed to converge, though it is not guaranteed to converge to the global optimum defined in Equation (19.7). Various tricks can be used both to speed up the iterations and to achieve lower minima [863].

Where MCD concentrates on identifying the core, the minimum volume enclosing ellipsoid (MVEE) algorithm concentrates on the periphery of the data. In contrast to Equation (19.7), the aim is to optimize

$$\min_{\mu,R} \det(R) \quad \text{where } (\mathbf{x}_n - \mu)^T R^{-1}(\mathbf{x}_n - \mu) \leq 1 \text{ for all } n \quad (19.8)$$

Unlike the optimization in Equation (19.7), the optimization here is convex and can be efficiently performed, using Khachiyan's algorithm [489], possibly including some of the further improvements that have since been suggested [522, 1048].

Although robustness against outliers and sensitivity to the periphery are seemingly opposite requirements, practical anomaly detection actually wants both. For datasets in which a very small number of samples are truly outliers (or are truly anomalies), we *do* not want to include these samples in our characterization of the background. But absent these outliers, we do want to identify where the tail of background distribution is, and that requires attention to the samples on the periphery [1029].

Figure 19.3 illustrates this tension between attention to the periphery and robustness to outliers by showing four algorithms lined up along two axes. All four algorithms seek ellipsoidal contours for the data, and all four can be expressed with an equation of the form in Equation (19.4); what differs is the methodology for estimating μ and R. As Figure 19.3 suggests, it is possible to combine

MCD and MVEE to create a new algorithm, called MVEE-h, that identifies a minimum volume ellipsoid that fully encloses not all, but most (in particular, a subset $\mathscr{H}$) of the data [394].

Other approaches have also been suggested for modifying the sample covariance in a way that better respects the points on the periphery. One family of such approaches uses the sample covariance to define eigenvectors, but modifies the eigenvalues in each of these eigenvector directions. This is consistent with the observation, noted by several authors [1027, 42, 10], that tails tend to be heavier in some directions than others. The approach taken by Adler-Golden [10] was based on the observation that the heavy-tailed distribution has different properties in different eigenvector directions. A model is fit for each of these directions (but it's only one-dimensional so the fit is generally stable and robust), and this leads to a modified anomaly detector:

$$\mathscr{A}(\mathbf{x}) = \sum_i \left(a_i |x_{(i)}|\right)^{p_i}, \tag{19.9}$$

where $x_{(i)} = \lambda_i^{-1/2} \mathbf{u}_i^T (\mathbf{x} - \mu)$ is the ith whitened coordinate of $\mathbf{x}$. Here, where $\mathbf{u}_i$ and λ_i are the ith eigenvector and eigenvalue, respectively, and a_i and p_i are obtained from the model fit for each component. If $a_i = 1$ and $p_i = 2$, then this is equivalent to Equation (19.4). A related approach has also been suggested, in which $\mathscr{A}(\mathbf{x}) = \sum_i x_{(i)}^2 / \sigma_i^2$ and σ_i is based on inter-percentile difference instead of variance [1029].

19.4 Subspace

For many datasets, the covariance matrix usually exhibits a wide range of eigenvalues, and the smallest values correspond to directions that can be projected out of the data with minimal loss in accuracy. Indeed, it is often the case that data can be accurately represented by a lower-dimensional plane (or manifold [864, 1009, 41, 597, 1209]), and there are often advantages to analyzing the data in that lower-dimensional space. For anomaly detection, however, one has to be extra careful.

Qualitatively speaking, there are two kinds of anomalies: "in-plane" anomalies and "out-of-plane" anomalies. The in-plane anomalies are unusual with respect to the distribution that is projected onto the lower-dimensional high-variance subspace ("the plane"), and thus these tend to be large magnitude samples. The out-of-plane anomalies are unusual in that they are far from the plane, where "far" refers to distances larger than the smallest eigenvalues. Thus an out-of-plane anomaly can be a lower magnitude sample, and if the data sample is projected into the subspace, it may lose its anomalousness.

A simple measure of out-of-plane anomalousness is Euclidean distance to the subspace [818] (or, in more sophisticated cases, to the manifold [610]). The subspace RX (SSRX) algorithm effectively computes a Mahalanobis distance to the subspace [907]. In practice this is achieved by projecting to a dual subspace (that is, projecting *out* the high-variance directions) and then performing standard RX in that space.

The qualitative difference between the high-variance and low-variance directions has led to a variety of Gaussian/Non-Gaussian (G/NG) models for high-dimensional distributions. In these models, the low-variance directions are modeled as Gaussian, but the high-variance directions are modeled with simplex-based [1016] or histogram-based [1045] distributions. This enables more sophisticated models to be employed, but because they are only used for a few high-variance directions, the complexity is bounded, and the curse of dimensionality is ameliorated.

Another approach based on projection to a lower-dimensional space was proposed by Kwon et al. [528]; here the projection operator is based on eigenvalues of a matrix that is the difference of two covariance matrices, one computed from an inner window (centered at the pixel under test in a moving window scenario) and one from an outer window (an annulus that surrounds the inner window and provides local context).

19.5 Kernels

19.5.1 Kernel Density Estimation

Given that the aim of anomaly detection is to estimate the background distribution $p_b(\mathbf{x})$, one of the most straightforward estimators is the kernel density estimator, or Parzen windows [770] estimator:

$$p_b(x) = (1/N) \sum_{n=1}^{N} \kappa(\mathbf{x}, \mathbf{x}_n),\qquad(19.10)$$

where the sum is over all points in the dataset, and where κ is a kernel function that is integrable and is everywhere non-negative. A popular choice is the Gaussian radial basis kernel,

$$\kappa(\mathbf{x}, \mathbf{x}_i) = \frac{1}{\sqrt{2\pi\sigma^d}} \exp\left(\frac{\|\mathbf{x} - \mathbf{x}_i\|^2}{2\sigma^2}\right).\qquad(19.11)$$

Equation (19.11) requires the user to choose a "bandwidth" σ that characterizes, in some sense, the range of influence of each point. Since density can vary widely over a distribution, variable and data-adaptive bandwidth schemes have been proposed [661, 662].

In the limit as bandwidth goes to zero, the anomalousness at $\mathbf{x}$ is dominated by the $\kappa(\mathbf{x}, \mathbf{x}_i)$ associated with the $\mathbf{x}_i$ that is closest to $\mathbf{x}$. Indeed, the anomalousness in that case is equivalent to that distance. An anomaly detector based on distance to the nearest point has been proposed [65], though with an additional step that uses a graph-based approach to eliminate a small fraction (typically 5%) of the points to be used as $\mathbf{x}_i$. An updated variant was later proposed [66] that included normalization, subsampling, and a distance defined by the average of the distances to the third, fourth, and fifth nearest points.

19.5.2 Feature Space Interpretation: The "Kernel Trick"

A particularly fruitful (if initially counter-intuitive) interpretation of kernel functions is as dot products in a (usually higher-dimensional) feature space.

Let $\phi(\mathbf{x})$ be a function that maps $\mathbf{x}$ to some feature space. Typically ϕ is non-linear, and the map is to a feature space that is of higher dimension than $\mathbf{x}$. Scalar dot product in this feature space can be expressed as (again, typically non-linear) functions of the values in the original data space. That is:

$$\kappa(\mathbf{r}, s') = \phi(\mathbf{r})^T \phi(s').\qquad(19.12)$$

The "kernel trick" is the observation that even though the function ϕ and the feature space are presumed to "exist" in some abstract mathematical sense, we do not actually need to use ϕ, as long as we have the kernel function κ. A popular choice is the Gaussian kernel

$$\kappa(\mathbf{r}, s') = \exp\left(-\frac{\|\mathbf{r} - s'\|^2}{2\sigma^2}\right),\qquad(19.13)$$

but many options are available. Polynomial kernels, for example, are of the form $\kappa(\mathbf{r}, s') = (c + \mathbf{r}^T s')^d$ for some polynomial dimension d. More general radial-basis kernels are scalar functions of the scalar value $\|\mathbf{r} - s'\|^2$; functions that are more heavy-tailed than the Gaussian have been proposed for this purpose [923].

This enables us re-derive the Parzen window detector from a different point of view. Given our data, $\{\mathbf{x}_1, \ldots, \mathbf{x}_N\}$, we first map to the feature space: $\{\phi(\mathbf{x}_1), \ldots, \phi(\mathbf{x}_N)\}$. In this feature space we define the centroid

$$\mu_\phi = \frac{1}{N} \sum_{n=1}^{N} \phi(\mathbf{x}_n)\qquad(19.14)$$

and we define anomalousness as distance to the centroid in this feature space.

$$
\begin{aligned}
\mathscr{A}(\mathbf{x}) &= \|\phi(\mathbf{x}) - \mu_\phi\|^2 \\
&= (\phi(\mathbf{x}) - \mu_\phi)^T (\phi(\mathbf{x}) - \mu_\phi) \\
&= \phi(\mathbf{x})^T \phi(\mathbf{x}) - 2\phi(\mathbf{x})^T \mu_\phi + \mu_\phi^T \mu_\phi
\end{aligned}
\tag{19.15}
$$

We observe that first term $\phi(\mathbf{x})^T \phi(\mathbf{x}) = \kappa(\mathbf{x}, \mathbf{x})$ is constant for radial basis kernels, that the third term is also constant, and that

$$
\phi(\mathbf{x})^T \mu_\phi = \frac{1}{N} \sum_{n=1}^{N} \phi(\mathbf{x})^T \phi(\mathbf{x}_n) = \frac{1}{N} \sum_{n=1}^{N} \kappa(\mathbf{x}, \mathbf{x}_n).
\tag{19.16}
$$

This leads to

$$
\mathscr{A}(\mathbf{x}) = \text{constant} - \frac{2}{N} \sum_{n=1}^{N} \kappa(\mathbf{x}, \mathbf{x}_n),
\tag{19.17}
$$

which is a negative monotonic transform of the density estimator $(1/N) \sum_{n=1}^{N} \kappa(\mathbf{x}, \mathbf{x}_n)$, and therefore equivalent to anomaly detection based on Parzen windows density estimation. The power of kernels in this case is that a seemingly trivial anomaly detector (Euclidean distance to the centroid of the data) in feature space maps back to a more complex data-adaptive anomaly detector in the data space.

The power of this feature-space interpretation of kernels is that it enables us to derive other expressions for anomaly detection, starting with very simple models in feature space that are then mapped back to more sophisticated data-adaptive anomaly detectors in the data space.

For instance, instead of Euclidean distance to the centroid μ_ϕ, consider a more periphery-respecting model that uses an adaptive center $\mathbf{a}_\phi$ that is adjusted to minimize the radius of the sphere that encloses all of the data (see Figure 19.4). That is,[3]

$$
\min_{r, \mathbf{a}_\phi} r^2 \quad \text{subject to: } \|\phi(\mathbf{x}_n) - \mathbf{a}_\phi\|^2 \leq r^2
\tag{19.18}
$$

or more generally, that *mostly* encloses the data:

$$
\min_{r, \mathbf{a}_\phi, \xi} r^2 + c \sum_n \xi_n
\tag{19.19}
$$

$$
\text{subject to: } \|\phi(\mathbf{x}_n) - \mathbf{a}_\phi\|^2 \leq r^2 + \xi_n
\tag{19.20}
$$

$$
\text{and: } \xi_n \geq 0,
\tag{19.21}
$$

which is equivalent to Equation (19.18) in the large c limit (which forces the "slack" variables ξ_n to zero). This optimization leads to the support vector domain decomposition (SVDD) anomaly detector [1004, 50], which has the form

$$
\mathscr{A}(\mathbf{x}) = \|\phi(\mathbf{x}) - \mathbf{a}_\phi\|^2 = \text{constant} - \sum_n a_n \kappa(\mathbf{x}, \mathbf{x}_n)
\tag{19.22}
$$

where the scalar coefficients a_n are positive and sum to 1. This is very much like the kernel density estimator for anomaly detection in Equation (19.17), but it puts uneven weight on the points in the dataset. The SVDD is very similar to the support vector machine for one-class classification [916], and has the property that $a_n = 0$ for points deep in the interior of the distribution.

In keeping with the general strategy of mapping data to kernel space, and applying anomaly

---

[3]Another way of expressing the centroid μ_ϕ is as the solution to the minimization of the average squared radius: $\mu_\phi = \operatorname{argmin}_\mu \sum_n \|\phi(\mathbf{x}_n) - \mu\|^2$; by comparison, we can say $\mathbf{a}_\phi$ is the solution to the minimization of the maximum squared radius: $\mathbf{a}_\phi = \operatorname{argmin}_\mathbf{a} \max_n \|\phi(\mathbf{x}_n) - \mathbf{a}\|^2$. We can interpret Equation (19.19) as the minimization of a "soft" maximum.

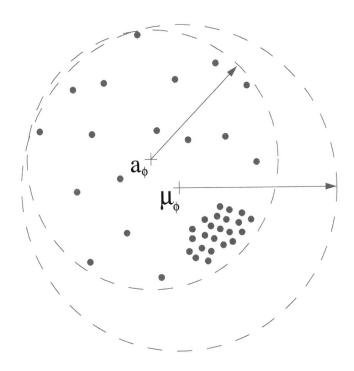

Figure 19.4: Adaptive center versus data centroid. Data samples in feature space are indicated by black dots. Here, $\mu_\phi = (1/N) \sum_{n=1}^{N} \phi(\mathbf{x}_n)$ is the centroid of the data, and a_ϕ is the adaptive center that enables the data to be enclosed by a smaller circle.

detection in this kernel space, we can also "kernelize" the RX algorithm. Here, given the data is mapped to kernel space $\{\phi(\mathbf{x}_1), \ldots, \phi(\mathbf{x}_N)\}$, we need to compute a mean and covariance matrix. The mean μ_ϕ is defined in Equation (19.14); to compute the covariance, we write

$$R_\phi = \sum_n [\phi(\mathbf{x}_n) - \mu_\phi][\phi(\mathbf{x}_n) - \mu_\phi]^T \tag{19.23}$$

A key step in the computation of RX anomalousness is the inversion of this covariance matrix. But R_ϕ has bounded rank (it is at most $n-1$), and depending on the dimension of the feature space, may not be invertible.

In Cremers et al. [226], this problem was addressed by regularizing to covariance, so that $R_\phi + \lambda I$ was inverted, where λ was taken to be a small but nonzero value. In Kwon and Nasrabadi [529], the pseudoinverse was taken. The effect of the pseudoinverse is to project data (in the feature space) to the in-sample data plane, but this projection can be problematic for anomaly detection [1028]. Anomalies are different from the rest of the data, and this difference will be suppressed by projection back into the in-sample data plane.

Indeed, another kernelization that can be effective is the kernel subspace anomaly detector [437, 720]. Here, principal components analysis is performed in the feature space, and a subspace is defined that includes the first few principal components. Anomalousness is defined in terms of the distance to this subspace.

19.6 Change

For the anomalous *change* detection problem, the aim is to find interesting differences between two images, taken of the same scene, but at different times and typically under different viewing conditions [908]. There will be some differences that are pervasive – e.g., differences due to overall calibration, contrast, illumination, look-angle, focus, spatial misregistration, atmospheric or even seasonal changes – but there may also be changes that occur in only a few pixels. These rare changes potentially indicate something truly changed in the scene, and the idea is to use anomaly detection to find them. But our interest is in pixels where *changes* between the pixels are unusual, not so much in unusual pixels that are "similarly unusual" in both images. Informally speaking, we want to learn the "patterns" of these pervasive differences, and then the changes that do not fit the patterns are identified as anomalous.

An important precursor to anomalous change detection is the co-registration of the two images. We say that images are registered if corresponding pixels in the two images correspond to the same position in the scene. Registering imagery is a nontrivial task, yet misregistration is one of the main confounds in change detection [1014, 675, 1035, 1107]. In what follows, let $\mathbf{x}$ and $\mathbf{y}$ refer to corresponding pixels in two images.

19.6.1 Subtraction-Based Approaches to Anomalous Change Detection

The most straightforward way to look for changes in a pair of images is to subtract them, $e = \mathbf{y} - \mathbf{x}$, and then to restrict analysis to the difference image e [138]. Simple subtraction, although it has the advantage of being simple, has the disadvantage that it folds in pervasive differences along with the anomalous changes.

Most anomalous change detection algorithms are based on subtracting images, but involve transforming the images to make them more similar. For instance, the chronochrome [910] seeks a linear transform of the first image to make it as similar as possible (in a least squares sense) to the second image. That is, it seeks L so that $\|\mathbf{y} - L\mathbf{x}\|^2$, averaged over the whole image, is minimized. To simplify notation, we will assume means have been subtracted from $\mathbf{x}$ and $\mathbf{y}$, and define the covariance matrices $X = \langle \mathbf{x}\mathbf{x}^T \rangle$, $Y = \langle \mathbf{y}\mathbf{y}^T \rangle$, and $C = \langle \mathbf{y}\mathbf{x}^T \rangle$. The linear transform that minimizes the least square fit of $\mathbf{y}$ to $L\mathbf{x}$ is given by $L = CX^{-1}$. Now the subtraction that is performed is $e = \mathbf{y} - L\mathbf{x}$, and this reduces the effect of pervasive differences on e while still "letting through" the anomalous changes. Note that there is an asymmetry in the chronochrome; by swapping the role of $\mathbf{x}$ and $\mathbf{y}$, and seeking L' to minimize $\|e = \mathbf{x} - L'\mathbf{y}\|^2$, one obtains a different anomalous change detector. Clifton [212] proposed a neural network version of chronochrome, in a non-linear function $\mathcal{L}(\mathbf{x})$ is chosen to minimize $e = \mathbf{y} - \mathcal{L}(\mathbf{x})$, with the aim of even further suppressing the pervasive differences.

A more symmetrical approach, which is sometimes called covariance equalization [911, 912] or whitening/de-whitening [664], transforms the data in both images before it subtracts them: $e = Y^{-1/2}\mathbf{y} - X^{-1/2}\mathbf{x}$. It can be shown [1013] that this is related to canonical coordinate analysis and to Nielsen's multivariate alteration detection (MAD) algorithm [735] and to a "total least squares" change detection algorithm [1030]. One reason there are so many variants is that there is not a unique whitening transform: if U is an arbitrary orthogonal matrix, then $UX^{-1/2}$ will also whiten the $\mathbf{x}$ samples. That is: if $\widehat{\mathbf{x}} = UX^{-1/2}\mathbf{x}$, then we say that $\widehat{\mathbf{x}}$ is whitened because $\langle \widehat{\mathbf{x}}\widehat{\mathbf{x}}^T \rangle = I$.

In all of these algorithms, a vector-valued difference, e, is produced. For anomaly detection, we need a scalar valued measure of anomalousness, and the RX formula provides the most straightforward way to achieve that. Let $E = \langle ee^T \rangle$ be the covariance matrix of the residuals, and then take the anomalousness to be

$$\mathcal{A}(e) = e^T E^{-1} e \qquad (19.24)$$

As a particular example, we can write Equation (19.24) for the chronochrome detector. Here $e = \mathbf{y} - CX^{-1}\mathbf{x}$, and

$$\mathcal{A}(\mathbf{x}, \mathbf{y}) = (\mathbf{y} - CX^{-1}\mathbf{x})^T [Y - CX^{-1}C]^{-1}(\mathbf{y} - CX^{-1}\mathbf{x}) \qquad (19.25)$$

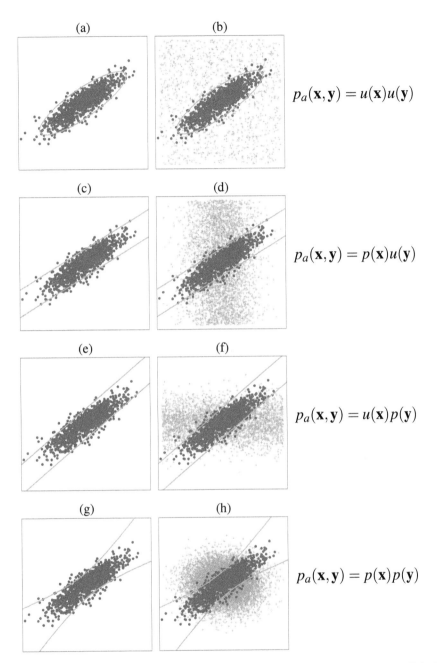

Figure 19.5: Four anomalous change detectors, derived from four distinct and explicit definitions of anomalous change. Here $\mathbf{x}$ is represented by the horizontal axis and $\mathbf{y}$ is the vertical axis. The left panels show the non-anomalous data (sampled from a correlated Gaussian distribution) and the boundaries outside of which are the points that the detectors consider anomalous changes. The panels to the right include samples drawn from $p_a(\mathbf{x}, \mathbf{y})$, the model for the distribution of anomalous changes. (a,b) RX detector is obtained from "straight" anomaly detection; (c,d) Chronochrome detector optimized for $\mathbf{x} \to \mathbf{y}$ changes; (e,f) Chronochrome optimized for $\mathbf{y} \to \mathbf{x}$ changes; and (g,h) hyperbolic anomalous change detector.

19.6.2 *Distribution-Based Approaches to Anomalous Change Detection*

While subtracting (suitably transformed) images is an intuitively plausible way to look for anomalous changes, we can take more of a machine learning point of view and treat the problem as one of two-class classification, where the two classes are pervasive differences and anomalous changes. If we can write down expressions for the underlying distributions of these two classes, then their ratio will be an optimal detector of anomalous changes.

The distribution for the pervasive differences is just the distribution of the data itself. We can call this $p_b(\mathbf{x},\mathbf{y})$ to indicate that it is the "background" distribution. What we need is a generative model for anomalous changes; that is, a distribution $p_a(\mathbf{x},\mathbf{y})$ that describes what we mean when we speak of anomalous changes. With that in hand, our anomaly detector is given by the likelihood ratio

$$\mathscr{L}(\mathbf{x},\mathbf{y}) = \frac{p_a(\mathbf{x},\mathbf{y})}{p_b(\mathbf{x},\mathbf{y})}. \tag{19.26}$$

The simplest choice, following our experience with straight anomaly detection, is a uniform distribution. We can write this $p_a(\mathbf{x},\mathbf{y}) = u(\mathbf{x})u(\mathbf{y})$. This is not an unreasonable choice, but it does not put any particular emphasis on *change*. A pair of pixels $(\mathbf{x},\mathbf{y})$ that are "similarly anomalous" in both images (e.g., the pixel might be unusually bright in both images) do not particularly indicate that something has changed at that pixel position in the scene.

But there are alternative models $p_a(\mathbf{x},\mathbf{y})$ that correspond to different notions of what an anomalous change is. For instance, if we are thinking specifically of changes $\mathbf{x} \to \mathbf{y}$, where $\mathbf{x}$ is a kind of "reference" image, and $\mathbf{y}$ is the new image that might contain the changes of interest, then $p_a(\mathbf{x},\mathbf{y}) = p_b(\mathbf{x})u(\mathbf{y})$. This corresponds to a non-anomalous $\mathbf{x}$ and an anomalous $\mathbf{y}$. It is interesting to note that using this definition of anomaly in Equation (19.26) leads to $\mathscr{L}(\mathbf{x},\mathbf{y}) = 1/p_b(\mathbf{y}|\mathbf{x})$, so that anomalousness of $\mathbf{y}$ varies inversely with the *conditional* probability density of $\mathbf{y}$. As with the chronochrome,[4] there is an asymmetry in this model of anomalous change; its mirror image is the situation in which $\mathbf{y}$ is considered the reference image and it is $\mathbf{y} \to \mathbf{x}$ changes that are of particular interest; here, $p_a(\mathbf{x},\mathbf{y}) = u(\mathbf{x})p_b(\mathbf{y})$.

Another choice for $p_a(\mathbf{x},\mathbf{y})$ has also been suggested [1031]. Here, $p_a(\mathbf{x},\mathbf{y}) = p_b(\mathbf{x})p_b(\mathbf{y})$, and $\mathbf{x} \to \mathbf{y}$ and $\mathbf{y} \to \mathbf{x}$ changes are treated equally. The informal interpretation is that unusual changes are pixel pairs $(\mathbf{x},\mathbf{y})$ which are collectively unusual, but individually normal. That is, the $\mathbf{x}$ pixel value is typical for the x-image, and the $\mathbf{y}$ pixel value is typical for the y-image, but the $(\mathbf{x},\mathbf{y})$ pair is unusual. If we use this model for anomalies in Equation (19.26), and take a logarithm, we obtain

$$\log \mathscr{L}(\mathbf{x},\mathbf{y}) = \log p_b(\mathbf{x}) + \log p_b(\mathbf{y}) - \log p_b(\mathbf{x},\mathbf{y}) \tag{19.27}$$

an expression for anomalousness that looks like negative mutual information of $\mathbf{x}$ and $\mathbf{y}$.

In the case of Gaussian $p_b(\mathbf{x},\mathbf{y})$, Equation (19.27) reduces to a quadratic expression in $\mathbf{x}$ and $\mathbf{y}$ that has hyperbolic contours (see Figure 19.5(g,h)). Experiments with real and simulated anomalous changes in real imagery indicated that this hyperbolic anomalous change detection (HACD) generally outperformed the subtraction-based anomaly detectors [1013].

A further advantage of the distribution-based approach is that the distribution needn't be Gaussian. Indeed, we can take a purely nonparametric view, and treat the problem of distinguishing pervasive differences from anomalous changes as a machine learning classification. Steinwart et al. [967] used support vector machines for just this purpose. But a simpler approach, that has also proven effective, is to consider a parametric distribution, but one slightly more general than the Gaussian. The class of elliptically-contoured (EC) distributions are, like the Gaussian, primarily parameterized by a mean vector and covariance matrix, but do not share the sharp e^{-r^2} tail of the Gaussian. Heavy-tailed EC distributions have been suggested for hyperspectral imagery in general [634], and for anomalous change detection in particular [909, 1034]. Although EC distributions do not affect the "straight anomaly detector" shown in Figure 19.5(a,b), they do generalize chronochrome and

---

[4]In fact, the chronochrome can be obtained from this model in the special case that $p_b(\mathbf{x},\mathbf{y})$ is Gaussian [1032].

hyperbolic anomalous change detectors in a way that can lead to improved performance [1034]. Kernelization of the EC-based change detector has also been shown to be advantageous [590].

19.6.3   *Further Comments on Anomalous Change Detection*

The description here treats pixel pairs as independent samples from an unknown distribution. But there is a lot of spatial structure in imagery, and further gains can be made by incorporating spatial aspects along with the spectral [480, 1020]. The description here also considers only pairs of images; often there are more than two images, and these algorithms can be extended to that case [11, 1023], though this approach may not be optimal for sequences of images (e.g., anomalous activities in video) where the order of the images in the sequence matters. Another issue that arises in remote sensing is that the anomalous targets may be subpixel in extent, which leads to a different optimization problem [1015].

Anomalous change detection is a problem that is particularly well matched to remote sensing imagery, and in that context, a variety of practical issues have been discussed [296, 297]. One of the biggest of these issues is misregistration, when the images don't exactly line up (and they never *exactly* line up). Although the effects of misregistration can to some extent be learned from the pervasive differences it creates in image pairs, it is still one of the main confounds to change detection [1014, 675]. Gains can be made by explicitly adapting the change detection algorithm to be more robust to misregistration error [1035, 1107].

19.7   Conclusion

Anomaly detection is seldom a goal in its own right. It is the first step in a search for data samples that are relevant, meaningful, or – in some sense that depends on where the data came from and what they are being used for – interesting.

The mystical "by definition undefined" aspect of anomaly detection mostly derives from the ambiguity of what one means by "interesting" and this has led to a wide variety of *ad hoc* anomaly detection algorithms, justified by hand-waving arguments and validated (if at all) by anecdotal performance on imagery with a statistically inadequate number of pre-judged anomalies.

By employing a framework in which anomalies are in fact well-defined, as samples drawn from some broad and flat distribution, anomaly detection algorithms can be objectively tested, and improvements can be confidently constructed.

Within this framework, many of the tools that have been developed for signal processing, machine learning, and data analytics in general, can be brought to bear on the detection of anomalies. These range from the venerable Gaussian distribution to kernels and subspaces (and kernelized subspaces!), and invoke the usual issues in underfitting and overfitting data.

The technical challenge of anomaly detection is not usually the anomalies themselves, but with characterizing what can be a complex and highly structured background. Since most anomaly detection scenarios require a low false alarm rate, it is out on the periphery of this background where the modeling is emphasized. This is something of a challenge, since the data density is much lower there. The modeling, however is discriminative, not generative, which means that the aim is not to model the distribution *per se*, but to find the boundary that separates the non-anomalous data from the anomalies.

Chapter 20

Anomalies in Microstructures

Stephen Bricker
University of Dayton

Craig Przybyla, Jeffrey P. Simmons
Air Force Research Laboratory

Russell Hardie
University of Dayton

20.1 Introduction

It is hypothesized that cracks in a failed specimen, such as seen in Figure 20.1, can be linked to component lifetimes via local fiber microstructure anomalies. Traditional statistical analysis of fibrous microstructure either focuses on averaged statistics of individual fibers (like fiber packing) or on fiber tow (woven bundle) statistics. Anomalies, however, hold little to no impact on averaging statistics and often completely escape statistical analysis.

In this chapter, we focus on the application of a standard anomaly detection technique to feature fields of material microstructure. Before we can apply anomaly detection techniques, we must represent what is physically happening in the microstructure mathematically. To do this we develop features of the microstructure that highlight the collective behavior of fiber neighborhoods, while avoiding averaging out the effect of any anomalies. Specifically, fiber orientation and fiber orientation field gradient are utilized to represent the microstructure at a scale greater than that of individual fibers but with greater detail than standard modeling techniques. Anomalies on this scale are hypothesized to shed insight into the crack growth pathways through fiber tows.

An anomaly as described in the previous chapter is "an observation that deviates in some way from the background clutter." Anomaly detection is generally approached as an unsupervised classification technique and thus does not require a pre-separated set of data. Because anomalous events have unknown distributions, anomaly detection is an appropriate approach to finding abnormal events in a material microstructure. The only a priori information necessary lies in the formation of a feature set representation of the microstructure that is suspected of holding information relating to the desired properties. Here, the construction of an appropriate feature set is approached from a mathematical basis and used to detect microstructural anomalies.

20.2 Features of the Local Fiber Microstructure

Textile-based fiber-reinforced composites are composed of a series of fiber groups, called tows, consisting of several hundred fibers woven into textile sheets. These sheets are then filled with a matrix material via one of several processing methods which ideally results in solid, dense matrix encasing the fibers and fiber tows. Most fiber weaves feature fiber tows woven perpendicular to one another and unbroken throughout the width of the material. When the composite material is imaged via serial sectioning, fibers viewed on planes perpendicular to the plane of the textile sheet appear

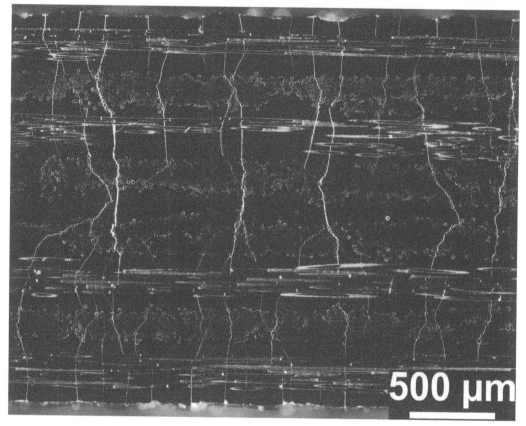

Figure 20.1: Cracks of a failed sample highlighted by dye penetrant and UV lighting. Fiber coatings also fluoresce under UV lighting.

as ellipses. Figure 20.2 shows the top and bottom slice of a sample, with fiber orientation illustrated by cylinders. The plot shows fibers of different tows plotted in 3D.

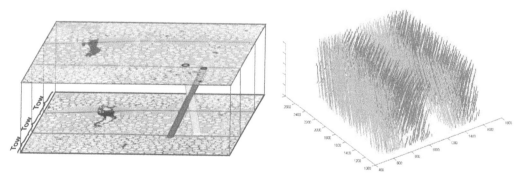

Figure 20.2: Fiber composite structure. Fibers in neighboring tows are oriented perpendicular to one another.

The resulting stack of images features groups of fibers (tows) with similar orientation that bend around one another according to the weave. Because the fibers are physically constrained within the fiber bundle, the fiber orientations do not differ significantly between fibers within a bundle.

Finding an appropriate feature for capturing the desired information about a material is a challenging problem. Here, we define and develop two feature fields for use with anomaly detection. The features, orientation and orientation gradient, are chosen because they describe the local fiber behavior of the microstructure. From a materials performance perspective, a ceramic matrix composite (CMC) research goal is to determine what local fiber microstructures produce preferential damage initiation sites, and it is hypothesized that fiber orientation and fiber orientation gradient may influence damage progression in the microstructure. Before a feature field for fiber microstructure can be developed, the fibers must be represented mathematically. Steps in the fiber extraction process include stitching and aligning images [788], image segmentation [219], ellipse identification [802], and fiber position tracking. Tracking of the fibers is dealt with in detail in Chapter 8. The end result of the mentioned preprocessing steps is a set of fiber positions for each serial section that is associated with the corresponding fibers between sections.

After a digital representation of the fibers has been constructed, the behavior of fiber orientation and fiber neighborhood interaction can be examined. Most microstructural analysis and modeling techniques focus on the behavior of entire fiber tows. Analysis of fiber neighborhoods is the goal of this chapter because the neighborhood behavior characterizes the microstructure on a scale larger than the individual fibers, but smaller than that of conventional statistical techniques. These meso-scale properties are believed to hold a large impact on the formation and propagation of damage in fibrous composites.

20.2.1   Orientation Field

The orientation field of fibers in the composite can be treated like a velocity field in a flowing fluid. Any point in the field can be modeled as a fluid streamer whose position varies smoothly along its length. Though the fibers are not *flowing* through the material, the equations of motion describing the path of a particle in a fluid adequately describe the path of a fiber. The position vector, $\mathbf{r} = (x, y, z)$, of the particle and its derivative, $\frac{\partial}{\partial \tau}\mathbf{r}$ describe the orientation of the fiber, analogous to instantaneous velocity of the particle. The parameter τ refers to a length traveled along the fiber. The fiber centers extracted from each serial section function as discrete samples of the continuous orientation vector field.

The sample representation is in the "Eulerian space," whereas the fibers are more easily described in "Lagrangian space." A subtle point to be distinguished here is that while the serial section data is in an orthogonal coordinate system that is readily converted to (x, y, z) coordinates, the fiber orientation is described by the distance traveled along each fiber. This section describes the transformations between these two coordinate systems and will be used as a foundation for the computation of orientation gradients, which will be the basis for meso-scale anomaly detection.

Lagrangian coordinates are often used to describe fluid flows and conservation laws as they apply to a specific particle in a fluid [798]. In the case of an embedded fiber, Lagrangian coordinates move along the length of the fiber as it moves in space. Quantitatively, a specific point on a fiber is denoted by $\boldsymbol{\rho} = (\xi, \eta, \tau)$ in a Lagrangian coordinate system, where ξ and η pinpoint a particular fiber, and τ indexes length traveled along the fiber path.

20.2.1.1   Description of the Orientation Field

Finding the orientation, $\frac{\partial}{\partial \tau}\mathbf{r}$, is not trivial because the serial section data is collected in an orthogonal reference frame, which is converted to Cartesian coordinates, (x, y, z). Because orientation is defined in Lagrangian coordinates and the data is in Eulerian coordinates, it is necessary to develop the transformation between the two systems. To accomplish this, following standard methods of multivariate calculus [223], we define the mapping between (ξ, η, τ) to (x, y, z) in terms of implicitly defined functions as follows:

$$
\begin{aligned}
x &= \Psi(\xi,\eta,\tau); & \xi &= F(x,y,z); \\
y &= \Phi(\xi,\eta,\tau); & \eta &= G(x,y,z); \\
z &= Z(\xi,\eta,\tau); & \tau &= H(x,y,z).
\end{aligned}
\tag{20.1}
$$

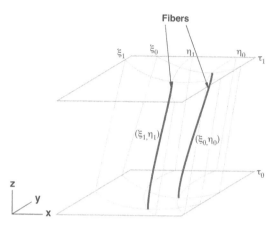

Figure 20.3: Representation of Lagrangian coordinate space for fibers.

For individual fibers the ξ and η coordinates remain constant while τ changes along the length of the fiber. Figure 20.3 shows the non-linear behavior of the Lagrangian space as it relates to Eulerian coordinates. The transformation to Lagrangian space via the functions F, G, and H can be expanded into a Taylor series about some reference point, (ξ_i, η_i, τ_i),

$$
\begin{aligned}
\xi &= \xi_i + \frac{\partial F}{\partial x}\Delta x + \frac{\partial F}{\partial y}\Delta y + \frac{\partial F}{\partial z}\Delta z + O(|\,\Delta \mathbf{r}\,|^2); \\
\eta &= \eta_i + \frac{\partial G}{\partial x}\Delta x + \frac{\partial G}{\partial y}\Delta y + \frac{\partial G}{\partial z}\Delta z + O(|\,\Delta \mathbf{r}\,|^2); \\
\tau &= \tau_i + \frac{\partial H}{\partial x}\Delta x + \frac{\partial H}{\partial y}\Delta y + \frac{\partial H}{\partial z}\Delta z + O(|\,\Delta \mathbf{r}\,|^2),
\end{aligned}
\tag{20.2}
$$

where $O(|\,\Delta \mathbf{r}\,|^2)$ represents the higher powers of the Taylor series, and i indexes some point in space.

The parameters Δx, Δy, and Δz represent changes in position within Euler space from the expansion point i: $\Delta x = x - x_i$, $\Delta y = y - y_i$ and $\Delta z = z - z_i$. The three equations can be combined in matrix form

$$
\begin{bmatrix} \Delta \xi \\ \Delta \eta \\ \Delta \tau \end{bmatrix}
=
\begin{bmatrix}
\frac{\partial F}{\partial x} & \frac{\partial F}{\partial y} & \frac{\partial F}{\partial z} \\
\frac{\partial G}{\partial x} & \frac{\partial G}{\partial y} & \frac{\partial G}{\partial z} \\
\frac{\partial H}{\partial x} & \frac{\partial H}{\partial y} & \frac{\partial H}{\partial z}
\end{bmatrix}
\begin{bmatrix} \Delta x \\ \Delta y \\ \Delta z \end{bmatrix}
+ O(|\,\Delta \mathbf{r}\,|^2),
\tag{20.3}
$$

where $\Delta \xi = \xi - \xi_i$, $\Delta \eta = \eta - \eta_i$, and $\Delta \tau = \tau - \tau_i$. This can be simplified by using the symbols, $\Delta \boldsymbol{\rho} = [\Delta \xi, \Delta \eta, \Delta \tau]^T$ and $\Delta \mathbf{r} = [\Delta x, \Delta y, \Delta z]^T$ to describe the corresponding change in position for Lagrangian and Eulerian coordinates, respectively. The matrix defining the coordinate transformation is a Jacobian matrix by neighbor and will be represented as $\mathbb{J}_{F,G,H}$.

Fibers embedded in the matrix are rigid and it can be assumed that the fiber does not have significant curvature between sections and the residual term, $O(|\Delta r|^2)$, may be neglected. Utilizing these substitutions and assumptions, Equation (20.3) becomes

$$\Delta\boldsymbol{\rho} = \mathbb{J}_{F,G,H}\Delta\mathbf{r}. \tag{20.4}$$

The coordinate transformation defined by Equation (20.4) holds true as long as the determinant of the Jacobian is non-zero, and the fibers are continuous. This is justified by assuming that (1) fibers do not join, so that each fiber i has its own unique pair of coordinates, (ξ_i, η_i) for a given τ_i, and (2) there are no breaks in the fibers. The first is virtually guaranteed by the process, which does not allow fibers to join, and the second is typically correct. A break in a fiber would result in an anomaly condition.

Since traveling along a fiber is represented by changing τ, each fiber can be uniquely identified by its (ξ, η) coordinates. The track of a fiber can be determined by indexing τ down its length and holding both ξ and η constant. To convert this to measurable coordinates, the Lagrangian coordinates of the fiber must be converted to the Eulerian coordinates of the laboratory. To accomplish this, Equation (20.4) is rearranged to solve for $\Delta\mathbf{r}$,

$$\mathbb{J}_{F,G,H}^{-1}\Delta\boldsymbol{\rho} = \Delta\mathbf{r}, \tag{20.5}$$

where $\mathbb{J}_{F,G,H}^{-1}$ is the inverse of the Jacobian transformation matrix. Mathematically, $\mathbb{J}_{F,G,H}^{-1} = \mathbb{J}_{F,G,H}^{T}$ [223] and can be solved for by inverting the Lagrangian vector and right multiplying both sides of the equation.

The change in position from Equation (20.5) can be used to approximate the orientation for small changes in τ. The orientation, $\frac{\partial}{\partial\tau}\mathbf{r}$, is found by taking the limit of $\Delta\mathbf{r}/\Delta\tau$ as $\Delta\tau$ goes to zero. Replacing $\Delta\mathbf{r}$ with its Lagrangian counterpart from Equation (20.5), the limit becomes

$$\lim_{\Delta\tau\to 0}\frac{\Delta\mathbf{r}}{\Delta\tau} = \lim_{\Delta\tau\to 0}\frac{\mathbb{J}_{F,G,H}^{-1}\Delta\boldsymbol{\rho}}{\Delta\tau}. \tag{20.6}$$

Because the Jacobian does not depend on $\Delta\tau$, it can be moved outside the limit while $\Delta\tau$ is distributed to the components of $\Delta\boldsymbol{\rho}$, yielding the general equation for orientation,

$$\frac{\partial\mathbf{r}}{\partial\tau} = \mathbb{J}_{F,G,H}^{-1}\lim_{\Delta\tau\to 0}\begin{bmatrix}\frac{\Delta\xi}{\Delta\tau}\\\frac{\Delta\eta}{\Delta\tau}\\1\end{bmatrix} = \mathbb{J}_{F,G,H}^{-1}\begin{bmatrix}\frac{\partial\xi}{\partial\tau}\\\frac{\partial\eta}{\partial\tau}\\1\end{bmatrix}, \tag{20.7}$$

for any arbitrary change in ξ and η. For a particular fiber, i, $\Delta\xi_i$ and $\Delta\eta_i$ are zero. The orientation of a fiber can then be described as,

$$\mathbf{V} = \frac{\partial\mathbf{r}_i}{\partial\tau} = \mathbb{J}_{F,G,H}^{-1}[0,0,1]^T, \tag{20.8}$$

the last column of the inverse Jacobian. Because orientation is only calculated for individual fibers, the rest of the Jacobian is not explored.

20.2.1.2 *Computational Simplification of the Orientation Field*

Due to serial sectioning and ellipse detection, the orientation field is only sampled at fiber locations for discrete increments in z. In the general case, no assumptions should be made about the orientations of the fibers. Here, however, to reduce expressions to two dimensions we assume fibers intersect the sectioning plane at the same angle, $\theta = 45°$. Under this constraint, all fibers advance by the same length, τ, between serial sections. Figure 20.4 details how the individual $[x, y, z]$ components of $\mathbf{r}$ can be defined in terms of τ. The vector,

$$\mathbf{r} = \begin{bmatrix} \tau \sin \theta \cos \phi \\ \tau \sin \theta \sin \phi \\ \tau \cos \theta \end{bmatrix} \tag{20.9}$$

can then be used to define the orientation of the fiber,

$$\frac{\partial \mathbf{r}}{\partial \tau} = \begin{bmatrix} \sin \theta \cos \phi \\ \sin \theta \sin \phi \\ \cos \theta \end{bmatrix}, \tag{20.10}$$

where $\cos \theta$ is a constant for all fibers. In practice, the x and y components of orientation are calculated by assuming that a fiber remains linear over a stretch of several serial sections. Zenith angle, θ, remains constant at $45°$ for all fibers, resulting in $\sin \theta = \cos \theta = \sqrt{2}/2$. The remaining components $\cos \phi$ and $\sin \phi$ can be approximated for each fiber by

$$\cos \phi \approx \frac{x}{P}$$
$$\sin \phi \approx \frac{y}{P}, \tag{20.11}$$

where P is the projection of $\mathbf{r}$ into the (x, y)-plane. For $\theta = 45°$, the magnitude of vector P is the same as z. This allows Equation (20.10) to be written as

$$\frac{\partial \mathbf{r}}{\partial \tau} \approx \begin{bmatrix} \frac{1}{\sqrt{2}} \frac{x}{z} \\ \frac{1}{\sqrt{2}} \frac{y}{z} \\ \frac{1}{\sqrt{2}} \end{bmatrix}, \tag{20.12}$$

which can be easily evaluated for a section of fiber. To reduce noise in the computation of orientation, x/z and y/z are computed from several slices about the current slice instead of between single slices. While the measured values theoretically must be multiplied by $\sin \theta$, the orientation direction does not change when the vector is normalized by the change in z, so the scaling is ignored. In the general case, $\sin \theta$ and $\cos \theta$ are not equal and must be included in the orientation calculation.

20.2.1.3   Color Visualization of the Orientation

Visualization of the orientation field is achieved through the use of the HSV color model. The HSV color space is parameterized by three parameters; hue, saturation, and value. Hue is useful in representing vector fields because it changes with angle as seen in Figure 20.5(a). Saturation controls the vibrancy or amount of color, while value adjusts the lightness/darkness. Typical color visualization uses fully saturated color to denote scalar values, ignoring two parameters for conveying information.

The orientation field is visualized by assigning hue to each fiber in the segmented image according to its angle of orientation. The saturation is indicative of magnitude of vector shift between slices. While value is constant for visualization of the orientation field, all three parameters are utilized for visualization of the orientation gradient. Tow groupings and shift direction are intuitively represented, and deviations from normal tow behavior become immediately clear. Figure 20.5(c) shows fibers in a real sample colored via the HSV model. Fibers that deviate the most from tow orientation tend to be along the tow boundaries where fibers are less constrained during weaving.

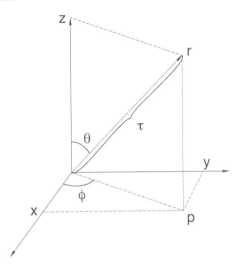

Figure 20.4: Vector diagram for the individual components of **r**.

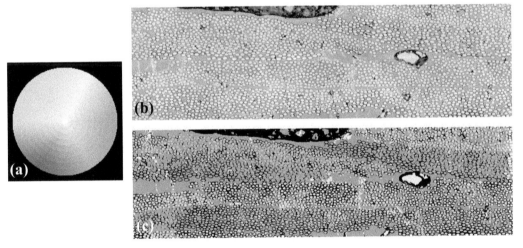

Figure 20.5: (a) Hue and saturation vector colormap used for visualization of the orientation field. Hue represents direction of fiber orientation(ϕ), while saturation represents magnitude (τ); (b) SiC matrix composite with SiC fiber reinforcement; (c) SiC/SiC composite segmented and colored by fiber orientation; e.g., blue indicates motion to the left and red indicates motion to the right.

20.2.2   Orientation Gradient Field

The features of the velocity field that describe neighborhood behavior collectively, such as eddies, boundaries, or other non-uniformity in the flow are characterized by the gradient of the velocity field. Assuming a smoothly varying orientation field, the gradient of the orientation can be used to describe how the orientation is changing with respect to each of the Eulerian coordinates, (x, y, z).

In practice, the orientation gradient cannot be directly calculated. Instead, a Taylor series expansion of orientation,

$$\Delta \frac{\partial \mathbf{r}}{\partial \tau} = \nabla \frac{\partial \mathbf{r}}{\partial \tau} \cdot \Delta \mathbf{r} + O(|\Delta \mathbf{r}|^2), \tag{20.13}$$

is used to express the orientation gradient in terms of change in orientation, $\Delta \partial \mathbf{r}/\partial \tau$, and change in position, $\Delta \mathbf{r}$. The difference in orientation between two points, $\Delta \partial \mathbf{r}/\partial \tau$, corresponds to the same three-dimensional difference in position as $\Delta \mathbf{r}$, related through the orientation gradient. Again, the residual term, $O(|\Delta \mathbf{r}|^2)$, can be neglected due to the assumption of linearity. Equation (20.13) can be represented between two arbitrary points, $\mathbf{r}_0$, a reference point, and $\mathbf{r}_1$. By replacing $\Delta \partial \mathbf{r}/\partial \tau$ and $\Delta \mathbf{r}$ with the resulting components, the general equation for the Taylor series becomes

$$
\frac{\partial \mathbf{r}_1}{\partial \tau} - \frac{\partial \mathbf{r}_0}{\partial \tau} = \nabla \frac{\partial \mathbf{r}_0}{\partial \tau} \cdot \begin{bmatrix} x_1 - x_0 \\ y_1 - y_0 \\ z_1 - z_0 \end{bmatrix}.
\tag{20.14}
$$

which can be utilized to solve for the gradient using measured values of $\Delta \mathbf{r}$ and $\Delta \partial \mathbf{r}/\partial \tau$. The orientation gradient at $\mathbf{r}_0$ can be expressed fully as

$$
\nabla \frac{\partial \mathbf{r}}{\partial \tau} = \begin{bmatrix} \frac{\partial}{\partial x}\left(\frac{\partial \Psi}{\partial \tau}\right) & \frac{\partial}{\partial y}\left(\frac{\partial \Psi}{\partial \tau}\right) & \frac{\partial}{\partial z}\left(\frac{\partial \Psi}{\partial \tau}\right) \\ \frac{\partial}{\partial x}\left(\frac{\partial \Phi}{\partial \tau}\right) & \frac{\partial}{\partial y}\left(\frac{\partial \Phi}{\partial \tau}\right) & \frac{\partial}{\partial z}\left(\frac{\partial \Phi}{\partial \tau}\right) \\ \frac{\partial}{\partial x}\left(\frac{\partial Z}{\partial \tau}\right) & \frac{\partial}{\partial y}\left(\frac{\partial Z}{\partial \tau}\right) & \frac{\partial}{\partial z}\left(\frac{\partial Z}{\partial \tau}\right) \end{bmatrix},
\tag{20.15}
$$

a $[3 \times 3]$ matrix defining how the Lagrangian components of orientation change with respect to changes in the Eulerian coordinates. The gradient can then be broken into symmetric and anti-symmetric components according to the equation

$$
\begin{aligned}
A &= \frac{1}{2}(A + A^T) + \frac{1}{2}(A - A^T) \\
&= A_{sym} + A_{anti},
\end{aligned}
\tag{20.16}
$$

where the symmetric part of the matrix, A_{sym}, contains the linear deformation components of the matrix, and the anti-symmetric part, A_{anti}, contains the rotational components. Splitting the matrix into symmetric and anti-symmetric parts allows it to intuitively represent the physical behavior of the transformation. The symmetric component of the orientation gradient, then, describes how the fibers are collectively moving towards or away from one another, as well as shearing past each other.

20.2.2.1 Geometric Simplifications of the Orientation Gradient

The S200 sample features measurements of the continuous orientation field only at the fiber positions for increments in the z-plane. The Taylor series of orientation can then be evaluated for any pair of fiber centers in the sample. Because the Taylor series is a linear approximation, only fiber centers that are near one another are used in calculations. For computational ease, differences in position and orientation are evaluated for fiber pairings within a single plane.

Measurement of the gradient is made in two steps. First, the linear approximation of fiber orientation is calculated, as in Section 20.2.1.2. The only change between fibers, with constant $\theta = 45°$, is $\cos\phi$ and $\sin\phi$, which can be approximated using Equation (20.12). Second, the linear approximation of the gradient is computed using measured values of position and the calculated values of linear orientation. Here, the vector for orientation is represented,

$$
\frac{\partial \mathbf{r}}{\partial \tau} = \begin{bmatrix} \frac{1}{\sqrt{2}}\frac{x}{z} \\ \frac{1}{\sqrt{2}}\frac{y}{z} \\ \frac{1}{\sqrt{2}} \end{bmatrix} = \begin{bmatrix} V_x \\ V_y \\ V_z \end{bmatrix},
\tag{20.17}
$$

to simplify the expression of following equations. Utilizing the substitution from Equation (20.17), the left-hand side of Equation (20.14) can be replaced, yielding

$$
\begin{bmatrix} V_{x_i} - V_{x_j} \\ V_{y_i} - V_{y_j} \\ V_{z_i} - V_{z_j} \end{bmatrix} = \nabla \frac{\partial \mathbf{r}}{\partial \tau} \begin{bmatrix} x_i - x_j \\ y_i - y_j \\ z_i - z_j \end{bmatrix},
\tag{20.18}
$$

where index, i, denotes evaluation at a reference fiber, and j, evaluation at some nearby fiber center in the same plane.

As previously stated, for the sample presented here, the zenith angle is assumed to be the same for all fibers, $\theta = 45°$. This results in $V_{z_i} = V_{z_j}$, causing the third row of $\Delta \partial \mathbf{r}/\partial \tau$ to be zero. Evaluating the orientation gradient in a single plane results in all measurements in a section having the same z, causing the third row of $\Delta \mathbf{r}$ to also reduce to zero. Computation of the gradient can be expanded to include differences between fibers in different slices resulting in a 3×3 gradient. Making all computations within a single slice reduces the computations to dealing with only two dimensions, though it makes the computed gradient noisier. The 3×3 representation is more accurate because it incorporates changing orientation in all directions rather than solely in plane. Because the z component, V_z, of orientation is constant, the derivative is zero, making the third row of the gradient matrix zero. This leads to the reduced expression,

$$
\begin{bmatrix} V_{x_i} - V_{x_j} \\ V_{y_i} - V_{y_j} \\ 0 \end{bmatrix} = \begin{bmatrix} \frac{\partial}{\partial x}(V_{x_i}) & \frac{\partial}{\partial y}(V_{x_i}) & \frac{\partial}{\partial z}(V_{x_i}) \\ \frac{\partial}{\partial x}(V_{y_i}) & \frac{\partial}{\partial y}(V_{y_i}) & \frac{\partial}{\partial z}(V_{y_i}) \\ 0 & 0 & 0 \end{bmatrix} \begin{bmatrix} x_i - x_j \\ y_i - y_j \\ 0 \end{bmatrix},
\tag{20.19}
$$

where, upon multiplication, the third column of the gradient disappears due to the zero in $\Delta \mathbf{r}$. The result is in an effective 2D orientation gradient Taylor series,

$$
\begin{bmatrix} V_{x_i} - V_{x_j} \\ V_{y_i} - V_{y_j} \end{bmatrix} = \begin{bmatrix} \frac{\partial}{\partial x}(V_{x_i}) & \frac{\partial}{\partial y}(V_{x_i}) \\ \frac{\partial}{\partial x}(V_{y_i}) & \frac{\partial}{\partial y}(V_{x_i}) \end{bmatrix} \begin{bmatrix} x_i - x_j \\ x_i - y_j \end{bmatrix},
\tag{20.20}
$$

that is used to calculate the in-plane orientation gradient for fibers.

The gradient can be calculated by replacing the values of $\Delta \mathbf{r}$ and $\Delta \partial \mathbf{r}/\partial \tau$ with known data associated with a fiber and its neighbor. Calculation of the gradient based on a single neighbor, however, is very noisy because fibers provide sparse sampling of the orientation field, and the Taylor series assumes small changes in location. Computation of the gradient is smoothed by utilizing a neighborhood of fibers surrounding a given fiber center in the computation.

20.2.2.2   Color Visualization of the Orientation Gradient

The orientation gradient is visualized according to the eigenvectors and eigenvalues of the symmetric portion of the gradient, $\hat{\mathbf{G}}_{sym}$. The gradient field is defined at all points, with fiber positions representing a neighborhood calculation. To capture the idea of neighborhood calculations, the gradient is visualized with a colored Voronoi diagram. Each cell in the Voronoi is centered on a fiber and assigned a hue based on the direction of the symmetric gradient's primary eigenvector. The saturation of each cell is determined by the difference in eigenvalues of $\hat{\mathbf{G}}_{sym}$ and value by the sum of eigenvalues.

The fibers within a tow are ideally parallel with one another and move past each other along smooth boundaries. The expected result is a jump discontinuity of shear between tows, and gradients of zero within tows. Because the gradient is calculated for a finite neighborhood size, the gradient

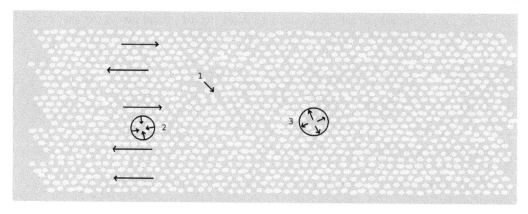

Figure 20.6: Synthetic data used for visualization of known microstructure. Anomaly 1 is a fiber of cross-tow orientation. Anomaly 2 features a region of uniform fiber orientation towards a center point (contraction). Anomaly 3 features uniform fiber orientation away from center (expansion).

should smoothly disappear with distance from a tow boundary. Testing these simplified expectations is achieved with the synthetic data (phantom) shown in Figure 20.6. A single fiber is replicated and pasted to a hexagonally packed grid to form a tow. Similar tows are placed one on top of another, and a uniform horizontal velocity is then assigned to all fibers in a tow, with neighboring tows receiving equal and opposite velocities. Horizontal arrows are placed in the center of each tow denoting direction of movement, while other arrows are used to identify inserted anomalies. A single fiber is given a cross-tow velocity, denoted by a diagonal arrow, and two neighborhoods are subjected to uniform contraction and expansion, denoted by circles and arrows. The cross-tow fiber displaces fibers in the immediate vicinity, resulting in neighborhood expansion preceding the fiber and contraction following the fiber. Figure 20.7 shows the colored Voronoi tesselation for the image shown in Figure 20.6. Figure 20.8 shows the velocity gradient for the same portion of the real sample as Figure 20.5(b). Again, tow boundaries follow expected trends, though horizontal tow edges feature large neighborhoods of expansion and contraction.

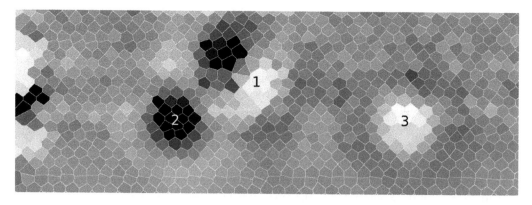

Figure 20.7: Orientation gradient representation of the phantom data.

20.2.3   Estimation of the Orientation Gradient Field

The 2D Taylor series from Equation (20.20) is computed for every fiber, $i = 1, 2, 3 \ldots I$. Noise reduction can be approached through the addition of neighboring points to make an over-determined set of equations. A neighborhood is defined around each fiber, i, to include every fiber within a given

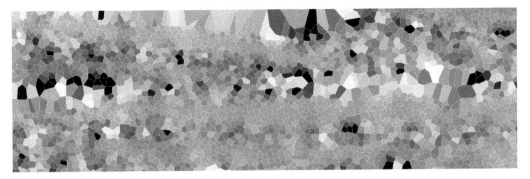

Figure 20.8: Orientation gradient representation of the real microstructure presented in 20.5(b).

radius. Each fiber has a different number of neighbors, indexed by $j = 1, 2, 3 \ldots J_i$, where J_i is the number of neighbors of i. Equation (20.20) can be written as a transpose and expanded to include all neighboring fibers,

$$
\begin{bmatrix}
\Delta V_x(i,1) & \Delta V_y(i,1) \\
\Delta V_x(i,2) & \Delta V_y(i,2) \\
\vdots & \vdots \\
\Delta V_x(i,J_i) & \Delta V_y(i,J_i)
\end{bmatrix}
=
\begin{bmatrix}
\Delta x(i,1) & \Delta y(i,1) \\
\Delta x(i,2) & \Delta y(i,2) \\
\vdots & \vdots \\
\Delta x(i,J_i) & \Delta y(i,J_i)
\end{bmatrix}
\begin{bmatrix}
\frac{\partial V_{x_i}}{\partial x} & \frac{\partial V_{y_i}}{\partial x} \\
\frac{\partial V_{x_i}}{\partial y} & \frac{\partial V_{y_i}}{\partial y}
\end{bmatrix}
$$
$$
\mathbf{B}_i = \mathbf{A}_i \mathbf{G}_i, \tag{20.21}
$$

where $\Delta V_x(i,j) = V_{x_i} - V_{x_j}$ and $\Delta V_y(i,j) = V_{y_i} - V_{y_j}$. Parameter $\mathbf{B}_i$ is the matrix of in-plane orientation change between all $j = 1, 2, 3 \ldots J_i$ neighbors of i and fiber i. The parameter $\mathbf{A}_i$ is the change in position matrix between fiber i and its J_i neighbors, and $\mathbf{G}_i$ is the orientation gradient at fiber i. Both $\hat{\mathbf{A}}_i$ and $\hat{\mathbf{B}}_i$ are known and can be solved for $\hat{\mathbf{G}}_i$ as an overdetermined set of linear equations via least squares, $\hat{\mathbf{G}}_i = (\hat{\mathbf{A}}_i^T \hat{\mathbf{A}}_i)^{-1} \hat{\mathbf{A}}_i^T \hat{\mathbf{B}}_i$. The gradient then, contains information on how the orientation is changing with position in a single (x,y)-plane. The gradient, $\hat{\mathbf{G}}_i$, can then be split into its rotational and linear components according to Equation (20.16). In planar laminates, the rotational component of the gradient is assumed to be minimal, thus the symmetric gradient, $\hat{\mathbf{G}}_{sym}$, is the focus of further analysis. More complex layup geometries may require the analysis of anti-symmetric components of the gradient alongside the symmetric gradient. The symmetric gradient itself represents a combination of possible scaling or shear in the x and y directions. In the case of pure shear, the transformation matrix in

$$
\begin{bmatrix} \hat{x} \\ \hat{y} \end{bmatrix}
=
\begin{bmatrix} a & s_1 \\ s_2 & b \end{bmatrix}
\begin{bmatrix} x \\ y \end{bmatrix}, \tag{20.22}
$$

exhibits a diagonal equal to one and s_1 and s_2 are equal and non-zero, where the axis of shear is always an eigenvector of the matrix. For bidirectional shear, the primary shear direction corresponds to the eigenvector associated with the larger eigenvalue, and the other eigenvector corresponds to the secondary shear direction. While shear lies on the off-diagonal of the transformation matrix and directly correlates to eigenvectors, expansion and contraction are represented by the magnitudes a and b in the diagonal. In the case of expansion or contraction in a given direction, the corresponding coefficient is greater than or less than one, respectively, and the eigenvectors lie along the axes of the original coordinate system. In the event of multiple factors, such as often encountered in real samples, the calculated eigenvectors are a combination of bidirectional shear and scaling.

20.3   Anomaly Detection

To utilize the methods presented in Section 20.2, measurements of any field can be represented as a feature vector

$$\mathbf{S}_i = [S_{k_1}, S_{k_2}, \ldots, S_K]^T, \tag{20.23}$$

where $S_{k_1}, S_{k_2}, \ldots, S_K$ can be any measurable features. Most anomaly detection problems can be approached as a data classification problem where the data can be separated into one of two classes: normal or anomalous. For all data classification methods, a model for each of the classes is optimized to produce the minimum number of expected classification errors. Typically, the model is generated based on a pre-divided set of training data and then applied to remaining unknown data to make classifications.

In the case of anomaly detection, particularly in materials science, anomalies are not known and cannot be labeled prior to modeling, thus unsupervised approaches must be considered. Unsupervised classifiers seek to fit a model to the bulk or background of the data, and to use the background model to identify anomalies. An assumption of this method is that the anomalies represent a negligible portion of the data, and that the resulting model is a fit of only the background (normal) behavior. This assumption holds true for parametric classifiers and large sample sizes, though non-parametric classifiers often incorporate anomalous behavior into the model and changing the anomaly classification boundary. This is due to the fact that non-parametric models such as Parzen density estimates and k-nearest neighbor estimates feature local probability density measures that are easily influenced by clusters of anomalies [339]. For this reason, non-parametric methods are not considered for modeling the local microstructure. Nonetheless, even with parametric models, if the anomaly contamination is too high when creating the model, any abnormal behavior will be included in the normal model and not be classified as an anomaly. For the S200 sample presented here, the data clusters densely into separate tows indicating the prevalence of background data. Because of the unknown nature of the microstructure, it is unknown what the anomaly to normal ratio is, though tight clustering suggests high prevalence of normal data.

20.3.1   Gaussian Mixture Modeling

A common simplifying assumption in modeling is that the underlying distribution is Gaussian. While this can be hard to verify, though due to the presence of multiple fiber tows in the sample

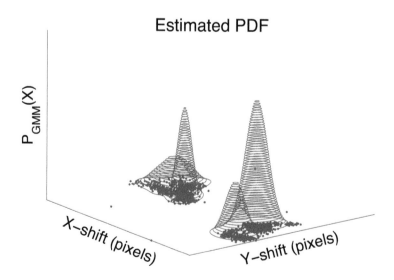

Figure 20.9: Gaussian mixture model probability density function of the fiber orientation field.

resulting in a multi-modal distribution, a mixture of Gaussians was necessary to model data accurately. In the case of all fiber tows being perfectly aligned in the weave plane, two Gaussians would be sufficient for modeling. While this condition is almost met, there is enough out of plane misalignment between tows to merit additional Gaussian components. In practice, one Gaussian per tow is sufficient for accurate models. The Gaussian mixture model (GMM) is defined as

$$p(\mathbf{S}) = \sum_{m=1}^{M} w_m \mathcal{N}_m(\mathbf{S}; \boldsymbol{\mu}_m, \boldsymbol{\Sigma}_m),$$ (20.24)

where $p(\mathbf{S})$ is the probability density (pdf) function of the modeled feature vector. Estimating the total number of components, M can be computationally expensive and, as previously stated, each tow can be modeled with a single Gaussian. The parameter w_m is the weight associated with each Gaussian, and

$$\mathcal{N}_m(\mathbf{S}; \boldsymbol{\mu}_m, \boldsymbol{\Sigma}_m) = (2\pi)^{-k/2} |\boldsymbol{\Sigma}_m|^{-1/2} \exp\left[-\frac{1}{2}(\mathbf{S} - \boldsymbol{\mu}_m)^T \boldsymbol{\Sigma}_m^{-1} (\mathbf{S} - \boldsymbol{\mu}_m)\right],$$ (20.25)

refers to each Gaussian component. The parameters $\boldsymbol{\mu}_m \in \mathbb{R}^K$ and $\boldsymbol{\Sigma}_m \in \mathbb{R}^{K \times K}$ are the mean and covariance of the mth component, respectively, and K is the length of the feature vector. Estimation of the model is accomplished via the expectation maximization (EM) algorithm [219]. Anomalies are then classified based on the probability density function (PDF) calculated by the EM algorithm. Figure 20.9 shows an example of a Gaussian mixture model fit to a feature vector $\mathbf{S} = [V_x(x,y), V_y(x,y)]^T$. The height at a given point specifies its likelihood of occurrence. An arbitrary threshold can be chosen, and adjusted to yield the desired number of anomalies. A given likelihood threshold, however, lacks physical meaning. Instead a probability bound is chosen to capture a certain percentage (98%) of the normal data. A fine grid is created and the likelihood values calculated at all points. Iterative integration is then applied to find the likelihood threshold that captures the appropriate percentage of the pdf. Each sample in the real feature vector is then evaluated for its likelihood and compared to the threshold. A sample is declared anomalous if its likelihood is less than the threshold.

20.3.2 Anomalies of the Microstructure

Results of the GMM anomaly detection are displayed by numbering points atop the field visualization that have been labeled anomalous. Figure 20.10(a) shows the anomaly detection applied to the orientation field, or feature vector $\mathbf{S} = [V_x, V_y]$. The fibers found by the Gaussian mixture model correspond well with the anomalies inserted into the phantom.

Figure 20.10(b) shows anomalies based on the feature vector $\mathbf{S} = [\lambda_1 - \lambda_2, \lambda_1 + \lambda_2]$ where λ_1 and λ_2 are the eigenvalues of the symmetric gradient. Anomalies inserted into the phantom are correctly classified by the Gaussian model. Additionally, anomalies corresponding to neighborhood behavior, such as expansion and contraction, are identified more accurately by means of the velocity gradient than the velocity field. Noise is more prevalent in the gradient field due to the neighborhood computation based on a relatively small number of samples. The noisy gradient field results in classifications of point anomalies where no anomaly exists in the phantom. The number of point anomalies can be reduced by increasing the number of neighborhood fibers used in computation, or by introducing a more restrictive threshold to the anomaly detector.

True anomalies are present in multiple layers due to the rigidity of the fibers. To emphasize the consistency of anomalies, anomalies are clustered based on average neighborhood distance, and colored according to the number of layers in which they are present. Blue signifies the presence of an anomaly while red indicates high consistency through multiple layers. Figure 20.11 shows the grouped anomalies for the gradient field. The grouped anomaly figure exhibits three main anomaly groupings. Anomalies within the real microstructure center on regions where neighboring tows separate due to the weave. Often the separation of tows as a result of weave architecture results in

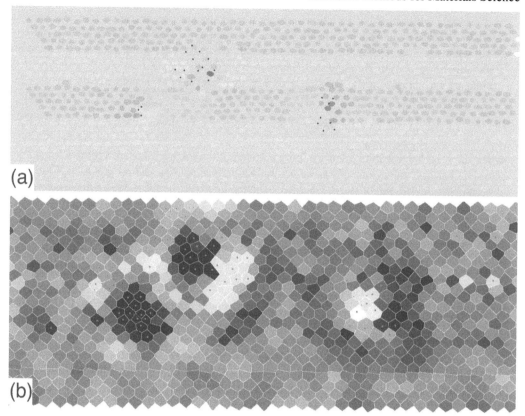

Figure 20.10: Anomalies of the contrived (phantom) microstructure. (a) Marked anomalies of the orientation field. (b) Marked anomalies of the orientation gradient.

Figure 20.11: Anomaly groupings based on neighborhood and depth consistency. Color saturation between tow boundaries indicates high shear in the gradient, while light or dark regions indicate expansion and contraction, respectively. Colors corresponding to the legend indicate presence of an anomaly through multiple layers.

matrix dense regions which open into a pore. Figure 20.11 features an anomaly on the right side where the end of a tow has created space for a pore. The anomaly visualization indicates high shear between the tows, as well as expansion as the edge of the tow recedes, leaving the open space of the pore. In the case of the anomaly grouping on the left-center of the image, a matrix dense region is seen preceding the presence of a pore.

Before grouped anomalies can be confidently called real microstructural anomalies, the fiber extraction and tracking must be examined for all layers contributing to the anomaly classification. Fiber location errors, such as fiber labels swapping during tracking, result in abnormal fiber tracks that the detector classifies as anomalies. While these false fiber tracks are indeed anomalies, they do

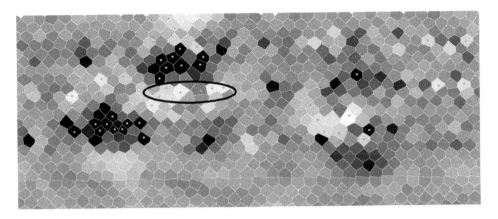

Figure 20.12: Anomalies (marked as red dots) of the orientation gradient of the phantom. Anomalies within the ellipse consist of both high neighborhood shear and expansion.

not correspond to anything real in the microstructure and should be eliminated from consideration in material property correlation.

After verifying the presence of a true anomaly, it is helpful in understanding the microstructure to know why a region was classified as an anomaly. This is accomplished by matching anomaly points on both the orientation gradient image as well as the plot of the feature vector **S** such as with Figures 20.12 and 20.13. Each anomaly within Figure 20.12 corresponds to a point in Figure 20.13. It can be seen that the anomalies lying below zero on the vertical axis correspond to contraction anomalies, and that those above zero are expansion anomalies. Anomalies from the cross-tow fiber exhibit expansion in front of the fiber and contraction behind, though some of the numbered anomalies can also be attributed to high magnitude of shear. Pairing the anomaly plot with the visualization of the orientation gradient field enables clear determination of why points are anomalous.

It can be seen in Figures 20.12 and 20.13 that the anomalies can be split into three general regions denoted by the dashed lines. Anomalies that lie above the bulk of the data (horizontal line) can be attributed to expansion. The points corresponding to the compression anomaly can be seen below the horizontal axis of the plot, as expected for compression. The stray fiber features expansion anomalies as neighboring fibers open to accommodate the fiber and compression anomalies as they close in behind the fiber. Though the indicated points are not particularly close to one another with respect to shear, they all exhibit similar expansion. Interestingly, the stray fiber also results in a number of fibers with high shear that lie in the rightmost region in the anomaly plot.

Similar regions can be defined for the S200 sample to aid in understanding the behavior of the local microstructure. Figures 20.14 and 20.15 show the anomalies of the orientation gradient on the gradient image and feature vector plot. The circled region in the orientation gradient image corresponds to the tightly clustered points circled in the anomaly plot. This region of contraction corresponds to the edge of a receding tow, leaving a pore between neighboring tows.

Anomalies based on the velocity gradient are valuable for their potential in predicting the emergence of certain microstuctural features. Due to the stiffness (local linearity) of the fibers, anomalies of the orientation gradient highlight emergence of either dense or sparse microstructures before they are within the field of view. Because destructive techniques are used during imaging, it is beneficial for experimentation purposes to know a defect is present before it is destroyed. Upon detection of the suspected defect, destructive imaging can be halted for experimentation to see if damage corresponds to anomalies of the orientation gradient.

20.4   Conclusion

Fibers within a composite microstructure drastically affect the properties of the material, making them a key component in accurate characterization. Because the fibers are constrained by one another prior to matrix infiltration, fibers tend to feature similar orientation within individual fiber tows. The fiber orientation field was developed analogous to a fluid velocity field where fibers were homologous to fluid streamers. The resulting representation of embedded fibers was used to develop the gradient of the orientation field, characterizing the collective behavior of small fiber neighborhoods. The orientation gradient field was shown to be related to the expansion/compression and skew distortion of the fiber neighborhood with changes in sample depth (slice number).

Visualization of both orientation and orientation gradient fields was achieved via the HSV colormodel, accurately representing the behavior of the microstructure with a single snapshot. While noise is present due to the discrete nature of the collected data, fiber neighborhoods generally exhibit similar behavior and large changes that correspond with the underlying microstructure.

Anomaly detection was developed and performed via Gaussian mixture modeling on the orientation gradient. The detected anomalies pinpoint areas theorized to be of interest in determining overall material strength. The anomalies identified by the Gaussian mixture model were verified with synthetic data to correspond to regions of high neighborhood expansion/contraction and skew. Within the true data, the anomalous regions were found to lie along fiber tow boundaries and commonly near the emergence of pores.

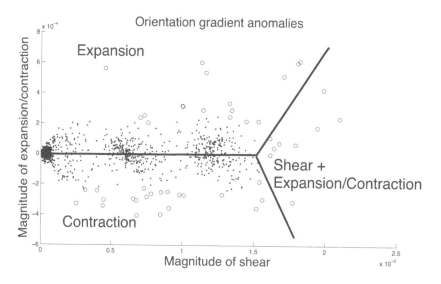

Figure 20.13: Anomalies (marked as circles) corresponding to the anomalies of the orientation gradient. Anomalies within the ellipse from Figure 20.12 correspond to circles in the shear+expansion/contraction region.

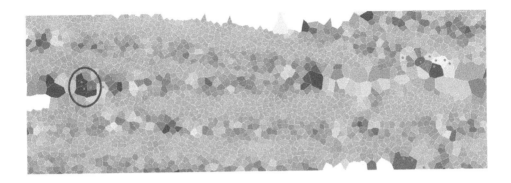

Figure 20.14: Anomalies displayed on the orientation gradient of the S200 sample. Anomalies within the circled region correspond to neighborhood contraction near emergence of a tow.

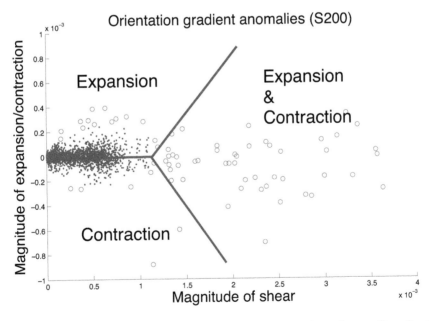

Figure 20.15: Anomalies corresponding to the anomalies of the orientation gradient for the S200 sample. Dark-colored anomalies in Figure 20.14 correspond to circles within the labeled contraction region.

Part VII

Sparse Methods

Sparsity was described in qualitative detail in the Introduction to this book and more quantitatively described in Chapter 5, in a general sense. This chapter gives some more specific applications of sparse methods that are currently in development. Research involving compressed sensing (CS) is probably the largest single application of sparse methods, but there are many other opportunities for this class of analysis in imaging work.

We chose to lead off this part of the book with a chapter by Willett, which gives an overview with applications of sparse methods for denoising applications. Essentially, these methods take advantage of the fact that features in an image are repeated many, many times (e.g., an interface has local characteristics that are repeated multiple times throughout an image). This allows a model of the denoised image to be constructed. Romberg then introduces the topic of CS in a broad way with some very specific applications of the concepts towards real applications. These methods rely on the fact that unsampled parts of the image may be reconstructed from a knowledge of what has been sampled. Nadakuditi et al. describe dictionary methods, a current method for implementing sparse algorithms, and outline how this may be used in microscopy. Finally, Larson et al. describe their current efforts in developing sparse sampling methods for electron microscopy.

Chapter 21

Denoising Methods with Applications to Microscopy

Rebecca Willett
University of Chicago

21.1 Introduction

Materials scientists and engineers collect a plethora of data they use to analyze different material properties and synthesis methods. Tools such as scanning electron microscopy (SEM), transmission electron microscopy (TEM), focused ion beam (FIB) microscopy, electron backscatter diffraction (EBSD) microscopy, and others all provide materials scientists with physical measurements critical to inferring material structure and properties. However, this data is typically corrupted by unknown errors, referred to as noise, that can confound the inference process. Broadly speaking, *denoising* is the process of removing these errors and estimating the "true" underlying intensity image or spectrum image. In this chapter, we examine different classes of methods used to remove noise from two-dimensional images and three-dimensional spectral images. In particular, we explore the landscape of popular and state-of-the-art denoising methods, examine their mathematical foundations, and provide insight into the relative advantages and disadvantages of various methods.

When the amount of noise is low (as in Figure 21.1(b)), we may be inclined to forego denoising and rely instead upon a visual inspection of raw data; after all, the human visual system is adept at inferring structure from noisy images. However, even small amounts of noise can result in significant errors when solving inverse problems (as discussed in Chapters 23 and 22). The insights and models used in denoising methods often provide the foundation for stable inverse problem solvers; hence understanding the fundamentals of image denoising can lead to more accurate solutions to other image analysis problems.

When the amount of noise is high (as in Figure 21.1(c) or Figure 21.1(d)), denoising can significantly improve our ability to interpret images. Many automated and quantitative image analysis tools are quite sensitive to noise. For more quantitative analyses it is often necessary to remove the noise before subsequent processing because noise impedes segmentation, estimating lengths, areas, or volumes of structures, and pattern recognition. For instance, imagine we wish to determine whether a doping compound bonded as expected within a synthesized material. "Eyeballing" noisy data leads to subjective and unreliable inference; in contrast, the denoising process exploits prior knowledge of material structure, image properties, and noise statistics that lead to more objective and reliable inference.

To keep the discussion as general as possible, we do not differentiate among two-dimensional images, three-dimensional spectral images, or other generalized notions of "images." Rather, we refer to all of these datasets as "images" and let N denote the total number of samples, pixels, or voxels in the image.[1] Our interest is in a true intensity image, denoted x, which we represent as a vector of N real numbers (i.e., $x \in \mathbb{R}^N$). The ith pixel of an image x is denoted x_i. The image

---

[1]For simplicity of presentation, we use the word "pixel" in a general sense throughout this chapter, so that it could refer to a voxel or other sample depending on the context.

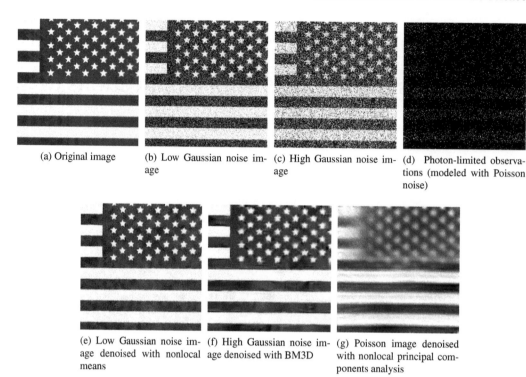

Figure 21.1: Example image and low-noise and high-noise Gaussian observations and Poisson observations.

x typically corresponds to a physical property of interest; the quantity x_i could correspond to, for instance, the reflectivity, transmittance, or fluorescence of a sample at location i.

The central challenge of image denoising is that we are unable to observe x directly; rather, for each pixel i we observe

$$y_i = x_i + \varepsilon_i,$$

where $\varepsilon_i \triangleq y_i - x_i$ is the noise or error in the measurement; we write $y = x + \varepsilon$ (for $y, \varepsilon \in \mathbb{R}^N$) to represent our observation model for all pixels in an image simultaneously. **Our goal is to estimate x as accurately as possible given the noisy observations y.** Denoised examples of the noisy images in Figure 21.1(b-d) are displayed in Figure 21.1(e-g); this chapter describes these and other denoising methods. Throughout the text, we measure the accuracy of the denoised image $\hat{x}$ using the "mean squared error" (MSE), defined as

$$\text{MSE}(\hat{x}, x) \triangleq \frac{1}{N} \sum_{i=1}^{N} (\hat{x}_i - x_i)^2. \tag{21.1}$$

In general, the estimation of x is only feasible if we place modeling assumptions of x and ε that allow the two to be distinguished. Often, these assumptions amount to stochastic models for the noise vector ε (e.g., corresponding to Gaussian or Poisson noise) and more geometric models for the image x (e.g., corresponding to smoothness or redundancy). In this chapter, we elaborate further on those models and the methods that leverage those models. We examine major classes of models and methods that are at the heart of many popular and effective image processing tools; however, our list is not exhaustive. Rather than encompassing all image denoising methods, the aim of this chapter is to review key concepts that are ubiquitous in much of image denoising.

21.1.1 Organization of Chapter

This chapter is organized as follows. Section 21.2 describes common models of image structure or geometry and sources of noise and their associated statistical models. Section 21.3 describes a widely-used class of denoising methods based on penalizing or regularizing a maximum likelihood estimator; examples of this approach are the widely-used wavelet thresholding and total variation estimators. We follow this with a discussion of both linear and adaptive kernel-based methods in Section 21.4, including Gaussian smoothing, the bilateral filter, and non-local means. At the heart of non-local means lie small, localized "patches" of image pixels; other patch-based methods, such as BM3D and non-local principal components analysis (PCA), are described in Section 21.5. Finally, Section 21.6 showcases several examples of image denoising in action, both for simulated and real materials science data.

21.2 Image and Noise Models

This section details common models for image geometry and structure and statistical models for image noise. These models are essential to understanding key concepts underlying modern image denoising methods.

21.2.1 Noise Models

Noise in microscopy data and other digital imagery typically arises during the image acquisition process. Many images of interest in materials science are acquired using photodetectors (e.g., charge-coupled device (CCD) arrays or photon-counting photodiodes) that measure the intensity of reflected, emitted, or transmitted light across a material. These intensity measurements can be corrupted by a number of natural processes [63], including

- **thermal noise (also known as Johnson/Nyquist noise),** caused by the thermal motion of charge carriers in the photodetector circuitry,
- **black body radiation** from the Earth and Sun,
- **shot noise,** caused by the random arrivals of discrete electrons or photons at the photodetectors and associated with "counting statistics,"
- **generation-recombination noise,** caused by thermal fluctuations in semiconductor carrier densities.

In general we consider noise as *unpredictable* measurement errors; in contrast, systematic measurement errors are predictable and can easily be subtracted from data. Because of the unpredictable nature of noise, we often consider it as arising from a stochastic model. The Gaussian and Poisson stochastic models capture the majority of noise present in microscopy data.

21.2.2 Gaussian Noise Model

Gaussian noise is by far the most commonly used noise model across application domains. There are two key reasons for this: (a) it is an accurate model of noise measurements collected from a number of physical sensors; (b) it facilitates simple mathematical analysis that lends itself to computationally fast algorithms. Under the Gaussian noise model, we assume that all N of the ε_i's are independent and identically distributed according to the Gaussian distribution

$$p(\varepsilon_i) = \frac{1}{\sqrt{2\pi\sigma^2}} e^{-\frac{\varepsilon_i^2}{2\sigma^2}},$$

where $\sigma^2 > 0$ is the variance of the noise. Given an image estimate $\hat{x}$, we can compute the negative log likelihood of the measurements y under this model as

$$-\log p(y|\hat{x}) = \sum_{i=1}^{N} \left[\frac{1}{2} \log(2\pi\sigma^2) + \frac{1}{2\sigma^2}(y_i - \hat{x}_i)^2 \right] \propto \|y - \hat{x}\|_2^2,$$

where $\|a\|_2^2 \triangleq \sum_{i=1}^{N} a_i^2$ is the squared Euclidean norm. Seeking an estimate of x which minimizes the negative log likelihood of y under the Gaussian noise model is equivalent to finding an estimate $\hat{x}$ which minimizes the sum of squared residuals (i.e., the least-squares fit). Generally, this estimate must be constrained or regularized (as described later in the chapter) to yield accurate estimates.

21.2.3 Poisson Noise Model

Shot noise, also referred to as photon noise or Poisson noise, arises in low-light settings in which we only observe a small number of photons for each pixel in an image; these models are often evaluated using counting statistics. This phenomenon occurs, for instance, if photons are emitted from a sample after its atoms are excited by an electron beam. In many settings using a high electron beam power to produce more photons can modify the sample, so low-power beams are used and the resulting photon counts are small. Spectral imaging can also be contaminated by photon noise when a fixed number of measured photons are divided among a large number of spectral bands.

To better understand photon noise, imagine focusing a microscope on a point in a sample and observing photons transmitted from that location after it is probed with an electron beam. If we could collect photons for an infinite amount of time, we might observe on average, say, 3.58 photons per second. In any given second, however, we receive a random quantity of photons, say 3 or 10. The number of photons observed in a given second is a random quantity well-modeled by the Poisson distribution. If we were to double the power of the electron beam used to probe our sample, then the average number of observed photons would also double; more generally, the photon arrival rate is proportional to the electron beam power. The signal-to-noise ratio (SNR) associated with Poisson noise is the square root of the average photon arrival rate (e.g., $\sqrt{3.58}$), so we might be tempted to increase the SNR of our data by simply increasing the power of our electron beam. However, high-power electron beams can damage samples, so we are often forced to cope with low SNR data.

We would like to estimate this average photon arrival rate based on the number of photons collected during a single second (or other integration time) for each pixel in a scene. Let T correspond to the overall "intensity" of our measurements; for instance, in a transmission electron microscopy system, T would correspond to the instantaneous electron beam power integrated over the entire sensing time. We may model the observed photon counts as follows:

$$y \sim \text{Poisson}(Tf) \qquad \text{or} \qquad y_i \sim \text{Poisson}(Tx_i) \,\, \forall i. \tag{21.2}$$

That is, y_i is the observed number of photons at detector element i, and Tx_i is the average photon arrival rate associated with pixel i in the scene. In general, small T corresponds to the so-called "photon-limited regime" and is very common in materials applications where large electron beam power can lead to damaged samples and long dwell or sensing times can lead to motion artifacts from heating or vibration effects.

The Poisson distribution has several features that distinguish it from the Gaussian distribution. First, the mean value of each pixel (e.g., Tx_i in (21.2)) is equal to the variance, as illustrated in Figure 21.2. In fact, for very large T we can approximate the distribution of y_i as a Gaussian with mean Tx_i and variance Tx_i – though this approximation is highly inaccurate for small T. Thus, unlike the Gaussian noise setting described above, the Poisson noise setting models image-dependent noise with variance that changes across the image. Furthermore, an arbitrary weighted combination of Poisson random variables is not itself Poisson; this is in contrast to Gaussian random variables, where arbitrary weighted combinations are also Gaussian. This perhaps subtle mathematical fact

(a) Photon-limited snapshot of ramp image

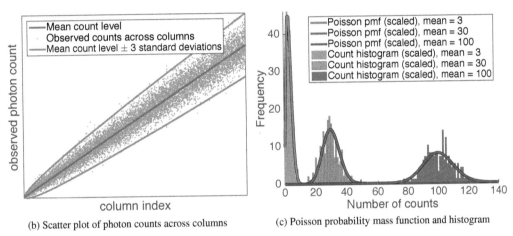

(b) Scatter plot of photon counts across columns | (c) Poisson probability mass function and histogram

Figure 21.2: Characteristics of Poisson noise. (a) Poisson observations of a ramp intensity image. (b) Scatter plot of pixel photon counts across column indicies. Poisson noise has higher variance where the underlying true intensity image is brighter. This means the noise standard deviation is a function of the mean, or that Poisson noise is image-dependent noise. (c) Poisson probability mass functions and count histograms for three different mean intensity levels, highlighting the mean-dependent variance and asymmetry.

about Gaussians underlies much image denoising theory and practice; since it does not extend to Poisson data, that theory and practice breaks down when applied to images corrupted by Poisson noise. *These are the main reasons that applying off-the-shelf image processing software designed for Gaussian noise to Poisson images yields poor estimates with undesirable artifacts.*

The Poisson model above implies that the likelihood of observing count y_i given intensity x_i is

$$p(y_i|x_i) = \frac{e^{-Tx_i}(Tx_i)^{y_i}}{y_i!} \qquad (21.3)$$

for each y_i independently. Given an image estimate $\hat{x}$, we can compute the negative log likelihood of our measurements y under this model as

$$-\log p(y|\hat{x}) \propto \sum_{i=1}^{N} [T\hat{x}_i - y_i \log(T\hat{x}_i)]. \qquad (21.4)$$

As we will see later, we can search for image estimates $\hat{x}$ which minimize this expression (while perhaps constraining $\hat{x}$ to lie in some model class). That strategy would be analogous to the least square fit associated with the Gaussian noise model, but rather than using the squared error metric, we used the negative Poisson log likelihood to explicitly account for Poisson noise. An example of a photon-limited image is depicted in Figure 21.1(d).

Variance stabilizing transforms   As mentioned above, if $y_i \sim \text{Poisson}(Tx_i)$, then the expected value $\mathbb{E}[y_i] = \text{var}(y_i) = Tx_i$, where $\mathbb{E}$ denotes expected value. Thus as the true image intensity x_i varies with location i, so does the variance of the noise. Unfortunately, many standard image denoising methods implicitly assume that all noisy pixels in the image have the same variance.

To address this challenge, one common approach to handling Poisson data is to use the Anscombe transform [27] or other variance stabilizing transforms (*cf.* [325, 336]). The basic idea is the following. Let

$$z_i \triangleq 2\sqrt{y_i + \frac{3}{8}};\tag{21.5}$$

it can be shown that z_i can be modeled approximately as a draw from a Gaussian distribution with mean

$$2\sqrt{Tx_i + \frac{3}{8}} - \frac{1}{4\sqrt{Tx_i}}$$

and variance one. The accuracy of this Gaussian approximation of the Anscombe-transformed observations is better when the number photon counts are relatively high [118, 1190]. Thus, while the original data had non-uniform variance corresponding to variations in the image x, the Anscombe-transformed data has nearly uniform variance, making it more amenable to off-the-shelf image denoising tools. Specifically, one can apply the tranformation in (21.5) to all the noisy observed pixels, denoise the z_i's to get $\hat{z}_i$'s with off-the-shelf image denoising methods (e.g., BM3D [232], Section 21.5), and then invert the transform to get an estimate $\hat{x}$, using either an algebraic inverse of $2\sqrt{Tx_i + \frac{3}{8}}$,

$$\hat{x}_i = \frac{1}{T}\left[\left(\frac{\hat{z}_i}{2}\right)^2 - \frac{3}{8}\right],$$

or a statistically unbiased inverse [623],

$$\hat{x}_i = \begin{cases} \frac{1}{4}\hat{z}_i^2 + \frac{1}{4}\sqrt{\frac{3}{2}}\hat{z}_i^{-1} - \frac{11}{8}\hat{z}_i^{-2} + \frac{5}{8}\sqrt{\frac{3}{2}}\hat{z}_i^{-3} - \frac{1}{8}, & \hat{z}_i > 2\sqrt{3/8} \\ 0, & \text{otherwise} \end{cases}.$$

However, when photon counts are very low these approaches may suffer, as shown later in this chapter. Recent work has proposed more sophisticated uses of the Anscombe transform for Poisson image denoising in a coarse-to-fine framework [38].

21.2.4   Image Models

The key to effective image denoising is that the properties of noise, especially as modeled using the Gaussian or Poisson noise distributions described above, are distinct from the properties of the underlying image of interest. This distinction is often immediate. For example, a simple model of images is that neighboring pixels are likely to be similar (i.e., the image is *smooth*, as in Figure 21.3(a)), and noise drawn from Gaussian or Poisson distributions is unlikely to exhibit this property. In the below sections, we examine Gaussian smoothing, which explicitly exploits image smoothness. At low noise levels, this approach can produce visually appealing results and is popular in many contexts.

However, the smoothness model does not capture the presence of edges, boundaries, and discontinuities that arise in many images. An alternative and more flexible model for images is *piecewise*

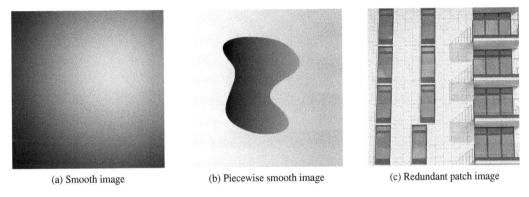

(a) Smooth image   (b) Piecewise smooth image   (c) Redundant patch image

Figure 21.3: Example images illustrating three common models of image structure and geometry.

smoothness – i.e., the image is smooth in most locations, but there is a small number of discontinuities that should be preserved by a denoising scheme, as in Figure 21.3(b). This chapter describes several methods which are designed for effective operation under this model.

The final model we consider reflects *redundancy* across an image. For instance, if we consider all $p \times p$ contiguous blocks of pixels, called patches, of an image, we might expect that collections of patches would exhibit some structure; for instance, large groups of patches may all be highly similar to one another. Commonality among patches is immediately obvious in some images (like Figure 21.3(c)) but far less apparent in more "natural" images. More general models of redundancy among patches has led to the current state-of-the-art methods for image denoising, as we see in Section 21.5. Note that smooth images and piecewise smooth images often contain redundancy among patches.

21.3 Denoising via Regularized Maximum Likelihood Estimation

A common approach to image denoising is to find an image which is a good fit to the data (as measured by the Gaussian or Poisson log likelihood) and which is *well-regularized*. Well-regularized images are typically good fits to the modeling assumptions in Section 21.2.4 – smooth, piecewise smooth, exhibit some form of redundancy, etc. A regularization function or *regularizer* $r(x)$ measures the mismatch between an image model and a candidate image estimate x. Regularizers are intimately connected to Bayesian prior probability models; in particular, given a prior probability distribution $p(x)$, we can choose the regularizer to be the negative log prior likelihood: $r(x) = -\log p(x)$.

In these settings, we seek an estimate by solving the following optimization problem:

$$\hat{x} = \operatorname*{argmin}_{\tilde{x}} \; -\log p(y|\tilde{x}) + \lambda r(\tilde{x}) \tag{21.6}$$

where $\lambda > 0$ is a tuning parameter. When $\lambda r(x)$ is a negative log prior probability, then the estimator in (21.6) is a *maximum a posteriori* (MAP) estimator. We may also impose constraints on the optimization problem, such as that all the pixel intensities are non-negative. The $-\log p(y|\tilde{x})$ term can be applied in the context of either the Gaussian or Poisson noise models, and for these models $-\log p(y|\tilde{x})$ is a convex function of $\tilde{x}$. Thus if $r(\tilde{x})$ is also a convex function, the optimization problem in (21.6) can be solved using convex solvers (*cf.* [409, 1154, 86]). The following subsections cover several common examples of different regularizers and the corresponding estimators.

21.3.1 Tikhonov Regularization

One classical regularizer is $r(x) = \|\Phi x\|_2^2$, where Φx is the result of applying a Laplacian filter to the image x, as illustrated in Figure 21.4(b). The Laplacian filter computes second-order differences

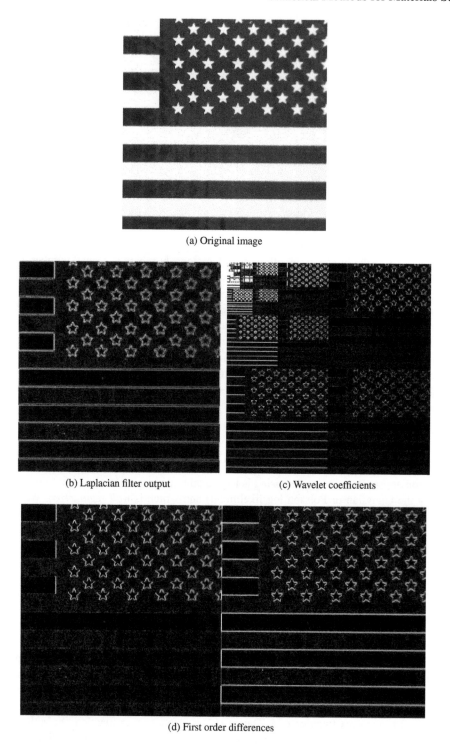

(a) Original image

(b) Laplacian filter output

(c) Wavelet coefficients

(d) First order differences

Figure 21.4: Φx for various Φ. (a) Original image x. (b) The output of a Laplacian filter applied to x, used in Tikhonov regularization. (c) Haar wavelet coefficients of x, used in sparsity regularization and wavelet denoising (see Section 21.3.2). (d) The horizontal and vertical first-order differences in x, used in total variation denoising. (b-d) Show only magnitudes and have been contrast enhanced for visibility.

among neighboring pixels in the image, returning values close to zero in nearly constant regions and returning large-magnitude values near large gradients, edges, and boundaries [373]. The regularizer $\|\Phi x\|_2^2 = \sum_{i=1}^{N} (\Phi x)_i^2$ measures the energy in the Laplacian of x and hence is a measure of the roughness of x. This approach is typically referred to as *Tikhonov regularization*. In the case of Gaussian noise with variance σ^2, the optimization problem in (21.6) amounts to

$$\hat{x} = \underset{\tilde{x}}{\text{argmin}} \, \|y - \tilde{x}\|_2^2 + \lambda \sigma^2 \|\Phi \tilde{x}\|_2^2$$

$$= (I + \lambda \sigma^2 \Phi^\top \Phi)^{-1} y,$$

which can be computed in closed-form without iterative optimization tools. Empirically, however, these estimators often either blur edges (for large λ) or exhibit noise artifacts (for small λ), and hence are often outperformed by the sparsity regularized and total-variation estimators described below.

21.3.2   *Sparsity and Wavelet Denoising*

The regularizer $r(x) = \|\Phi x\|_1 = \sum_{i=1}^{N} |(\Phi x)_i|$ can be used to measure the sum of the absolute values of the coefficients of x in some orthonormal basis $\Phi^\top$. The idea generalizes beyond bases to frames and redundant dictionaries of image features. The sum of absolute values used here (as opposed to the sum of squares used in Tikhonov regularization) promotes an estimate $\hat{x}$ with a sparse set of coefficients, in which most basis coefficients or elements of Φx are zero-valued; hence, this approach is typically referred to as *sparsity regularization*.

For example, consider the wavelet transform coefficients of an image x. Wavelets form an orthonormal basis for images, with different wavelet basis functions capturing features of images at different scales and locations. The wavelet coefficients of the flag image have been arranged in an image and are shown in Figure 21.4(c). In that figure, wavelet coefficients corresponding to the finest level of detail are displayed in the lower left, lower right, and upper right quadrants, representing vertical, horizontal, and diagonal edge information, respectively. The upper left quadrant can again be split into four quadrants, where the coefficients corresponding to the second-finest level of detail are displayed in the lower left, lower right, and upper right quadrants. This pattern is repeated for about $\log N$ levels of detail, where N is the number of pixels in the image. An accessible introduction to wavelets is in [145]; for the purposes of this chapter, note that wavelet coefficients have many zero-valued elements for images that have many smooth surfaces and boundaries. Therefore, wavelet coefficients of images can exhibit the sparsity promoted by this regularizer. Wavelets are frequently used in this context, resulting in piecewise smooth image estimates.

In the case of Gaussian noise with variance σ^2, the optimization problem in (21.6) amounts to

$$\hat{x} = \underset{\tilde{x}}{\text{argmin}} \, \|y - \tilde{x}\|_2^2 + \lambda \sigma^2 \|\Phi \tilde{x}\|_1. \tag{21.7}$$

This optimization problem can be solved in closed form. Let $\theta = \Phi y$ be the noisy coefficients of the observed image y. Then let

$$\hat{\theta} = \underset{\tilde{\theta}}{\text{argmin}} \, \|\theta - \tilde{\theta}\|_2^2 + \lambda \sigma^2 \|\tilde{\theta}\|_1$$

be the denoised basis coefficients; we can compute each element of $\hat{\theta}$ independently; for $i = 1, \ldots, N$,

$$\hat{\theta}_i = \underset{\tilde{\theta}_i}{\text{argmin}} (\theta_i - \tilde{\theta}_i)^2 + \lambda \sigma^2 |\tilde{\theta}_i| = \text{SoftThreshold}(\theta_i, \lambda \sigma^2) \triangleq \text{sign}(\theta_i)(|\theta_i| - \lambda \sigma^2)_+ \tag{21.8}$$

where

$$\text{sign}(a) = \begin{cases} -1, & a < 0 \\ 1, & \text{otherwise} \end{cases} \quad \text{and} \quad (a)_+ = \begin{cases} a, & a > 0 \\ 0, & \text{otherwise} \end{cases}.$$

Given the denoised basis coefficients, the denoised image can be computed as

$$\hat{x} = \Phi^\top \hat{\theta}.$$

Note that this approach is only viable when Φ is an orthonormal matrix (e.g., a wavelet basis). The soft thresholding operation in (21.8) can equivalently be expressed as

$$\text{SoftThreshold}(\theta_i, \lambda\sigma^2) = \begin{cases} \theta_i + \lambda\sigma^2, & \theta_i < \lambda\sigma^2 \\ 0, & |\theta_i| \leq \lambda\sigma^2 \\ \theta_i - \lambda\sigma^2, & \theta_i > \lambda\sigma^2. \end{cases}$$

Essentially, this operator maps noisy wavelet coefficients to zero if their magnitude is smaller than the threshold $\lambda\sigma^2$ and shifts them by $\lambda\sigma^2$ towards zero otherwise.

Note that the sparsifying basis Φ (e.g., the wavelet basis) is selected independently of the data; however, the ideas here can be leveraged in *dictionary learning* (Chapter 23) contexts where we wish to choose a good sparse representation from the data itself. For more general sparsity regularization in non-orthonormal, redundant dictionaries, iterative methods must be used to solve the optimization problem in (21.7).

The effect of using wavelet soft thresholding for a noisy Poisson observation is shown in Figure 21.5. Note that the soft thresholding operation was derived with Gaussian noise models in mind and hence is a poor fit to many Poisson denoising problems. While it is often used this way because of its speed and simplicity, the accuracy problems are evident in Figure 21.5(c). One approach to addressing this challenge is to first apply the Anscombe transform (see Section 21.2.3) to the noisy image, then apply wavelet soft thresholding, and then apply either the algebraic or unbiased inverse Anscombe transform. The results of this approach are displayed in Figure 21.5(d-e); while this approach can be effective at high photon count levels, it suffers from artifacts in very photon-limited settings (i.e., small T in (21.2)).

There exist several methods aimed at addressing these challenges in the context of Poisson noise (*cf.* [94, 1047, 503, 505, 891, 28, 487, 504, 586, 89]). One such approach adaptively bins neighboring pixels corresponding to similar intensities and performs smoothing in each bin separately. These steps are completed with explicit dependence on the Poisson likelihood and are consistent with the core principles underlying wavelet and multiresolution analysis [1145, 1144]. The data-adaptive binning gives these estimators the capability of spatially varying the resolution to automatically increase the smoothing in very regular regions of the image and to preserve detailed structure in less regular regions (e.g., near edges and boundaries). Fast and practical tree-based algorithms effectively address both model accuracy for piecewise smooth images and computational efficiency challenges, nearly achieve theoretical lower bounds on the best possible error convergence rates, and perform very well in practice. This approach is demonstrated in Figure 21.5(f-g), in which very few photon observations are used to reconstruct an underlying intensity with high accuracy using adaptive binning and constant model fits within each bin.

21.3.3 *Total Variation*

The total variation regularizer [182, 72] measures how much an image varies across pixels, so that a highly textured or noisy image has a large TV seminorm,[2] while a smooth or piecewise smooth image would have a relatively small TV seminorm. Total variation regularization is often a useful alternative to wavelet-based regularizers, which are also designed to be small for piecewise smooth images but can result in spurious large, isolated wavelet coefficients and related image artifacts.

---

[2]A seminorm is a norm that can have zero value for non-zero vectors; for instance, the TV seminorm of the vector $\begin{bmatrix} 2,2,2,2 \end{bmatrix}$ is zero because the signal never changes.

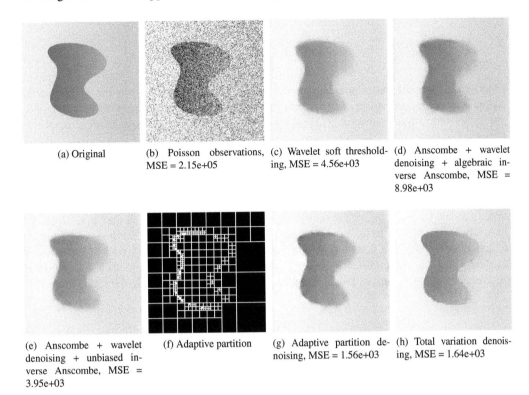

(a) Original

(b) Poisson observations, MSE = 2.15e+05

(c) Wavelet soft thresholding, MSE = 4.56e+03

(d) Anscombe + wavelet denoising + algebraic inverse Anscombe, MSE = 8.98e+03

(e) Anscombe + wavelet denoising + unbiased inverse Anscombe, MSE = 3.95e+03

(f) Adaptive partition

(g) Adaptive partition denoising, MSE = 1.56e+03

(h) Total variation denoising, MSE = 1.64e+03

Figure 21.5: Denoising Poisson observed images via wavelet denoising and total variation. (a) Original image. (b) Poisson observation with an average of 3.84 photons per pixel. (c) Result of wavelet soft thresholding, which is poorly suited to Poisson noise. (d) Result of applying the Anscombe transform, wavelet soft thresholding, and the algebraic inverse Anscombe transform. (e) Result of applying the Anscombe transform, wavelet soft thresholding, and the unbiased inverse Anscombe transform. (f) Adaptive partition associated with the image in (a), with smaller bins corresponding to image regions with less homogeneity. (g) Result of using multiscale adaptive photon binning, designed for Poisson noise. (h) Result of total variation denoising using the Poisson log likelihood.

The (anisotropic) total variation seminorm is defined as

$$r(x) = \|\Phi x\|_1 = \|x\|_{\mathrm{TV}} \triangleq \sum_{k=1}^{\sqrt{N}-1} \sum_{l=1}^{\sqrt{N}} |x_{k,l} - x_{k+1,l}| + \sum_{k=1}^{\sqrt{N}} \sum_{l=1}^{\sqrt{N}-1} |x_{k,l} - x_{k,l+1}|,$$

where Φx contains first-order pixel differences and where we use a slight abuse of notation by using 2D pixel indices instead of vector indices by assuming that $x \in \mathbb{R}^N$ is a square $\sqrt{N} \times \sqrt{N}$ image. This highlights the fact that the TV seminorm is simply a measure of the magnitude of all vertical and horizontal first-order differences, which are illustrated in Figure 21.4(d). This property makes TV especially well-suited for image denoising and inverse problems for piecewise constant images.

In the case of Gaussian noise with variance σ^2, the typical optimization problem in (21.6) amounts to

$$\hat{x} = \underset{\tilde{x}}{\mathrm{argmin}} \quad \|y - \tilde{x}\|_2^2 + \lambda \sigma^2 \|\tilde{x}\|_{\mathrm{TV}}. \tag{21.9}$$

There is no closed-form solution to this optimization problem (as there was in the case of wavelet denoising), but work by Beck and Teboulle [72] and others present fast iterative algorithms for solving (21.9). In the Poisson setting, we seek to solve the TV-regularized problem

$$\hat{x} = \underset{\tilde{x}}{\text{argmin}} \quad \sum_{i=1}^{n} [T\tilde{x}_i - y_i \log(T\tilde{x}_i)] + \lambda \|\tilde{x}\|_{\text{TV}} \tag{21.10}$$

which requires additional convex optimization machinery [409]. An example of this method in action is displayed in Figure 21.5(g). Note that total variation denoising is widely used as a central component[3] of many inverse problem solvers (e.g., deblurring and tomographic reconstruction, Chapter 5).

21.4   Kernel Denoising Methods

Kernel denoising methods focus less explicitly on the regularized likelihood formulation in Section 21.3, and instead rely on the central ideal that neighboring pixels in the "true" scene tend to have similar intensities. In light of this image model, if we average neighborhoods of similar pixels together, we should get a good estimate of the true intensity x and remove a lot of the noise. Kernel methods essentially determine which pixels belong in the same neighborhoods and how they should be averaged to remove noise. In general, kernel smoothing estimates have the following form:

$$\hat{x}_i = \sum_{j=1}^{N} w_{i,j} y_j; \tag{21.11}$$

that is, the estimate of pixel i is a weighted sum of the noisy pixels $\{y_j\}$. The function determining the weights is referred to as the kernel. With *linear* kernel estimators, the weights $\{w_{i,j}\}$ are independent of the observed data, while with *non-linear* kernel estimators (e.g., the bilateral filter), the weights are selected *data dependently*.

Many image denoising algorithms can be interpreted as estimators of this form with particular choices of weights [716, 140, 95, 191, 192, 31, 686]. Choosing the weights for this estimator is a challenging task that determines how much noise is removed and how much image structure is preserved. In this section, we describe three popular choices to illustrate some of the key tradeoffs that must be considered when choosing weights.

21.4.1   *Linear Smoothing*

A particularly popular choice for the weights corresponds to a Gaussian filter; i.e.,

$$w_{i,j} = e^{-d_{i,j}^2/h^2}, \tag{21.12}$$

where $d_{i,j}$ is the spatial distance between pixels. The parameter h^2 determines the width of the Gaussian filter; wider filters remove more noise but also blur boundaries in piecewise smooth images. Figure 21.6(b) illustrates weights associated with the Gaussian filter. For denoising pixel i in the center of the image, weights $w_{i,j}$ are larger (shown as brighter) when they are closer to the center pixel.

When only moderate amounts of noise are present, a small Gaussian filter may be effective. Its effectiveness, the simplicity of its implementation, and its interpretability have all contributed to the filter's lasting popularity. However, as noise levels increase, larger values of h become necessary and boundaries become increasingly blurred. This is illustrated in Figure 21.7(b), (g), and (l).

The linear Gaussian kernel is well-known to blur edges, and Figure 21.8(a) illustrates why. In that image, all the pixels within the red square are contributing (via the average in (21.11)) to the estimate of the red pixel in the center. In other words, pixels which are close to the center pixel

---

[3]Specifically, total variation denoising is often used as a proximal operator [765].

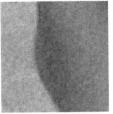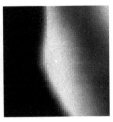

(a) Fragment of image to be denoised. (b) Linear kernel weights (c) Yaroslavsky / bilateral filter kernel weights (d) nonlocal means weights

Figure 21.6: Non-linear kernel smoothing illustration (Gaussian noise). Consider estimating the true intensity of the center pixel in (a). A kernel estimate would compute a weighted sum of all the pixels in (a). The relative value of the weights for different types of kernels are depicted in (b)-(d), where brighter pixels correspond to higher weights. In (b), the weights correspond to a linear Gaussian kernel and depend only on the spatial proximity of each pixel to the center. In (c), the weights are a product of the spatial kernel in (b) and the similarity of each pixel to the center pixel, measured by the absolute difference in noisy pixels. In (d), the weights are a product of the spatial kernel in (b) and the similarity of each pixel to the center pixel, measured by the similarity of a small patch around the center pixel to be denoised and a small patch around each pixel in the noisy data in (a). These images reproduced with permission from [31].

being denoised but on the opposite side of an edge in the image receive a large weight, so that our denoised pixel is a weighted sum of both dark and light pixels instead of just the dark pixels on the same side of the boundary as our pixel of interest. If we decrease our kernel bandwidth, then we reduce the number of light pixels used to compute our estimate and achieve less blurring, but simultaneously increase our sensitivity to noise.

This tradeoff can be mitigated to some extent by leveraging more sophisticated prior models of image structure to choose better weights. However, if the weights must be chosen independently of the data, there is no mechanism for having weights depend on the location of boundaries or discontinuities in the image. For this capability, we must turn to non-linear methods.

21.4.2 Ideal Weights

As described in the previous section, the linear Gaussian kernel is well known to blur edges because, in the context of Figure 21.8(a), bright pixels on one side of a boundary get averaged with dark pixels on the other side of the boundary to yield biased (blurry) estimates of pixels near boundaries. Ideally, we would like to estimate our center pixel of interest using an average of the noise pixels outlined in red in Figure 21.8(b) – i.e., pixels that an "oracle" tells us are on the same side of a boundary as the pixel to be denoised. It can be shown that this oracle estimator is optimal [31].

In Figure 21.7(c,h,m), we simulate these ideal weights by choosing

$$w_{i,j} = e^{-d_{i,j}^2/h^2} \mathbb{1}\{|x_i - x_j| \le h'\}, \tag{21.13}$$

where h, h' are bandwidth parameters and

$$\mathbb{1}(a) = \begin{cases} 1 & \text{if } a \text{ is true,} \\ 0 & \text{otherwise} \end{cases}.$$

In other words, when we wish to compute a denoised estimate of pixel i, we use a weighted sum of neighboring pixels where the weight for neighbor pixel j is large if two conditions are met: (a)

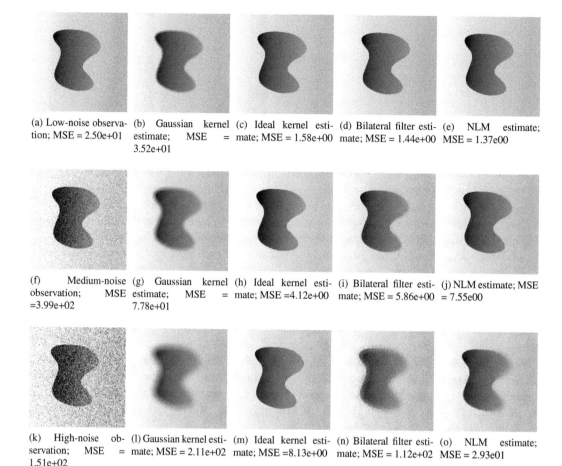

(a) Low-noise observation; MSE = 2.50e+01   (b) Gaussian kernel estimate; MSE = 3.52e+01   (c) Ideal kernel estimate; MSE = 1.58e+00   (d) Bilateral filter estimate; MSE = 1.44e+00   (e) NLM estimate; MSE = 1.37e00

(f) Medium-noise observation; MSE =3.99e+02   (g) Gaussian kernel estimate; MSE = 7.78e+01   (h) Ideal kernel estimate; MSE =4.12e+00   (i) Bilateral filter estimate; MSE = 5.86e+00   (j) NLM estimate; MSE = 7.55e00

(k) High-noise observation; MSE = 1.51e+02   (l) Gaussian kernel estimate; MSE = 2.11e+02   (m) Ideal kernel estimate; MSE =8.13e+00   (n) Bilateral filter estimate; MSE = 1.12e+02   (o) NLM estimate; MSE = 2.93e01

Figure 21.7: Comparison of linear and non-linear kernel estimation methods in the presence of Gaussian noise. These images reproduced with permission from [31].

pixels i and j are spatially close and (b) true intensity pixel values x_i and x_j are similar. Note that this estimator could never be used in practice, since it depends on the true intensity image x. However, it provides a benchmark of how well we might hope to do with other, more practical estimators, both from an empirical and theoretical perspective. In fact, both the bilateral filter (Section 21.4.3) and nonlocal means (Section 21.4.4) can be interpreted as computing empirical estimates of the ideal weights in (21.13).

21.4.3   Bilateral Filters

The key to the efficacy of the bilateral filter is that, when the noise level is sufficiently low, the bilateral filter weights are an effective proxy for the ideal weights in Section 21.4.2. That is, the bilateral filter is accurately accounting for whether each pair of pixels are on the same side of an image boundary or on opposite sides. As a result, at low noise levels the bilateral filter is optimal [31] for images consisting of smooth surfaces separated by smooth boundaries.

The core ideas of bilateral filtering were introduced by Yaroslavsky [1172] and independently by Lee [554], and more modern variants include SUSAN [950] and bilateral filtering [1049]. The basic idea is to choose the weight $w_{i,j}$ in (21.11) to be larger when pixels i and j are more similar. In the linear smoothing considered above, similarity was estimated via the spatial proximity of two pixels;

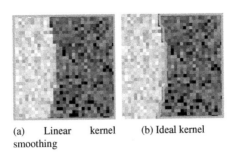

(a)   Linear   kernel   (b) Ideal kernel
smoothing

Figure 21.8: To estimate the center pixel in this image, we compute an average of all the pixels inside the red boundary. (a) Illustration of the effect of a linear kernel, in which pixels are included in the average regardless of their luminosity, resulting in blurry image estimates. (b) Illustration of an "ideal" kernel in which only pixels on the same side of the boundary as the center pixel are included in the average. It can be shown that the ideal kernel is optimal for denoising images consisting of smooth surfaces separated by smooth boundaries. However, this ideal kernel is generally not computable. Methods like the bilateral filter and nonlocal means attempt to estimate the ideal kernel from the noisy data. These images reproduced with permission from [31].

with the bilateral filter, similarity is estimated via both the spatial proximity and the luminosity of two pixels. A key element of this method is that the weights $\{w_{i,j}\}$ are selected in a *data-dependent* manner, making the estimator inherently non-linear.

Several different expressions for the weights reflect this approach; for example, consider weights in (21.11) of the form

$$w_{i,j} = e^{-d_{i,j}^2/h^2} \mathbb{1}\{|y_i - y_j| \le h'\},\tag{21.14}$$

where h, h' are bandwidth parameters. Analogous to the ideal weights in Section 21.4.2, the weight for neighbor pixel j is large if two conditions are met: (a) pixels i and j are spatially close and (b) observed noisy pixel values y_i and y_j are similar. The first term in this expression controls the dependence of the weights on the spatial proximity between a pair of pixels while the second term controls the weight's dependence on the photometric or luminosity proximity. As in classical kernel smoothing (see Section 21.4), h plays the role of spatial bandwidth; the new parameter h' is a photometric bandwidth.

Bilateral filter weights are illustrated in Figure 21.6(c). Clearly pixels spatially far from the center pixel receive low weight, just as with the linear Gaussian kernel. In addition, pixels which are much brighter than the center pixel, such as those to the left of the boundary, receive low weight; as a result, in practice we observe much less blurring around boundaries in denoised images. Further observe that the weights shown in Figure 21.6(c) appear somewhat "noisy" – pixels on the right side of the boundary and near the center pixel being estimated sometimes receive very low weights. In practice, this effect is exacerbated in the presence of strong noise, making this method most effective in low-noise environments.

It is clear that when the noise variance is low, then $|x_i - x_j| \approx |y_i - y_j|$ and hence the ideal weights in (21.13) are close to the bilateral filter weights in (21.14); in this setting the bilateral filter is highly effective. In contrast, when the noise variance is high, $|x_i - x_j|$ may be very far from $|y_i - y_j|$, so the bilateral filter weights may be very far from the ideal weights, making the bilateral filter ineffective. This effect is reflected in the empirical results in Figures 21.7(d,i,n).

21.4.4   Nonlocal Means (NLM)

The bilateral filter described above is very effective at low noise levels, but its performance degrades as the noise level increases. One explanation for this effect is that pixels on the same side of the

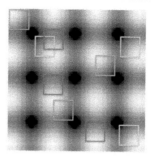

Figure 21.9: Image patches in a synthetic image generated with materials science applications in mind. Note that the red patches share similar structure despite being spatially distant; similarly for the green and blue patches. (These ideas extend naturally to spectral images and three-dimensional patches.)

boundary as the pixel being denoised may have very different observed luminosities even if the true underlying pixel intensities are similar; this effect is illustrated by the "noisiness" of the weights in Figure 21.6(c).

Nonlocal means generalizes the idea of including photometric proximity information in the choice of weights [140], but measures photometric intensity in a way that is more robust to noise than the absolute pixel difference used by the bilateral filter in (21.14). The basic idea here is that the photometric similarity used to compute weights is based on *patches* of pixels and is much less sensitive to noise. As was the case with the bilateral filter, the weights $\{w_{i,j}\}$ associated with NLM are selected in a *data-dependent* manner, making the estimator non-linear. In fact, bilateral filters are a special case of patch-based methods in which each patch contains only a single pixel.

A generic description of NLM is the following. Let $h_P > 0$ be a bandwidth parameter and let P_i be the patch of width h_P centered at i, i.e.,

$$P_i = \{j : \tilde{d}_{i,j} \leq h_P\} \tag{21.15}$$

for a measure of spatial distance $\tilde{d}_{i,j}$. (The distance measure is usually chosen so that the patch P_i is a square or cube; i.e., if (i_x, i_y) denotes the coordinate of the ith pixel and likewise for (j_x, j_y), then we might use $\tilde{d}_{i,j} = \max(|i_x - j_x|, |i_y - j_y|)$.) Let $\mathbf{y}_{P_i} = (y_j : j \in P_i)$ be the vector of noisy pixels within patch P_i. We consider overlapping patches: all P_i for $i = 1, \ldots, N$.[4] A subset of the overlapping patches in an image is illustrated in Figure 21.9. Because of the *redundancy* inherent in this image, large subsets of patches (e.g., the red patches) share similar spatial structure. These similarities are exploited by patch-based methods.

With this notation, we could choose nonlocal means weights according to

$$w_{i,j} = e^{-d_{i,j}^2/h^2} \mathbb{1}\{\|\mathbf{y}_{P_i} - \mathbf{y}_{P_j}\|_2 \leq h'\}, \tag{21.16}$$

where h, h' are bandwidths, as before. The photometric similarity is based on the Euclidean distance between the patches (as vectors) around pixels i and j.[5]

Consider the NLM weights shown in Figure 21.6(d). In contrast to the weights associated with the linear Gaussian kernel in Figure 21.6(b), the NLM weights are largest on the right side of the boundary and deemphasize the brighter pixels on the left side of the boundary, similar to the bilateral

---

[4]For simplicity of presentation, we ignore boundary effects. Patches close to the boundary can be formed by padding the original image.

[5]In [140], emphasis is placed on $h = \infty$, so that spatial proximity is ignored. [140] used spatial proximity only as a numerical parameter to solve a computational issue. However, later works (*cf.* [884, 1210]) have shown that spatial proximity can improve NLM performance. Also, other functions of $\|\mathbf{y}_{P_i} - \mathbf{y}_{P_j}\|_2$ besides $\mathbb{1}(\cdot \leq h')$ can be used to measure photometric similarity; indeed, a Gaussian kernel is more common in practice. We focus on the version in (21.16) for illustrative purposes.

filter weights in Figure 21.6(c). In contrast to the bilateral filter weights, the NLM weights are much less sensitive to noise, since a single noisy pixel is used only as a member of a patch and hence has less direct influence over any individual weight.

It can be shown [31] that the NLM weights in (21.16) are close to the ideal weights in (21.13) even at noise levels where the bilateral filter weights in (21.14) are far from the ideal weights. In other words, when we compute $\|\mathbf{y}_{P_i} - \mathbf{y}_{P_j}\|$ in homogeneous or smooth regions, we get a close approximation to $\sqrt{p}|x_i - x_j|$, where p is the number of pixels in a patch, because we average over noisy pixels to compute the distance between patches, reducing sensitivity to noise. This effect is reflected in the empirical results in Figure 21.7(e,j,o). An additional example comparing the linear Gaussian kernel, oracle estimates, bilateral filtering, and nonlocal means is in Figure 21.14.

21.5   Patch-Based Methods for Image Denoising

In the previous section, we explored nonlocal means as a denoising method based on the similarity among *patches* of pixels. This core idea can be generalized in many ways and is a key element of many modern image processing methods, including BM3D [232] and dictionary learning (see Chapter 23).

If our observed noisy image y has N pixels, arranged in an $N_1 \times N_2$ grid, then a patch of y is a collection of pixels defined as in Section 21.4.4 via

$$P_i = \left\{ j : \tilde{d}_{i,j} \le h_P \right\} \qquad \text{and} \qquad \mathbf{y}_{P_i} = (y_j : j \in P_i)$$

for a patch sidelength parameter h_P. For a d-dimensional image (conventionally $d = 2$), the vector $\mathbf{y}_{P_i}$ of pixels in patch P_i has $p = (2h_P + 1)^d$ elements.[6]

All of the patch-based denoising methods described in this section have a similar structure that is illustrated in Figure 21.10. Roughly speaking, nonlocal means, nonlocal principal components analysis (PCA), and BM3D, which are all described in this chapter, have this basic structure at their heart:

1. To denoise pixel y_i, choose patch P_i of pixels centered around location i; denote this $\mathbf{y}_{P_i}$.

2. Find all patches similar to $\mathbf{y}_{P_i}$ in the noisy image.

3. Feed this collection of noisy patches into a "patch denoising" routine.

   - For nonlocal means, the patch denoising method computes a weighted average of the patches, where patches that are more similar to $\mathbf{y}_{P_i}$ receive higher weight. This weight generally depends on $\|\mathbf{y}_{P_i} - \mathbf{y}_{P_j}\|_2^2$, and is illustrated in Figure 21.6(d), where brighter pixels correspond to pixels receiving higher weight in the weighted average.

   - For nonlocal PCA, the patches are vectorized and concatenated into a matrix, and PCA is applied to this matrix to remove noise from the patches.

   - For BM3D, the patches are stacked to form a 3D cube, and this cube is denoised with a combination of wavelet thresholding (see Section 21.3.2) and Wiener filtering [373].

4. The center pixel of the denoised patch P_i is used as the estimate $\hat{x}_i$.

Note that this sketch is intended to provide helpful intuition about patch-based denoising, but it omits many important details of the individual algorithms.[7] The details of nonlocal means were described in Section 21.4.4; nonlocal PCA and BM3D are detailed below.

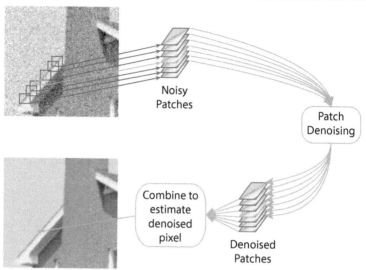

Noisy
Patches

Patch
Denoising

Combine to
estimate
denoised
pixel

Denoised
Patches

Figure 21.10: Illustration of a generic patch-based image denoising method. To denoise the pixel y_i (marked with a red dot in the noisy image), we examine a patch of pixels centered around location i (marked with a red square, denoted P_i in the text). We then find similar patches in the noisy image (marked with blue squares). This group of patches is then fed into a "patch denoising" routine; the nature of the patch denoising routine is essentially what differentiates methods such as nonlocal means, nonlocal PCA, and BM3D. The denoised patches are used to estimate $\hat{x}_i$. This image reproduced with permission from [1171].

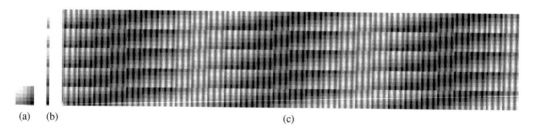

(a)   (b)   (c)

Figure 21.11: Illustration of matrix representation of image patches for the image in Figure 21.9. (a) One highlighted patch from Figure 21.9. (b) Vector representation of patch in (a). (c) Concatenated collection of vector representations of all patches. (These ideas extend naturally to spectral images and three-dimensional patches.) These images reproduced with permission from [1171].

21.5.1   Principal Components Analysis

The vector/matrix representation of patches used in BM3D and nonlocal means is depicted in Figure 21.11. Given the collection of vectorized patches described above, we can concatenate them into a single $p \times N$ matrix as in Figure 21.11; we denote this matrix Y. One simple approach to

---

[6]Patches can have unequal sidelengths; this is particularly common in spectral imaging, where we might wish to have a patch cover a small spatial area but a large number of spectral bands. Our discussion focuses on patches with equal sidelengths for simplicity of presentation.

[7]NLM can be interpreted in this framework, where "combining" denoised patches amounts to selecting the center pixel; however, the patch denoising and patch recombination steps are typically implemented as one step for computational efficiency. BM3D implements this cycle twice, with two different patch denoising methods.

denoising y is to perform principal components analysis (PCA) on Y (*cf.* [335, 100] for a nice introduction to PCA for data analysis). The underlying idea is that we wish to find some small number $r \ll \min(N, p)$ of *representative patches* such that each image patch can be accurately approximated as a weighted sum of those representatives. Note that in nonlocal means, we also would choose a patch and then denoise it using a weighted sum of representative patches. However, the key difference is that with NLM, the representative patches (a) could be different for each patch being denoised, (b) must be similar to the patch being denoised (as measured by Euclidean distance), and (c) assigned nonnegative weights only. In contrast, with the PCA approach, the representative patches (a) are the same for every patch to be denoised, (b) are a diverse (indeed orthogonal) collection of patches that may, by themselves, be quite dissimilar from a patch being denoised, and (c) can be assigned arbitrary weights. For example, imagine denoising an image of a checkerboard and consider a target patch which is white on the left half and black on the right half. NLM would only leverage other patches with this structure, while PCA would *also* leverage patches which are black on the left and white on the right, the negative of the target patch.

PCA is computed by first calculating the singular value decomposition of the matrix Y, yielding orthogonal matrices U and V and diagonal matrix Σ such that $Y = U\Sigma V^\top$, then setting all but the r largest diagonal elements in Σ (i.e., all but the r largest singular values of Y) to zero to produce $\hat{\Sigma}$, and finally setting $\hat{X} = U\hat{\Sigma}V^\top$. The r columns of U corresponding to the non-zero rows of $\hat{\Sigma}$ are the representative patches, and the corresponding r rows of $\hat{\Sigma}V^\top$ are the weights assigned to those representative patches for each of the N patches to be denoised. From here, the denoised patches in $\hat{X}$ can be recombined to estimate each denoised pixel $\hat{x}_i$. For instance, we can find all the patches which contain pixel i, and average the denoised value of pixel i from all those patches to produce $\hat{x}_i$.

This approach is simple, interpretable, and can be computed quickly on large datasets. Its impact on a materials science dataset is illustrated in Figure 21.17. Smaller r yields a model with fewer representative patches, fewer degrees of freedom, and hence estimates with higher bias but less sensitivity to noise. Unfortunately, for many real-world materials science problems, r must be somewhat large relative to N and p to accurately capture the diversity of features in an image; that is, the true intensity image patches lie in a relatively high dimensional subspace. As a result, the denoising performance of this method is rarely optimal. We will see two alternatives below which extend this model and yield more accurate image estimates.

21.5.2 Nonlocal PCA

In the PCA approach described in Section 21.5.1, finding r representative patches amounted to finding a single r-dimensional subspace of $\mathbb{R}^p$ so that all N noisy patches with p pixels were close to this subspace. A conceptually simple yet powerful generalization of the subspace model is that the patches lie in a *union of subspaces*, illustrated in Figure 21.12. Conceptually, this means that instead of finding *one* set of r representative patches for the entire image, as in Section 21.5.1, we instead cluster the patches into K groups and find a *different* set of r/K representative patches for each group. In both cases, we have a total of r representative patches, but the second model requires that each patch be a weighted sum of only r/K representatives instead of all r. The second approach is called *nonlocal PCA* because we perform PCA on collections of patches which are not spatially localized.

More specifically, similar to the model in Section 21.5.1, we estimate $\hat{X} = UV^\top$, where $U \in \mathbb{R}^{p \times r}$ contains r representative patches, each with p pixels, and $V^\top \in \mathbb{R}^{r \times N}$ contains r-dimensional vectors of weights for each of the N patches in the intensity image. In contrast to Section 21.5.1, however, (a) the representative patches in U are not necessarily orthogonal to one another, and (b) the weights in $V^\top$ are not arbitrary. The columns of U can be divided into K disjoint sets (corresponding to K subspaces), each containing r/K representative patches. The r-dimensional weight vector associated with a given patch will similarly have K subsets of r/K rows, corresponding to r/K weights for each of K subspaces. If a patch lies in the union of subspaces, then the corresponding column of $V^\top$ of weights will have nonzero weights for *only one* of the K subsets of rows, and

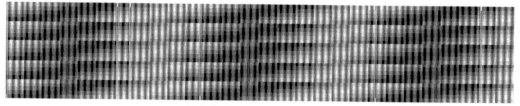

(a) Collection of vectorized patches

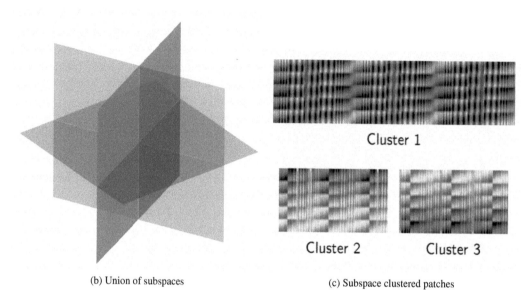

(b) Union of subspaces   (c) Subspace clustered patches

Figure 21.12: Illustration of a union of subspaces. We model each patch in our spectral image as lying in such a union. Some patches are highlighted with colored lines to illustrate their subspace membership. If we knew a priori which patches lie in which subspaces, then we could estimate the subspaces using principal components analysis (PCA) or a variant that accounts for non-Gaussian noise. Our approach estimates these subspaces without that prior information about subspace membership.

the weights corresponding to the remaining $K - 1$ subspaces will be zero-valued. This idea is illustrated in Figure 21.13. Thus, even though in this setting both the PCA and nonlocal PCA models use the same number of representative patches, the latter model constrains the weights that can be used to combine patches and produce a denoised patch. This constraint limits the number of degrees of freedom in the nonlocal PCA model, which in turn allows us to increase r without an unmanageable increase in the variance (i.e., sensitivity to noise) of the estimates.

If we knew a priori which patches lie in which subspaces, then we could estimate the subspaces using a principal components analysis (PCA) or a variant that accounts for non-Gaussian noise (e.g., [216, 587]). The nonlocal PCA approach estimates these subspaces without that prior information about subspace membership.

These ideas are closely related to *subspace clustering* [769, 302, 790] and piecewise linear image models [1182, 1183]. Optimal algorithms for subspace clustering are a topic of current research, but most involve iteratively solving an inverse problem for each patch in the image, which can be time consuming for large-scale materials science datasets. A suboptimal but fast alternative that has been studied primarily in the context of Poisson image denoising has the following steps:

1. Divide image into overlapping patches.

2. Cluster patches into K groups. In the context of Poisson noise, we use a variant of k-means

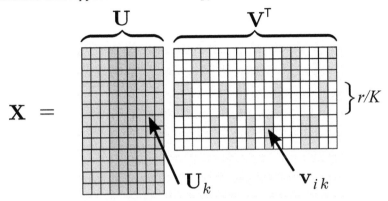

Figure 21.13: Matrix factorization associated with a union of subspaces model. The matrix U has $K = 3$ groups, each with $r/K = 3$ columns corresponding to three representative patches per group or nine total representative patches. U_k is the set of representative patches for the kth group. $v_{i,k}$ is the set of weights for the ith patch projected onto the kth subspace. Note that each patch has nonzero weights for only *one* of the K subspaces.

clustering with Poisson Bregman divergence to measure similarity of patches rather than the standard Euclidean distance measure [51].

3. Perform PCA on each cluster of patches to find an r/K-dimensional patch subspace. In the context of Poisson noise, we do this by searching for the subspaces which minimize the negative Poisson log-likelihood in (21.4), as described in [216, 886].

4. For each of the N noisy patches, estimate r/K subspace coefficients. In the context of Poisson noise, we do this by minimizing the negative Poisson log-likelihood plus a sparsity regularizer corresponding to the sum of the absolute values of the subspace coefficients [409]. (Recall sparsity regularization was discussed in Section 21.3.)

5. Recombine the denoised patches to form the final image estimate.

The nonlocal Poisson principal components analysis (PCA) denoising method described above performs steps such as noisy image patch clustering, subspace estimation, and projections of patches onto subspaces *using routines based on optimizing objective functions built around the Poisson noise model*. As a result, this approach naturally accounts for the non-uniform noise variance [886] and leads to strong empirical performance, as illustrated in the experiments in Section 21.6.2. The impact of this approach on materials science data is illustrated in Figure 21.17. Code for this method is available online.[8]

21.5.3   BM3D

BM3D [232] is widely considered to be representative of the state-of-the-art in Gaussian image denoising, with fast code available online.[9] Like nonlocal means and non-linear PCA, BM3D is based on patches of image pixels. Refer again to Figure 21.10. The first step is to extract groups of similar patches from the noisy image, as discussed before; this is referred to as *block matching* (the "BM" in BM3D). BM3D then denoises the stack of noisy patches using 3D wavelet thresholding (similar to what was discussed in Section 21.3.2, but applied to a 3D stack of image patches rather than the full image). These are used to form an intermediate denoised image. Next, groups of similar patches are extracted from this intermediate image, and corresponding groups of patches are extracted from

---

[8]http://josephsalmon.eu/code/index_codes.php?page=NLPCA
[9]http://www.cs.tut.fi/~foi/GCF-BM3D/

the original noisy image. Thus we have two groups of patches corresponding to the same spatial locations in the image: one from the original noisy image, and one from the intermediate denoised image. This latter group is used to estimate the power spectral density of the patch group, which is in turn used to denoise the group of noisy patches using Wiener filtering [373]. The effect of this procedure on an image contaminated with a high level of Gaussian noise is displayed in Figure 21.1(f) and on real materials science data in Figure 21.17.

There has been some interest in adapting BM3D to images contaminated with Poisson noise (*cf.* [677] for an example in the context of materials science). The most common approach is to first apply the Anscombe transform to the noisy image (as described in Section 21.2.3), then feed this transformed image into BM3D, and finally apply the inverse Anscombe transform. Examples of this procedure are illustrated in Section 21.6. More recent work has considered specialized variants of BM3D for materials science data that exploit underlying periodicities [677]. This type of estimator is still under active development, with recent coarse-to-fine approaches [38, 637] yielding particularly promising results.

21.6   Examples

21.6.1   *Comparison of Linear Gaussian Smoothing, Bilateral Filtering, and Nonlocal Means*

Figure 21.14 depicts the result of an experiment on synthetic data similar to Figure 21.7. The same image is corrupted with varying levels of Gaussian noise, and denoised using the ideal weights (21.13) from Section 21.4.2, bilateral filter weights from (21.14) from Section 21.4.3, and nonlocal means weights (21.16) from Section 21.4.4. This example highlights effects described above, such as the efficacy of the ideal weights (not computable in practice), the similar performance of the ideal weights and the bilateral filter at low noise levels, and the improvement in accuracy with NLM relative to the bilateral filter as noise levels increase.

21.6.2   *Examples of Poisson Image Denoising on Simulated Data*

Figure 21.15 displays comparisons of several algorithms for denoising Poisson images on simulated data. In this experiment, we compare BM3D combined with the Anscombe transform (described in Section 21.5.3) with the NLPCA method (described in Section 21.5.2); more extensive comparisons are available in [885]. In general, we see that both using the Anscombe transform (from Section 21.2.3) combined with BM3D (from Section 21.5.3) and using Poisson NLPCA (from Section 21.5.2) effectively capture redundancies inherent in the underlying true intensity image x, even when the noisy observations y are composed of very small numbers of photon counts. Note that faint ridges in the top row of stars in the flag image are accurately recovered by NLPCA even when they are barely perceptible in the original noisy data, and that mean squared errors are reduced by two orders of magnitude by denoising with tools which explicitly account for both image models (Section 21.2.4) and noise models (Section 21.2.1).

21.6.3   *Application to Electron Microscopy Spectrum Images*

In this section, we consider the application of PCA and NLPCA to denoising spectrum images acquired via energy dispersive X-ray spectroscopy (EDS). These results originally appeared in [1171]. Here, we demonstrate the nonlocal Poisson PCA method described in Section 21.5.2 to EDS spectrum images. Specifically, we used a Ca-doped $Nd_{2/3}TiO_3$ sample acquired at atomic resolution in a scanning transmission electron microscope and sought to generate a spectrum image from the noisy data that could subsequently be used to create composition maps, phase maps, or maps of variations in electronic structure. This data must be collected with a low-power electron beam (i.e., low "dose") because increasing the dose through the dwell time increases distortions in the image caused by time-dependent instabilities of the sample and the microscope, and increasing the dose

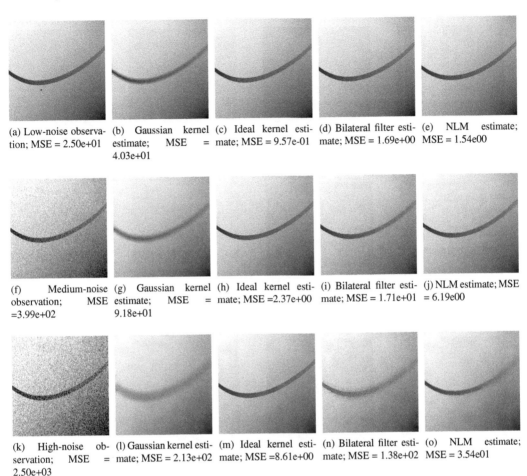

(a) Low-noise observa-
tion; MSE = 2.50e+01

(b) Gaussian kernel
estimate;   MSE =
4.03e+01

(c) Ideal kernel esti-
mate; MSE = 9.57e-01

(d) Bilateral filter estimate; MSE = 1.69e+00

(e)   NLM   estimate;
MSE = 1.54e00

(f)   Medium-noise
observation;   MSE
=3.99e+02

(g) Gaussian kernel
estimate;   MSE =
9.18e+01

(h) Ideal kernel esti-
mate; MSE =2.37e+00

(i) Bilateral filter estimate; MSE = 1.71e+01

(j) NLM estimate; MSE
= 6.19e00

(k)   High-noise   ob-
servation;   MSE   =
2.50e+03

(l) Gaussian kernel esti-
mate; MSE = 2.13e+02

(m) Ideal kernel esti-
mate; MSE =8.61e+00

(n) Bilateral filter estimate; MSE = 1.38e+02

(o)   NLM   estimate;
MSE = 3.54e01

Figure 21.14: Comparison of linear and non-linear kernel estimation methods with Gaussian noise.
These images reproduced with permission from [31].

through the beam power increases the probe size and reduces resolution. As a result, the data has
a very low signal-to-noise ratio (SNR) and the noise can be modeled accurately with a Poisson
distribution.

We first apply post-acquisition non-rigid registration [1170] to a series of spectrum images and
then sum the registered spectrum images to achieve a moderate increase in SNR. Next, the (still
noisy) spectrum image was integrated into seven elemental bands (corresponding to the characteris-
tic X-ray signals for O Kα, Ca Kα, Ti Kα, Ti Kβ, Nd Lα, Nd Lβ, and Nd Lβ4). This reduces the
size of the EDS spectrum dataset to $240 \times 244 \times 7$ voxels. The corresponding seven noisy images
are displayed in Figure 21.16. These images are dominated by photon noise, which we model using
a Poisson distribution.

We next apply several denoising schemes to this noisy spectrum image dataset. Figure 21.17
shows the result of applying four different denoising methods to this dataset. The first row shows
the result of applying conventional PCA to the matrix of noisy patches; a large amount of noise
remains in the estimate and key structures are not visible. The second row shows the result of
applying NLPCA (described above) to the same dataset, yielding much more noise removal and
clearly revealing the atomic lattice in the Ca, Ti, and Nd maps. The third row shows the result of
applying non-local means to this dataset, which results in significant artifacts that hide the lattice

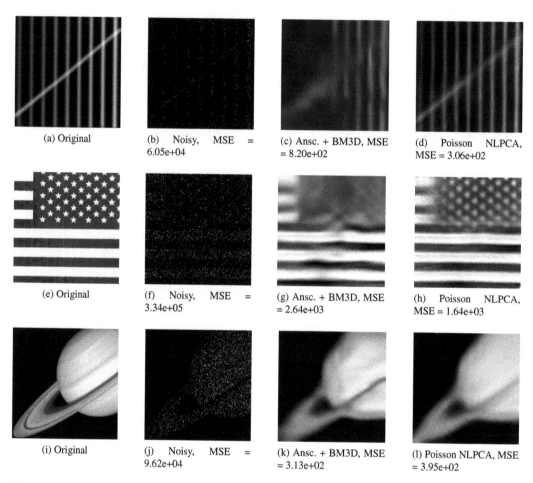

Figure 21.15: Denoising images contaminated with Poisson noise, corresponding to very low photon count data. Top row: Ridges image corrupted with Poisson noise with maximum true intensity $\max_i f_i = 0.1$. Second row: Flag image corrupted with Poisson noise with maximum true intensity $\max_i f_i = 0.1$. Third row: Saturn image corrupted with Poisson noise with maximum true intensity $\max_i f_i = 0.2$. For parameter settings, see [886]. These images reproduced with permission from [886].

structure. The final row shows the result of applying BM3D to this dataset, which again exhibits significant artifacts. The NLPCA result can be used to make more reliable inference about the material sample – for example, that Ca occupies the Nd sublattice, not the Ti sublattice.

21.7 Conclusion

The denoising methods described in this chapter are far from an exhaustive treatment of the plethora of denoising methods in the literature. However, the methods described are representative of major classes of methods and are intended to provide readers with a basic understanding of the underlying principles and ideas. As mentioned in the introduction, image denoising plays two key roles in understanding materials science microscopy images: (a) it facilitates more accurate segmentation and quantitative image analysis, and (b) it plays a central role in inverse problem solvers used in deblurring and tomographic reconstruction, (see Chapter 5). All the methods described in this chapter navigate the complex interplay between models of image structure and geometry and models of noise in the observations. It is only by carefully considering both image and noise models that we can achieve strong empirical results. While the majority of the image denoising literature focuses on Gaussian noise models, Poisson noise is prevalent in materials science. The unique statistical properties of Poisson noise can strongly influence the accuracy of various image denoising methods, but recent and emerging algorithms for Poisson image denoising successfully account for these statistics and have significant potential in the context of materials science.

21.8   Acknowledgments

Thank you to Mr. Anthony Wang for simulations and experiments in support of this chapter.

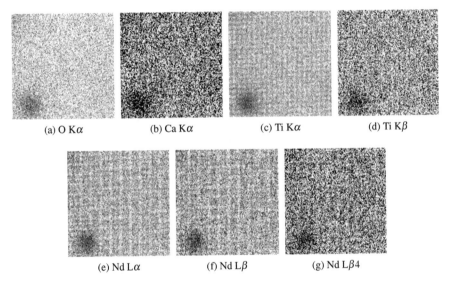

    (a) O Kα (b) Ca Kα (c) Ti Kα (d) Ti Kβ

    (e) Nd Lα (f) Nd Lβ (g) Nd Lβ4

Figure 21.16: Calcium-doped neodymium titanate (a perovskite ceramic). Raw data courtesy of Thomas Slater and Sarah Haigh at University of Manchester. Noise is modeled as Poisson. Non-rigid alignment and averaging were performed using the method in [1170]. This experiment was first reported in [1171].

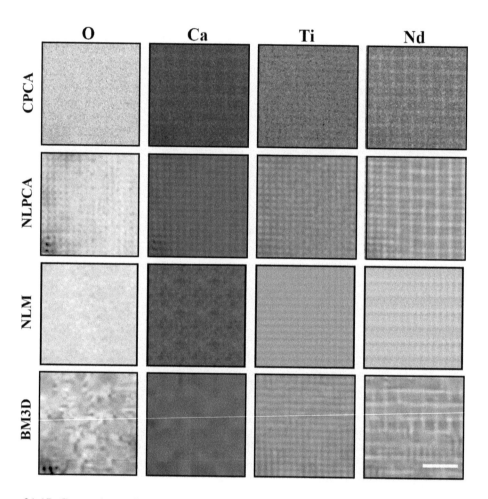

Figure 21.17: Comparison of conventional PCA (corresponding to patches in a single subspace), nonlocal PCA (corresponding to a union of subspaces model), NLM and BM3D applied to experimental data from Figure 21.16. Elemental maps from left to right display the integrated intensity from O Kα, Ca Kα, Ti Kα+Ti Kβ, Nd Lα+Nd Lβ+Nd Lβ4 peaks. Scale bar is 2nm and applies to all panels. The atomic lattice is clearly visible in the Ca, Ti and Nd maps estimated by nonlocal PCA, showing, for example, that Ca occupies the Nd sublattice, not the Ti sublattice. This experiment was first reported in [1171].

Chapter 22

Compressed Sensing Theory and Algorithms for Imaging Applications

Justin Romberg
Georgia Institute of Technology

22.1 Introduction

In this chapter, we give a brief overview of the theory and methods for compressed sensing (CS), particularly as it pertains to problems in imaging. The CS framework is very general in that it applies to any imaging system that can be abstracted into a finite-dimensional linear inverse problem: $\mathbf{y} = \mathbf{A}\mathbf{x}_0 + \text{noise}$. Given the (noisy) indirect observations $\mathbf{y}$ are indirect of our image, we wish to recover (or estimate) $\mathbf{x}_0$ as accurately and stably as possible. We can think of the entries of $\mathbf{y}$ as readings we take off of one or more sensors, the $\mathbf{x}_0$ as a discrete representation of the image we are targeting (e.g., pixel values or basis coefficients), and the matrix $\mathbf{A}$ as a (linear) model for our acquisition device. The number of rows M in $\mathbf{A}$ are the total number of measurements we make of the image; the number of columns N is the number of variables we wish to estimate from these readings. In many cases, N can be interpreted as the target resolution of the reconstruction.

Solving systems of linear equations is perhaps the most fundamental problem in all of applied mathematics, and plays a key role in almost every computational imaging system. The compressed sensing literature developed over the last decade has managed to give us new insight into this old problem. Classically, analyzing the solution to a linear system of equations revolves around the singular values of $\mathbf{A}$ — we can stably solve an $M \times N$ system when $\mathbf{A}$ has N non-zero singular values whose magnitudes do not vary too dramatically. This of course requires the number of equations M to be at least as large as the number of variables N that we are trying to estimate.

The main message of the compressed sensing literature is that if the target vector $\mathbf{x}_0$ is *sparse*, then $\mathbf{y} = \mathbf{A}\mathbf{x}_0 + \text{noise}$ can be meaningfully solved even when $M \ll N$. This has caused a philosophical shift in how we think about the requirements for an imaging system. The dominant factor in determining the number of measurements we need for a faithful reconstruction is not the target resolution N, but rather the complexity of the image.

We will start our exposition with a simulated imaging experiment that illustrates this message.

22.1.1 An Imaging Experiment

Consider the following imaging experiment. There is an image[1] $f(\mathbf{s})$ that we would like to acquire; to be concrete, we use the one shown in Figure 22.1(a) for the purposes of discussion. We measure this image using an instrument which takes indirect observations of $f(\mathbf{s})$ — instead of observing samples of the image, say that we instead have samples of its 2D spatial Fourier transform. A single

---

[1]We are using the bold $\mathbf{s}$ as the argument to stress that f might be naturally indexed by more than one coordinate. In the examples in this chapter, $\mathbf{s} \in \mathbb{R}^2$, but the general principles hold for objects in any dimension.

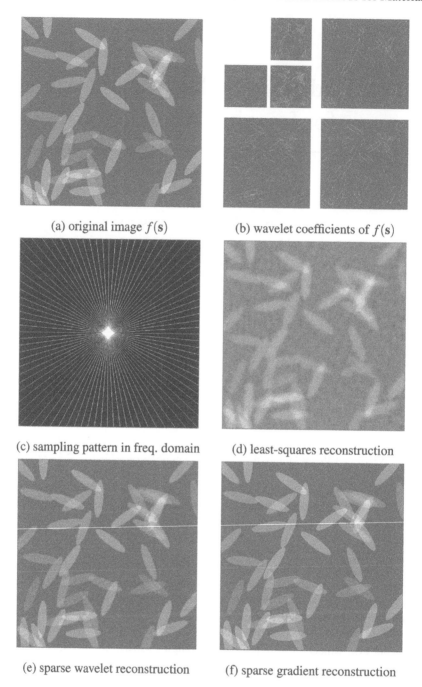

(a) original image $f(\mathbf{s})$ (b) wavelet coefficients of $f(\mathbf{s})$

(c) sampling pattern in freq. domain   (d) least-squares reconstruction

(e) sparse wavelet reconstruction   (f) sparse gradient reconstruction

Figure 22.1: A numerical imaging experiment. The original image in part (a), which contains $512 \times 512 = 262,144$ pixels, can be described by a very small number of wavelet coefficients, shown at two different scales (comprising 94% of the coefficients) in part (b). Part (c) illustrates the locations that the Fourier transform of the image is sampled; there are $M = 21,977$ total sample locations samples along 44 radial lines passing through $\omega = \mathbf{0}$. The number of frequency samples we are given is less than 10% of the number of pixels we are aiming to reconstruct. Part (d) shows the classical least-squares reconstruction, part (e) shows the reconstruction with an ℓ_1 penalty on the wavelet coefficients, and part (f) shows the reconstruction with an ℓ_1 penalty on the gradient of the image.

sample at spatial frequency ω_m can be modeled mathematically as

$$y_m = \int_{\mathbf{s} \in \mathbb{R}^2} f(\mathbf{s}) e^{-j\omega_m^{\mathrm{T}} \mathbf{s}} \, d\mathbf{s} + \text{noise}. \tag{22.1}$$

An acquisition would consist of observing some number M of such samples.

What determines how well we can reconstruct $f(\mathbf{s})$ from the y_m? Of course, the particulars of the measurement scenario matter a great deal, especially how many measurements were taken, what frequencies ω_m were used for those measurements, and the size of the errors introduced by the noise. Classical theory tells us that if there is a small amount of noise and the samples were taken at reasonably dense locations inside the region $[-\Omega, \Omega]$ in the Fourier domain, then we can expect to recover $f(\mathbf{s})$ with resolution on the order of $\sim 1/\Omega$.

Statements of this kind, however, do not account for one and perhaps equally important aspect of the problem: *structure* in the image $f(\mathbf{s})$. If we have a priori information about $f(\mathbf{s})$, and this information can be described using mathematics in a tractable way, then it may be possible to reconstruct a far more accurate version of $f(\mathbf{s})$ than the classical theory suggests. Over the past decade, an rich theory and an extensive set of tools have been developed to take advantage of *sparsity*, described more below, in recovery problems of this general type. Taking advantage of the sparsity in $f(\mathbf{s})$ can make up for missing measurements, super-resolve images beyond the bandwidth of the measurement device, and mitigate the effects of noise.

We say that an image is *sparse* if it can be transformed in such a way that almost all of the energy in the image concentrates onto a small number of transform coefficients. For example, while the simple image in Figure 22.1(a) has energy spread out over a significant fraction of the spatial domain; if we look at the image in the *wavelet domain*, then only a small fraction of the coefficients are significant. The wavelet transform, like the more familiar discrete Fourier transform and discrete cosine transform, is lossless in that we can move back and forth between the image and its wavelet coefficients without losing any information. The wavelet coefficients tell us how to build up the image by taking local gradients at different scales — as most of the image consists of "flat" regions, very few of these coefficients are "active." The wavelet coefficients for an image can themselves be naturally arranged into 2D arrays of different sizes spatially indexed in the same manner as the original image; each of these arrays is called a *subband*. Six subbands of the wavelet transform for our test image are depicted in Figure 22.1(b); this represents $15/16 \approx 94\%$ of the transform. The coefficients are imaged so that coefficients that are zero (or close to zero) are blue, while large coefficients are white. The arrays in Figure 22.1(b) are mostly blue, with just a small amount of white. This is an illustration of what we mean by sparsity in the wavelet domain.

We can now compare two image reconstructions, one which ignores this sparse structure in the wavelet domain, and one which takes advantage of it. The measurements $\mathbf{y}$ come from sampling the Fourier transform in the star pattern shown in Figure 22.1(c). In this stylized numerical experiment, we approximate this process by taking a pixelated version of the image (512×512) and then extracting the appropriate part of its 2D discrete Fourier transform.

The first reconstruction, shown in Figure 22.1(d), comes from solving a simple least-squares problem, discussed in detail in Section 22.1.3 below. This reconstruction finds an image that is consistent with the measurements while having an energy which is not too large. Controlling the energy in the reconstruction *regularizes* the solution, and is necessary as there are fewer measurements than unknowns — there are many possible images that match the measurements exactly, and many of those have extremely large energy; the regularization term weeds these undesirable solutions out. We notice that while we can make out the general characteristics of the original image, there are many high-frequency artifacts.

The second reconstruction, shown in Figure 22.1(e), comes from solving an optimization program that has a similar form but with one key difference: the regularizer that penalizes energy has been replaced with one that promotes sparsity. Specifically, we use the ℓ_1 norm, as discussed in Sections 22.2 and 22.3.3. We are now finding an image that has as few wavelet coefficients active as possible while still being consistent with the measurements. This seemingly small change in the reconstruction process leads to a dramatic difference in the quality of the image produced. Even though the number of measurements we make of the image is an order of magnitude smaller than the number of pixels that we are reconstructing (around $22,000$ versus $262,400$), we are able to produce an extremely high-quality image of comparable resolution to the original.

Taking this example one step further produces a result that is even more interesting. The image in Figure 22.1(a) is piecewise constant, meaning that its gradient is *exactly* sparse; local differences in the image are non-zero only around the small number of edges. If instead of penalizing sparsity in the wavelet domain, we penalize the sparsity of the gradient, we recover[2] the original image *exactly*. Again, this is in spite of there being over 10x as many unknowns as there are observations.

The numerical experiment in Figure 22.1 uses synthetic data (and noise-free measurements) to make the main points above clear. But what makes this example work is that the transform coefficients of the image concentrate onto a small set. This type of sparsity model has a long history in signal and image processing, and in fact lies at the core of modern audio and video compression algorithms. In fact, the measurement and sparsity models used in this experiment are very similar to those used in magnetic resonance imaging (see [607], for example).

The phenomena illustrated in this numerical experiment can be abstracted into more mathematical terms: while underdetermined systems of linear equations do not have unique solutions in general, they may very well have unique sparse solutions. This statement has formed the basis for the body of work now known as "compressed sensing" that has been developed over the past 10–12 years. In what follows, we will give an overview of the main theoretical results from this field, as well as an introduction to the algorithms used for sparse recovery.

22.1.2 *Mathematical Formulation*

The theory of compressed sensing is most easily understood in the context of linear algebra. We observe $\mathbf{y} = \mathbf{A}\mathbf{x}_0 + \text{noise}$, and want to estimate $\mathbf{x}_0$. In other words, we want to invert the action of $\mathbf{A}$, even when there are fewer equations than unknowns. This is possible when the target $\mathbf{x}_0$ obeys a specific structural assumption: it is *sparse*, meaning that it contains a relatively small number of important (i.e., large magnitude) entries, and a large number of insignificant (small magnitude) entries.

The first step is to pose the imaging or acquisition problem using the language of linear algebra. There are two steps in doing this: modeling the acquisition device as a linear operator that maps a function $f(\mathbf{s})$ of a continuous spatial variable $\mathbf{s}$ to a finite set of measurements, and then deciding on a basis in which to represent $f(\mathbf{s})$. The basis representation serves a dual purpose. First, it gives us a natural way to discretize the target by parameterizing $f(\mathbf{s})$ by a finite set of coefficients. Second, if properly chosen, it gives us a simple model for the structure in $f(\mathbf{s})$: we will assume images of interest can be closely approximated by combining very few basis functions.

22.1.2.1 *Linear Measurements*

We start by reviewing a standard technique for discretizing a linear inverse problem using a basis expansion. Our unknown object (signal, image, intensity map, etc.) is a function $f(\mathbf{s})$ of one or more continuous variables that we encapsulate in the vector $\mathbf{s}$. Our measurements consist of evaluations of a finite number of known *linear functionals* of $f(\mathbf{s})$,

$$y_m = \mathcal{L}_m(f) + \text{noise}, \quad m = 1, \ldots, M.$$

It is only a slight abuse of terminology to say that this means we observe integrals of f against a series of kernels $h_m(\mathbf{s})$:

$$\mathcal{L}_m(f) = \int f(\mathbf{s}) h_m(\mathbf{s}) \, d\mathbf{s}. \tag{22.2}$$

The examples below illustrate this formulation in several familiar scenarios.

---

[2]The image was recovered by solving (22.24) in Section 22.4 with $\tau_1 = 0$ and τ_2 very small.

Sampling. If we are measuring samples of $f(\mathbf{s})$ at locations $\mathbf{s}_1, \ldots, \mathbf{s}_M$, then $\mathcal{L}_m(f) = f(\mathbf{s}_m)$, and we can take the h_m above to be Dirac delta functions at these locations, $h_m(\mathbf{s}) = \delta(\mathbf{s} - \mathbf{s}_m)$.

Tomography. In tomographic imaging applications, we observe integrals along lines of the target $f(\mathbf{s})$, which are samples of the Radon transform. In this case, the $h_m(\mathbf{s})$ are delta ridges along lines at different offsets and orientations.

Frequency measurements. In certain radar and magnetic resonance imaging applications, we can model each observation as a sample of the continuous-time (or space) Fourier transform $\hat{f}(\omega)$ of $f(\mathbf{s})$. This is the type of measurement we used for the numerical experiment in Section 22.1.1; an observation at (time or spatial) frequency ω_m is given by (22.1). This has the precise form of (22.2) with $h_m(\mathbf{s}) = e^{-j\omega_m^{\mathrm{T}}\mathbf{s}}$.

Convolutions. Another common measurement model is that we observe samples of the convolution of $f(\mathbf{s})$ with a known template $h(\mathbf{s})$. In active sonar or radar imaging, a transmitter sends out a pulse, this pulse interacts with the environment, and returns convolved with the unknown reflectivity field. In coded aperture imaging [379], the sensor array samples the unknown image convolved with one or more diverse point-spread functions. In these cases, the h_m are shifts of a flipped template h, $h_m(\mathbf{s}) = h(\mathbf{s}_m - \mathbf{s})$, where the $\mathbf{s}_m$ are the locations at which the convolution is sampled.

22.1.2.2 Bases for Discretization and Sparsity

As an estimate of the object $f(\mathbf{s})$ is *computed* from the measurements, it will also have to be represented by a finite list of numbers. A *basis representation* is a standard mathematical tool for moving from the continuum to the discrete. We will assume that our f of interest can be written as a linear combination of a fixed and finite number of basis functions $\psi_1(\mathbf{s}), \ldots, \psi_N(\mathbf{s})$:

$$f(\mathbf{s}) = \sum_{n=1}^{N} x_n \psi_n(\mathbf{s}). \tag{22.3}$$

The basis functions are again functions of continuous-valued variables in $\mathbf{s}$, but since they are fixed, the decomposition above allows us to capture $f(\mathbf{s})$ with the finite (but possibly large) list of numbers $x_1, \ldots, x_N$. We are of course introducing an implicit modeling assumption here by asserting that $f(\mathbf{s})$ is in the linear space of dimension N spanned by the ψ_n; in practice, equality in (22.3) will hold only approximately. But by choosing the ψ_n carefully and making N large enough, we can generally have approximations that are arbitrarily close.

The synthesis formula (22.3) is a linear generative model. Its effectiveness depends only on the properties of $f(\mathbf{s})$ and is completely independent of the way the object is observed. Unlike the h_m in the measurement process (22.2), we have complete control over the choice of the ψ_n. There are a number of considerations when choosing a basis, the foremost of which is that the objects of interest can be closely approximated by a linear combination of the basis elements — making the space sufficiently dense, though, is usually just a matter of selecting the appropriate dimension N.

Along with closely approximating the object of interest, a good basis representation should do it with as few terms as possible. In the expansion (22.3), we would like a very small percentage of the N coefficients to be significant. If every object of interest has significant coefficients concentrated on the same small subset, then of course we can tighten up the model to include only these terms, effectively reducing the dimension N. The model underlying compressive sensing is slightly different: each object can be closely approximated using a small subset of the $\psi_n(\mathbf{s})$, but this subset is different for every $f(\mathbf{s})$. We want the $\{\psi_n(\mathbf{s})\}$ to provide a **sparse representation** for objects of interest.

It is easy to imagine sensing problems where the target is sparse. Sparsity in the spatial domain (taking ψ_n as pixels, say) is a common model for astronomical images, where the energy is concentrated around a small number of points in space. It also accurately describes the differences between video frames; if there are only a few slowly moving objects, only the pixels around their edges will be active in the frame differences. But sparsity is a far broader phenomenon than concentration in time or space. There are many interesting objects of interest whose energy is distributed more or less evenly in the time or spatial domain, but the energy concentrates on a small number of terms in a basis expansion. For example, signals and images that are smooth except at a small number of singularities (or along a small number of edges), tend to be very sparse in the wavelet domain [626]. This idea is at the core of many image compression [942] and denoising algorithms [271].

Examples of natural basis expansions include:

Pixels. Perhaps the simplest discretization technique is to reconstruct an $f(\mathbf{s})$ piecewise constant. This corresponds to making the $\psi_n(\mathbf{s})$ indicator functions over regions which tile the domain. For example, when $\mathbf{s} \in [0,1]^2$ and the $\psi_n(\mathbf{s})$ are indicator functions over squares, the model (22.3) for $f(\mathbf{s})$ is that it is a pixelated image. As N gets larger, the size of these pixels gets smaller, and the approximation becomes more accurate.

Splines. A slightly more sophisticated model is that the object $f(\mathbf{s})$ is piecewise polynomial with a certain number of continuous derivatives. B-spline functions form a convenient basis for this space, and are equipped with efficient computational algorithms. The classic reference for these types of bases is [1115], and [1070, 1071] are surveys for how splines are used in signal processing.

Local Fourier Bases. In many applications in signal and image processing, objects are naturally described by their frequency content over successive windows of time. This type of basis representation might be thought of as a piecewise trigonometric polynomial model, where intervals in time or space are built up using localized sinusoids at different frequencies. There are many variations on this theme. The simplest is to partition the domain of the object into blocks, and use independent Fourier-like representations over each of these blocks; this representation is central to the JPEG image compression standard [775]. Other variations include the *lapped orthogonal transform* [630], used in audio encoding standards, which carefully allow the subdomains to overlap, and the basis functions to be globally smooth.

Wavelets. Wavelet bases build up $f(\mathbf{s})$ in a multiscale manner. The basis functions can be partitioned into nested dyadic scales, with each successive scale adding details that effectively double resolution from all previous scales. The basis functions are also local in time/space, making the representation more robust to local singularities: edges and jumps in signals and images affect only a small portion of the transform coefficients. Roughly speaking, wavelet basis functions are locally oscillatory and act like local edge detectors. A thorough introduction to the theory and practice of using wavelet transforms in signal and image processing can be found in [237, 626].

Dictionary Learning. Sparse bases can also be learned from data. If we have many examples of $f(\mathbf{s})$ from the class of interest, we can search for a fixed basis in which they are all sparse. Techniques for solving this problem are called *dictionary learning* algorithms [751, 15]. Training a sparse basis from examples involves solving a complicated and computationally demanding optimization program, but this can be done once and the result then applied to many problem instances. In fact, there is a chapter on dictionary learning for problems in materials elsewhere in this volume [819].

22.1.2.3 *The Linear Acquisition Model*

With the basis chosen (22.3) and the measurements defined (22.2), the discrete model is in place. We can now write the problem in matrix form, as each of the measurements is a finite linear combination of the expansion coefficients,

$$
\begin{aligned}
y_m = \int f(\mathbf{s}) h_m(\mathbf{s}) \, d\mathbf{s} &= \int \left(\sum_{n=1}^{N} x_n \psi_n(\mathbf{s}) \right) h_m(\mathbf{s}) \, d\mathbf{s} \\
&= \sum_{n=1}^{N} x_n \left(\int h_m(\mathbf{s}) \psi_n(\mathbf{s}) \, d\mathbf{s} \right) \\
&= \sum_{n=1}^{N} A_{m,n} x_n,
\end{aligned}
$$

where

$$
A_{m,n} = \int h_m(\mathbf{s}) \psi_n(\mathbf{s}) \, d\mathbf{s}.
$$

We can collect the weights $A_{m,n}$ for each of these linear combinations, formed as above from the continuous-time (or -space) inner products between the measurement functions $h_m(\mathbf{s})$ and basis functions $\psi_n(\mathbf{s})$, into an $M \times N$ matrix $\mathbf{A}$, giving us the fundamental model

$$
\mathbf{y} = \mathbf{A}\mathbf{x}.
$$

Above, we have collected the M measurements $y_1, \ldots, y_M$ into $\mathbf{y} \in \mathbb{R}^M$, the (unknown) basis expansion coefficients $x_1, \ldots, x_N$ into $\mathbf{x} \in \mathbb{R}^N$, and the $M \times N$ continuous-time (or -space) inner products between the measurement functions $h_m(\mathbf{s})$ and basis functions $\psi_n(\mathbf{s})$ into the matrix $\mathbf{A}$.

The acquisition problem is now firmly in the realm of linear algebra. As we will start to appreciate in the next section, this allows us to make general statements about our ability to recover the expansion coefficients $\mathbf{x}$ (and hence the object of interest $f(\mathbf{s})$) from $\mathbf{y}$ based on well-defined properties of the matrix $\mathbf{A}$ and our model for $\mathbf{x}$.

22.1.3 Classical Least Squares Recovery

With the linear model in place, and our problem set in the framework of linear algebra, we can now apply general computational tools to recover $\mathbf{x}_0$ from $\mathbf{y} = \mathbf{A}\mathbf{x}_0 + \text{noise}$. Classically, the starting point for this problem is *least-squares*, where we solve

$$
\underset{\mathbf{x}}{\text{minimize}} \ \|\mathbf{y} - \mathbf{A}\mathbf{x}\|_2^2. \tag{22.4}
$$

This problem has a closed-form solution[3] that is linear in the measurements $\mathbf{y}$

$$
\hat{\mathbf{x}}_{\text{ls}} = (\mathbf{A}^{\mathsf{T}}\mathbf{A})^{-1}\mathbf{A}^{\mathsf{T}}\mathbf{y}. \tag{22.5}
$$

The eigenvalues of $\mathbf{A}^{\mathsf{T}}\mathbf{A}$ (or equivalently the singular values of $\mathbf{A}$) completely determine how well we can recover $\mathbf{x}_0$ from $\mathbf{y}$ using (22.5). If $\mathbf{A}^{\mathsf{T}}\mathbf{A}$ is well-conditioned in that all of its eigenvalues are within a reasonable factor of one another, then $(\mathbf{A}^{\mathsf{T}}\mathbf{A})^{-1}$ is well-defined, and the recovery will be stable in the presence of noise. Geometrically, the smallest and largest eigenvalues of $\mathbf{A}^{\mathsf{T}}\mathbf{A}$ give us an upper and lower bound on how the action of $\mathbf{A}$ affects the distances between vectors:

$$
\lambda_{\min} \|\mathbf{x}_1 - \mathbf{x}_2\|_2^2 \leq \|\mathbf{A}\mathbf{x}_1 - \mathbf{A}\mathbf{x}_2\|_2^2 \leq \lambda_{\max} \|\mathbf{x}_1 - \mathbf{x}_2\|_2^2, \quad \text{for all } \mathbf{x}_1, \mathbf{x}_2 \in \mathbb{R}^N. \tag{22.6}
$$

If $\lambda_{\min}$ and $\lambda_{\max}$ are not too different from one another, then $\mathbf{A}$ approximately preserves the distances between all pairs of vectors in $\mathbb{R}^N$. In some sense, this means that vectors are as easy to identify from their image through $\mathbf{A}$ as they are in the ambient space.

---

[3]For readers not familiar with Equation (22.5), I recommend the classic exposition in [975, Chapter 3] and the more modern discussion in [976, Chapter 8].

A direct consequence of (22.6) is that if we observe $\mathbf{y} = \mathbf{A}\mathbf{x}_0 + \boldsymbol{e}$, where $\boldsymbol{e}$ is an unknown error vector, then the recovery error will obey

$$\|\hat{\mathbf{x}}_{ls} - \mathbf{x}_0\|_2^2 \leq \frac{1}{\lambda_{min}} \|\boldsymbol{e}\|_2^2, \tag{22.7}$$

where λ_{min} is the smallest eigenvalue of $\mathbf{A}^T\mathbf{A}$. In other words, the energy in the worst-case recovery error is proportional to the noise energy, where the multiplicative constant depends on the smallest singular values of $\mathbf{A}$. When $\boldsymbol{e}$ is *random*, then an average case reconstruction error is also straightforward to derive (see, for example, [906, Chapter 9]). In particular, if the entries of $\boldsymbol{e}$ are uncorrelated zero-mean normal random variables with variance σ^2, then the expected recovery error obeys

$$E[\|\hat{\mathbf{x}}_{ls} - \mathbf{x}_0\|_2^2] = \left(\frac{1}{N} \sum_{n=1}^{N} \frac{1}{\lambda_n} \right) M\sigma^2, \tag{22.8}$$

where the λ_n are the eigenvalues of $\mathbf{A}^T\mathbf{A}$. So the average reconstruction error is controlled by the average of the inverse eigenvalues.

Of course, if the rank of $\mathbf{A}$ is less than N (which is certainly the case if there are fewer measurements than unknowns, $M < N$) or more generally if $\mathbf{A}^T\mathbf{A}$ is very poorly conditioned, then the solution to (22.4) is not unique and is unstable. A standard technique for stabilizing the solution is by adding a small energy term,

$$\underset{\mathbf{x}}{\text{minimize}} \ \|\mathbf{y} - \mathbf{A}\mathbf{x}\|_2^2 + \tau \|\mathbf{x}\|_2^2. \tag{22.9}$$

The solution is still linear in $\mathbf{y}$: $\hat{\mathbf{x}} = (\mathbf{A}^T\mathbf{A} + \tau\mathbf{I})^{-1}\mathbf{A}^T\mathbf{y}$, but now it is always unique. The addition of a multiple of the identity has the effect of "lifting up" all of the eigenvalues so that they are at least τ, which stabilizes both (22.7) and (22.8).

22.2   Principles of Sparse Recovery

When $\mathbf{A}$ is underdetermined, there are no longer uniform guarantees for recovering arbitrary $\mathbf{x}_0 \in \mathbb{R}^N$. In this case, $\mathbf{A}^T\mathbf{A}$ has (possibly many) eigenvalues that are equal to zero, the inverse of $\mathbf{A}^T\mathbf{A}$ does not exist, and the solution to the optimization program (22.4) is not unique. Regularizing least-squares, as in (22.9), makes the solution unique and stable. But the recovery will essentially contain only the part of $\mathbf{x}_0$ that is in the span of the rows of $\mathbf{A}$ — we recover only what we measure, and everything else is cut out.

The central message in body of work that has come to be known as *compressed sensing* is that if the solution $\mathbf{x}_0$ is structured, then we can estimate it with surprising accuracy from underdetermined measurements. The theory has two central components. First, the matrix $\mathbf{A}$ must somehow be compatible with the type of structure we are expecting in the solution. There are many different notions of compatibility, but the most straightforward is that $\mathbf{A}$ *stably embeds* sparse vectors into the subspace spanned by its rows. Rather than ask that all of our vectors are contained in the rows space of $\mathbf{A}$, we ask only that their projections onto this space are identifiable. The second component is a computational framework, complete with mathematical guarantees of accuracy, for performing the recovery. We will see in Section 22.3 that there is a wide variety of algorithms with similar recovery guarantees. Which of the these algorithms is the "best" depends heavily on the application.

One of the sparse recovery techniques that we will look at below, and perhaps the one which has received the most attention in the literature, is to replace the ℓ_2 regularization term in (22.9) with an ℓ_1 penalty:

$$\underset{\mathbf{x}}{\text{minimize}} \ \|\mathbf{y} - \mathbf{A}\mathbf{x}\|_2^2 + \tau \|\mathbf{x}\|_1. \tag{22.10}$$

This seemingly small change has dramatic consequences. The ℓ_1 term above can be viewed as a regularization term that encourages sparsity in the solution. For a fixed energy (ℓ_2 norm), vectors which are concentrated on a small subset have much smaller ℓ_1 norm than vectors which are diffuse.

The conditions under which (22.10) and its variations can effectively recover sparse vectors was a subject of intense study in the latter half of the 2000s. The main results of this work are probably best understood in parallel to the classical least-squares results from the last section. When $\mathbf{A}$ is underdetermined, meaning it has a non-trivial null space, then $\lambda_{min} = 0$ in (22.6), and multiple vectors map to the same set of measurements. In this situation, of course, we will not be able to recover an arbitrary vector from its image through $\mathbf{A}$. If we are interested in recovering sparse vectors, though, we only need a weaker condition on $\mathbf{A}$. In particular, if we are interested in recovering a $\mathbf{x}_0 \in \mathbb{R}^N$ that has at most S non-zero components, then ℓ_1 regularization and other sparse recovery algorithms are effective if $\mathbf{A}$ preserves the pairwise differences between such vectors. We ask that there is a $0 < \delta < 1$ such that

$$(1-\delta)\|\mathbf{x}_1 - \mathbf{x}_2\|_2^2 \le \|\mathbf{A}\mathbf{x}_1 - \mathbf{A}\mathbf{x}_2\|_2^2 \le (1+\delta)\|\mathbf{x}_1 - \mathbf{x}_2\|_2^2, \quad \text{for all } S\text{-sparse } \mathbf{x}_1, \mathbf{x}_2 \in \mathbb{R}^N. \quad (22.11)$$

This condition has come to be known at the *restricted isometry property* [168] for $\mathbf{A}$. The analogy with (22.6) is clear, but instead of (22.11) being a statement about the range of singular values of $\mathbf{A}$, it is a statement about the range of singular values for all submatrices of $\mathbf{A}$ of a certain size. If we form a $M \times 2S$ matrix $\mathbf{A}_\Gamma$ by selecting any subset of $2S$ columns from $\mathbf{A}$, then the condition (22.11) means that all of the eigenvalues of $\mathbf{A}_\Gamma^T \mathbf{A}_\Gamma$ are between $1 - \delta$ and $1 + \delta$.

If (22.11) holds for δ small enough, then there are mathematical guarantees about how well we can recover a sparse vector in the presence of noise. For example, if we observe $\mathbf{y} = \mathbf{A}\mathbf{x}_0 + \mathbf{e}$, where now $\mathbf{x} \in \mathbb{R}^N$ is S-sparse and $\mathbf{A}$ is underdetermined, then for an appropriate choice of τ, the solution $\hat{\mathbf{x}}_1$ to (22.10) will obey

$$\|\hat{\mathbf{x}}_1 - \mathbf{x}_0\|_2^2 \le \frac{\text{Const}}{1-\delta}\|\mathbf{e}\|_2^2. \quad (22.12)$$

This parallels the least-squares result (22.7), where in place of the smallest eigenvalue of $\mathbf{A}^T\mathbf{A}$, we have the lower bound in (22.11) (times a small constant). Variants of (22.10) have also been used to produce average case results [169]. If the entries of $\mathbf{e}$ are independent and normally distributed with zero mean and variance σ^2, then the expected recovery error will obey

$$\mathrm{E}[\|\hat{\mathbf{x}}_1 - \mathbf{x}_0\|_2^2] \le \text{Const}_\delta \cdot S\sigma^2,$$

where the constant now depends on the isometry constant δ. The effect of generic errors, then, is proportional to the number of active components in the target; the sparser the signal is, the more accurate the recovery will be. Least-squares, on the other hand, has an expected recovery error (see (22.8)) that is the same for all targets $\mathbf{x}_0$.

22.2.1 Sparse Embeddings

Above, we interpreted the property (22.11) as meaning that sparse signals are about as far apart from one another after being mapped through $\mathbf{A}$ as they were in the original ambient space. Another way to say this is that $\mathbf{A}$ *stably embeds* the set of $\mathbf{x} \in \mathbb{R}^N$ that are sparse in the lower-dimensional space $\mathbb{R}^M$.

The question remains as to what kinds of matrices obey (22.11). We start with two concrete examples. Consider

$$\mathbf{A} = \frac{1}{2}\begin{bmatrix} 1 & -1 & 1 & -1 & -1 & -1 & -1 \\ -1 & 1 & -1 & 1 & -1 & -1 & 1 \\ 1 & 1 & 1 & -1 & -1 & 1 & 1 \\ 1 & -1 & -1 & 1 & 1 & -1 & -1 \end{bmatrix}$$

In this case, the restricted isometry property (22.11) does not hold for $S = 1$. To see this note that columns 2 and 7 are linear multiples of one another, as are columns 3 and 4. This means, for instance, that

$$\mathbf{A}\mathbf{x}_1 - \mathbf{A}\mathbf{x}_2 = \mathbf{0}, \quad \text{for } \mathbf{x}_1 = \begin{bmatrix} 0 \\ \alpha \\ 0 \\ 0 \\ 0 \\ 0 \\ 0 \end{bmatrix}, \; \mathbf{x}_2 = \begin{bmatrix} 0 \\ 0 \\ 0 \\ 0 \\ 0 \\ 0 \\ \alpha \end{bmatrix}, \; \alpha \in \mathbb{R}.$$

That is, we have found two one-sparse vectors that have the same image through $\mathbf{A}$, and so the lower bound in (22.11) cannot hold. For our second example, consider:

$$\mathbf{A} = \frac{1}{2} \begin{bmatrix} 1 & -1 & 1 & 1 & 1 & -1 & 1 \\ 1 & 1 & -1 & 1 & 1 & -1 & -1 \\ 1 & 1 & 1 & -1 & 1 & 1 & -1 \\ 1 & 1 & 1 & 1 & -1 & 1 & 1 \end{bmatrix}.$$

In this case, (22.11) holds with $S = 1$ for $\delta = 1/2$. That this is true is not evident simply by looking at the matrix; to get this value of δ, we extracted every combination of two columns (there are 21 of them), and then looked at the conditioning of each of the 4×2 submatrices. All of these submatrices have singular values between $\sqrt{1/2}$ and $\sqrt{3/2}$.

Unlike (22.6), where the upper and lower bounds can be computed using a singular value decomposition, finding the smallest δ such that (22.11) holds becomes computationally intractable [502, 49] when M, N, and S are only moderately large. The brute-force search used in the example above cannot be fundamentally improved on, and its complexity is exponential in S. (If we were to attempt to calculate δ for $N = 100$, $M = 50$, and $S = 10$, it would require computing the SVD of 4.7 trillion 50×20 matrices.) The constant δ can be bounded in terms of the maximum off-diagonal entry in $\mathbf{A}^T \mathbf{A}$ [273], but these bounds are usually too loose to be of practical interest.

Despite the difficulty in verifying (22.11), there are many *random* constructions for $\mathbf{A}$ where we know that (22.11) will hold with probability very close to 1. For example, if we created an $M \times N$ random matrix by choosing each entry to be 1 or -1 with equal probability, then after normalizing by $1/\sqrt{M}$, this matrix will obey (22.11) with probability *extremely* close to 1 when

$$M \geq \text{Const} \cdot S \log(N/S). \tag{22.13}$$

This means that the number of rows in $\mathbf{A}$ (which represents the number of equations in our system, or the number of linear measurements we have taken of $\mathbf{x}_0$) can be significantly less than the number of variables N we are trying to recover, as long it is only slightly larger than the sparsity S of the vectors we are interested in recovering. This same result holds when the entries in the matrix are generated from any one of a number of distributions (other examples include entries that are distributed normally or uniformly) [56, 168, 270, 1102].

Numerically, for reasonable values of (N, M, S) (each on the order of 10^5, say), we can typically recover an S-sparse vector from $M \approx 4S$ measurements for a variety of different types of $\mathbf{A}$ (see the early computational experiments in [166], or the more recent and extensive ones in [279, 700]). In fact, when $\mathbf{A}$ contains independent and identically distributed Gaussian entries, the recovery behavior is completely understood, with the theory matching the numerical experiments exactly [278].

One way to interpret these results for random matrices is that "most" matrices of a certain size obey (22.11) for vectors of a certain sparsity — if we select one at random, it will almost surely work. This is interesting, but of course in practice the matrix $\mathbf{A}$ is determined by physics; it is a model for the instrumentation we are using to observe the object of interest. There are indeed imaging applications where random mixing is a good model for the measurements [710, 30, 283, 353, 834, 848], but in general, the measurements will have definitive structure. Fortunately, there are similar embedding guarantees for matrices that have *structured randomness* that is more amenable to physical implementation. These (theoretical) results still rely on injecting some kind of controlled randomness into the measurement system, but this is done in such a way that it is physically implementable.

Qualitatively, the properties we look for in a good compressed sensing matrix are the opposite of the sparse structure we use as a model for the vector. Matrices that allow sparse recovery have rows which are diverse, and their entries are not concentrated on a small set. The intuition for this is that if every row $\mathbf{a}_m$ of $\mathbf{A}$ is spread out, then each measurement $y_m = \langle \mathbf{x}_0, \mathbf{a}_m \rangle$ will contain information about all of the entries of $\mathbf{x}_0$ — since we do not know which entries of $\mathbf{x}_0$ are important a priori, collecting information about all of them allows us to triangulate the positions of the significant terms in as few measurements as possible. Put another way, if $\mathbf{x}_0$ is concentrated on a small set, we want $\mathbf{A}\mathbf{x}_0$ to be diffuse.

Below, we give a few examples of this type of measurement system, including some discussion of how they arise in applications.

Subsampled Fourier. Signals which are concentrated on small sets in the time domain (sparse) tend to be spread out in the frequency domain. In fact, the discrete uncertainty principles in [277, 275, 165] show that the energy concentrated on generic subsets of a certain size will contain about the same amount of energy for (most) sparse signals.

A consequence of this phenomena is that (most) subsampled Fourier matrices are efficient sparse embeddings. To state this result precisely, let $\mathbf{F}$ be the $N \times N$ discrete Fourier matrix, $F[m,n] = e^{-j2\pi mn/N}$, $0 \leq m, n \leq N - 1$. We measure the sparse signal by observing its Fourier coefficients on a set of locations indexed by $\Omega \subset \{0, \ldots, N-1\}$; the size of this set is $|\Omega| = M$. Certain choices of this subset Ω will be better than others. For example, if M divides N evenly and Ω contains the equally spaced indexes $\Omega = \{N/M, 2N/M, \ldots, N\}$, then the well-known aliasing effect will make components in $\mathbf{x}_0$ at locations n and $n + M$ indistinguishable from one another. Aliasing, however, is the exception and not the rule. If $M \gtrsim S \log^4 N$ frequencies are chosen uniformly at random, then $\mathbf{A} = M^{-1/2}\mathbf{F}_\Omega$ obeys (22.11) with high probability [870, 168]. The only difference between this result and the independent entry result in (22.13) is the log factor. The increased factor in the Fourier case appears to be an artifact of the analysis; numerically, we are able to recover a sparse vector from roughly the same number of randomly selected Fourier coefficients as inner products against random vectors.

Measuring an object of interest against a series of sinusoids at different frequencies is a scenario encountered in many different imaging applications. Techniques for recovering a sparse vector from limited frequency measurements have influenced magnetic resonance imaging [607], stepped frequency radar [402], and seismic imaging [580].

Convolutions. Another type of compressive sensing system that can by physically implemented is convolution with a fixed, diverse kernel followed by subsampling. If the kernel we convolve the (sparse) signal or image with is random in nature, then we can dramatically subsample the convolution without sacrificing resolution; this is demonstrated with theory in [416, 822, 858, 511], using numerical examples in [639, 642], and in hardware in [35, 455, 622, 1112]. Moreover, if this convolution is combined with certain kinds of random subsampling schemes, then it is *universal* in that it is equally effective for sparsity in any orthobasis $\{\psi_n\}$ [858]. Similar architectures have been proposed for increasing the field-of-view of a camera by superimposing multiple images onto the same sensor array, and then treating their untangling as an underdetermined linear inverse problem

[859, 640, 641]. This same principle has also been used to accelerate forward simulations in seismic imaging [727, 428].

Modulate and Integrate. We can interpret convolution with a random kernel as a *modulation* in the frequency domain. Compressive sampling can also be realized by modulating the signal in the time/spatial domain, low pass filtering, and then sampling on a coarse uniform grid [1058, 691]. Mathematically, this corresponds to a pointwise multiplication of the vector $\mathbf{x}_0$ with a random (or pseudorandom) code, and then summing the result over blocks. This type of compressed sensing architecture has been applied to radar pulse detection [1181] and video capture [889, 1094].

22.3   Algorithms for Sparse Recovery

Now that we have seen several examples of underdetermined systems that produce different images for every different sparse vector they take as input, we turn our attention to algorithms for finding a sparse solution to a system of equations. The basic problem is that we are given $\mathbf{y} \in \mathbb{R}^M$ and an $M \times N$ matrix $\mathbf{A}$, and we want to find a $\hat{\mathbf{x}} \in \mathbb{R}^N$ with as few non-zeros as possible such that $\mathbf{y} \approx \mathbf{A}\hat{\mathbf{x}}$. Many techniques for solving this problem have been developed recently, and we do not have space to review them all. Instead, we will focus on three popular methods chosen because 1) they are easy to understand, and 2) they form the basis for many more advanced techniques.

The sparse recovery algorithms presented below come with various mathematical performance guarantees for both their run time and their ability to produce an accurate estimate. As these results, and the conditions under which they hold, are complicated to state exactly, we point the interested reader to the references provided. Basically, these algorithms perform extremely well when $\mathbf{A}$ is the type of sparse embedding described in Section 22.2.1, and tend to produce competitive results even when this condition does not (or is not known to) hold.

To close, in Section 22.3.4 below we briefly discuss image reconstruction algorithms that exploit more subtle structural properties of the target. Although theoretical results for these techniques are for the most part very limited, moving to models beyond sparsity can significantly improve the quality of the reconstructed image.

22.3.1   *Orthogonal Matching Pursuit*

The most straightforward sparse recovery algorithm is probably *orthogonal matching pursuit* (OMP). It was introduced into the mainstream consciousness of the signal and image processing community in [627, 242], analyzed in the context of sparse approximation in [1057], analyzed in the context of structured (noiseless) recovery in [1056, 821, 240], and stable recovery in [328, 1199] which include theoretical performance guarantees of the type (22.12).

The algorithm is exceptionally simple — as can be seen in the pseudocode in Algorithm 22.1 — and is a benchmark for small problems. Roughly speaking, the algorithm builds up the support of the vector one element at a time by seeing which column in $\mathbf{A}$ is most correlated with the current *residual*, the portion of the measurements $\mathbf{y}$ not yet explained. (Variants which add multiple columns to the support also exist, see for example the StOMP algorithm in [280] and ROMP in [726].) Its simplicity makes OMP a natural starting point for solving sparse recovery algorithms that have a small number of variables.

The computational complexity of OMP to build an S-term approximation is $O(S^2 M)$, and the constants hidden in the $O(\cdot)$ notation are small. The complexity is dominated by step 4 in Algorithm 22.1, which is effectively the Gram–Schmidt algorithm for finding an orthogonal basis for the vectors which have been selected into the representation so far. The speed and simplicity of OMP are nearly impossible to beat when S is small (on the order of 10s) and the matrix $\mathbf{A}$ is created and handled explicitly. However, OMP becomes computationally intractable when S grows to even moderate values. In a typical large-scale imaging experiment, where N, M, and S are on the order of 10^4 to 10^6, using OMP is out of the question. From a performance standpoint, OMP does not per-

---

Algorithm 22.1: Orthogonal matching pursuit (OMP) for approximating a M vector $\mathbf{y}$ using a linear combination of a small number of columns from the $M \times N$ dictionary $\mathbf{A}$.

Require: data $\mathbf{y}$, system matrix $\mathbf{A}$ with columns $\mathbf{A}_1, \ldots, \mathbf{A}_N$, tolerance ε
Ensure: $\hat{\mathbf{x}}$
1: **Intialize:** set the estimate $\mathbf{x}_0 = 0$, and residual $\boldsymbol{r}_0 = \mathbf{y}$
2: **repeat** while incrementing k
3:     $\gamma_k = \arg\max_{1 \le n \le N} |\mathbf{A}_n^{\mathrm{T}} \boldsymbol{r}_k|$ $\triangleright$ Find the most closely correlated column with $\boldsymbol{r}_k$
4:     $\mathbf{u}_k' = \mathbf{A}_{\Gamma_k} - \sum_{\ell=0}^{k-1} (\mathbf{A}_{\gamma_k}^{\mathrm{T}} \mathbf{u}_\ell) \mathbf{u}_\ell$
5:     $\mathbf{u}_k = \mathbf{u}_k' / \|\mathbf{u}_k'\|_2$ $\triangleright$ The $\{\mathbf{u}_k\}$ are an orthobasis for the space the estimate $\mathbf{x}_k$ lives in
6:     $\mathbf{x}_{k+1} = \mathbf{x}_k + (\boldsymbol{r}_k^{\mathrm{T}} \mathbf{u}_k) \mathbf{u}_k$ $\triangleright$ Update the signal estimate
7:     $\boldsymbol{r}_{k+1} = \boldsymbol{r}_k - (\boldsymbol{r}_k^{\mathrm{T}} \mathbf{u}_k) \mathbf{u}_k$ $\triangleright$ Update the residual
8: **until** $\|\boldsymbol{r}_{k+1}\|_2 \le \varepsilon$ or $k \ge k_{\max}$
9: $\hat{\mathbf{x}} = \mathbf{x}_{k+1}$

---

form as well as variational techniques based on ℓ_1 minimization or approximate message passing (see, for example, [446] for a detailed performance comparison in the context of channel estimation for underwater communications).

In short, OMP is extremely efficient for finding very sparse solutions to small systems of equations, but it does not scale (either computationally or in terms of performance) to large problems.

22.3.2 Iterative Hard Thresholding

Given a direct observation of a vector $\mathbf{x}_0$, finding the best sparse approximation (in a least-squares sense) is a simple task. We formulate this problem as follows: given $\mathbf{x}_0$, solve

$$\underset{\mathbf{x}}{\text{minimize}} \; \|\mathbf{x}_0 - \mathbf{x}\|_2^2 \quad \text{subject to} \quad \|\mathbf{x}\|_0 = S. \tag{22.14}$$

Here, $\|\mathbf{x}\|_0$ is the number of non-zero terms[4] in $\mathbf{x}$, and so S is the desired level of sparsity. To solve (22.14), we take $\mathbf{x}_0$, retain its S largest terms, and set the rest to zero.

We can also solve (22.14) in Lagrange form. We set up an unconstrained problem that balances the distance to $\mathbf{x}_0$ against the number of non-zero terms,

$$\underset{\mathbf{x}}{\text{minimize}} \; \|\mathbf{x}_0 - \mathbf{x}\|_2^2 + \tau \|\mathbf{x}\|_0. \tag{22.15}$$

The parameter τ above adjusts the trade off between finding an $\mathbf{x}$ that matches $\mathbf{x}_0$ (small τ), and finding an $\mathbf{x}$ that has only a small number of non-zero components (large τ). As both of the terms in the objective are separable, the solution can be found term-by-term; it is given by applying a *hard thresholding operator* to $\mathbf{x}_0$:

$$\hat{x}[n] = \begin{cases} x_0[n], & |x_0[n]| \ge \sqrt{\tau}, \\ 0, & |x_0[n]| < \sqrt{\tau}, \end{cases} \quad n = 1, \ldots, N. \tag{22.16}$$

So solving (22.15) simply requires us to apply a non-linearity to each entry of $\mathbf{x}_0$.

When working from indirect observations, $\mathbf{y} \approx \mathbf{A}\mathbf{x}_0$, the analogs of (22.14) and (22.15) become intractable. While we just saw that (22.14) has a straightforward solution, solving

$$\underset{\mathbf{x}}{\text{minimize}} \; \|\mathbf{y} - \mathbf{A}\mathbf{x}\|_2^2 \quad \text{subject to} \quad \|\mathbf{x}\|_0 = S, \tag{22.17}$$

---

[4]Despite the notation, the number of non-zero terms is not a valid norm on $\mathbb{R}^N$. The notation $\|\cdot\|_0$ is historical, and is unfortunately now standard.

exactly is an NP hard problem [721] for general $\mathbf{A}$. We can, however, attempt to solve with an iterative algorithm that works by making a guess at the entire support of the target vector, testing how well this support explains the measurements $\mathbf{y}$, then adjusting this guess by seeing which elements of the support that were not included can reduce the residual the most. This iteration is known as *iterative hard thresholding* (IHT); this is a somewhat classical algorithm which was popularized in the sparse recovery community by two papers [128, 108, 109] which showed how to use it effectively and analyzed its performance. (The reference [128] in particular illustrates the application of IHT to several problems in imaging.)

The IHT algorithm can be captured with one line. We use $H_S(\mathbf{v})$ to denote the operator that implements (22.16) above, taking a vector $\mathbf{v}$ and zeroing out all but the largest (in magnitude) S elements. Then starting with $\mathbf{x}_0 = 0$, we iterate over k with

$$\mathbf{x}_{k+1} = H_S(\mathbf{x}_k + \mathbf{A}^{\mathrm{T}}(\mathbf{y} - \mathbf{A}\mathbf{x}_k)). \tag{22.18}$$

Conceptually, at each iteration we are taking the residual $\mathbf{y} - \mathbf{A}\mathbf{x}_k$, correlating with all the columns of $\mathbf{A}$ (by applying $\mathbf{A}^{\mathrm{T}}$), adding the result to the current estimate, then keeping only the S largest terms. Of course, our estimate will have exactly S non-zero terms in it. There is of course no guarantee that this procedure finds the solution to (22.17) in the general case. However, if $\mathbf{A}$ is a sparse embedding as in (22.11), then this procedure does come with guarantees for recovering vectors which are sufficiently sparse [109].

Empirically, the active support (i.e., the locations of the S largest elements) tends to go through dramatic changes in the first few iterations of IHT, but if a sparse solution does indeed exist, the algorithm settles on the correct support within 10s of iterations, especially for small-scale algorithms. Notice that even if the correct S elements have been chosen, (22.18) might still iterate to make corrections on the support.

Each iteration of IHT is inexpensive; the cost is almost exactly the cost of one application of $\mathbf{A}$ and one application of $\mathbf{A}^{\mathrm{T}}$. In general, these matrix multiplications have a computational complexity of $O(MN)$. But if there are fast, implicit algorithms for applying $\mathbf{A}$ and its transpose, IHT can take direct advantage of this.

For small problems with truly sparse solutions (N and M on the order of 10^3 and S on the order of 10^2), IHT converges very quickly, typically taking 10s or 100s of iterations. For larger problems, and for problems in which the support size is relatively large (more than 10% of N) or there is significant measurement noise, the convergence can be much slower, sometimes taking 1000s of iterations. In addition, its performance in terms of reconstruction quality on this type of problem is not competitive when compared against state-of-the-art variational techniques.

There is a method which is in some sense a combination of OMP and IHT called *compressive sampling matching pursuit* (CoSaMP) [724] that is widely used in the sparse recovery community. The iterations in the algorithm, pseudocode for which can be found in Algorithm 22.2, are similar in structure to IHT in that they choose a candidate support by correlating against the residual, but CoSaMP refines the estimate on the support by additionally optimizing the coefficient sequence on this support (i.e., solving a least-squares problem). It adjusts the support globally at each iteration like IHT, but maintains an "optimal" coefficient sequence on that support like OMP.

Because of this more careful choice of expansion coefficients on the support, CoSaMP tends to converge in many fewer iterations than IHT (and IST). But each iteration is more expensive — along with an application of $\mathbf{A}$ and $\mathbf{A}^{\mathrm{T}}$, a $3S \times 3S$ system of equations is solved at every step. In theory, solving this system has a computational complexity of $O(MS^2)$, making it intractable for large problems. But if we have an implicit, fast algorithm for applying $\mathbf{A}$, this substep can be solved using conjugate gradients (or any other method for solving a system of linear equations). Moreover, the matrices $\mathbf{A}_\Gamma$ in Algorithm 22.2, which are $M \times 3S$ matrices formed by extracting the $3S$ columns of $\mathbf{A}$ indexed by Γ, are generally very well conditioned, meaning that an iterative solver can get a very good approximate solution in a very few steps.

For medium- to large-scale imaging applications, CoSaMP tends to give better quality solutions

---

Algorithm 22.2: CoSaMP algorithm for approximating an M vector $\mathbf{y}$ using a linear combination of a small number of columns from the $M \times N$ dictionary $\mathbf{A}$.

Require: data $\mathbf{y}$, system matrix $\mathbf{A}$, target sparsity S, tolerance ε
Ensure: $\hat{\mathbf{x}}$
1: **Intialize:** set the estimate $\mathbf{x}_0 = 0$, and residual $\boldsymbol{r}_0 = \mathbf{y}$
2: **repeat** while incrementing k
3:     $\mathbf{u} = \mathbf{A}^\mathsf{T} \boldsymbol{r}_k$
4:     $\Gamma' =$ locations of largest $2S$ elements in $\mathbf{u}$
5:     $\Gamma = \Gamma' \cup \text{support}(\mathbf{x}_k)$ $\triangleright$ merge Γ' into the support of $\mathbf{x}_k$
6:     $\mathbf{x}_{k+1}|_\Gamma = (\mathbf{A}_\Gamma^\mathsf{T} \mathbf{A}_\Gamma)^{-1} \mathbf{A}_\Gamma^\mathsf{T} \mathbf{y}$ $\triangleright$ solve least-square on the support Γ
7:     $\mathbf{x}_{k+1}|_{\Gamma^c} = 0$ $\triangleright$ set the values off Γ to 0
8:     $\boldsymbol{r}_{k+1} = \mathbf{y} - \mathbf{A}\mathbf{x}_{k+1}$ $\triangleright$ update the residual
9: **until** $\|\boldsymbol{r}_k\|_2 \leq \varepsilon$ or $k \geq k_{\max}$

---

than IHT, and does tend to converge more quickly despite its more expensive iterations. In general, the reconstructions are not quite as good as those based on state-of-the-art variational methods, but these relative performances are application dependent.

22.3.3 Sparse Recovery Using ℓ_1 Minimization

Perhaps the most studied technique for finding a sparse vector $\hat{\mathbf{x}}$ that explains a set of linear measurements is through ℓ_1 minimization. That solving a least-squares problem with an ℓ_1 regularized (such as (22.10) above) results in a solution that has a small number of non-zero terms is a classical heuristic. Programs such as these have been used for decades in model selection in statistical learning [1044], solving inverse problems [207, 890] in geophysics, high-resolution channel and direction of arrival estimation in radar [337, 338], and decoupling disparate phenomena in signal processing [275, 386, 962].

Algorithms for sparse recovery based on ℓ_1 minimization are in some sense more agnostic than the iterative support selection algorithms discussed in previous sections. The algorithms do not explicitly seek out sparse vectors that explain the measurements, but rather simply find the minimum of a functional under given constraints. That these functionals and constraints are convex means that there are many ways in which the optimization programs can be solved. Because of this, ℓ_1 solvers have benefited from decades of research in convex optimization.

In this section, we will look at one particular type of ℓ_1 minimization algorithm that is easy to motivate, is related to the sparse recovery algorithms in the previous sections, and forms the basis for many of the best modern solvers. We start by looking at ℓ_1 minimization problems when we have direct observations; that is, there is no matrix $\mathbf{A}$ in the way. In Section 22.3.2 above, we saw that the sparse approximation problem (22.15) could be solved by component-wise hard thresholding as in (22.16). The analogous ℓ_1 problem,

$$\underset{\mathbf{x}}{\text{minimize}} \ \|\mathbf{x}_0 - \mathbf{x}\|_2^2 + \tau \|\mathbf{x}\|_1, \tag{22.19}$$

has a similarly straightfoward solution: we can compute the solution simply by applying a *soft thresholding* operator to $\mathbf{x}_0$:

$$\hat{x}[n] = \begin{cases} x_0[n] - \tau, & x_0[n] > \tau, \\ 0, & |x_0[n]| \leq \tau, \\ x_0[n] + \tau, & x_0[n] < -\tau. \end{cases} \tag{22.20}$$

Algorithm 22.3: Iterative soft thresholding algorithm for solving the ℓ_1 minimization program in (22.21). For simplicity, we use a fixed stepsize γ, but in practice this might be chosen judiciously from step to step.

Require: data $\mathbf{y}$, system matrix $\mathbf{A}$, parameter τ, tolerance ε
Ensure: $\hat{\mathbf{x}}$
 1: **Intialize:** set the estimate $\mathbf{x}_0 = 0$, and residual $\mathbf{r}_0 = \mathbf{y}$
 2: **repeat** while incrementing k
 3:      $\mathbf{u} = \mathbf{A}^\mathrm{T} \mathbf{r}_k$
 4:      $\mathbf{x}_{k+1} = \mathrm{prox}_\gamma(\mathbf{x}_k - \gamma \mathbf{u})$ ▷ Soft thresholding (22.20) using $\gamma\tau$ as the threshold
 5:      $\mathbf{r}_{k+1} = \mathbf{y} - \mathbf{A}\mathbf{x}_{k+1}$
 6: **until** $\|\mathbf{x}_{k+1} - \mathbf{x}_k\|_2 / \|\mathbf{x}_k\|_2 \le \varepsilon$ or $k \ge k_{\max}$

Soft thresholding plays a critical role in solving the problem that we are really interested in: recovering $\mathbf{x}_0$ from indirect observations $\mathbf{y} \approx \mathbf{A}\mathbf{x}_0$. Unlike the ℓ_0 problem in (22.17), the ℓ_1 minimization program

$$\underset{\mathbf{x}}{\text{minimize}} \ \|\mathbf{y} - \mathbf{A}\mathbf{x}\|_2^2 + \tau\|\mathbf{x}\|_1 \tag{22.21}$$

remains perfectly tractable. The program above is an unconstrained convex program, and although it is not smooth (the ℓ_1 norm is not differentiable), it can be solved in a systematic way using a *proximal method* [218, 764]. Instead of iteratively descending on the functional in the optimization program above, proximal methods start from a *fixed point condition*: it is a fact [218] that minimizers $\mathbf{x}^*$ of (22.21) obey, for every $\gamma > 0$,

$$\mathbf{x}^* = \mathrm{prox}_\gamma\left(\mathbf{x}^* - \gamma\mathbf{A}^\mathrm{T}(\mathbf{A}\mathbf{x}^* - \mathbf{y})\right),$$

where the prox operator is

$$\mathrm{prox}_\gamma(\mathbf{z}) = \underset{\mathbf{x}}{\mathrm{argmin}} \ \|\mathbf{x} - \mathbf{z}\|_2^2 + \gamma\tau\|\mathbf{x}\|_1,$$

which we have already seen we can evaluate quickly using soft thresholding. Finding an $\mathbf{x}^*$ that satisfies the fixed point condition, and hence solves (22.21), can be done with the iteration

$$\mathbf{x}_{k+1} = \mathrm{prox}_{\gamma_k}\left(\mathbf{x}_k - \gamma_k\mathbf{A}^\mathrm{T}(\mathbf{A}\mathbf{x}_k - \mathbf{y})\right), \tag{22.22}$$

for a sequence of carefully chosen "stepsizes" $\{\gamma_k\}$. This iteration can be interpreted in two steps: we take a gradient step to decrease the $\|\mathbf{y} - \mathbf{A}\mathbf{x}\|_2^2$ term, and then soft threshold the result. As such, algorithms with this basic form are referred to as *iterative soft thresholding* (IST). While they are guaranteed to converge for a small enough fixed step size [238, 323], significant gains in convergence speed can be realized through judicious choices of the γ_k [1155, 99]. They can also be accelerated by adding extrapolation terms inside the prox operator, with a small multiple of $(\mathbf{x}_k - \mathbf{x}_{k-1})$ added to the first $\mathbf{x}_k$ [73]. This has no effect on the computational complexity, but can result in significantly faster convergence.

Pseudo-code for ℓ_1 minimization based on iterative soft thresholding is shown in Algorithm 22.3. It is exceedingly simple and scalable. The cost of each iteration is dominated by the application of $\mathbf{A}$ and $\mathbf{A}^\mathrm{T}$. If we have fast methods for applying these operators (which we might if our acquisition matrix is sparse or can be broken down into Fourier transforms and/or convolutions), then these iterations will be equally fast.

We close this section by noting that there are other competitive techniques for performing ℓ_1 minimization that are not based on IST; interested readers can see the works [1168, 1177, 74].

22.3.4   Beyond Sparsity: Recovery Algorithms for Alternative Structure

Compressive recovery is possible because we know the target of interest has special structure: it is sparse. To close this section, we make brief mention of recovery algorithms for signals with other kinds of structure. In [460, 420], a Bayesian framework was developed to take advantage of persistence across scale, a structural property enjoyed by images in the wavelet domain. An alternative Bayesian perspective, which learns the model from the data itself, can be found in [1149]. In [55], a version of the CoSaMP algorithm from Section 22.3.2 above was introduced for signals obeying a more structured "union of subspaces" model, of which sparsity is one example; related work for images where important coefficients appear in clusters can be found in [181]. Another powerful modeling framework, based on approximate message passing [276], that produces state-of-the-art image recovery with a computationally efficient algorithm can be found in [953].

The recent activity in the machine learning community on deep learning [551] has led to an entirely different, data driven way to find structural models for images. These models give us a new way to regularize ill-posed image reconstruction problems [1165, 268, 116], replacing sparsity with a neural-net based geometric model that is learned from examples.

Neural networks have also led to new reconstruction algorithms. Notice that the iterations in (22.18) and (22.22) consist of a linear operator applied to the current estimate followed by a non-linearity. Neural nets mimic this structure with layers consisting of pointwise non-linearities connected by weighted edges. To train the network, a large corpus of measurements is presented along with its corresponding "true" images, and the weights are adjusted so that the network maps the measurements to images that are as close to the true images as possible. Training the neural net (fitting the weights to data), then, can be interpreted as optimizing the linear operators in (22.18), (22.22) so that the reconstructions are as faithful as possible after a small number of "iterations" (layers in the networks). The resulting reconstruction algorithms are extremely fast, and produce state-of-the-art results [385, 195, 466, 707].

Finally, there is a parallel theory and suite of algorithms for recovering *matrices* that are low rank from a small number of linear observations. Original work on this problem can be found in [833, 164], and an overview of low-rank matrix recovery from limited measurements can be found in [239].

22.4   Numerical Example: Computed Tomography

In this section, we give a real-world example of how sparse regularization can make a measurable difference in the quality of the reconstructed image. Our example uses one of the most important and broadly studied computational imaging techniques: tomographic reconstruction [473, 723]. The particular image we are using comes from medicine, but the problem of reconstructing an image from projections arises in fields including geophysics, materials science, and archaeology.

As we have discussed above, in order to apply our sparse reconstruction framework, there are two things that have to be codified: the measurement system and the manner in which we are capturing the (expected) structure in the reconstruction. In order to get state-of-the-art reconstruction results, we depart slightly from the general approach outlined in Section 22.1.2 in two different ways. First, we will use different representations to discretize the measurement operator and model the sparsity; we will describe this further below. Second, we will use an optimization framework similar to that discussed in Section 22.3.3, but we will combine two different regularizers to model the structure of the image. The first type of regularizer is a sparsity penalty in the wavelet domain, similar to what was described in the numerical example in Section 22.1.1. However, rather than using the (convex) ℓ_1 norm for this penalty, we use the ℓ_p norm

$$\|\mathbf{x}\|_p = \left(\sum_{n=1}^{n} |x[n]|^p \right)^{1/p},$$

for some $p < 1$ ($p = 1/2$, to be specific). This has the advantage of in some ways being a better model for sparsity, as the quantity above approaches the number of non-zero terms in $\mathbf{x}$ as $p \to 0$. It has the disadvantage that it is a non-convex function of $\mathbf{x}$. The main consequence of this is that we do not have any guarantees of finding the *global* minimizer of the associated optimization program ((22.24) below). Nevertheless, iterative algorithms with structure very similar to Algorithm 22.3 above tend to produce good results in practice [186, 1110].

Along with this sparsity penalty in the wavelet domain, we will consider a similar type of penalty on the gradient of the image, known as the *total variation* (TV). If $\mathbf{x}$ is an $N_1 \times N_2$ image, then

$$\|\mathbf{x}\|_{\mathrm{TV}} = \sum_{i=1}^{N_1-1} \sum_{j=1}^{N_2-1} |\nabla \mathbf{x}(i,j)| = \sum_{i=1}^{N_1-1} \sum_{j=1}^{N_2-1} \sqrt{(x[i+1,j] - x[i,j])^2 + (x[i,j+1] - x[i,j])^2}.$$

The TV norm might be roughly described as the ℓ_1 norm (sum of magnitudes) of the gradient of $\mathbf{x}$; adding it as a penalty encourages the resulting image to have as few oscillations as necessary while still explaining the observations [872].

Our measurement operators $\mathcal{L}_m$ take integrals of the image along lines parameterized by an angle θ_m and an offset r_m from the center of the image,

$$\mathcal{L}_m(f) = \int f(\mathbf{s}) \delta(\mathbf{s}^{\mathrm{T}} \eta_m - r_m) \, d\mathbf{s}, \quad \eta_m = \begin{bmatrix} -\sin \theta_m \\ \cos \theta_m \end{bmatrix}, \quad m = 1, \ldots, M. \tag{22.23}$$

We discretize the image by representing it using pixels, i.e., as a linear superposition of indicator functions

$$f(\mathbf{s}) = \sum_{i=1}^{N_1} \sum_{j=1}^{N_2} x_0[i,j] \psi_{i,j}(\mathbf{s}), \quad \psi_{i,j}(\mathbf{s}) = \begin{cases} 1, & \mathbf{s} \in [(i-1)/N_1, i/N_1] \times [(j-1)/N_2, j/N_2] \\ 0, & \text{otherwise.} \end{cases}$$

The entries in the measurement matrix $\mathbf{A}$, then, are the length of the line (θ_m, r_m) corresponding to the row index that passes through the pixel (i,j) corresponding to the column index. Above, we find it convenient to arrange the pixel values in a two-dimensional array; however, when we write $\mathbf{A}\mathbf{x}_0$, we understand it as an $M \times N_1 N_2$ matrix acting on these coefficient rasterized into a vector of length $N_1 N_2$.

Given $\mathbf{y} = \mathbf{A}\mathbf{x}_0 + \text{noise}$, we recover the image by solving the optimization program

$$\underset{\mathbf{x}}{\text{minimize}} \, \|\mathbf{y} - \mathbf{A}\mathbf{x}\|_2^2 + \tau_1 \|\mathbf{W}^{\mathrm{T}} \mathbf{x}\|_p^p + \tau_2 \|\mathbf{x}\|_{\mathrm{TV}}, \tag{22.24}$$

for $p = 1/2$ and some fixed values of τ_1, τ_2. Above, $\mathbf{W}$ is a undecimated complex dual-tree wavelet transform [432, 925]; this representation is overcomplete in that $\mathbf{W}$ has more columns than rows. This overcompleteness means that penalizing $\|\mathbf{W}^{\mathrm{T}} \mathbf{x}\|_p^p$ is subtly different than penalizing $\|\alpha\|_p^p$ with the constraint that $\mathbf{W}\alpha = \mathbf{x}$. For an in-depth discussion of using overcomplete non-orthogonal expansions for sparse recovery, see [299, 170].

In the experiment[5] shown in Figure 22.2, the "ground truth" image is measured by taking line integrals at 45 equally spaced angles, with 729 offsets at each angle, for a total of $M = 32,805$ measurements. The image is reconstructed at a resolution of $512 \times 512 = 264,144$ pixels. The classical least-squares reconstruction is shown in Figure 22.2(b); as we can see, the undersampling results in heavy oscillations in the reconstruction. The solution to (22.24) is shown in Figure 22.2(c). Not only have many of the spurious oscillations been removed, but the structure in the region of interest is much clearer.

Images of the type shown in Figure 22.2(a) certainly have the type of sparse structure that we know how to exploit using the tools of compressed sensing. However, it is not known if the

---

[5]This numerical result comes from [1185]. A full description of how to solve (22.24) can also be found in that reference.

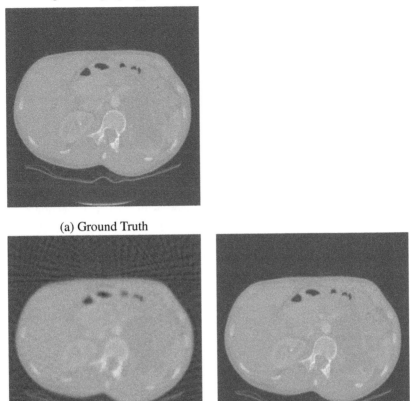

(a) Ground Truth

(b) Least-Squares   (c) Solution to (22.24)

Figure 22.2: Numerical simulation of a tomographic reconstruction. (a) A cross-section of a human kidney at high resolution. Line integrals as in (22.23) of this image were computed numerically at $M = 32,805$ different orientations. (b) The least-squares reconstruction, computed by solving (22.9) with $\tau = 0.04$, of the image on a 512×512 grid of pixels. The undersampling causes significant oscillatory artifacts, and interesting features have been blurred. (c) Sparse reconstruction computed by solving (22.24) with $p = 0.5, \tau_1 = 0.04, \tau_2 = 10^{-5}$. By taking advantage of the structure in the image, we are able to sharpen many of the important details.

measurement operator obeys the condition (22.11). When the number of samples (offsets) taken per angle θ is large though, as it is in this case, the collected data is equivalent to sampling the spatial Fourier transform of the image along lines passing though the origin (this is known as the *Radon-slice theorem*, see [723]). The measurements, then, have the qualitative characteristics of being global and diverse, the same characteristics we look for in compressed sensing systems.

22.5 Summary

Scientists and engineers working in computational imaging have long made use of prior information in image reconstruction algorithms. Having the target image adhere to a certain mathematical structure can make inverse problems better posed, and can make up for deficiencies in the acquisition process (e.g., sensor noise or undersampling). Compressed sensing gives us a solid algorithmic framework for handling sparsity, perhaps the most common model in image processing. What is more, these algorithms come equipped with theoretical guarantees and mathematical theory for when they are effective.

In this chapter, we have given an overview of what the theory of compressed sensing says, and discussed how it might apply to problems in computed imaging. Our discussion started in earnest in Section 22.1.2 with a clear description of how different imaging problems can be cast into the language of linear algebra. While the classical least-squares reconstruction is almost completely characterized by the singular values of the system matrix, we saw in Section 22.2 how the conditions for sparse recovery can be much weaker — they rely not on the system itself being well-conditioned, but only on the systems formed by subsets of the columns. It is this fact that makes exact sparse recovery possible even when there are fewer equations than there are unknowns.

That sparse recovery is possible from underdetermined measurements is certainly interesting, but having a suite of algorithms available (along with provable guarantees for their performance) is what has allowed compressed sensing to connect theory with practice. In Section 22.3, we reviewed a number of core algorithms whose performance is very well understood. These should serve as a starting point for the practitioner; their particulars can be adapted to specific applications of interest. We closed that section by discussing current research on modeling methods that go beyond sparsity.

Chapter 23

Dictionary Methods for Compressed Sensing: Framework for Microscopy

Saiprasad Ravishankar
University of Michigan

R. Rao Nadakuditi
University of Michigan

23.1 Introduction

Techniques exploiting sparsity have become extremely popular in various imaging and image processing applications in recent years. These techniques typically use the sparsity of images or image patches in a transform domain or dictionary to compress [638], denoise, or restore/reconstruct images. Compressed sensing (CS) is a method that enables accurate reconstruction of images from a few measurements by exploiting the sparsity of the images in a known transform domain or dictionary. CS has been recently used in applications such as materials science microscopy and enables dose reduction and accelerated data acquisition. In this chapter, we focus on adaptive dictionary methods and discuss the approach of blind compressed sensing (BCS), which aims to reconstruct images in the scenario when a good sparse model for the image is unknown a priori. BCS allows the dictionary to be adaptive to the underlying images, leading to sparser image representations and better image reconstructions from few measurements. In the following, we briefly review the synthesis dictionary model and its learning, CS, and BCS.

23.1.1 Synthesis Model and Sparsity Measures

The well-known synthesis model suggests that a signal $z \in \mathbb{R}^n$ is approximately a linear combination of a small subset of atoms or columns of a dictionary $D \in \mathbb{R}^{n \times K}$, i.e., $z = D\alpha + e$ with $\alpha \in \mathbb{R}^K$ sparse, and e is assumed to be a small modeling error or approximation error in the signal domain [137]. We say that $\alpha \in \mathbb{R}^K$ is sparse if $\|\alpha\|_0 \ll n$. The ℓ_0 "norm" counts the number of non-zero entries in a vector. Since different signals may be approximately spanned by different subsets of columns of the dictionary D, the synthesis model is also referred to as a union of subspaces model [1103, 301]. When $n = K$ and D is full rank, it is a basis. When $K > n$, D is called an overcomplete dictionary.

The ℓ_0 "norm" (not a norm because it is invariant to non-trivial vector scaling or in other words violates the homogeneity condition) function $\|\alpha\|_0 = \sum_{i=1}^{K} 1_{\{\alpha_i \neq 0\}}$ with α_i the ith entry of $\alpha \in \mathbb{R}^K$ and $1_{\{\alpha_i \neq 0\}}$ the indicator function of $\alpha_i \neq 0$ is a straighforward indicator of sparsity and measures the proportion of non-zeros in a vector. However, this function is non-convex as well as discontinuous. Hence, alternative sparsity promoting functions are also popular. Often the ℓ_0 "norm" is replaced with its (tightest) convex relaxation, the ℓ_1 norm $\|\alpha\|_1 = \sum_{i=1}^{K} |\alpha_i|$. Minimizing the ℓ_1 norm subject to linear constraints (a convex problem) is known to promote sparsity. Other sparsity promoting functions include $\|\alpha\|_p^p = \sum_{i=1}^{K} |\alpha_i|^p$ for $0 < p < 1$ (ℓ_p norms). On the other hand, the well-known ℓ_2 norm (with $\|\alpha\|_2^2 = \sum_{i=1}^{K} |\alpha_i|^2$) or the ℓ_∞ norm ($\|\alpha\|_\infty = \max_i |\alpha_i|$) do not measure sparsity.

23.1.2   Synthesis Sparse Coding

For a given signal z and dictionary D, the process of finding a sparse representation α involves solving the well-known synthesis sparse coding problem. This problem is to minimize $\|z - D\alpha\|_2^2$ subject to $\|\alpha\|_0 \leq s$, where s is some given sparsity level. The synthesis sparse coding problem is NP-hard (non-deterministic polynomial-time hard) in general [721, 241]. Numerous algorithms [771, 628, 377, 408, 194, 290, 725, 233] including greedy and relaxation algorithms have been proposed for sparse coding. While some of these algorithms are guaranteed to provide the correct solution under certain conditions, these conditions are often restrictive and violated in applications. Moreover, these sparse coding algorithms typically tend to be computationally expensive in large-scale settings.

23.1.3   Dictionary Learning

In recent years, the data-driven adaptation of synthesis dictionaries, called dictionary learning, has been investigated in several works [752, 304, 15, 1167, 618]. Dictionary learning has shown promise in several applications including compression, denoising, inpainting, deblurring, demosaicing, super-resolution, and classification [298, 139, 620, 14, 621, 619, 595, 813, 506, 464]. It has also been demonstrated to be useful in inverse problems such as those in tomography [574], and magnetic resonance imaging (MRI) [824, 825, 1121, 447].

Given a collection of training signals $\{z_j\}_{j=1}^N$ that are represented as columns of the matrix $Z \in \mathbb{R}^{n \times N}$, the dictionary learning problem is often formulated as follows [15]:

$$\min_{D,B} \|Z - DB\|_F^2 \text{ s.t. } \|b_j\|_0 \leq s \ \forall j, \|d_k\|_2 = 1 \forall k. \tag{23.1}$$

Here, d_k and b_j denote the kth and jth columns of the dictionary $D \in \mathbb{R}^{n \times K}$ and sparse code matrix $B \in \mathbb{R}^{K \times N}$, respectively, and s denotes the maximum sparsity level (non-zeros in representations b_j) allowed for each training signal. The columns of the dictionary are constrained to have unit norm to avoid the scaling ambiguity [387]. Variants of Problem (23.1) include replacing the ℓ_0 "norm" with an ℓ_1 norm or an alternative sparsity criterion, or enforcing additional properties (e.g., incoherence [60, 813]) for the dictionary D, or solving an online version (where the dictionary is updated sequentially as new training signals arrive) of the problem [618].

23.1.4   Why Are There Many Dictionary Learning Algorithms?

The dictionary learning problem can be formulated in a variety of ways using (23.1), its variants, etc. Several learning algorithms exist for Problem (23.1) or its variants [304, 15, 1167, 867, 943, 618, 754, 882, 949, 877, 924, 717, 52] that typically alternate in some form between a *sparse coding step* (updating B), and a *dictionary update step* (updating D). Some of these algorithms (e.g., [15, 949, 924]) also partially update B in the dictionary update step. A few recent methods attempt to solve for D and B jointly in an iterative fashion [811, 417]. The K-SVD method [15] has been particularly popular in numerous applications [298, 620, 799, 139, 824, 825]. However, Problem (23.1) is highly non-convex and NP-hard, and most dictionary learning approaches lack proven convergence guarantees. Moreover, the algorithms for (23.1) tend to be computationally expensive (particularly alternating-type algorithms), with the computations usually dominated by the synthesis sparse coding step.

Some recent works [958, 12, 33, 1166, 52, 13] have studied the convergence of (specific) synthesis dictionary learning algorithms. Bao et al. [52] find that their method, although a fast proximal scheme, denoises less effectively than the K-SVD method [298]. Many of these recent works use restrictive assumptions (e.g., noiseless data, etc.) for their convergence results.

In this chapter, we consider a recent efficient ℓ_0 dictionary learning method [830, 829] called Sum of OUter Products DIctionary Learning (SOUP-DIL). The approach models the training dataset as an approximate sum of sparse rank-one matrices (or outer products), and then uses a simple and exact block coordinate descent approach to estimate the unknowns. The sparse coding step in the method uses a form of thresholding, and the dictionary atom update step involves a matrix-vector product that is computed efficiently. Apart from being efficient, the SOUP-DIL method enjoys good theoretical convergence properties [830], and performs well in applications such as image denoising.

23.1.5 Compressed Sensing

In the context of imaging, the recent theory of compressed sensing (CS) [166, 270, 167] (see also [316, 131, 315, 1100, 132, 349, 1173, 130] for the earliest versions of CS for Fourier-sparse signals and for Fourier imaging) enables accurate recovery of images from much fewer measurements than the number of unknowns. In order to do so, it requires that the underlying image be sufficiently sparse in some transform domain or dictionary, and that the measurement acquisition procedure be incoherent, in an appropriate sense, with the transform. However, the image reconstruction procedure for CS is non-linear.

The image reconstruction problem in CS is typically formulated as follows:

$$\min_{x} \|\Psi x\|_0 \quad \text{s.t. } Ax = y. \tag{23.2}$$

Here, $x \in \mathbb{R}^p$ is a vectorized representation of the image to be reconstructed, and $y \in \mathbb{R}^m$ denotes the measurements. The operator $A \in \mathbb{R}^{m \times p}$ (with $m \ll p$ typically) is the sensing matrix or measurement matrix. The matrix $\Psi \in \mathbb{R}^{t \times p}$ is a sparsifying transform (typically chosen as orthonormal). The aim of Problem (23.2) is to find the image satisfying the measurement equation $Ax = y$, that is the sparsest possible in the Ψ-transform domain. Since, in CS, the measurement equation $Ax = y$ represents an underdetermined system of equations, an additional model (such as the sparsity model) is needed to accurately estimate the underlying image.

When Ψ is orthonormal, Problem (23.2) can be rewritten as

$$\min_{z} \|\alpha\|_0 \quad \text{s.t. } A\Psi^T \alpha = y, \tag{23.3}$$

where we used the substitution $\Psi x = \alpha$, and $(\cdot)^T$ denotes the matrix transpose operation. Similar to the synthesis sparse coding problem, Problem (23.3) too is NP-hard, with Ψ^T denoting the synthesis dictionary for the image x. When the ℓ_0 "norm" in (23.2) is replaced with its convex relaxation, the ℓ_1 norm [272], the following convex problem is solved to reconstruct the image when the CS measurements are noisy [607, 274]:

$$\min_{x} \|Ax - y\|_2^2 + \lambda \|\Psi x\|_1. \tag{23.4}$$

In Problem (23.4), the ℓ_2 penalty for the measurement fidelity term can also be replaced with alternative penalties such as a weighted ℓ_2 penalty, depending on the physics of the imaging process and the statistics of the measurement noise.

Recently, such sparsity-driven reconstruction techniques have been applied to imaging modalities such as magnetic resonance imaging (MRI) [607, 608, 187, 1060, 494, 807, 808], computed tomography (CT) [193, 200, 571], electron microscopy [24], and positron emission tomography (PET) imaging [1080, 624], demonstrating high-quality reconstructions from a reduced set of measurements. Such compressive measurements are highly advantageous and accelerate the applications or help reduce radiation dose. Inverse problems such as inpainting (where an image is reconstructed from a subset of measured pixels) can also be viewed as CS problems.

23.1.6 Compressed Sensing with Adaptive Dictionaries

While conventional CS techniques utilize fixed analytical dictionaries or sparsifying transforms such as wavelets [625], finite differences, DCT, or contourlets [264], to reconstruct images, in this chapter, we consider using adaptive or data-driven dictionaries for reconstruction. Such dictionaries can provide sparser representations of images or image patches and could enable potentially better reconstructions. For example, the adaptive dictionary could be learned or trained from a dataset of images related to the one(s) to be reconstructed. An alternative even more adaptive approach is called blind compressed sensing (BCS) [824, 825, 366, 583, 1121], where the underlying sparse model is assumed unknown a priori, and the goal is to simultaneously reconstruct the underlying image(s) as well as the dictionary from highly undersampled measurements. Thus, BCS fully exploits the observed measurements (information) and enables the dictionary model to be adaptive to the specific data under consideration.

Recent works on BCS [824, 828] have demonstrated the usefulness of dictionary or transform-based blind compressed sensing for MRI, even in the case when the undersampled measurements corresponding to (only) a single image are provided. In the latter case, the overlapping patches (rectangular blocks) of the underlying image are assumed to be sparse in a dictionary (or alternatively a transform [826]), and the (unknown) patch-based dictionary that is typically much smaller in size than the image, is learned directly from the limited measurements.

BCS techniques have been shown to provide better image reconstruction quality compared to compressed sensing methods that utilize a fixed sparsifying transform or dictionary [824, 583, 1121]. However, dictionary-based BCS algorithms such as DLMRI [824] tend to be computationally expensive. In this chapter, we focus on fast BCS algorithms that rely on recent efficient dictionary optimization techniques such as SOUP-DIL [830].

23.1.7   Organization

The BCS problem formulations and algorithms are described in Section 23.2 and Section 23.3, respectively. In Section 23.4, we present preliminary experiments illustrating the usefulness of the BCS approach for materials science microscopy. Section 23.5 summarizes our conclusions.

23.2   BCS Problem Formulations

The CS image reconstruction Problem (23.4) can be viewed as a particular instance of the following constrained regularized inverse problem, with $\zeta(x) = \lambda \|\Psi x\|_1$, $v = 1$, and $\mathscr{S} = \mathbb{R}^p$:

$$\min_{x \in \mathscr{S}} v \|Ax - y\|_2^2 + \zeta(x) \tag{23.5}$$

However, CS image reconstructions employing fixed, non-adaptive image models may suffer from artifacts at high undersampling factors [824]. BCS allows the sparse model to be directly adapted to the object(s) being imaged. For example, the overlapping patches of the underlying image(s) may be assumed to be sparse in an adaptive dictionary model. Here, we use the following patch-based dictionary learning regularizer [830] within Problem (23.5) along with $\mathscr{S} = \mathbb{R}^p$:

$$\zeta(x) = \min_{D,B} \sum_{j=1}^{N} \left\| P_j x - D b_j \right\|_2^2 + \lambda^2 \|B\|_0 \text{ s.t. } \|d_k\|_2 = 1 \, \forall \, k, \ \|B\|_\infty \le L. \tag{23.6}$$

The resulting synthesis dictionary-based BCS formulation is as follows:

$$(\text{P0}) \ \min_{x,D,B} v \|Ax - y\|_2^2 + \sum_{j=1}^{N} \left\| P_j x - D b_j \right\|_2^2 + \lambda^2 \|B\|_0 \text{ s.t. } \|d_k\|_2 = 1 \, \forall \, k, \ \|B\|_\infty \le L.$$

Here, $v > 0$ and $\lambda > 0$ are weights, and $P_j \in \mathbb{R}^{n \times p}$ represents the operator that extracts a patch (e.g., a $\sqrt{n} \times \sqrt{n}$ 2D patch, or a 3D patch, etc.) as a vector $P_j x \in \mathbb{R}^n$ from the image x. A total of N overlapping patches are used. The synthesis model allows each patch $P_j x$ to be approximated by a linear combination $D b_j$ of a small number of columns from a dictionary $D \in \mathbb{R}^{n \times K}$, where

$b_j \in \mathbb{R}^K$ is sparse. The columns of the learned dictionary (represented by $d_k, 1 \leq k \leq K$) in (P0) are additionally constrained to be of unit norm in order to avoid the scaling ambiguity [387, 830]. The dictionary, and the image patch, are assumed to be much smaller than the image(s) ($n, K \ll p$) in (P0). Problem (P0) thus enforces the N (a typically large number) overlapping image patches to be sparse in some dictionary D, which can be considered as a strong yet flexible prior on the underlying image.

We use $B \in \mathbb{R}^{K \times N}$ to denote the matrix that has the sparse codes of the patches b_j as its columns, and $\|B\|_0$ counts the number of non-zeros in B. The $\|B\|_0$ penalty in (P0) helps keep the total number of non-zeros (or aggregate sparsity) in the sparse codes small. Problem (P0) also has a constraint on $\|B\|_\infty \triangleq \max_{1 \leq j \leq N} \|b_j\|_\infty$ that limits the maximum magnitude of the elements of matrix B. Such a constraint is needed because the function within the minimization in (23.6) is non-coercive [830]. The ℓ_∞ constraints prevent pathologies that could theoretically arise (e.g., unbounded algorithm iterates) due to the non-coercive objective. In practice, we set L very large, and the constraint is typically inactive.

Problem (P0) is to learn a patch-based synthesis sparsifying dictionary, and simultaneously reconstruct the image from measurements. An alternative to (P0) is obtained by replacing the $\|B\|_0$ penalty with the scale-varying $\|B\|_1$ penalty that sums the magnitudes of the elements of B. This results in the following formulation, where $\eta > 0$ is a weight:

$$(\text{P1}) \quad \min_{x,D,B} \nu \|Ax - y\|_2^2 + \sum_{j=1}^N \|P_j x - D b_j\|_2^2 + \eta \|B\|_1 \quad \text{s.t.} \quad \|d_k\|_2 = 1 \, \forall \, k.$$

We do not need the constraint on $\|B\|_\infty$ in (P1) because the $\|B\|_1$ penalty is a coercive function.

23.3   BCS Algorithms

Here, we discuss efficient block coordinate descent algorithms for the BCS problems (P0) and (P1) [829]. Our algorithms alternate between updating the dictionary and sparse codes (D, B) (*dictionary learning step*), and updating x (*image update step*). In the following, we describe these steps in detail.

23.3.1   Dictionary Learning Step for (P0)

Minimizing (P0) with respect to (D, B) leads to the following non-convex problem, where $X \in \mathbb{R}^{n \times N}$ is a matrix with the patches $P_j x$ as its columns, and $C \triangleq B^T \in \mathbb{R}^{N \times K}$:

$$(\text{P2}) \quad \min_{D,C} \|X - DC^T\|_F^2 + \lambda^2 \|C\|_0 \quad \text{s.t.} \quad \|d_k\|_2 = 1 \, \forall \, k, \quad \|C\|_\infty \leq L.$$

Denoting the columns of C by $\{c_k\}_{k=1}^K$, we have $\|C\|_0 = \sum_{k=1}^K \|c_k\|_0$ and $DC^T = \sum_{k=1}^K d_k c_k^T$ (a sum of K rank-one matrices or outer products). Substituting these into (P2), we adopt a block coordinate descent method to estimate the unknown variables [830]. For each k ($1 \leq k \leq K$), we perform two steps. First, we solve (P2) with respect to c_k keeping all the other variables fixed, which we call the *sparse coding step*. Once c_k is updated, we solve (P2) with respect to d_k keeping all other variables fixed, which we call the *dictionary atom update step* or *dictionary update step*. We describe the sparse coding and dictionary atom update steps in detail in the following.

23.3.1.1   Sparse Coding Step

Minimizing (P2) with respect to c_k leads to the following non-convex problem, where $E_k \triangleq Y - \sum_{j \neq k} d_j c_j^T$ is a fixed matrix based on the most recent estimates of all other atoms and coefficients:

$$\min_{c_k} \left\| E_k - d_k c_k^T \right\|_F^2 + \lambda^2 \|c_k\|_0 \quad \text{s.t.} \quad \|c_k\|_\infty \leq L. \tag{23.7}$$

The following proposition (cf. [830] for a proof) provides the solution to Problem (23.7), where the hard-thresholding operator $H_\lambda(\cdot)$ is defined as

$$(H_\lambda(b))_i = \begin{cases} 0, & |b_i| < \lambda \\ b_i, & |b_i| \geq \lambda \end{cases} \tag{23.8}$$

with $b \in \mathbb{R}^N$, and the subscript i indexes vector entries. We assume $L > \lambda$ and let 1_N denote a vector of ones of length N. The operation "$\odot$" denotes element-wise multiplication, sign$(\cdot)$ computes the signs of the elements of a vector, and $z = \min(a, u)$ for vectors $a, u \in \mathbb{R}^N$ denotes the element-wise minimum operation.

Proposition 1. *Given $E_k \in \mathbb{R}^{n \times N}$ and $d_k \in \mathbb{R}^n$, and assuming $L > \lambda$, a global minimizer of the sparse coding problem (23.7) is*

$$\hat{c}_k = \min\left(\left|H_\lambda\left(E_k^T d_k\right)\right|, L 1_N\right) \odot \text{sign}\left(H_\lambda\left(E_k^T d_k\right)\right). \tag{23.9}$$

The solution is unique if and only if $E_k^T d_k$ has no entry with a magnitude of λ.

23.3.1.2   Dictionary Atom Update Step

Minimizing (P2) with respect to d_k leads to the following non-convex problem:

$$\min_{d_k} \left\| E_k - d_k c_k^T \right\|_F^2 \quad \text{s.t.} \quad \|d_k\|_2 = 1 \tag{23.10}$$

Proposition 2 (see [830] for proof) provides the closed-form solution [866] for (23.10).

Proposition 2. *Given $E_k \in \mathbb{R}^{n \times N}$ and $c_k \in \mathbb{R}^N$, a global minimizer of the dictionary atom update problem (23.10) is*

$$\hat{d}_k = \begin{cases} \dfrac{E_k c_k}{\|E_k c_k\|_2}, & \text{if } c_k \neq 0 \\[2mm] v, & \text{if } c_k = 0 \end{cases} \tag{23.11}$$

where v can be any unit ℓ_2 norm vector (e.g., the first column of the $n \times n$ identity matrix). The solution is unique if and only if $c_k \neq 0$.

23.3.2   Dictionary Learning Step for (P1)

Minimizing (P1) with respect to (D, B) leads to the following non-convex problem, where $X \in \mathbb{R}^{n \times N}$ is again the matrix with the patches $P_j x$ as its columns, and $C \triangleq B^T$:

$$(P3) \quad \min_{D,C} \left\| X - DC^T \right\|_F^2 + \eta \|C\|_1 \quad \text{s.t.} \quad \|d_k\|_2 = 1 \, \forall \, k. $$

Here, $\|C\|_1 = \sum_{k=1}^K \|c_k\|_1$.

Similarly as for (P2), we use a block coordinate descent method to update the unknown variables in (P3) (cf. the OS-DL method in [878]). For each k ($1 \leq k \leq K$), we perform a sparse coding step to update c_k and a dictionary atom update step to estimate d_k, with all other variables kept fixed in each step. The dictionary atom update step is identical to the one for (P2). The sparse coding step [878] is discussed in the following.

Minimizing (P3) with respect to c_k leads to the following problem, where $E_k \triangleq Y - \sum_{j \neq k} d_j c_j^T$ is as before fixed based on the most recent values of the respective variables:

$$\min_{c_k} \left\| E_k - d_k c_k^T \right\|_F^2 + \eta \|c_k\|_1 \tag{23.12}$$

The following proposition states the solution to Problem (23.12), where the soft-thresholding operator $S_{\eta/2}(\cdot)$ is defined as

$$
\left(S_{\eta/2}(b)\right)_i = \begin{cases} b_i - \frac{\eta}{2}, & b_i \geq \frac{\eta}{2} \\ 0, & -\frac{\eta}{2} < b_i < \frac{\eta}{2} \\ b_i + \frac{\eta}{2}, & b_i \leq -\frac{\eta}{2} \end{cases} \tag{23.13}
$$

with $b \in \mathbb{R}^N$, and the subscript i indexes vector entries.

Proposition 3. *Given $E_k \in \mathbb{R}^{n \times N}$ and $d_k \in \mathbb{R}^n$, the unique global minimizer of the sparse coding problem (23.12) is*

$$
\hat{c}_k = S_{\eta/2}\left(E_k^T d_k\right). \tag{23.14}
$$

23.3.3 Image Update Step

In this step, we solve Problem (P0) or (P1) for the image x, with the other variables kept fixed. This involves solving the following optimization problem:

$$
\min_x \nu \|Ax - y\|_2^2 + \sum_{j=1}^{N} \|P_j x - Db_j\|_2^2 \tag{23.15}
$$

This is a least squares problem, whose solution satisfies the following normal equation:

$$
\left(\sum_{j=1}^{N} P_j^T P_j + \nu A^T A\right) x = \sum_{j=1}^{N} P_j^T D b_j + \nu A^T y. \tag{23.16}
$$

Here, $\sum_{j=1}^{N} P_j^T D b_j$ can be computed using simple patch-based operations. The matrix $\sum_{j=1}^{N} P_j^T P_j$ is diagonal and its (diagonal) entries correspond to the number of patches that overlap each pixel in the image. If periodically positioned, overlapping image patches with a patch stride of 1 pixel are used, and the patches that overlap the image boundaries "wrap around" on the opposite side of the image, then $\sum_{j=1}^{N} P_j^T P_j = nI$, where n is the number of pixels in an image patch and I is the identity matrix. In several applications (e.g., image inpainting), the matrix $A^T A$ is also diagonal or easily diagonalizable, in which case the optimal solution in (23.16) can be found directly and efficiently. More generally, the solution to (23.16) can be found using iterative techniques such as conjugate gradients (CG).

23.3.4 Overall Algorithms

Figure 23.1 shows the dictionary-based BCS Algorithms A0 and A1 for Problems (P0) and (P1), respectively. The algorithms assume that an initial estimate (x^0, D^0, B^0) for the variables is provided. For example, the initial image estimate could be set to $x^0 = A^\dagger y$ with $A^\dagger$ denoting the pseudo-inverse of A, and the initial sparse coefficients could be set to zero, and the initial dictionary could be a known analytical dictionary such as the DCT. When $c_k^t = 0$, setting $d_k^t = v$ in the algorithms could also be replaced with other (equivalent) settings such as $d_k^t = d_k^{t-1}$ or setting d_k^t to a random unit norm vector. Such settings have been observed to work well in practice. A random ordering of the atom/sparse coefficient updates in Figure 23.1 (i.e., random k sequence) also works in practice in place of cycling in the order 1 through K every iteration. One could also alternate several times between the sparse coding and dictionary atom update steps for each k, or cycle multiple times over the (update) sequence 1 through K for each iteration t in Figure 23.1. These alternatives would involve increased computation.

BCS Algorithms A0 and A1

Inputs :   measurements y obtained with sensing matrix A, parameters λ, ν, η, upper bound L, and number of iterations J.

Outputs :   reconstructed image x^J, learned dictionary D^J, and learned patch sparse code matrix B^J.

Initial Estimates: $\left(x^0, D^0, B^0\right)$ and $C^0 = \left(B^0\right)^T$.

For $t = 1 : J$ **Repeat**

1. Form the matrix X with $P_j x^{t-1}$ as its columns.

2. **For** $k = 1 : K$ **Repeat**

    (a) $C = \left[c_1^t, ..., c_{k-1}^t, c_k^{t-1}, ..., c_K^{t-1}\right].$
    $D = \left[d_1^t, ..., d_{k-1}^t, d_k^{t-1}, ..., d_K^{t-1}\right].$

    (b) **Sparse coding:**

$$g^t = X^T d_k^{t-1} - CD^T d_k^{t-1} + c_k^{t-1} \tag{23.17}$$

$$c_k^t = \min\left(\left|H_\lambda\left(g^t\right)\right|, L1_N\right) \odot \operatorname{sign}\left(H_\lambda\left(g^t\right)\right) \quad \text{(For Algorithm A0)} \tag{23.18}$$

$$c_k^t = S_{\eta/2}\left(g^t\right) \qquad \text{(For Algorithm A1)} \tag{23.19}$$

    (c) **Dictionary atom update:**

$$h^t = Xc_k^t - DC^T c_k^t + d_k^{t-1}\left(c_k^{t-1}\right)^T c_k^t \tag{23.20}$$

$$d_k^t = \begin{cases} \frac{h^t}{\|h^t\|_2}, & \text{if } c_k^t \neq 0 \\ \nu, & \text{if } c_k^t = 0 \end{cases} \tag{23.21}$$

    End

3. $C^t = [c_1^t, ..., c_K^t]$. Set $B^t = \left(C^t\right)^T$.

4. Update x^t by solving (23.16) directly or by CG with fixed (D^t, B^t).

End

Figure 23.1: The BCS Algorithms A0 and A1 for Problems (P0) and (P1), respectively. Superscript of t denotes the iterates in the algorithms. The vectors g^t and h^t above are computed very efficiently using sparse operations.

An advantage of the BCS algorithms in Figure 23.1 compared to prior dictionary-based BCS algorithms such as DLMRI [824] is that the sparse coding step in Figure 23.1 is performed efficiently by hard or soft thresholding type operations, whereas in DLMRI, an NP-hard sparse coding problem is repeatedly solved in an approximate manner using greedy (and often expensive) algorithms such as orthogonal matching pursuit (OMP). A detailed theoretical convergence analysis of Algorithms A0 and A1 is presented in [829].

23.4   Numerical Experiments

23.4.1   Framework for Electron Microscopy

We present some preliminary results illustrating the usefulness of blind compressed sensing for electron microscopy. Electron microscopes are used in biology, geology, and materials science for imaging and structural or compositional analysis of various materials, and provide higher image resolutions than optical microscopes. Anderson et al. [24] recently demonstrated the usefulness of compressed sensing for accelerating data acquisition in scanning electron microscopes (SEM). A sparse sampling strategy was employed for SEM, and the CS image reconstruction method effectively filled in the missing bits (or pixels) of the image. Stevens et al. [968] also investigated CS for electron microscopy, but employed a variant of blind compressed sensing called Bayesian compressive sensing. However, this approach for image reconstruction was also observed to be quite slow [968]. Instead, we apply the efficient blind compressed sensing methodology presented in this chapter to obtain high-quality image reconstructions from CS SEM measurements. Although we illustrate the potential of blind compressed sensing for SEM, the technique could also be used with CS in other applications such as transmission electron microscopy (TEM), atomic force microscopy (AFM), etc.

In our experiments, we consider the publicly available SEM images from the Dartmouth database [442], and simulate CS by sampling a subset of the pixels of the images (i.e., the operator A in (P0) is a diagonal matrix with ones and zeros). Such an approach has been adopted in prior work [968]. Algorithm A0 is used to reconstruct (i.e., inpaint) the full high-resolution images from CS measurements. We have observed that Algorithm A1 also works well in practice. However, a complete exploration of the alternative Algorithm A1 or other variants of the presented methods is left for future work.

Algorithm A0 for (P0) was executed with a 64×20 ($n = 64$, $K = 20$) dictionary learned for 8×8 maximally overlapping image patches, and the parameters J (number of iterations) and v were set to 100 and 10^7, respectively. The parameter λ that sets the threshold during sparse coding was decreased from 0.3 to 0.05 (final value) during the first 90 iterations using a logarithmic interval spacing, and this led to faster convergence of the algorithm (compared to a fixed λ setting).[1] The initial image x^0 in the algorithm was obtained by setting the missing pixel intensities to zero, and the sampled pixel intensities to their measured values (i.e., $x^0 = A^\dagger y$). The initial sparse coefficients were set to zero, and the initial dictionary was a 64×20 DCT matrix generated similarly as in [830].

We measure the quality of image reconstructions in the experiments using the well-known peak signal-to-noise ratio (PSNR) metric that is computed between the true reference and the reconstructed images. We express PSNR in decibels (dB). We compare the image reconstruction quality achieved using blind compressed sensing (Algorithm A0) with that obtained by keeping the dictionary D fixed to its initial setting during the algorithm iterations. The latter case corresponds to a conventional (non-adaptive) reconstruction. All simulations ran on MATLAB® R2013a using an Intel Core i5 CPU at 2.5 GHz and 4 GB memory, and employing a 64-bit Windows 7 operating system.

23.4.2   Results and Discussion

The reference images used in our simulations are shown in Figure 23.2 and are labeled Image 1 and Image 2, respectively. We consider various levels of subsampling of these images. Table 23.1 lists the image reconstructions PSNRs obtained by employing Algorithm A0 along with the PSNRs for the initial images (x^0) and for the reconstructions obtained by keeping the dictionary D fixed (to initial setting) in Algorithm A0. BCS using Algorithm A0 clearly achieves significantly better

---

[1] A similar strategy was also adopted in a prior BCS work [827].

Figure 23.2: Reference images (normalized by peak intensity): (a) an excised 430×450 block (Image 1) of a scanning electron microscope (SEM) image available from the Dartmouth public SEM dataset [442]; and (b) a 500×500 block (Image 2) of an SEM image from the Dartmouth SEM dataset [442]. These are biological images.

Table 23.1: PSNR values in decibels (dB) for the initial image (x^0), the final BCS reconstruction (with 64×20 dictionary), and for the reconstruction obtained by keeping the dictionary fixed to the initialization (i.e., no learning) in the algorithm. The results are shown for Image 1 and Image 2 for various percentages of sampled (measured) pixels.

Image	Method	5%	10%	20%	30%	50%	80%
1	Initial image	8.3	8.6	9.1	9.7	11.1	15.1
	fixed D	25.3	27.8	30.2	31.9	34.8	40.0
	BCS	27.4	32.3	34.6	35.6	37.9	42.2
2	Initial image	7.8	8.0	8.5	9.1	10.5	14.5
	fixed D	24.7	27.0	29.1	30.7	33.4	39.0
	BCS	28.2	31.6	33.5	35.0	37.4	42.3

image reconstruction PSNRs compared to non-adaptive or conventional CS (fixed D) in Table 23.1. This illustrates the potential of a data-driven approach like BCS.

Figure 23.3 shows the initial image estimates and final BCS reconstructions at various undersampling factors for Images 1 and 2. The BCS reconstructions retain key image features even with tenfold undersampling of the pixels. Algorithm A0 is able to obtain such reconstructions from limited measurements by learning good sparse signal models for the images. The algorithm had a runtime of about 7 minutes on average in Table 23.1 for reconstructing the 500×500 Image 2. We expect lower runtimes with code optimization and C/C++ implementations.

Figure 23.4 shows the dictionaries learned using BCS for Images 1 and 2 at fivefold pixel undersampling (i.e., when 20% of the image pixels are measured). Both dictionaries display features specific to the SEM images, and are very different from the initial DCT dictionary (Figure 23.4 (c)). Although the learned dictionaries here have only 20 columns, they provide highly sparse representations for the reconstructed image patches (an average of 2 non-zero coefficients per patch for both test images), and may prevent over-fitting. The data thus effectively lives in very low-dimensional subspaces. We observed that dictionaries with even more columns provided some improvements in reconstruction quality for small undersampling factors (i.e., when many image pixels are measured).

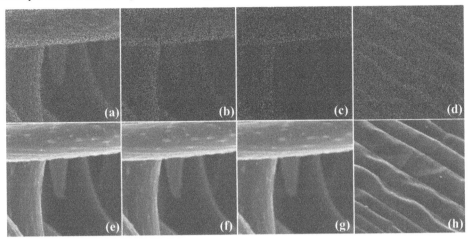

Figure 23.3: Initial images (x^0) for the BCS algorithm at various undersampling factors: (a) twofold (50% measured pixels) undersampling (Image 1); (b) fivefold undersampling (Image 1); (c) tenfold undersampling (Image 1); and (d) fivefold undersampling (Image 2). Final BCS reconstructions at various undersampling factors: (e) twofold undersampling (Image 1); (f) fivefold undersampling (Image 1); (g) tenfold undersampling (Image 1); and (h) fivefold undersampling (Image 2). The initial images are obtained by setting the missing pixel intensities to zero.

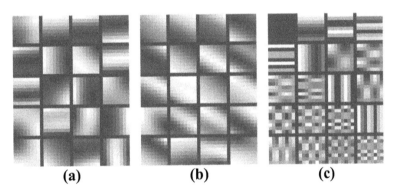

Figure 23.4: Dictionary models (64×20): (a) dictionary obtained from Image 1 with fivefold undersampling (20% measured pixels); (b) dictionary obtained from Image 2 with fivefold undersampling; and (c) the initial dictionary in the algorithm. The columns of the dictionaries are shown as 8×8 patches. For each dictionary atom (vector), we subtracted out the minimum value and then normalized the result by the maximum value for display.

23.5 Conclusion

Data-driven methods have received increasing attention in imaging and image processing applications in recent years. In this chapter, we considered data-driven image recovery techniques based on adaptive dictionaries, and investigated the methodology called blind compressed sensing that enables joint estimation of the sparse signal model and the image from undersampled measurements. Our preliminary experiments illustrated some potential for BCS for electron microscopy compared to conventional CS involving fixed image models. Since dictionary learning-based image reconstruction approaches including BCS are part of a nascent research field, we expect more breakthroughs in this area in coming years. Data-driven sampling strategies could further improve the image quality and acceleration factors achieved by CS in various microscopy applications.

Chapter 24

Sparse Sampling in Microscopy

Kurt Larson
Sandia National Laboratories

Hyrum Anderson
Endgame, Inc.

Jason Wheeler
Sandia National Laboratories

A fundamental goal of microscopy is capturing images, which in the digital age generally means representing the variable spatial properties of a specimen as an array of pixels with different values. The operator of a microscope chooses pixel spacing—the physical spacing of samples on a specimen—as a parameter of the image collection process. On many microscopes pixel spacing is derived from other user-selected parameters such as field of view and pixels per image. The relationships between signal structure and sufficient sampling were worked out by Nyquist, Shannon, and others [746, 928]. For electron microscopes in which the pixel spacing is greater than the electron beam spot size, or optical microscopy above the diffraction limit, the pixel spacing is equivalent to half the dimension of reliably detectable features. In practice, the operator often chooses a pixel spacing that is significantly smaller than half the minimum dimension, as many pixels per structural feature reduces the impact of noise, improves visual interpretation, and significantly improves algorithms for tasks such as segmentation that may be used for automated image interpretation.

Due to variable structures in most specimens and user-selected spatial oversampling, most images collected in microscopy have pixel values that are correlated spatially. Generally, neighboring pixels are likely to have similar intensity values, although other spatial correlations exist. Such images are sparse or compressible; they can be represented as a linear combination of elements in a basis in fewer values than the number of pixels. Often, the most compact sparse representation may be a transformation of the original image (for example, a gradient image). For compressible images, the Nyquist and Shannon signal detection criteria may not be necessary to completely describe the signal. Sparse sampling and reconstruction, also known as compressed sensing (CS), was mathematically formalized by Candes, Donoho, and others in the mid-2000s [172, 171, 270]. CS describes the situations and methods for which a sparse representation of the domain of interest can be exploited, and provides formal performance guarantees for some situations. One goal of a CS microscope can be the acquisition of the specimen structure in fewer measurements than the number of pixels in the image. The opportunities are broader, however. For the designer, CS provides an alternative framework for thinking about sensing systems. The potential benefits of CS in any particular application vary depending on the intent and creativity of the designer. For domains and systems which have constraints on the sampling process—such as one or more of cost, power, size, bandwidth, dose, speed, etc.—design utilizing CS may provide surprising or even radically disruptive performance compared to traditional approaches. The well-known single-pixel camera, in addition to being a demonstration of CS, could be an alternative imaging design in wavelengths for which detector arrays are difficult to fabricate [285]. Because CS is computationally simple to encode, an on-orbit reprogramming of the Herschel Space Observatory is possible for transmis-

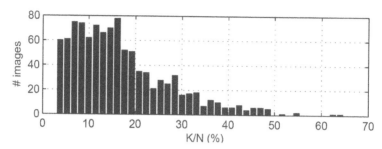

Figure 24.1: The sparsity of 1022 various electron microscopy images, where sparsity is measured by counting the number of block-DCT coefficients K that account for at least 99.75% of the image energy and dividing by the number N of pixels. The average sparsity is 17%, with half of all images less than 15% sparse, and three-quarters less than 20% sparse.

sion of high-bandwidth data without lossy compression artifacts, while at the same time improving resolution [58].

In this chapter, we consider the collection of sparse samples in electron microscopy, either by modification of the sampling methods utilized on existing microscopes, or with new microscope concepts that are specifically designed and optimized for collection of sparse samples (i.e., a compressive sensing electron microscope (CSEM)). Electron microscopy is well suited to compressed or sparse sampling due to the difficulty of building electron microscopes that can accurately record more than one electron signal at a time.

24.1 Motivations for Sparse Sampling in Electron Microscopy

Electron microscope images are usually smooth and are often compressed via block-DCT (discrete cosine transform) or wavelet compression schemes using JPEG (Joint Photographic Experts Group) or JPEG-2000 standards, respectively, while still maintaining high image fidelity. To assess image sparsity of typical electron microscopy images, we gathered 1022 electron microscopy images (SEM (scanning electron microscope), TEM (transmission electron microscope), and E-SEM (environmental SEM)) from the public domain Dartmouth gallery [1126]. The images are of a variety of biological, geological, and materials specimens. To standardize analysis, we excised the center 512×512 pixels of each image to remove banners and rescaled images to $[0, +1]$ grayscale values. For each 512×512 image, we computed the sparsity K by counting the number of coefficients in the block-DCT domain (32×32 blocks) that cumulatively accounted for at least 99.75% of the total coefficient energy. A histogram of the results is shown in Figure 24.1, demonstrating that most of these images are highly compressible even when losing only 0.25% of the image energy.

Sparse sampling in electron microscopy has been considered for dose reduction, improving 3D (three-dimensional) reconstructions and accelerating data acquisition. In optical microscopy, sparse sampling biological fluorescence microscopes [981] and holographic sparse sampling microscopes have been developed [126, 645].

24.2 Sparse Sampling and Reconstruction

CS allows signal acquisition and reconstruction using few measurements. Whereas the Shannon–Nyquist condition provides recoverability guarantees for band limited signals provided a sufficiently dense sampling, CS theory provides recoverability guarantees for sparse or compressible signals, provided that a sufficient number of non-adaptive measurements are taken. Romberg, in Chapter 22, discusses reconstruction of signals from sparse data.

In CS, a measurement vector y is acquired according to the linear model

$$y = Ax + n, \tag{24.1}$$

where A is an $M \times N$ matrix (e.g., M measurements of an N pixel image) that can be decomposed into $A = \Phi\Psi$, and n is additive noise. The "sensing matrix" Φ represents the measurements employed in the system and Ψ is a compression basis or dictionary suitable to the domain (such as wavelets or block-DCT). Thus, Ψx is the signal or image decomposed into a vector x with only K significant coefficients. For example, x could represent the wavelet coefficients of an image with the wavelets captured in the basis Ψ; or x could, itself, represent a sparse image (e.g., a "starfield") with $\Psi = I$ (no compression basis). In compressed sensing, the measurements are non-adaptive, in that A does not depend on x in any way. In the context of electron microscopy, the sensing matrix Φ is some set of illumination patterns that can be projected onto the sample by the electron microscope, where the signal for each pattern, measured by an electron detector, is collected as a single entry in the measurement vector y.

Since $M \ll N$, the transformation from x to y is a dimensionality reduction or compression, that in general, loses information. However, since only K elements of x are nonzero, recovering x from (noiseless) measurements y could, in theory, be achieved through NP-hard (non-deterministic polynomial-time hard) combinatorial search for the support of x using only $M \geq 2K$ appropriately defined measurements. Indeed, if x is a vector with only K nonzero elements, then a brute force solution of the non-convex problem

$$\min_{x} ||x||_0 \quad \text{subject to } Ax = y, \tag{24.2}$$

could recover x, where $||x||_0$ counts the number of nonzero elements in x. However, that problem is computationally intractable even for very small N.

Although many alternative methods have been proposed to reconstruct the signal x from measurements y, basis pursuit—a computationally tractable ℓ_1 inverse problem that is essentially a convex relaxation of the combinatorial recovery problem—is the most common approach that enjoys broad theoretical guarantees. Basis pursuit for noiseless measurement solves

$$\min_{x} ||x||_1 \quad \text{subject to } Ax = y, \tag{24.3}$$

where $||x||_1 = \sum_{i=1}^{N} |x_i|$, and the constraint is replaced by $||Ax - y||_2^2 \leq \varepsilon^2$ for noisy measurements. Myriad efficient methods exist to solve basis pursuit in polynomial time.

Ideally, compressive sensing measurements are "wholistic" or global and each measurement represents a response that interrogates a broad spatial area (for two-dimensional (2D) reconstructions) or volumetric area (for 3D reconstructions). Compressive sensing theory does not automatically promise a benefit if the number of conventional measurements is simply reduced. The greatest benefit from CS could derive from collecting measurements of the specimen in non-traditional ways. If less-than-ideal benefit is acceptable, however, any measurement method can potentially be used. A direct approach to trying compressive sensing in microscopy is to simply choose a sparsifying basis and collect samples using the normal measurement methods, just fewer of them. In scanning methods such as SEM and atomic force microscopy (AFM) where each pixel is individually measured, by this approach the measurement matrix is simply a subset of the identity matrix in which many rows representing individual pixel measurements are removed (not measured). This yields a "starfield" sampling pattern. Simply reducing the number of point samples might be expected to yield less promising results than alternative multi-point sampling strategies could.

24.3 Sparse Sampling in Transmission Electron Microscopy

Sparse sampling in electron microscopy has been most extensively reported in transmission electron microscopy (TEM), described in this volume by Venkatakrishnan's Chapter 6. TEMs direct a beam of nominally 100-300 keV electrons through a thin specimen and measure the signal emitted from the side opposite the illumination. In ordinary TEM, an array of detectors collects an image from a wide-field beam. For sparse sampling, variations of scanning TEM (STEM) are typically used. In STEM, the electron probe is scanned across the specimen, and the detector measurement is recorded as a function of probe location. Many modes of operation have been devised for TEM. Single scans are performed to explore two-dimensional material structures at sub-nm resolution, usually with the beam axis aligned with a crystallographic axis. To perform three-dimensional tomography, many 2D scans are completed with various rotations and tilts of the specimen. Goris et al. mention some of the modes of TEM that can be used for tomographic reconstruction [376]. Each traditional 2D tilt-rotation measurement is wholistic or global and suited for CS in the sense that it integrates across the thickness of the sample. In traditional weighted back-projection reconstruction as well as sparse reconstruction, the set of tilt-rotation STEM measurements provides the constraint for the inferred 3D structure. Compressive sensing has been simulated and experimentally demonstrated in TEM to recover 2D and 3D structure. The earliest explorations emphasized better recovery of structure with fewer scans. Reduction in electron dose for sensitive specimens is also identified as a potential benefit and has been explored.

CS may have been first applied to electron microscopy in 2007 for cryo-TEM which seeks to reconstruct 3D structure from one 2D image containing many specimen copies in different orientations [493]. A novel L_1 reconstruction method that converges using a "sliding mode" technique was developed to reconstruct virus structure in experimental and simulated data. In simulated data reconstruction was shown for fewer views. Reconstructions were better than those acquired using conventional weighted back-projection. Subsequent developments in simulated cryo-TEM data include a technique for improved SNR and resolution by promoting sparsity in the wavelet domain and exploiting Toeplitz properties of the forward-projection and back-projection operators [1106, 1117].

Veeraraghaven et al. [1095] used CS computed tomography in serial section TEM to simultaneously achieve z-resolution of 5 nm and reduce the number of tilted sample images from tens or hundreds to about five. Thirty consecutive, 50-nm thick sections of fly larva brain were imaged at 4 nm x, y resolution for five tilts. Utilizing prior knowledge in the domain of neural tissue, a sparsifying basis was constructed using planar local basis functions in a variety of orientations as well as basis functions to represent uniform intensity and blob image primitives. Segmentation of neurites was improved compared to non-tomographic methods for the cost of four extra tilt-rotation images, and the speed of acquisition was much faster than traditional computed tomography which uses many more tilt-rotation images, or focused-ion beam (FIB) sectioning. Demonstrating a different custom basis for reconstruction, Hu et al. [445] used an unsupervised learning method to infer a domain-specific sparsifying 3D basis from a high-resolution training set collected using FIB. Using the learned basis, they demonstrate on 15 slices of fly larvae neural tissue that the number of STEM tomographic tilt-rotations could be reduced to five while maintaining high resolution using a sparse reconstruction technique. A drawback was noted that the images require very precise alignment prior to sparse reconstruction. A number of potential mathematical enhancements of the reconstruction process were noted.

Goris et al. [376] used a total variation minimization (TVM) sparse reconstruction technique and compared it to the simultaneous iterative reconstruction technique (SIRT) algorithm for tomographic reconstruction of dispersed Ag nanoparticles, PbSe-CdSe core shell particles, and a Si needle with Pb inclusions. The three specimens were imaged in different TEM microscopes with different techniques. For all three specimens, the TVM sparse reconstruction suffered less from the missing wedge of projections due to the limited orientations possible in a TEM system, required fewer projections, yielded better segmentation, and demonstrated higher detection probability for small features. Subsequently, using experimental nanoparticle data and synthetic phantoms, others demonstrate reduced elongation in the direction of the missing wedge, improved segmentation, improved resolution or detection of small features, reduced artifacts, and 3D reconstruction with fewer samples, compared to SIRT and weighted back-projection [547, 1039].

Motivated principally to reduce the dose for specimens that might otherwise be altered or destroyed, Binev et al. [97] develop the application of compressed sensing to TEM for 2D scanning and 3D tomographic recovery, and validate concepts via numerical simulations. Stevens et al. [968] use Bayesian dictionary learning to infer a domain-specific basis for STEM, which is used to recover effective images from approximately 5% of the pixels (20×) reduction. Béché et al. [71] implement a fast shutter mechanism in a 2D scanning TEM to enable arbitrary, random illumination of pixels without modifying the scanning mechanism. Images with 50% reduction in samples (50% dose reduction) maintained image quality across all specimens tested; images with 80% reduction had variable results depending on specimen and imaging factors.

Although CS may yield the greatest benefit with alternative illumination strategies, we are unaware of simulation or experiments for TEM in modifying the beam (for example, utilizing STEM with a beam with a spot size broader than one pixel) or utilizing multiple beams in 2D or 3D TEM to implement sparse sampling, although it has been conceived as a potential approach to reducing dose in TEM sampling [97].

24.4   Sparse Sampling in Scanning Surface Microscopy

The traditional SEM or atomic force microscope (AFM) samples each pixel independently to generate an image, and it is natural to consider whether sparse sampling could be applied to these instruments. It is fairly easy to conceptualize and model sparse sampling techniques that could capture sufficient information for reliable compressive recovery in fewer samples for data acquisition with a reduced dose. For example, one could randomly sample a subset of the desired pixels and utilize sparse reconstruction to recover the image. A high-speed shutter would be useful to precisely select pixels and exposure times, such as has been recently created for TEM [70] and are commercially available for SEMs.

To speed up imaging, however, an alternative faster than scanning every pixel equally is necessary. Dynamic processes in scan electronics preclude arbitrarily sampling individual pixels at high speed. Simple robust methods to recover an accurate image with fewer measurements, such as iterative sub-sampling, in practice yield a process that takes more time than sampling all of the pixels at the desired resolution. An additional important consideration is that CS favors measurements of a global nature — rather than a reduced set of point measurements.

In both SEM and AFM, electrical lags in both electrostatic and electromagnetic components exist and the beam or tip does not come to rest at a commanded position until some finite time after the command. In the case of AFM, the tip assembly also has physical inertia. SEM and AFM typically compensate for these dynamical lags by moving slowly or by allowing the system to come into stable motion prior to imaging; in SEM this occurs beyond the edges of the desired image. Essentially, extra pixels outside of the desired image area are scanned and thrown away until the beam speed becomes constant. A consistent image is created by replicating identical scan commands for each row of an image, some of which is discarded. Since the dynamical motion is highly dependent on the specific piece part values of amplifiers and other electrical components, each SEM or AFM needs to be calibrated whenever hardware is replaced so that electrical control signals can be correlated to real dimensions. One implication of the electrical and mechanical dynamics is, in most scans, even after the transients have died out, the probe is constantly lagging the continuously changing commanded positions, and the measurements represented as a pixel in an image are actually collected by a probe in motion sampling a linear section of the specimen. Electrical systems with electromagnetic components cannot avoid transient and steady-state dynamical responses. Even electrostatically-scanning SEMs have lag dynamics because although power supplies with fast switching and large voltage range suitable for SEMs are possible, they are expensive and not normally used. When these system dynamics are taken into consideration, many concepts for collecting data sparsely in existing real microscopes are likely to be no faster, or even slower than sampling every single pixel at the desired signal-to-noise ratio. This may be acceptable for exploratory R&D or dose reduction applications, but is a significant challenge in applications emphasizing speed.

24.5 Sparse Sampling in Atomic Force Microscopy

AFM uses a sensing probe on a cantilever arm to detect variations in the interaction of the probe and specimen as a function of probe location. The sensing probe and property measured have many variations [2], which affect the rate of data acquisition. Interest in CS in AFM is aimed at increasing speed of acquisition or reduced sample interaction to reduce damage. Monitoring the progress of in-situ surface or biochemical reactions emphasizes speed of acquisition or revisit rates.

Song et al. developed a sparse sampling procedure and reconstruction method for AFM which uses a fixed sampling pattern in which the tip traces a square spiral [954]. The sampling ratio (sampled pixels to total pixels) is not specified. They use a total-variation norm to reconstruct the image, which shows possible aliasing due to the highly structured sampling methodology. Because their sampling process is highly structured and has appreciable gaps, the method is not generalizable to all samples.

Andersson and Pao randomly chose 20% of the pixels from a complete AFM dataset to simulate random-pixel sampling in AFM [25], and also tried a concentric-square sampling method similar to that used by Song et al. [954]. Pure L_1 reconstruction was poor for both methods. Much better results were obtained using total variation minimization and Delaunay triangulation. The authors conclude that CS could be used advantageously in AFM when sample interaction needs to be minimized, or for speedup when very slow measurements, such as force mapping, are being conducted.

A fast and arbitrarily controlled (or random) probe motion and image reconstruction technique has been developed which could be applicable to AFM [23]. This method would require inversion of accurate models of AFM tip motion such as those developed by Chang et al. for use in a feedback-controlled variable scanning process [184].

24.6 Sparse Sampling in Scanning Electron Microscopy

Anderson et al. demonstrate a sparse imaging method in SEM, in which the electron probe measured a subset of locations on the specimen with varying illumination times [23]. The dynamics of the scanning process were modeled and compensated for to allow a CS reconstruction of the image. Each detector measurement encodes an integrated response over many pixels as the beam moves transiently to commanded locations. By this method, recovery of an image similar to the normal SEM image was possible in about 1/3 of the total imaging time. In addition to speed, a dose-minimization motivation is applicable to use of CS in SEM, in which certain biological or dielectric materials may be damaged or exhibit charging artifacts if the dose rate is too high.

24.7 Compressed Sensing in Multi-Beam Electron Microscopes

Volumetric and wide-area imaging at resolution of a few nm is being sought in neuroscience and materials science. In neuroscience, imaging the approximately 500 mm^3 volume of a mouse brain [1104] with 4 nm pixel pitch and 30 nm tissue slices [576] would generate 10^{18} voxels. The Intelligence Advanced Research Projects Activity (IARPA) Circuit Analysis Tools program sought a 400× increase in the speed of imaging a 1 cm^2 area at 2.75 nm resolution (13.2 trillion pixels) for microcircuit analysis [381], which was not realized by the program [382]. The IARPA Rapid Analysis of Various Emerging Nanoelectronics (RAVEN) program seeks to image all layers of a 1 cm^2-area microcircuit with 14 nm minimum feature size within 25 days [383]. In microcircuit fabrication, which must keep pace with an ever-shrinking minimum feature size, the capability to routinely inspect large areas with ˜nm resolution, which is well beyond all optical techniques, is becoming necessary for process development and defect detection. Statistical characterizations of structural heterogeneity benefit greatly from wide-area imaging and 3D reconstructions, e.g., natural materials such as geologic samples, or engineered materials after failure, chemical degradation, or aging. These applications demonstrate that near-nm resolution imaging at sustained rates over large areas is scientifically and commercially valuable. For many applications, imaging at sustained rates of hundreds of millions of pixels per second could be sufficient; in the case of neuroscience, sustained imaging at many billions of pixels per second is necessary to even conceive that large volumes of brain tissue could be imaged.

Collecting surface images of samples by SEM is ultimately limited in speed by the single-pixel-at-a-time nature of the scanning process. Even as detectors become faster, and system engineering innovations reduce time required for beam dynamics and stage movement, the maximum attainable speed at a given resolution is limited by one or more of (1) the physics of material contrast, (2) electron repulsion and lens aberrations which limit beam current in the column for a given spot size, and (3) sample dose tolerances limiting the beam intensity per unit area on the sample. The fastest scanning electron microscopes operate in specialized applications such as microelectronics inspections at approximately 100 million pixels per second. Compared to the necessary speeds, the single beam systems are not fast enough to collect the desired quantities of data in a reasonable amount of time. Parallel collection of pixels is necessary.

In the microcircuit fabrication domain, parallel imaging can be implemented in a footprint the size of the wafer, so widely separated scanning systems can be built which simplifies parallel imaging. In biology and materials science, with unique small samples, the parallel imaging must be achieved in a much smaller area. Development of electron microscopes that can image in parallel in a small area is difficult. In response to an incident beam, electrons scatter from the sample by elastic and inelastic processes with a broad range of energies. The broad range of electron energies creates challenging conditions for imaging electron optics. For this reason, standard scanning electron microscopes use a simple potential field in the chamber to attract low-energy, inelastically-scattered (secondary) electrons to a single detector, one pixel at a time. If used for multiple illumination beams, this detection paradigm mixes the electrons emitted from the multiple incident beams. Wide-field or parallel imaging optics utilizing complex lens assemblies have been devised for specialized applications as in LEEM [68, 1008]. The Zeiss MultiSEM uses innovative illumination and imaging optics to scan many single-beam SEM images in parallel at landing energies less than 3 keV [484, 286]. Due to the complexity of their imaging optics, the LEEM and MultiSEM have significant constraints on electron landing energy which makes these systems specialized rather than general-purpose.

The 61-beam Zeiss MultiSEM 505 collects 1.22 billion samples per second allocated as approximately 200 electrons per pixel per sample ($^\sim$672 pA $\times$ 50 ns) with a minimum pixel spacing of 3 nm over a field of view about 100 μm wide [679, 629]. Some specimens would require more than one sample per pixel to generate sufficient material contrast in the collected image. Stage moves and other non-imaging processes occur between imaging intervals. With two samples per pixel and assuming 1 second for non-imaging process such as a stage move between imaging intervals, the sustained rate for large-area imaging using the MultiSEM in the neighborhood of 200 million pixels per second. Clearly, it is important to reduce the "down-time" for stage moves. Given a suitable characterization of the amplifier and electrostatic plate beam dynamics in the MultiSEM, a single-beam CS technique could be attempted with it [23, 21]. Since the scanning time in the current MultiSEM is a relatively small fraction of the elapsed time, using compressive sensing in the current system would not significantly increase its effective speed. Other benefits might be possible such as improved image quality for a given exposure time or mitigating sample charging by using an unstructured illumination pattern. If the stage move time in the MultiSEM becomes significantly smaller, in some domains the use of CS in it could conceivably increase its effective rate into the billions of pixels per second range.

The difficulties associated with using multiple detectors to image in parallel makes surface electron microscopy very appealing for CS approaches, which generally utilize one or a few detectors, each of which aggregates information sampled globally rather than locally. Illuminating an area of many μm^2 with a series of electron patterns that satisfy the global nature desired for sparse sampling is feasible, as various multi-beam electron illumination columns for mask writing or lithography applications have been developed in which illumination patterns with essentially arbitrary structure over the field-of-view can be generated. One uses MEMs apertures and deflection electrodes to create 262,144 electron beams with 20 nm or 10 nm spot size within an 82 μm $\times$ 82 μm write field [498]. The beams are individually programmable on/off, and the overall system is capable of addressing the write location of the beams within 0.1 nm. Another uses MEMs electron mirror/trap devices to create an array of 248 $\times$ 4096 adjacent beams (1,015,808 total) that can be switched on/off individually with beam location precision of about one nm [666]. Some of the technologies developed in the multi-electron beam lithography and photomask R&D could potentially be adapted or repurposed for CS electron microscopy.

Anderson et al. [22] model CS recovery in SEM with a single detector when multiple electron probes (beams) simultaneously illuminate the specimen. For a 50% samples-to-pixels ratio using simulated data with ~5% sparsity, binary (on-off) electron beam illumination patterns yielded minimum reconstruction error for sparse beam-pixel ratios of 0.1% to 6%; above and below these values, reconstruction error increased rapidly. For Bernoulli measurements with $-1, +1$ illumination, which in electron microscopy requires a method to estimate the mean response of the specimen, reconstruction error was low for all fill ratios between 0.25% and 50%. Simulated reconstruction by sampling images collected in SEM suggested that non-ideal noise in real electron microscopes may be an important limiter in CS. The authors also noted that the methodology could have introduced more noise than would occur in a real multi-beam microscope and that a less-than-optimal reconstruction method was used, both of which suggest the results are potentially pessimistic. Experiments in a multi-beam instrument were identified as necessary to fully understand the effect of stochastic sample interactions and detector responses in CS.

24.8   Theoretical Analysis of an Electron Column for a Multi-Beam CSEM

Here we develop a theoretical analysis of a hypothetical compressive sensing electron microscope (CSEM) to develop understanding of potential performance and challenges. First, the desired maximum diameter of the electron beam at the sample, also known as the spot size, is defined as 5 nm. Collimation is the angular spread of electrons in the beam and is defined as the spot size divided by the focal length of the objective lens. A value of 5 mm focal length is chosen for this analysis. This leads to a beam collimation requirement of 1×10^{-6} radians (1 micro-radian).

The multi-beam CSEM concept is illustrated in Figure 24.2. It uses an array of apertures to create a pattern of beamlets from an incident flood beam that is created by an electron gun and collimation lens. An array of micro-electro-mechanical (MEMs) devices over about 2 mm $\times$ 2 mm (4 mm^2) is assumed as a design starting point and seems feasible to construct, as similar devices have been developed for electron lithography [498]. A beam waist is a position in a microscope where the beam has a minimum dimension and angular divergence. The 4 mm^2 area of the aperture array is selected as a beam waist from which an emittance value could be derived. Emittance E is defined as $E^2 = A\omega$, where A is the area of the beam waist and ω is the solid angle of the beam, which is approximately the collimation squared. Therefore, the emittance of the flood beam entering the aperture array is 2×10^{-9} A/m^{2}sr. The beam current I is related to brightness and emittance by $I = BE^2$, where brightness is a property of the electron emitter. Schottky field-emission emitters have a brightness of about 5×10^{12} A/m^{2}sr. Thus, the beam current entering the 4 mm^2 aperture array is about 20 μA.

To continue the analysis, an assumption about how the aperture array divides the incident collimated flood beam into beamlets is necessary. We assume the aperture array passes 10% of the incident electrons as some unspecified number of beams. A design process iterating between electron column models and sparse reconstruction analysis would be necessary to determine the optimal number of beams for a real system. For the time being, the additional optics to create patterns and demagnify the beams onto a sample are ignored, which would also require significant iterative modeling to verify resolution and other performance attributes. Given these assumptions, about 2 μA of beam current could be used to create "sensing function" patterns for sparse sampling.

2 μA of beam current is equivalent to approximately 12,500 electrons per nanosecond. At ordinary detector quantum efficiencies (DQE), acceptable SNR for many imaging applications can

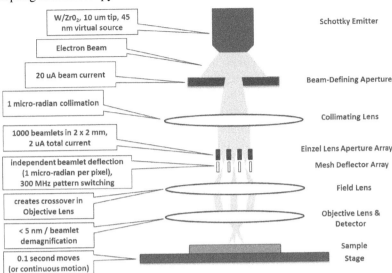

Figure 24.2: Conceptual schematic of a multi-beam CSEM electron column and stage. Major components are listed on the right, and attributes discussed subsequently in the text that could lead to high speed are listed on the left. This concept includes a single electron source that is split into many electron beams, each of which is individually steerable over some area of the sample. A set of illuminations with differing beam locations on the sample comprise the sensing functions for "global" sparse sampling. The detector response for each pattern is a single compressed measurement of the sample response over a large area of simultaneous, spatially-varying illumination. Some key subsystems of a CSEM such as high-speed pattern-generating electronics, data acquisition systems, and computational image recovery methods are not shown.

be generated from between 10^3 and 10^4 incident electrons per pixel. If 10^4 electrons per pixel are budgeted to generate SNR or contrast, a conservatively large value, the theoretical data rate for the multi-beam compressive microscope is 1.25 billion pixels per second. Specimens that require fewer electrons per pixel to generate sufficient material contrast, or a system that has a higher DQE, could achieve proportionately higher pixel rates. Furthermore, as illustrated in Figure 24.1, many surface microscopy images are sparse, so an additional increase in the pixel rate could potentially be achieved because compressive sensing can enable collection of fewer samples than pixels. This first-order estimate of a potential data rate in the billions of pixels per second range for a multi-beam CSEM exceeds the current state of the art in electron microscope data acquisition for most applications. Further investigation of the possible designs and engineering challenges of a multi-beam CSEM seem warranted.

24.9   Potential Embodiments of a Multi-Beam CSEM

In this section we explore potential embodiments of a multi-beam CSEM. As has been noted, multi-beam electron lithography systems have developed a variety of technologies that could be used in multi-beam compressive sensing [498, 666]. In lithography, a common approach to creating the exposure patterns is to assign each beam a fixed location in the array of beams, and turn the beam off (e.g., with an electrostatic deflection into a blanking plate or absorption) if exposure is not necessary at that location in the current exposure period. In areas where the lithographic pattern is sparse, most beams are off; where it is dense, most of the beams are on. The beam current or number of patches being exposed at any one time is job and location specific and is not a quantity of principal concern. The bandwidth or switching speed of the patterning system is related to switching beams on and off.

A similar approach could be taken for a compressive sensing microscope [538]. A contiguous array of many on–off switchable beams could be constructed, with each beam mapping to one location in the image to be reconstructed. Such an array could expose any number of pixels or patches in a contiguous region of the sample, depending on the operator's choice of which beams are on or off. A sequence of patterns could be specified as the sensing matrix. This sort of microscope could potentially borrow technology for many devices from electron beam lithography systems.

Many possible designs for compressive sensing microscopes could be made with on–off switching, and undoubtedly multi-beam compressive sensing could be tested and thoroughly explored in an on–off system. Such an embodiment may not be ideal for applications emphasizing imaging speed. An application interest in minimizing acquisition time emphasizes maximizing beam current to the sample, and raises electronics bandwidth management as a challenging engineering issue. In an on–off system, there are as many potential beams as pixels, and 50% (or more) of the beams are off during any sensing pattern illumination. One perspective of such a system is that the available electronics bandwidth is used to reduce beam current by interrupting beams that have already been formed. It thus seems like a sub-optimal approach to utilizing the available beam current and bandwidth. Rather than switching beams on or off, could a system be made to switch beams "here" or "there"? In other words, could the electronics bandwidth be used to modify illumination patterns without affecting the beam current to the sample? Approximately half of all pixels (or fewer) are not illuminated in any given sensing pattern. Instead of absorbing or deflecting 50% or more of the formed beams, a more efficient and therefore faster microscope might result if each beam could be steered in some way to select among two (or more) pixels (or patches), because in this manner many more pixels than beams would be measured in any sequence of patterns. This section explores the potential for compressive sensing microscopes with multiple steerable beams, or more precisely, beams that are switchable among a pre-determined set of locations. Each beam is always "on" and beam current to the specimen is constant or, in other words, maximized to drive speed up; what varies among the sensing patterns is the location each beam is illuminating. We devise two potential methods and explore the compressive sampling and reconstruction, engineering challenges, and potential speed of each.

24.9.1   Concept for Steerable Beams

Larson et al. [537, 538] explore using steerable beams to create the patterned illumination of the measurement matrix. The switching mechanism would be located in the electron column between the electron gun and the objective lens. It would use a system of microelectronic devices with two purposes. First, the devices create an array of beams from a collimated flood beam emerging from the electron gun. Second, the devices deflect (steer) the beams to their desired position on the sample. To be consistent with the preceding theoretical evaluation, the total area of the apertures is about 10% (or potentially more) of the array area, and the size of the apertures is inversely proportional to their number. The area not utilized for apertures would be used for circuitry such as steering electrodes, switching logic, routing, and power. The devices would need to be mechanically, thermally, and electrically stable in the environment of an electron column.

Each device or aperture would have x-axis and y-axis circuitry to control deflection in orthogonal directions. Conceptually, the circuitry design could be simple. Each aperture is divided into four quadrants: $-X, +X, -Y, +Y$. Voltage differential in the X elements provide X-axis deflection; similarly for Y. Grounding all elements allows a beam to pass undeflected. By applying voltage steps to the control elements, each beam can be deflected by pixel increments along the X and Y axes. With two voltages in each axis (e.g., $[0, +1]$), each beam could address four different pixels. With three voltages $[-1, 0, +1]$, each could address nine pixels. Thus, each beam can be "addressed" to a number of pixels that is the product of the available X and Y voltage increments. The number of pixels that can be patterned is some multiple of the number of beams. Depending on the number of pixel offsets allocated to each beam and the spacing of the switching apertures, a variety of exposure patterns and beam/pixel ratios can be created including ones in which pixels can be "exposed" by more than one beam at a time. The bandwidth is utilized to select a particular pattern rather than to reduce beam current. The bandwidth per pattern required by the switching mechanism is directly related to the number of beams, the number of pixel offsets allowed for each beam, and the electron exposure required to obtain a sufficient measurement from each pattern. We explore two varieties of this configuration here.

24.9.2 *Array of Correlated Steerable Beams*

To explore sparse sampling with a high ratio of beams to pixels, a square array is postulated with nominally 100×100 switching apertures, for ten thousand total beams. Given 2 μA total beam current to the sample from 10% of the area of a 4 mm^2 array, each beam aperture has a diameter of about 7 μm and passes 200 pA current. The circuit complexity and bandwidth required to independently switch this many beams in a small area is likely prohibitive, so a simplification is used. Each X-axis beam control electrode is linked to its neighbors along each row so that they all receive the same deflection offset command, and each Y-axis control electrode is similarly linked along each column [537]. The trade off for this significant simplification of circuitry is that the offsets along each row and along each column are correlated. All beams in a row are offset by the same number of pixels in the orthogonal Y direction, and all beams in a column are offset by the same number of pixels in X. Correlation in the sampling matrix results from this concept, which ought to reduce the efficiency of sparse sampling and require more samples for image reconstruction. The engineering benefits are that control circuitry can be placed external to the array, and the switching bandwidth is significantly reduced. This idea is conceptually similar to Toeplitz or circulant approaches that can be used to reduce control bandwidth and which can be nearly as good as random sampling [1178].

The deflection array and subsequent optics could be configured so the beams have non-overlapping address space. For example, if each control line has two settings $[0, +1]$, the beam centers (neutral position) would occur on every other pixel position and a 10,000 beam array could create patterns over forty thousand pixels, for $1/4$ beam-to-pixel fill ratio. Three setting controls would have beam centers on every third pixel index and interrogate ninety thousand pixels for $1/9$ fill, and so forth. The no-overlap configuration minimizes bandwidth and prevents a pixel from being illuminated by multiple beams simultaneously. A potential disadvantage of the no-overlap configuration is only one pixel within a single beam's address area can be illuminated at a time, which means that the sensing patterns cannot directly test the correlation of some adjacent pixels. This approach also makes the system more vulnerable to defective beamlets, which may manifest as "always off" or "unsteerable."

Alternatively, the deflection array can be configured to allow the address space of beams to overlap. The possible fill ratios are as in the no-overlap case, however the increased addressing allows for any two adjacent pixels to be illuminated in a sensing pattern as well as the condition of multiple beams illuminating one pixel at the same time. The bandwidth required for the overlap configuration is higher than the no-overlap case. A hardware deflection array built to accommodate an overlap configuration could be limited in software to run in the no-overlap mode.

This configuration of beam controls and signals can be described precisely. The i-th 2-dimensional $U \times V$ electron sensing pattern P_i can be written as the outer product of a vector that defines the rows $r_i \in \mathbb{R}^U$, and another that defines the columns $c_i \in \mathbb{R}^V$, as $R_i = r_i c_i^T$. This scenario is related to previous work in separable imaging operators for compressed sensing [284, 844]. Sensing patterns generated by one potential configuration of a 21×21 array of beams that addresses a 49×49 array of pixels is shown in Figure 24.3. A parameter-less implementation of basis pursuit called SPGL1 [1083, 1082] evaluates this sampling for randomly-generated images with 5%, 10%, and 15% sparsity over a range of sampling ratios (Figure 24.4).

24.9.3 *Array of Individually Steerable Beams*

Previously, it has been shown that sparse matrices can perform theoretically as well as dense Gaussian ensembles for compressed sensing recovery [363], when the degree (fill ratio) of the sensing

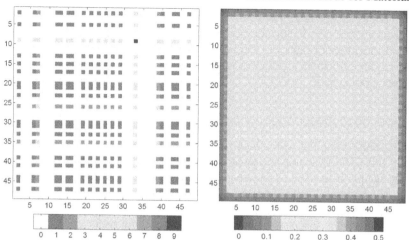

Figure 24.3: (left) One realization of a sampling pattern for an array of 47×47 beams, with a two-pixel offset between beam centers and a $-2, -1, 0, +1, +2$ steerable offset randomly selected and applied uniformly to each row and column. The beam-to-pixel ratio in this configuration is 25% and a pixel is illuminated by 0 to 9 beams in any realization, as indicated by the color bar. (right) The average illumination per pixel in a realization if offsets are selected uniformly from the distribution.

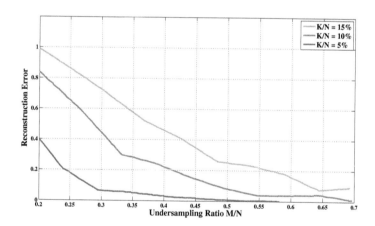

Figure 24.4: Reconstruction results for an array of correlated beams configured as in Figure 24.3. Acceptable reconstruction error of less than 0.1 is achieved for images with 5% or lower sparsity above approximately 0.3 under-sampling ratio.

matrix was very small. Sparse measurement matrices offer an advantage of efficient image recovery, since each iteration of the process becomes a simple multiplication by a sparse matrix. For recovering images that are canonically sparse, (e.g., starfield images), very efficient methods exist that are based on expander graphs [457].

The beam steering apparatus could be configured to create a very sparse fill ratio, on the order of 1 beam per thousand pixels. Assuming a system in which $1,000$ beams illuminate sensing functions over a million pixels, each beam needs a minimum of $\sqrt{1000} \approx 32$ addresses in X and Y. More addresses would allow for overlapping fields of illumination. In terms of bandwidth, five bits per beam for each X and Y control axis requires 32 different voltages and yields $(2^5)^2 = 1,024$ pixel ad-

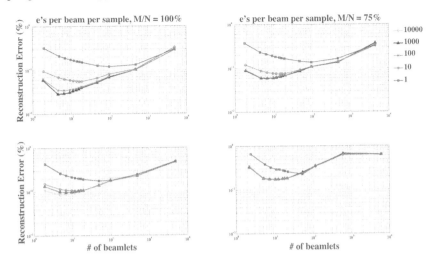

Figure 24.5: Reconstruction error as a function of the number of beams (beamlets) in a hypothetical CSEM for a set of 100×100 pixel images. Colored lines indicate the electrons per beam per sample. The four subplots test sampling ratio (samples to pixels). Global allocation means that each beam was assigned a random location in the image without restriction. Global allocation was uniformly slightly better than a local allocation scheme in which each beam was restricted to a 4×4 pixel patch (results not shown). Reconstruction error is smaller for sparse measurement matrices (few beams) rather than dense. Ten or fewer electrons per beam yield significantly poorer reconstruction, and 100 or more electrons per beam per sample provide indistinguishable results. Further investigation determined the knee in the curve, or threshold electron budget, to be about 40 electrons per beam per sample.

dresses per beam; six bits requires 64 voltages yielding $4,096$ positions. The potential performance of this system is discussed in the following section.

24.9.4   Managing the Electron Budget

The allocation of electrons among beams and compressive samples is a key design consideration of a fast compressive sensing electron microscope. For a given source and column design, electrons are the resource in scarce supply, and utilizing them efficiently is the key to achieving quality images in the shortest possible time.

There are several factors that bear consideration. First, since the system has multiple beams and electrons are few and discrete, an initial question is how many electrons must pass through each aperture to ensure that the sample is sufficiently interrogated by each beam in a sample? The time and location of an electron being emitted from a source is essentially a random process, and thus the aperture that an individual electron passes through is unknowable.

For the system concept described in Section 24.8, the emission of electrons from a Schottky source was modeled as a Poisson process, in which electrons are emitted at inexact time intervals in a normal distribution over the facet of the source. Thus, each aperture hole has an equivalent chance of passing the emitted electron. If the number of electrons is approximately equal to the number of holes, the statistics work out such that many holes will pass one electron, some will pass more than one, and some will pass zero. To assess the number of emitted electrons necessary to achieve statistically-equivalent illumination per beam, sparse image reconstruction accuracy as a function of the average number of electrons per aperture hole was assessed through computer simulation, while also varying the beam-to-pixel fill ratio (Figure 24.5). A sparse fill ratio was significantly superior to dense, and an average value of 40 or more electrons per hole was the "knee in the curve" above which additional illumination provided slight to negligible improvement in reconstruction accuracy for this source of noise. This method of quantifying noise is independent of the beam current in the system and the specimen being imaged and is thus a useful representation of the minimum number of electrons that must be transmitted to the sample, per measurement per beam, in a multi-beam CSEM. Since reduced-dose imaging is a motivation for compressive sensing, particularly in TEM, additional exploration of the minimum realizable electron dose per compressive sample would be useful for several applications.

Is there any benefit in going above the minimum electron budget? Stochastic processes in the interaction of sample and beams and detectors make this an issue that may be best answered experimentally. However, some insight from SEMs might be applicable. Many SEMs have a fixed sampling rate, 10 to 20 million samples per second on many modern machines. In a SEM with 200 pA beam current and 10 million samples per second, each measurement uses about 125 electrons. SNR in an image pixel is controlled by operator selections of beam current (electrons per sample) and averaging more than one sample into the reported pixel value. In the case of sample averaging, the SNR increases as the square root of the number of samples. Excellent SNR in the recorded image (aka material contrast) is often built up from many poor measurements.

In a multi-beam CSEM, responses from many pixels are aggregated into one compressive measurement. An open question is the necessary SNR of the individual pixel responses. Each pixel is measured many times across the set of compressive measurements. For example, a multi-beam CSEM with 1,000 beams, a million pixel image area, and 100,000 sparse measurements (measurements/pixels = 0.10) will represent each pixel in about 100 compressive measurements. If on the order of 100 electrons per pixel is allocated for each compressive measurement, each pixel's value is represented by approximately 10,000 electrons across the entire collection. Fundamentally, if few electrons per pixel are sufficient in each compressive sample, then fast collection of the compressive samples can be achieved. Conversely, if many electrons are required per pixel per compressive sample, fast imaging becomes challenging.

24.10   Multi-Beam CSEM Speed Estimates

Estimates of speed are developed in this section for a variety of potential embodiments. In Section 24.8, a conceptual column design for a multi-beam CSEM was developed suggesting that approximately 12,500 electrons per nanosecond could be delivered to the sample if distributed over many beams. Accepting the critical caveat that scientific uncertainty and engineering challenges exist, which are discussed in a following section, and assuming they can be solved, the necessary speed-related parameters are principally the electron budget per pixel per sample — which as previously discussed could be as low as about 40, but not likely lower — and the number of compressive samples to be collected, which depends theoretically on the sparsity of the specimen being imaged and practically on the optimization of the sampling matrices and image reconstruction methods. In the scenarios presented in Table 24.1, 100 electrons per beamlet per sample and a 10% ratio of measurements to pixels are assumed. Both of these are unverified in an actual multi-beam CSEM. Speed estimates are linear in total electron budget, electrons per beam per sample, and the compression ratio.

Scenarios 1 through 4 explore dense measurement matrices. Scenarios 1 and 2 explore the effect of [on–off] vs. steerable switching. Scenario 1 has 10,000 [on–off] switchable beamlets and a 10,000 pixel image, with 50% of the beamlets on at any one time, and half of the electron budget diverted to a collection plate. Scenario 2 has 5,000 steerable beamlets over two positions, and maintains the full electron budget on the sample for all patterns. Scenario 2 is twice as fast as Scenario 1. Scenarios 2 and 3 show that for a given pixel-to-beam ratio and total beam current, the system with fewer beams is faster. Scenario 3 recovers a 20,000 pixel image, and Scenario 4 recovers a 160,000 pixel image. Comparing Scenarios 3 and 4 shows that, all other factors being equal, the

Table 24.1: Calculations of speed for a hypothetical multi-beam compressive sensing electron microscope in which 2 nA of beam current (12,500 electrons per nanosecond) are distributed among the indicated number of beams.

Scenario	# Beams	PBR	μs/image	MP/s	PSR (MHz)	nA/Beam	MP/s Formula
1	10000	1 [on–off]	80	125	12.5	0.2	125
2	5000	2	40	250	25	0.4	250
3	10000	2	160	125	12.5	0.2	125
4	10000	16	1280	125	12.5	0.2	125
5	1000	1000	800	1250	125	2	1250
6	1000	500	400	1250	125	2	1250
7	500	1000	200	2500	250	4	2500
8	250	1000	50	5000	500	8	5000

Note: The exposure is assumed to be 100 electrons per beam (pixel) per compressive sample. The pixel to beam ratio (PBR) is simply the size of the recovered image divided by the number of beams used to collect it. The μs/image indicates the total time to collect samples necessary for reconstruction, assuming no switching latency. The MP per second is the data rate for a single image. The pattern switch rate (PSR) indicates the rate of compressive sample collection. The nA per beam indicates the beam current per beam. The Formula and Scenarios columns are discussed in the text.

systems have the same realizable speed. In fact, Scenarios 1, 3, and 4 each have 10,000 beams and are equally fast, even though the other system design parameters vary by large factors.

Scenarios 5 through 8 explore sparse measurement matrices. Scenarios 5 and 6 show that, like Scenarios 3 and 4, the ratio of the number of pixels in the image to the number of beams in the sensing system has no effect on the achievable speed. Scenarios 6 and 7 show that for images with the same number of pixels, the system with fewer beams will have a faster rate. Scenarios 7 and 8 also show that for a given image size to beam ratio, the system with fewer beams is faster.

The Switch Rate, MHz column (PSR) explores the engineering complexity of the Scenario, indicating the detector rate required and the maximum pattern switch rate for the system. This is an indicator of the bandwidth required by the system. Extremely careful electronics design or even fundamental advances in microelectronics would be required to achieve the performance some of these scenarios demand. The nA/Beam column indicates the physics complexity of the Scenario; in most electron microscopes, the resolution is inversely proportional to beam current.

The MP/s formula column indicates the value calculated by the formula A/BCD where A is the electron budget per second (12.5×10^{12}), B is the number of beams in the Scenario, C is the electrons per beamlet per sample (100), and D is the sparsity, or ratio of samples to pixels, 0.1. The units cancel

$$\frac{\frac{e}{sec}}{beam \frac{e}{beam \times sample} \frac{sample}{pixel}} = \frac{\frac{e}{sec}}{\frac{e}{pixel}} = \frac{pixel}{sec} \tag{24.4}$$

to a rate of pixel/sec. The values calculated by this formula match those displayed in the MP/s column. The electron budget (beam current) in the numerator indicates an unsurprising direct linear correlation with speed. Sparsity in the denominator is also expected, and shows that the microscope has an effective increase in speed, just by reducing the number of compressive measurements such as might occur over time with R&D into image reconstruction methods and improved sparse representations. The electron budget per beam per sample in the denominator is also not surprising. The electron dose per beam per sample has a minimum realizable value of around 40, and as discussed previously the necessary value is an open scientific question for compressive sensing in electron microscopy (it is also dependent on the specific detector system: electron capture and detector efficiency).

A somewhat non-intuitive aspects of the speed formula is that the speed of a multi-beam electron microscope implementing sparse sampling is inversely proportional to the number of beams and does not depend on the beam-to-pixel ratio. The key insight is the available beam current, in the numerator, ought to be allocated to as few beams as possible which is physics-limited for a domain of application. In a sense, the multi-beam compressive sensing system optimized for speed would use as much beam current per beam as would be used in the single beam of a SEM highly optimized for speed.

The intermediate term is the ratio of total electron budget to the average number of electrons each pixel receives in a compressive sample. For a fixed number of electrons per beam per measurement, the only variable in the denominator is the number of beams. Thus, the number of beams in the illumination pattern is a primary control on the number of times a pixel is sampled in a fixed number of compressive measurements, or in other words, the number of incident electrons each pixel receives in a given measurement matrix.

Conveniently, the number of pixels per image and the methods for selecting pixels in any sensing pattern are free variables that can be chosen optimally for the engineering constraints of the system or the mathematical constraints of the compressive sensing issues in a given domain of interest.

24.11   Scientific Challenges in a Multi-Beam CSEM

A variety of multi-beam electron microscopes exist for lithography applications, proving that various and essentially arbitrary illumination patterns can be created using multiple electron beams. In general, compressive reconstructions are more sensitive to noise in the measurement process than traditional imaging methods. The principal scientific challenges in a multi-beam CSEM are the extents to which sources of measurement uncertainty in the system dominate the compressed measurements. As compressive sensing in electron microscopy is lightly explored, the topics introduced in this section are intended to outline areas that would benefit from research and development, rather than to raise insurmountable objections to the concept.

A finite and relatively small number of electrons are used in electron microscopy to measure responses of the specimen. A well-optimized system is essentially signal-to-noise dominated in which the elapsed time to acquire an image is principally related to the number of electrons per pixel necessary to acquire the desired image quality. In a multi-beam system in which an array of apertures is used to create many beams from one source, each electron emitted can pass through at most one aperture hole. An immediate uncertainty is the number of electrons necessary to create sufficiently uniform illumination among the beams created by the aperture array. This uncertainty was evaluated by simulation and becomes negligible with 40 or more electrons per aperture per sample. Alternative methods for creating multiple beams, such as an array of sources or multiple illumination of a photocathode create variable emission that could be problematic in compressive sensing.

An associated uncertainty, already discussed, is the extent to which imprecise vs. precise measurement of each pixel value is necessary in each compressive sample. As in SEM and TEM, the secondary and backscatter electron response of the sample to the incident beam is a principal source of uncertainty and the principal driver of long illumination time to increase SNR, which improves as the square root of the illumination duration. An open scientific question is the necessary SNR or dwell in a CSEM collecting sparse samples of a global nature. This is related to CS theory but also to stochastic and noise processes in an electron microscope system. Clearly, if each pixel can be "undersampled" in compressive measurements, samples can be collected faster.

While not noise per se, beam deformation and spreading due to lens aberrations and electron repulsion require careful consideration. For example, to reduce electron repulsion effects, multi-beam systems being developed for lithography applications are often operated at electron acceleration voltages higher than the 0.5-30 keV typically used in surface electron microscopy [498, 666]. Methods for simulating these quantities in complex microscopes are mature and commercially available [952, 922]. For a given total beam current, electron repulsion would tend to increase as the number of beams decreases.

Clearly, a fast CSEM requires an efficient detector system which collects a large percentage of the electrons emitted from the sample. Chamber response is the emission of electrons (indirect back-scatter electrons (BSE) and secondary electrons) from surfaces struck by BSE emitted from the sample. Since it derives directly from BSE emitted by the sample, it can be considered as an indicator or amplification of primary BSE signal [369]. Chamber response electrons would be measurement noise in a CSEM only if the specific application requires exclusive collection of secondary electrons. In principal, detectors are a final potential source of noise, however most detectors in common use are relatively clean compared to the variability of specimen response.

The uncertainties and sources of noise discussed above are all common to existing electron microscopes. A real compressive sensing electron microscope may have additional uncertainties in the precise shapes and positions of the beams in the global illumination patterns, deriving from many sources. First, the shape of each beam will to some extent be unknown and varying across the array due to the discrete, stochastic nature of electrons as well as various engineering imperfections and optical aberrations in the illumination electron column. This will be true even in the case of long illumination times, careful calibration, and the potential use of anti-stigmation micro-optics on each beam.

The optimal size of the beams on the sample is an area for further study. The use of beams that are smaller than the intended pixels may help in reducing the effects of uncertainty of beam shape and position. Using beams larger than pixels, which would sample patches of pixels simultaneously while maintaining pixel precision in location, may have helped average out noise in pixel-domain sampling [780]. However, uncertainty in beam location is an explanation offered for degraded reconstruction in experiments on a SEM [23], which could occur for patch sampling in a multi-beam CSEM.

Another source of sampling matrix uncertainty is imperfect knowledge of the sample. Uneven topography will distort the sensing pattern. Sample tilt will induce skew distortions. The application of a CSEM favors flat samples in mountings that provide high precision orientation orthogonal to the axis of illumination.

A further source of sampling matrix uncertainty is charging in the sample. Under many imaging conditions, a sample illuminated by an electron beam develops an imbalanced charge near the incident beam. Surface charge (of either polarity) causes the incident beam to deflect from the intended position, and can affect the capture efficiency of detectors. Since a multi-beam CSEM envisions sampling globally over an area perhaps many microns wide, differential surface charge could develop which would cause spatially-variable distortion of the measurement patterns and possibly spatially-variable detection efficiency of the emitted electrons. The magnitude of the charging-induced distortions would be unknowable and potentially progressive over time depending on the illumination process and sample properties. Spatially-variable, time-progressive distortions would be a significant challenge in reconstruction. Specimens or specimen preparations resistant to charging such as carbon coating or active charge suppression methods such as gas injection or complementary illumination would be desirable for surface imaging applications of compressive sensing. Since sample charging is a local phenomenon, it could be that sparse measurement matrices such as $1/1000$ beam-to-pixel ratio would allow charge imbalance to dissipate as the beam landing locations jump around irregularly.

The devices that create the sensing patterns, like many multi-beam electron lithography systems, would probably use some sort of switchable or configurable microelectronics in the illumination column. The devices in this system are subject to degradation and possible failure, which would be another source of measurement matrix uncertainty. An aperture could become contaminated, affecting illumination and beam shape. Of potentially greater impact would be the failure of a device such that a beam that is supposed to vary in its location or intensity becomes stuck in one position. For long service life between major maintenance, the favored system would be resilient in device function, provide run-time knowledge of device state, and allow any of several beams to illuminate a location in the sensing matrix, so that failure of one device does not leave gaps in areal coverage over a series of patterns.

Methods to address image recovery in the presence of uncertain deviations from intended sensing patterns and other uncertainties have been considered [768]. It is likely that a CSEM would require reconstruction methods that accommodate some measurement matrix uncertainty.

In summary, the principal scientific questions for electron microscopy with sparse sampling and compressive reconstruction center on the impact of system noise and uncertainty on the compressive reconstruction process. It seems possible that experiments using one or more of the various existing multi-beam electron systems, adapted from electron beam lithography, could be conducted to explore and characterize the specific effects of each of these uncertainties. As these experiments would intend to answer issues of a mostly scientific nature, they could be conducted at much slower speed and much lower resolution than would be desired in an operational CSEM. A well-designed and executed demonstration of compressive sensing in a multi-beam electron microscope, exploring and answering the issues outlined in this section, would be an important milestone.

24.12   Engineering Challenges in a Multi-Beam CSEM

A scientific demonstration of compressive sensing in a multi-beam system could be undertaken at any resolution and speed, and thus any existing multi-beam system is a potential candidate for compressive sensing experiments that could significantly reduce scientific uncertainty. A compressive sensing microscope capable of sustained operation collecting imagery faster than existing alternatives at near-nanometer resolution is significantly more difficult. Some SEMs operate at near SNR-limited rates over large mosaics, and an emergent parallel scanning SEM could potentially be optimized to collect data over sustained rates more closely to its peak speed of greater than 1 billion samples per second. For a CSEM to go faster, significant engineering challenges exist in many of the necessary subsystems. This section identifies and briefly discusses some of the anticipated engineering issues that would need to be solved in a CSEM that is dramatically faster than alternative approaches. As in the scientific challenges section, the intent here is to open avenues of potential research rather than to raise objections to the possibility of a fast CSEM. Only a few people have examined the vast engineering landscape of a fast CSEM and feasible solutions may already exist for some or many of these challenges.

A simple but vital issue is focus and stability. How would a multi-beam CSEM be focused, and how would the focus be kept stable for a long period of time so that the sustainable data rate is high? Most likely, a scanning mode or setup mode would be necessary, in which the compressive sampling and reconstruction is bypassed to enable a slower and more direct imaging process. With fast enough reconstruction (potentially in hardware), an iterative scheme could be attempted.

Like the electron beam lithography industry, a multi-beam CSEM would likely incorporate micro-optic devices to create, shape, and control beams. Two principal engineering challenges are micro-optic bandwidth and fabrication. Bandwidth is the product of the number of independent controls (bits) times the switching speed of the array, yielding bits per second. It is not possible to route hundreds of thousands or millions of time-synchronized lines, each one carrying one of several possible voltages, into micro-optic devices laid out over a few square mm. Either only a few thousand (at most) deflection elements can be built, or some method of reducing routing bandwidth is necessary. A steerable beam concept might utilize on the order of a thousand independent deflection devices. To explore bandwidth, assume each deflection device requires X and Y deflection controls with one deflection voltage per control setting. If each beam has just two voltage settings (0 and $+1$), a thousand independent beams requires $4,000$ bits per pattern. A sampling frequency of 20 million hertz (50 ns dwell per sample) is common in existing SEMs. Such a CSEM would have a bandwidth of 80 billion bits per second. A different steerable beam concept might create a million beams arranged in a $1,000 \times 1,000$ array with some sort of bandwidth-reducing control system as discussed previously. A million beam array configured thus would use a similar number of bits per sample as the array of a thousand independent beams. Eighty billion bits per second is large for MEMs devices and thus bandwidth to the deflection array of micro-optic devices is a significant challenge. A vast worldwide community desires increased microelectronics bandwidth, however, so a current challenge becomes increasingly tractable over time.

The voltage controls on the micro-optics will require either CMOS logic integrated proximal to the devices, or a pattern switching method that is very low bandwidth. CMOS integration with MEMs structures has been performed in electron microscopes [498], however some embodiments of the CSEM could require MEMs-CMOS integration at a complexity not yet attempted. A bandwidth-reduction solution that reduced randomness in the patterns of beams would theoretically require more compressive samples and hence represent an engineering trade off between switching bandwidth and microscope speed. Similar to bandwidth, in this area the CSEM benefits from innovation in the broader microelectronics industry.

A second area of engineering challenges is in the objective lens of the microscope, which demagnifies the beam pattern from the micron-scale dimensions in the electron column to the nanometer scale dimensions necessary on the sample. As mentioned previously, approximately $2 \mu A$ of beam current could pass through the objective lens as $1,000$ or more individual beams. An IMAGE computer model of parallel beams passing through an existing electromagnetic objective lens maintained spot sizes on the sample of several nanometers suggesting that this requirement is reasonable [537]. However, contrary evidence is that the electron beam lithography industry shifts to high acceleration voltage (~50 keV) to reduce electron repulsion interactions.

A thorough canvassing of available detectors and their suitability for compressive sensing applications has not been conducted. Potential issues include detector placement, dynamic range, sensitivity, and speed. A detector placed at the bottom of the polepiece is sometimes used in SEMs, and would present a large capture area for CS measurements. The optimal detector configuration and location would be determined in formal design. As the compressive measurement patterns are created by many beams striking the sample simultaneously with high total current, there will be a large electron flux. The detector may need a dynamic range and sensitivity to be able to distinguish relatively small variability in signals. If relatively few beams and few electrons per beam per sample are sufficient, the sampling rate may exceed 100 million samples per second.

Alternative approaches to potential detector challenges may be possible. Noting that undersaturation compresses dynamic range and detector speed is inversely proportional to bit depth, sampling strategies and sparse recovery conditions have been developed for saturated responses and as few as 1 bit of detector depth [539, 456]. Independent of their potential as a path forward for detector challenges, these alternative strategies point to the value of "contrarian thinking" in designing a CSEM.

In compressive sensing, the measurements are somewhat independent of the back-end reconstruction. During the design and prototyping phases, reconstruction algorithms simply need to be "good enough" to provide quick iteration and useful feedback to designers, and evidence that the proposed system is competitive or economically justifiable. As system development proceeds, custom, domain-specific sparsifying representations can be developed, as has already been done in TEM. Improved reconstruction methods occur as domain-specific constraints are identified and mathematical challenges are worked out. These developments can only result in improved image quality and higher speed by requiring fewer measurements. Thus, relative to the scientific uncertainties and engineering challenges in a multi-beam CSEM, robust methods for image reconstruction appear as a lesser risk.

Compressive reconstruction can be computationally complex. Significant computational resources may be necessary to invert the compressive measurements into usable images. Since myriad approaches are possible, including custom hardware, the issues of computational complexity are regarded as an engineering challenge. In the case of using compressive sensing to collect large quantities of data, the engineering challenge of compressive reconstruction could be insignificant compared to the overall challenge of storing and automatically processing terabytes to petabytes or, someday, exabytes of pixel or voxel data to a compact representation of the desired information.

24.13   Conclusion

Sparse sampling and compressive reconstruction offer an alternative design space for electron microscopy, which is largely unexplored. The first applications of CS in electron microscopy used existing electron microscopes with reduced sampling and sparse reconstruction. This is most active in TEM, which integrates information tomographically in each measurement which is highly suited to CS. Identified benefits of CS in TEM include reduced number of tilt-rotation samples required, reduced overall dose to the specimen, improved resolution of the z-dimension (depth) in the reconstructed volume, and reduced impact of the "missing wedge" (inaccessible tilt-rotations) in reconstructed images. Fewer explorations of sparse sampling and reconstruction exist for AFM and SEM, which do not intrinsically measure integrative samples.

Engineering modification of existing electron microscopy technology for sparse sampling has been explored in a fast electrostatic beam shutter in TEM, alternative sampling patterns in AFM with sparse reconstruction, star-field sampling in SEM, and modeling and inverting SEM beam dynamics enabling knowledge of the multi-pixel sensing function associated with each measurement. The modification of beam shape in existing electron microscopes has not been explored.

Sparse sampling and reconstruction in multi-beam electron microscopes has been lightly explored from theoretical and conceptual design perspectives, chiefly motivated by the desire for fast collection of large quantities of imagery. The difficulty of wide-field imaging in electron microscopy makes multi-beam illumination with single-detector response a natural fit for compressive sensing and motivates development. The existence and advanced technological development of multi-beam electron lithography microscopes demonstrate the ability to create precise, repeatable, and essentially arbitrary electron beam patterns on a specimen. Furthermore, many of the beam-forming and control technologies developed for lithography appear highly suited to multi-beam compressive sensing. A conceptual design suggests that multi-beam electron microscopy with sparse sampling and reconstruction could collect upwards of billions of pixels per second, due to parallel illumination allowing high total beam current. Principal scientific challenges relate to uncertainties in noise processes and their effects in sparse reconstruction. Of the many engineering challenges, a principal concern is the micro-electronics bandwidth to create the sensing patterns. The theoretical/conceptual analysis suggests that a multi-beam CSEM optimized for speed would have as few beams as possible with maximal current in each beam subject to resolution and other issues, rather than fewer beams with lower beam current.

Acknowledgments

Sandia National Laboratories is a multi-program laboratory managed and operated by Sandia Corporation, a wholly-owned subsidiary of Lockheed Martin Corporation, for the U.S. Department of Energy's National Nuclear Security Administration under contract DE-AC04-94AL85000.

Appendix

A   List of Symbols for Chapters 6, 7, and 13

Symbols for Chapter 6

α_k	Detector gain associated with the measurements at tilt k
$\bar{I}$	Average scaled gain parameter
$\beta_{T,\delta}$	Generalized Huber function with cutoff parameter T and scale parameter δ
Λ_k	Diagonal matrix with entries proportional to the variance of the measurements at tilt k
ϕ	Vector of unknown calibration parameters
$\rho(.)$	Prior model potential function
σ_f	Prior model scaling parameter
σ_k	Noise variance parameter at tilt k
c_{HAADF}	MBIR cost-function for the HAADF-STEM tomography
A_k	$M \times N$ projection matrix for tilt k
c_{BF}	MBIR cost-function for BF tomography
d_k	Offset associated with the measurements at tilt k
f	Vector of unknown voxels
$f(x,y,z)$	Scatter coefficient at location (x,y,z)
g	Vector of all measurements
$G_k(u,v)$	Total number of electrons scattered into the HAADF detector at location (u,v) at tilt k
$h_i(u,v)$	Kernel which averages the electron flux over the area corresponding to the ith measurement
$I_0(u,v)$	Electron flux at location (u,v)
I_k	Scaled gain associated with the measurements at tilt k
K	Total number of views at which measurements are made
R_{θ_k}	Orthonormal rotation matrix corresponding to angle θ_k at tilt k
w_{ij}	Prior model weight factor between voxel i and j
Z	Prior model normalization parameter
[	D

Symbols for Chapter 7

V_η	Covariance of c
$\hat{r}_{\eta_0}(\mathbf{x},\mathbf{x}')$	Estimated value of V
$\hat{\bar{\rho}}_{\eta_0}(\mathbf{x})$	Estimated value of $\bar{c}$
$\bar{c}^\eta$	Mean of c
$\phi_\tau^{(\eta)}(\mathbf{x})$	Basis function used to form the electron scattering intensity
$\rho(\mathbf{x})$	Electron scattering intensity of an instance in the ηth class, represented as a linear combination of $N_c(\eta)$ basis functions of $\phi_\tau^{(\eta)}(\mathbf{x})$
θ	Projection direction and center location
$c \in \mathbb{R}^{N_c(\eta)}$	Vector of weights of basis functions, Gaussian random vector with mean $\bar{c}^\eta$ and covariance V_η
c_τ	Unknown weights of basis functions used to form the electron scattering intensity
$i \in \{1,\ldots,N_v\}$	Index for instances of the object
$j \in \{1,\ldots,N_T\}$	Index for the tilt series (recording several images at known relative projection directions)
L	Matrix to relate the unknown weight vector c and the image vector y
$N_c(\eta)$	Number of basis functions
N_T	Number of images taken of each object
N_v	Number of objects imaged
N_y	Number of pixels in each image
Q	Covariance of pixel noise of image vector y
$S^{(\eta)} \subset \mathbb{R}^3$	Support of the electron scattering intensity of an instance
$w_{i,j}$	Vector of pixel noise
[	E

Symbols for Chapter 13

Ψ	Euler characteristic. An topological invariant that relates the number of edges, faces, and vertices of shapes. Can be applied to analysis of percolation
$\aleph^{h_1}(x)$	Surface indicator function. Indicates whether a point is on the surface of a feature defined by local state h. Used for calculating various microstructure descriptors
$\beta_n(h)$	The nth Betti number. The Betti numbers parameterize the topological homology or connectivity of the microstructure
$\beta_n(h,A)$	Relative Betti number. Betti number calculated only for local state h. Can be used to calculate number of percolating paths in an image or 3D dataset
$\chi^{h_1}(x)$	Indicator function. 1 if $\mathbf{h}(x) = h_1$ and 0 otherwise. Used to calculate various microstructure descriptors
$\mathbf{h}(x)$	The microstructure local state vector. The set of relevant microstructure descriptors for a given application
$\mathbf{m}(x,h)$	The microstructure function. A stochastic process which generates realizations or observations of the microstructure
$\mathbf{x}$	A random or stochastic variable
$\mathscr{F}$	A set of events. Used to formally define a probability space
$\mathscr{P}$	The probability measure. Assigns probabilities to the events in $\mathscr{F}$
μ	The mean or expected value of a random variable
Ω	Probability sample space, a set of all theoretical outcomes of an experiment
ω	A single specific experimental outcome
σ^2	Variance. The expected square deviation of a random variable from its mean. Gives an indication of the scatter of observations
C_{ij}	Covariance matrix. Generalizes the concept of variance to random vectors. Expresses the dependencies between random variables
$d(X,Y)$	Generalized distance between random vectors. Expresses expected distance from X to Y in terms of the variance of $f(Y)$
$D_m(X,f(Y))$	The Mahalanobis distance. A generalization of the distance between point X and the probability distribution $f(Y)$. It is equivalent to expressing the distance from X to the mean of Y in terms of # of standard deviations of $f(Y)$
$E\{\cdot\}$	Expectation operator. Returns the expected or average value of the input
$f(\nabla h_1, \nabla h_1)$	Surface to surface correlation function. Expresses the probability of finding two internal surfaces or interfaces separated by a given distance
$f(\nabla h_1, h_1)$	Local state surface correlation. Expresses the probability of finding a local state a given distance from an internal surface or interface
$f(h,x)$	The first order probability density of the microstructure. Gives the volume fraction of local state h at spatial position x
$f_2(h_1, h_2 \mid r)$	The microstructure correlation function. Expresses the probability of finding the specified local states separated by a vector r
$f_{\mathbf{x}}(x)$	Probability distribution for random variable $\mathbf{x}$
$L(h \mid z)$	Lineal path function. Gives the probability that a line of length z will lie completely in local state h when randomly placed in the microstructure
$m(x,h,\omega)$	A specific microstructure relization, e.g. a micrograph

$p(h|z)$ Chord length distribution function. Gives the distribution of straight line distances between internal surfaces or interfaces

r_c Microstructure coherence length. Defines the length scale at which microstructure features exhibit spatial correlation or the minimum length scale for statistical analysis of the microstructure

R_{ij} Correlation matrix. Describes the relationship or dependencies between random variables or elements of a random vector

S_v^h Specific surface area per unit volume of local state h. Can be used to relate microstructure correlations to classical stereology

Bibliography

[1] Ossama B. Abouelatta. Classification of copper alloys microstructure using image processing and neural network. *Journal of American Science*, 9(6), 2013.

[2] Daniel Y. Abramovitch, Sean B. Andersson, Lucy Y. Pao, and Georg Schitter. A tutorial on the mechanisms, dynamics, and control of atomic force microscopes. In *American Control Conference, ACC'07*, pages 3488–3502. IEEE, 2007.

[3] Marc Achermann. Exciton-plasmon interactions in metal-semiconductor nanostructures. *J. Phys. Chem. Lett.*, 1(19):2837–2843, 2010.

[4] G. J. Ackland and A. P. Jones. Applications of local crystal structure measures in experiment and simulation. *Phys. Rev. B*, 73(5):054104, 2006.

[5] B. L. Adams, P. R. Morris, T. T. Wang, K. S. Willden, and S. I. Wright. Description of orientation coherence in polycrystalline materials. *Acta Metallurgica*, 35(12):2935–2946, 1987.

[6] Brent Adams, Stuart Wright, and Karsten Kunze. Orientation imaging: The emergence of a new microscopy. *Metallurgical and Materials Transactions A*, 24:819–831, 1993.

[7] Brent A. Adams. personal communication.

[8] Brent L. Adams, Xiang Carl Gao, and Surya R. Kalidindi. Finite approximations to the second-order properties closure in single phase polycrystals. *Acta Materialia*, 53(13):3563–3577, 2005.

[9] Brent L. Adams, Surya Kalidindi, and David T. Fullwood. *Microstructure sensitive design for performance optimization*. Butterworth-Heinemann, 2013.

[10] S. M. Adler-Golden. Improved hyperspectral anomaly detection in heavy-tailed backgrounds. *Proc. 1st IEEE Workshop on Hyperspectral Image and Signal Processing: Evolution in Remote Sensing (WHISPERS)*, 2009.

[11] S. M. Adler-Golden, S. C. Richtsmeier, and R.M. Shroll. Suppression of subpixel sensor jitter fluctuations using temporal whitening. *Proc. SPIE*, 6969:69691D, 2008.

[12] A. Agarwal, A. Anandkumar, P. Jain, and P. Netrapalli. Learning sparsely used overcomplete dictionaries via alternating minimization. *SIAM Journal on Optimization*, 26(4):2775–2799, 2016.

[13] A. Agarwal, A. Anandkumar, P. Jain, P. Netrapalli, and R. Tandon. Learning sparsely used overcomplete dictionaries. *Journal of Machine Learning Research*, 35:1–15, 2014.

[14] M. Aharon and M. Elad. Sparse and redundant modeling of image content using an image-signature-dictionary. *SIAM Journal on Imaging Sciences*, 1(3):228–247, 2008.

[15] M. Aharon, M. Elad, and A. Bruckstein. k-SVD: An algorithm for designing overcomplete dictionaries for sparse representation. *Signal Processing, IEEE Transactions on*, 54(11):4311–4322, 2006.

[16] Y. Aharonov and D. Bohm. Significance of electromagnetic potentials in the quantum theory. *Phys. Rev.*, 115:485–491, 1959.

[17] C. Aldrich, C. Marais, B. J. Shean, and J. J. Cilliers. Online monitoring and control of froth flotation systems with machine vision: A review. *International Journal of Mineral Processing*, 96(1):1–13, 2010.

[18] John Allison. Integrated computational materials engineering: A perspective on progress and future steps. *JOM Journal of the Minerals, Metals and Materials Society*, 63(4):15–18, 2011.

[19] Shaul Aloni, Virginia Altoe, Allard Katan, Florent Martin, and Miquel Salmeron. Scanned electron diffraction studies of self-assembled monolayers. *Microscopy and Microanalysis*, 18(S2):1600–1601, 2012.

[20] N. S. Altman. An introduction to kernel and nearest-neighbor nonparametric regression. *The American Statistician*, 46(3):175–185, 1992.

[21] H. Anderson, J. Helms, J. W. Wheeler, and K. W. Larson. Sparse sampling and reconstruction for electron and scanning probe microscope imaging. Technical report, U.S. Patent 9,093,249 B2, 2015.

[22] H. S. Anderson, Wheeler J. W., and K. W. Larson. Compressed sensing for fast electron microscopy. In *TMS 2014 Supplemental Proceedings*, pages 517–526, 2014.

[23] Hyrum S. Anderson, Jovana Ilic-Helms, Brandon Rohrer, Jason Wheeler, and Kurt Larson. Sparse imaging for fast electron microscopy. In *IS&T/SPIE Electronic Imaging*, pages 86570C–86570C–12. International Society for Optics and Photonics, 2012.

[24] Hyrum S. Anderson, Jovana Ilic-Helms, Brandon Rohrer, Jason Wheeler, and Kurt Larson. Sparse imaging for fast electron microscopy. In *Proc. SPIE*, volume 8657, pages 86570C–86570C–12, 2013.

[25] Sean B. Andersson and Lucy Y. Pao. Non-raster sampling in atomic force microscopy: A compressed sensing approach. In *American Control Conference (ACC)*, pages 2485–2490. IEEE, 2012.

[26] N. Andrei. Kolmogorov. Foundations of the Theory of Probability, 1950.

[27] F. J. Anscombe. The transformation of Poisson, binomial and negative-binomial data. *Biometrika*, 35:246–254, 1948.

[28] A. Antoniadis and T. Sapatinas. Wavelet shrinkage for natural exponential families with quadratic variance functions. *Biometrika*, 88(3):805–820, 2001.

[29] Relja Arandjelović and Andrew Zisserman. Three things everyone should know to improve object retrieval. In *Computer Vision and Pattern Recognition (CVPR), 2012 IEEE Conference on*, pages 2911–2918. IEEE, 2012.

[30] H. Arguello, H. F. Rueda, and G. R. Arce. Spatial super-resolution in code aperture spectral imaging. *Proceedings of the SPIE*, 8365:83650A, 2012.

[31] Ery Arias-Castro, Joseph Salmon, and Rebecca Willett. Oracle inequalities and minimax rates for non-local means and related adaptive kernel-based methods. *SIAM Journal on Imaging Sciences*, 5(3):944–992, 2012. https://doi.org/10.1137/110859403.

[32] S. Arivazhagan and L. Ganesan. Texture classification using wavelet transform. *Pattern Recognition Letters*, 24(9):1513–1521, 2003.

[33] S. Arora, R. Ge, and A. Moitra. New algorithms for learning incoherent and overcomplete dictionaries. In *Proceedings of the 27th Conference on Learning Theory*, volume 35, pages 779–806, 2014.

[34] M. Ashikhmin. Synthesizing natural textures. In *In I3D Conference 2001, pp. 217–226*, 2001.

[35] A. Ashok and M. Neifeld. Pseudorandom phase masks for superresolution imaging from subpixel shifting. *Applied Optics*, 46(12):2256–2268, 2007.

[36] E. A. Ashton. Multialgorithm solution for automated multispectral target detection. *Optical Engineering*, 38:717–724, 1999.

[37] K. M. Asl and J. Luo. Impurity effects on the intergranular liquid bismuth penetration in polycrystalline nickel. *Acta Materialia*, 60:149, 2012.

[38] L. Azzari and A. Foi. Variance stabilization for noisy+ estimate combination in iterative Poisson denoising. *IEEE Signal Processing Letters*, pages 1086–1090, 2016.

[39] L. Bachega, J. Theiler, and C. A. Bouman. Evaluating and improving local hyperspectral anomaly detectors. *IEEE Applied Imagery and Pattern Recognition (AIPR) Workshop*, 39, 2011.

[40] F. Bachman, R. Hielscher, and H. Schaeben. Texture analysis with mtex- free and open source software toolboc. *Solid State Phenomena*, 160:63–68, 2010.

[41] Charles M. Bachmann, Thomas L. Ainsworth, and Robert A. Fusina. Exploiting manifold geometry in hyperspectral imagery. *IEEE Trans. Geoscience and Remote Sensing*, 43:441–454, 2005.

[42] P. Bajorski. Maximum Gaussianity models for hyperspectral images. *Proc. SPIE*, 6966:69661M, 2008.

[43] Matthew L. Baker, Junjie Zhang, Steven J. Ludtke, and Wah Chiu. Cryo-EM of macromolecular assemblies at near-atomic resolution. *Nature Protocols*, 5(10):1697–1708, 2010.

[44] T. S. Baker, N. H. Olson, and S. D. Fuller. Adding the third dimension to virus life cycles: Three-dimensional reconstruction of icosahedral viruses from cryo-electron micrographs. *Microbiology and Molecular Biology Reviews*, 63(4):862–922, December 1999.

[45] Sinan Balci, Coskun Kocabas, Simge Ates, Ertugrul Karademir, Omer Salihoglu, and Atilla Aydinli. Tuning surface plasmon-exciton coupling via thickness dependent plasmon damping. *Physical Review B*, 86(23):235402, 2012.

[46] R. W. Balluffi, S. M. Allen, and W. C. Carter. *Kinetics of Materials*. John Wiley & Sons, 2005.

[47] S. Bals, C. F. Kisielowski, M. Croitoru, and G. Van Tendeloo. Annular dark field tomography in TEM. *Microscopy and Microanalysis*, 11:2118–2119, 2005.

[48] D. Balzani, L. Scheunemann, D. Brands, and J. Schröder. Construction of two-and three-dimensional statistically similar rves for coupled micro-macro simulations. *Computational Mechanics*, 54(5):1269–1284, 2014.

[49] A. S. Bandeira, E. Dobriban, D. G. Mixon, and W. F. Sawin. Certifying the restricted isometry property is hard. *IEEE Trans. Inform. Theory*, 59(6):2448–2450, 2013.

[50] A. Banerjee, P. Burlina, and C. Diehl. A support vector method for anomaly detection in hyperspectral imagery. *IEEE Trans. Geoscience and Remote Sensing*, 44:2282–2291, 2006.

[51] A. Banerjee, S. Merugu, I. S. Dhillon, and J. Ghosh. Clustering with Bregman divergences. *J. Mach. Learn. Res.*, 6:1705–1749, 2005.

[52] C. Bao, H. Ji, Y. Quan, and Z. Shen. L0 norm based dictionary learning by proximal methods with global convergence. In *IEEE Conference on Computer Vision and Pattern Recognition (CVPR)*, pages 3858–3865, 2014.

[53] Maya Bar Sadan, Lothar Houben, Sharon G. Wolf, Andrey Enyashin, Gotthard Seifert, Reshef Tenne, and Knut Urban. Toward atomic-scale bright-field electron tomography for the study of fullerene-like nanostructures. *Nano Letters*, 8:891–896, 2008.

[54] M. Baram, D. Chatain, and W. D. Kaplan. Nanometer-thick equilibrium films: The interface between thermodynamics and atomistics. *Science*, 332:206, 2011.

[55] R. G. Baraniuk, V. Cevher, M. F. Duarte, and C. Hegde. Model-based compressive sensing. *IEEE Trans. Inform. Theory*, 56:1982–2001, 2010.

[56] R. G. Baraniuk, M. Davenport, R. DeVore, and M. Wakin. A simple proof of the restricted isometry property for random matrices. *Constructive Approximation*, 28(3):253–263, 2008.

[57] Z. H. Barber, J. A. Leake, and T. W. Clyne. The doitpoms project: A web-based initiative for teaching and learning materials science. *Journal of Materials Education*, 29(1/2):7, 2007.

[58] Nicolas Barbey, Marc Sauvage, J.-L. Starck, Roland Ottensamer, and Pierre Chanial. Feasibility and performances of compressed sensing and sparse map-making with herschel/pacs data. *Astronomy & Astrophysics*, 527:A102, 2011.

[59] Montserrat Bárcena and Abraham J. Koster. Electron tomography in life science. *Seminars in Cell & Developmental Biology*, 20:920 – 930, 2009.

[60] D. Barchiesi and M. D. Plumbley. Learning incoherent dictionaries for sparse approximation using iterative projections and rotations. *IEEE Transactions on Signal Processing*, 61(8):2055–2065, 2013.

[61] Rizia Bardhan, Nathaniel K. Grady, Joseph R. Cole, Amit Joshi, and Naomi J. Halas. Fluorescence enhancement by au nanostructures: Nanoshells and Nanorods. *ACS Nano*, 3(3):744–752, 2009.

[62] K. Barmak, J. Kim, C.-S. Kim, W. E. Archibald, G. S. Rohrer, A. D. Rollett, D. Kinderlehrer, S. Taàsan, H. Zhang, and D. J. Srolovitz. Grain boundary energy and grain growth in Al films: Comparison of experiments and simulations. *Scripta Materialia*, 54(6):1059–1063, 2006.

[63] H. H. Barrett and K. J. Myers. *Foundations of Image Science*, pp. 1584. Wiley-VCH, 1, 2003.

[64] Gianni Bartolacci, Patrick Pelletier, Jayson Tessier, Carl Duchesne, Pierre-Alexandre Bossé, and Julie Fournier. Application of numerical image analysis to process diagnosis and physical parameter measurement in mineral processes part i: flotation control based on froth textural characteristics. *Minerals Engineering*, 19(6):734–747, 2006.

[65] B. Basener, E. Ientilucci, and D. Messinger. Anomaly detection using topology. *Proc. SPIE*, 6565:65650J, 2007.

[66] W. F. Basener and D. W. Messinger. Enhanced detection and visualization of anomalies in spectral imagery. *Proc. SPIE*, 7334:73341Q, 2009.

[67] W. F. Basener, E. Nance, and J. Kerekes. The target implant method for predicting target difficulty and detector performance in hyperspectral imagery. *Proc. SPIE*, 8048:80481H, 2011.

[68] E. Bauer. Low energy electron microscopy. *Reports on Progress in Physics*, 57(9):895, 1994.

[69] Herbert Bay, Tinne Tuytelaars, and Luc Van Gool. Surf: Speeded up robust features. In *Computer vision–ECCV 2006*, pages 404–417. Springer, 2006.

[70] A. Béché, B. Goris, B. Freitag, and J. Verbeeck. Development of a fast electromagnetic beam blanker for compressed sensing in scanning transmission electron microscopy. *Applied Physics Letters*, 108(9):093103, 2016.

[71] Armand Béché, Bart Goris, Bert Freitag, and Jo Verbeeck. Development of a fast electromagnetic shutter for compressive sensing imaging in scanning transmission electron microscopy. *arXiv preprint arXiv:1509.06656*, 2015.

[72] A. Beck and M. Teboulle. Fast gradient-based algorithms for constrained total variation image denoising and deblurring problems. *IEEE Transactions Image Processing*, 18(11):2419–34, 2009.

[73] A. Beck and M. Teboulle. A fast iterative shrinkage-thresholding algorithm for linear inverse problems. *SIAM J. Imaging Sci.*, 2(1):183–202, 2009.

[74] S. Becker, J. Bobin, and E. Candès. NESTA: A fast and accurate first-order method for sparse recovery. *SIAM J. Imaging Sci.*, 4(1):1–39, 2011.

[75] H. Beladi, Q. Chao, and G. S. Rohrer. Variant selection and intervariant crystallographic planes distribution in martensite in a Ti-6Al-4V alloy. *Acta Materialia*, 80:478–489, 2014.

[76] H. Beladi, N. T. Nuhfer, and G. S. Rohrer. The five-parameter grain boundary character and energy distributions of a fully austenitic high-manganese steel using three dimensional data. *Acta Materialia*, 70:281–289, 2014.

[77] H. Beladi and G. S. Rohrer. The relative grain boundary area and energy distributions in a ferritic steel determined from three-dimensional electron backscatter diffraction maps. *Acta Materialia*, 61(4):1404–1412, 2013.

[78] H. Beladi, G. S. Rohrer, A. D. Rollett, V. Tari, and P. D. Hodgson. The distribution of intervariant crystallographic planes in a lath martensite using five macroscopic parameters. *Acta Materialia*, 63:86–98, 2014.

[79] M. Beleggia, S. Tandon, Y. Zhu, and M. De Graef. Electron-optical phase shift of magnetic nanoparticles, part II: Polyhedral particles. *Phil. Mag. B*, 83:1143–1161, 2003.

[80] M. Beleggia and Y. Zhu. Electron-optical phase shift of magnetic nanoparticles, part I: Basic concepts. *Phil. Mag.*, 83:1045–1057, 2003.

[81] Alex Belianinov, Rama Vasudevan, Evgheni Strelcov, Chad Steed, Sang Mo Yang, Alexander Tselev, Stephen Jesse, Michael Biegalski, Galen Shipman, Christopher Symons, et al. Big data and deep data in scanning and electron microscopies: deriving functionality from multidimensional data sets. *Advanced Structural and Chemical Imaging*, 1(1):1–25, 2015.

[82] Yoshua Bengio. Deep learning of representations: Looking forward. In *Statistical Language and Speech Processing*, pages 1–37. Springer, 2013.

[83] Jerome Berclaz, Francois Fleuret, and Pascal Fua. Multiple object tracking using flow linear programming. In *IEEE International Workshop on Performance Evaluation of Tracking and Surveillance*, pages 1–8, 2009.

[84] Asbjørn Berge and Anne H. Schistad Solberg. Structured Gaussian components for hyperspectral image classification. *IEEE Trans. Geosci. Remote Sensing*, 44(11):3386–3396, November 2006.

[85] James G. Berryman. Relationship between specific surface area and spatial correlation functions for anisotropic porous media. *Journal of Mathematical Physics*, 28(1):244–245, 1987.

[86] Dimitri P. Bertsekas. Incremental gradient, subgradient, and proximal methods for convex optimization: A survey. *Optimization for Machine Learning*, 2010:1–38, 2011.

[87] J. Besag. Spatial interaction and the statistical analysis of lattice systems. *Journal of the Royal Statistical Society B*, 36:192–236, 1974.

[88] J. Besag. Towards Bayesian image analysis. 16(3):395–407, 1989.

[89] P. Besbeas, I. De Feis, and T. Sapatinas. A comparative simulation study of wavelet shrinkage estimators for Poisson counts. *Internat. Statist. Rev*, 72(2):209–237, 2004.

[90] Manish H. Bharati, J. Jay Liu, and John F. MacGregor. Image texture analysis: Methods and comparisons. *Chemometrics and Intelligent Laboratory Systems*, 72(1):57–71, 2004.

[91] A. Bhattacharyya. On a measure of divergence between two statistical populations defined by their probability distribution. *Bulletin of the Calcutta Mathematical Society*, 35:99–110, 1943.

[92] Peter J. Bickel and Elizaveta Levina. Regularized estimation of large covariance matrices. *Annals of Statistics*, 36(1):199–227, 2008.

[93] Peter J. Bickel and Bo Li. Regularization in statistics. *Test*, 15(2):271–344, 2006.

[94] A. Bijaoui and G. Jammal. On the distribution of the wavelet coefficient for a Poisson noise. *Signal Processing*, 81:1789–1800, 2001.

[95] R. C. Bilcu and M. Vehvilainen. Fast nonlocal means for image denoising. In Russel A. Martin, Jeffrey M. DiCarlo, and Nitin Sampat, editors, *Digital Photography III*, volume 6502, page 65020R. SPIE, 2007.

[96] Jeff A. Bilmes. A gentle tutorial of the EM algorithm and its application to parameter estimation for Gaussian mixture and hidden Markov models. Technical Report TR-97-021, Department of Electrical Engineering and Computer Science, University of California at Berkeley, April 1998.

[97] Peter Binev, Wolfgang Dahmen, Ronald DeVore, Philipp Lamby, Daniel Savu, and Robert Sharpley. *Compressed Sensing and Electron Microscopy*. Springer, 2012.

[98] C. Bingham. An antipodally symmetric distribution on the sphere. *The Annals of Statistics*, pages 1201–1225, 1974.

[99] J. Bioucas-Dias and M. Figueiredo. A new TwIST: Two step iterative shrinkage/thresholding algorithms for image restoration. *IEEE Trans. Image Proc.*, 16(12):2992–3004, 2007.

[100] C. M. Bishop. *Pattern Recognition and Machine Learning*. Springer, 2006.

[101] Sushmita Biswas, Jinsong Duan, Dhriti Nepal, Ruth Pachter, and Richard Vaia. Plasmonic resonances in self-assembled reduced symmetry gold nanorod structures. *Nano Letters*, 13(5):2220–2225, 2013.

[102] Sushmita Biswas, Jinsong Duan, Dhriti Nepal, Kyoungweon Park, Ruth Pachter, and Richard A. Vaia. Plasmon-induced transparency in the visible region via self-assembled gold nanorod heterodimers. *Nano Letters*, 13(12):6287–6291, 2013.

[103] Sushmita Biswas, Xiaoying Liu, Jeremy W. Jarrett, Dean Brown, Vitaliy Pustovit, Augustine Urbas, Kenneth L. Knappenberger, Paul F. Nealey, and Richard A. Vaia. Nonlinear chiro-optical amplification by plasmonic nanolens arrays formed via directed assembly of gold nanoparticles. *Nano Letters*, 15(3):1836–1842, 2015.

[104] Pendry B. J. Negative refraction makes a perfect lens. *Phys. Rev. Lett.*, 85:3966–3969, 2000.

[105] James Black, Tim Ellis, and Paul Rosin. Multi view image surveillance and tracking. In *Workshop on Motion and Video Computing*, pages 169–174, 2002.

[106] Samuel S. Blackman. *Multiple-Target Tracking with Radar Applications*. Dedham, MA, Artech House, Inc., 1986, 463 p., 1, 1986.

[107] A. Blake. Comparison of the efficiency of deterministic and stochastic algorithms for visual reconstruction. *IEEE Trans. Pattern Anal. Mach. Intell.*, 11:2–12, 1989.

[108] T. Blumensath and M. Davies. Iterative hard thresholding for compressed sensing. *Appl. and Comp. Harmonic Analysis*, 27(3):265–274, November 2009.

[109] T. Blumensath and M. E. Davies. Iterative thresholding for sparse approximations. *J. Fourier Analysis and Appl.*, 14:629–654, 2008.

[110] D. A. Boas, D. H. Brooks, E. L. Miller, C. A. DiMarzio, M. Kilmer, R. J. Gaudette, and Quan Zhang. Imaging the body with diffuse optical tomography. *IEEE Signal Processing Magazine*, 18:57–75, 2001.

[111] Thomas Böhlke. Application of the maximum entropy method in texture analysis. *Computational Materials Science*, 32:276 – 283, 2005.

[112] Roland Böhmer, Catalin Gainaru, and Ranko Richert. Structure and dynamics of monohydroxy alcoholsmilestones towards their microscopic understanding, 100 years after debye. *Physics Reports*, 545(4):125–195, 2014.

[113] C. F. Bohren and R. D. Huffman. *Absorption and Scattering of Light by Small Particles*. New York: John Wiley & Sons, Inc., 1998.

[114] J. S. D. Bonet. Multiresolution sampling procedure for analysis and synthesis of texture images. In *In SIGGRAPH 97*, pages 361–368, 1997.

[115] D. Bonn and D. Ross. Wetting transitions. *Reports on Progress in Physics*, 64(9):1085–1163, 2001.

[116] A. Bora, A. Jalal, E. Price, and A. G. Dimakis. Compressed sensing using generative models. arxiv:1703.03208, March 2017.

[117] Ramin Bostanabad, Anh Tuan Bui, Wei Xie, Daniel W Apley, and Wei Chen. Stochastic microstructure characterization and reconstruction via supervised learning. *Acta Materialia*, 103:89–102, 2016.

[118] J. Boulanger, C. Kervrann, P. Bouthemy, P. Elbau, J-B. Sibarita, and J. Salamero. Patch-based non-local functional for denoising fluorescence microscopy image sequences. *IEEE Trans. Med. Imag.*, 29(2):442–454, 2010.

[119] C. A. Bouman and K. Sauer. A generalized Gaussian image model for edge-preserving MAP estimation. 2(3):296–310, July 1993.

[120] C. A. Bouman and K. Sauer. A unified approach to statistical tomography using coordinate descent optimization. *IEEE Trans. on Image Processing*, 5(3):480 –492, March 1996.

[121] Charles A. Bouman. *Model Based Image and Signal Processing*. 2012.

[122] Y-Lan Boureau, Francis Bach, Yann LeCun, and Jean Ponce. Learning mid-level features for recognition. In *Computer Vision and Pattern Recognition (CVPR), 2010 IEEE Conference on*, pages 2559–2566. IEEE, 2010.

[123] Y-Lan Boureau, Jean Ponce, and Yann LeCun. A theoretical analysis of feature pooling in visual recognition. In *Proceedings of the 27th International Conference on Machine Learning (ICML-10)*, pages 111–118, 2010.

[124] Stephen Boyd and Lieven Vandenberghe. *Convex Optimization*. Cambridge University Press, 2004.

[125] G. Bradski. OpenCV. *Dr. Dobb's Journal of Software Tools*, 2000.

[126] David J. Brady, Kerkil Choi, Daniel L. Marks, Ryoichi Horisaki, and Sehoon Lim. Compressive holography. *Optics Express*, 17(15):13040–13049, 2009.

[127] A. Brahme, M. H. Alvi, D. Saylor, J. Fridy, and A. D. Rollett. 3D reconstruction of microstructure in a commercial purity aluminum. *Scripta Materialia*, 55(1):75–80, 2006.

[128] K. Bredies and D. A. Lorenz. Iterated hard shrinkage for minimization problems with sparsity constraints. *SIAM J. Sci. Comput.*, 30(2):657–683, March 2008.

[129] Michael D. Breitenstein, Fabian Reichlin, Bastian Leibe, Esther Koller-Meier, and Luc Van Gool. Robust tracking-by-detection using a detector confidence particle filter. In *International Conference on Computer Vision*, pages 1515–1522, 2009.

[130] Y. Bresler. Spectrum-blind sampling and compressive sensing for continuous-index signals. In *2008 Information Theory and Applications Workshop*, pages 547–554, 2008.

[131] Y. Bresler and P. Feng. Spectrum-blind minimum-rate sampling and reconstruction of 2-D multiband signals. In *Proc. 3rd IEEE Int. Conf. on Image Processing, ICIP'96*, pages 701–704, Sep 1996.

[132] Y. Bresler, M. Gastpar, and R. Venkataramani. Image compression on-the-fly by universal sampling in Fourier imaging systems. In *Proc. 1999 IEEE Information Theory Workshop on Detection, Estimation, Classification and Imaging*, page 48, Feb 1999.

[133] William L. Briggs et al. *The DFT: An Owner's Manual for the Discrete Fourier Transform*. SIAM, 1995.

[134] T. Ben Britton, Soran Birosca, Michael Preuss, and Angus J. Wilkinson. Electron backscatter diffraction study of dislocation content of a macrozone in hot-rolled Ti-6Al-4V alloy. *Scripta Materialia*, 62(9):639–642, 2010.

[135] H. G. Brokmeier. Neutron diffraction texture analysis. *Physica B: Condensed Matter*, 234:977–979, 1997.

[136] David B. Brough, Abhiram Kannan, Benjamin Haaland, David G. Bucknall, and Surya R. Kalidindi. Extraction of process-structure evolution linkages from x-ray scattering measurements using dimensionality reduction and time series analysis. *Integrating Materials and Manufacturing Innovation*, pages 1–13, 2017.

[137] A. M. Bruckstein, D. L. Donoho, and M. Elad. From sparse solutions of systems of equations to sparse modeling of signals and images. *SIAM Review*, 51(1):34–81, 2009.

[138] L. Bruzzone and D. F. Prieto. Automatic analysis of the difference image for unsupervised change detection. *IEEE Trans. Geoscience and Remote Sensing*, 38:1171–1182, 2000.

[139] Ori Bryt and Michael Elad. Compression of facial images using the K-SVD algorithm. *Journal of Visual Communication and Image Representation*, 19(4):270–282, 2008.

[140] A. Buades, B. Coll, and J-M. Morel. A review of image denoising algorithms, with a new one. *Multiscale Model. Simul.*, 4(2):490–530, 2005.

[141] Doryen Bubeck, David J. Filman, and James M. Hogle. Cryo-electron microscopy reconstruction of a poliovirus-receptor-membrane complex. *Nature Structural & Molecular Biology*, 12:615–618, July 2005.

[142] Benjamin Bunday. Hvm metrology challenges towards the 5nm node. *Proc. SPIE 9778, Metrology, Inspection, and Process Control for Microlithography*, 9778:97780E–97780E–34, 2016.

[143] H.-J. Bunge. *Texture Analysis in Materials Science: Mathematical Methods*. Butterworths, 1982.

[144] H. J. Bunge. Three-dimensional texture analysis. *International Materials Reviews*, 32(1):265–291, 1987.

[145] C. S. Burrus, R. A. Gopinath, and H. Guo. *Introduction to Wavelets and Wavelet Transforms: A Primer*. Prentice Hall, 1997.

[146] Peter R. Buseck, Rafal E. Dunin-Borkowski, Bertrand Devouard, Richard B. Frankel, Martha R. McCartney, Paul A. Midgley, Mihly Psfai, and Matthew Weyland. Magnetite morphology and life on mars. *Proceedings of the National Academy of Sciences*, 98(24):13490–13495, 2001.

[147] C. E. Caefer, J. Silverman, O. Orthal, D. Antonelli, Y. Sharoni, and S. R. Rotman. Improved covariance matrices for point target detection in hyperspectral data. *Optical Engineering*, 47:076402, 2008.

[148] C. E. Caefer, M. S. Stefanou, E. D. Nelson, A. P. Rizzuto, O. Raviv, and S. R. Rotman. Analysis of false alarm distributions in the development and evaluation of hyperspectral point target detection algorithms. *Optical Engineering*, 46:076402, 2007.

[149] J. W. Cahn. The impurity drag effect in grain boundary motion. *Acta Metallurgica et Materialia*, 10:789–798, 1962.

[150] J. W. Cahn. Critical point wetting. *Journal of Chemical Physics*, 66(8):3667–72, 1977.

[151] J. W. Cahn. Transition and phase equilibria among grain boundary structures. *Journal de Physique*, 43:C6, 1982.

[152] J. W. Cahn and D. W. Hoffman. Vector thermodynamics for anisotropic surfaces – ii. curved and faceted surfaces. *Acta Metallurgica*, 22(10):1205–1214, 1974.

[153] John W. Cahn and John E. Hilliard. Free energy of a nonuniform system. I. Interfacial free energy. *The Journal of Chemical Physics*, 28(2):258–267, 1958.

[154] JW Cahn and DW Hoffman. A vector thermodynamics for anisotropic surfaces – I. curved and faceted surfaces. *Acta Metallurgica*, 22(10):1205–1214, 1974.

[155] D. M. Cai, M. Gokhale, and J. Theiler. Comparison of feature selection and classification algorithms in identifying malicious executables. *Computational Statistics and Data Analysis*, 51:3156–3172, 2007.

[156] T. Tony Cai, Cun-Hui Zhang, and Harrison H. Zhou. Optimal rates of convergnece for covariance matrix estimation. *Annals of Statistics*, 38(4):2118–2144, 2010.

[157] P. G. Callahan. *Quantitative characterization and comparison of precipitate and grain shape in Ni-base superalloys using moment invariants*. PhD thesis, Carnegie Mellon University, 2012.

[158] P. G. Callahan and M. De Graef. Precipitate shape fitting and reconstruction by means of 3D Zernike functions. *Modeling and Simulations in Materials Science and Engineering*, 20:015003, 2012.

[159] P. G. Callahan and M. De Graef. Dynamical EBSD patterns Part I: Pattern simulations. *Microscopy and MicroAnalysis*, 19:1255–1265, 2013.

[160] P. G. Callahan, M. Groeber, and M. De Graef. Towards a quantitative comparison between experimental and synthetic grain structures. *Acta Materialia*, 111:242–252, 2016.

[161] H. B. Callen. *Thermodynamics and an Introduction to Thermostatistics, 2nd ed.* Wiley, 1985.

[162] Geoffrey H. Campbell, Thomas LaGrange, Judy S. Kim, Bryan W. Reed, and Nigel D. Browning. Quantifying transient states in materials with the dynamic transmission electron microscope. *Journal of Electron Microscopy*, 59(S1):S67–S74, 2010.

[163] N. A. Campbell. Robust procedures in multivariate analysis I: Robust covariance estimation. *Applied Statistics*, 29:231–237, 1980.

[164] E. Candès and B. Recht. Exact matrix completion via convex optimization. *Found. of Comput. Math.*, 9(6):717–772, 2009.

[165] E. Candès and J. Romberg. Quantitative robust uncertainty principles and optimally sparse decompositions. *Foundations of Comput. Math.*, 6(2):227–254, 2006.

[166] E. Candès, J. Romberg, and T. Tao. Robust uncertainty principles: Exact signal reconstruction from highly incomplete frequency information. *IEEE Trans. Inform. Theory*, 52(2):489–509, February 2006.

[167] E. Candès and T. Tao. Decoding by linear programming. *IEEE Trans. on Information Theory*, 51(12):4203–4215, 2005.

[168] E. Candès and T. Tao. Near-optimal signal recovery from random projections: Universal encoding strategies? *IEEE Trans. Inform. Theory*, 52(12):5406–5245, December 2006.

[169] E. Candès and T. Tao. The Dantzig selector: statistical estimation when p is much smaller than n. *Annals of Stat.*, 35(6):2313–2351, 2007.

[170] E. J. Candès, Y. C. Eldar, D. Needell, and P. Randall. Compressed sensing with coherent and redundant dictionaries. *Appl. and Comp. Harm. Analysis*, 31(1):59–73, 2011.

[171] Emmanuel J. Candès. Compressive sampling. *Proceedings of the International Congress of Mathematicians, Madrid, Spain*, III:20, 2006.

[172] Emmanuel J. Candès, Justin K. Romberg, and Terence Tao. Stable signal recovery from incomplete and inaccurate measurements. *Communications on Pure and Applied Mathematics*, 59(8):1207–1223, 2006.

[173] R. M Cannon, M. Rühle, M. J. Hoffmann, R. H. French, H. Gu, A. P. Tomsia, and E. Saiz. Adsorption and wetting mechanisms at ceramic grain boundaries. *Ceramic Transactions (Grain Boundary Engineering in Ceramics)*, 118:427–444, 2000.

[174] Patrick R. Cantwell, Ming Tang, Shen J. Dillon, Jian Luo, Gregory S. Rohrer, and Martin P. Harmer. Overview no. 152: Grain boundary complexions. *Acta Materialia*, 62:1–48, 2014.

[175] P. R. Cantwell, M. Tang, S. J. Dillon, J. Luo, GS Rohrer, and M. P. Harmer. Grain boundary complexions. *Acta Materialia*, 62:1–48, 2014.

[176] G. Cao and C. A. Bouman. Covariance estimation for high dimensional data vectors using the sparse matrix transform. *Advances in Neural Information Processing Systems*, 21:225–232, 2009.

[177] Barbara Caputo, Eric Hayman, Mario Fritz, and Jan-Olof Eklundh. Classifying materials in the real world. *Image and Vision Computing*, 28(1):150–163, 2010.

[178] A. P. Carleer, Olivier Debeir, and Eléonore Wolff. Assessment of very high spatial resolution satellite image segmentations. *Photogrammetric Engineering & Remote Sensing*, 71(11):1285–1294, 2005.

[179] M. J. Carlotto. A cluster-based approach for detecting man-made objects and changes in imagery. *IEEE Trans. Geoscience and Remote Sensing*, 43:374–387, 2005.

[180] George Casella and Roger L. Berger. *Statistical Inference*. Duxbury, Wadsworth Group, Pacific Grove, CA, 2nd edition, 2002.

[181] V. Cevher, M. F. Duarte, C. Hegde, and R. G. Baraniuk. Sparse signal recovery using Markov random fields. In *Proc. Workshop of Neural Information Processing Systems (NIPS)*, Vancouver, Canada, 2008.

[182] T. Chan and J. Shen. *Image Processing and Analysis: Variational, PDE, Wavelet, and Stochastic Methods*. Society for Industrial and Applied Mathematics, 2005.

[183] Kenneth Chang. Israeli scientist wins Nobel prize for chemistry. *New York Times*, 5, 2011.

[184] Peter I. Chang, Peng Huang, Jungyeoul Maeng, and Sean B. Andersson. Local raster scanning for high-speed imaging of biopolymers in atomic force microscopy. *Review of Scientific Instruments*, 82(6):063703, 2011.

[185] Shery L. Y. Chang, Christian Dwyer, Juri Barthel, Chris B. Boothroyd, and Rafal E. Dunin-Borkowski. Performance of a direct detection camera for off-axis electron holography. *Ultramicroscopy*, 161:90–97, 2016.

[186] R. Chartrand. Fast algorithms for nonconvex compressive sensing: MRI reconstruction from very few data. In *IEEE Int. Symp. on Biomedical Imag.: From Nano to Macro*, pages 262–265, 2009.

[187] R. Chartrand. Fast algorithms for nonconvex compressive sensing: MRI reconstruction from very few data. In *Proc. IEEE International Symposium on Biomedical Imaging (ISBI)*, pages 262–265, 2009.

[188] D. Chatain, V. Ghetta, and P. Wynblatt. Equilibrium shape of copper crystals grown on sapphire. *Interface Science*, 12(1):7–18, 2004.

[189] Ken Chatfield, Victor S. Lempitsky, Andrea Vedaldi, and Andrew Zisserman. The devil is in the details: An evaluation of recent feature encoding methods. In *BMVC*, volume 2, page 8, 2011.

[190] Ken Chatfield, Karen Simonyan, Andrea Vedaldi, and Andrew Zisserman. Return of the devil in the details: Delving deep into convolutional nets. *arXiv preprint arXiv:1405.3531*, 2014.

[191] P. Chatterjee and P. Milanfar. A generalization of non-local means via kernel regression. In *SPIE*, volume 6814, 2008.

[192] P. Chatterjee and P. Milanfar. Patch-based near-optimal image denoising. In *ICIP*, 2011.

[193] G. H. Chen, J. Tang, and S. Leng. Prior image constrained compressed sensing (piccs): A method to accurately reconstruct dynamic CT images from highly undersampled projection data sets. *Med. Phys.*, 35(2):660–663, 2008.

[194] S. S. Chen, D. L. Donoho, and M. A. Saunders. Atomic decomposition by basis pursuit. *SIAM J. Sci. Comput.*, 20(1):33–61, 1998.

[195] Y. Chen, W. Yu, and T. Pock. On learning optimized reaction diffusion processes for effective image restoration. In *Proc. IEEE Conf. Comp. Vision and Patt. Rec.*, pages 5261–5269, 2015.

[196] Yu-Hui Chen, Dennis Wei, Gregory Newstadt, Marc DeGraef, Jeffrey Simmons, and Alfred Hero. Parameter estimation in spherical symmetry groups. *IEEE Signal Processing Letters*, 22(8):1152–1155, 2015.

[197] R. Holland Cheng, Vijay S. Reddy, Norman H. Olson, Andrew J. Fisher, Timothy S. Baker, and John E. Johnson. Functional implications of quasi-equivalence in a $T = 3$ icosahedral animal virus established by cryo-electron microscopy and x-ray crystallography. *Structure*, 2:271–282, 15 April 1994.

[198] A. Chilkoti, T. Boland, B. D. Ratner, and P. S. Stayton. The relationship between ligand-binding thermodynamics and protein-ligand interaction forces measured by atomic force microscopy. *Biophysical Journal*, 69(5):2125–2130, 1995.

[199] J. H. Cho, H. M. Chan, M. P. Harmer, and J. M. Rickman. Influence of yttrium doping on grain misorientation in aluminum oxide. *Journal of the American Ceramic Society*, 81(11):3001–3004, 1998.

[200] K. Choi, J. Wang, L. Zhu, T.-S. Suh, S. Boyd, and L. Xing. Compressed sensing based cone-beam computed tomography reconstruction with a first-order method. *Med. Phys.*, 37(9):5113–5125, 2010.

[201] Ki-Young Choi and S. S. Kim. Morphological analysis and classification of types of surface corrosion damage by digital image processing. *Corrosion Science*, 47(1):1–15, 2005.

[202] Edwin K. P. Chong and Stanislaw H. Zak. *An Introduction to Optimization*, volume 76. John Wiley & Sons, 2013.

[203] Aritra Chowdhury, Elizabeth Kautz, Bülent Yener, and Daniel Lewis. Image driven machine learning methods for microstructure recognition. *Computational Materials Science*, 123:176–187, 2016.

[204] Haiso-Chiang Chuang, Mary L. Comer, and Jeff P. Simmons. Texture classification in microstructure images of advanced materials. In *Image Analysis and Interpretation, 2008. SSIAI 2008. IEEE Southwest Symposium on*, pages 1–4. IEEE, 2008.

[205] Radoslaw Martin Cichy, Aditya Khosla, Dimitrios Pantazis, Antonio Torralba, and Aude Oliva. Comparison of deep neural networks to spatio-temporal cortical dynamics of human visual object recognition reveals hierarchical correspondence. *Scientific Reports*, 6, 2016.

[206] Mircea Cimpoi, Subhransu Maji, and Andrea Vedaldi. Deep filter banks for texture recognition and segmentation. In *Proceedings of the IEEE Conference on Computer Vision and Pattern Recognition*, pages 3828–3836, 2015.

[207] J. F. Claerbout and F. Muir. Robust modeling of erratic data. *Geophysics*, 38:826–844, 1973.

[208] E. Clapeyron. Memoire sur la puissance motrice de la chaleur. *Jounal de l'École Polytechnique*, 1834.

[209] D. R. Clarke. On the equilibrium thickness of intergranular glass phases in ceramic materials. *Journal of the American Ceramic Society*, 70(1):15–22, 1987.

[210] David A. Clausi. An analysis of co-occurrence texture statistics as a function of grey level quantization. *Canadian Journal of Remote Sensing*, 28(1):45–62, 2002.

[211] Rudolf Clausius. *The Mechanical Theory of Heat: With Its Applications to the Steam-Engine and to the Physical Properties of Bodies*. J. van Voorst, 1867. This quote appears on p. 365.

[212] Chris Clifton. Change detection in overhead imagery using neural networks. *Applied Intelligence*, 18:215–234, 2003.

[213] R. L. Coble and R. M. Cannon. Current paradigms in powder processing, 1978 1978.

[214] Y. Cohen, Y. August, D. G. Blumberg, and S. R. Rotman. Evaluating sub-pixel target detection algorithms in hyper-spectral imagery. *J. Electrical and Computer Engineering*, 2012:103286, 2012.

[215] Eyal Cohen-Hoshen, Garnett W. Bryant, Iddo Pinkas, Joseph Sperling, and Israel Bar-Joseph. Excitonplasmon interactions in quantum dotgold nanoparticle structures. *Nano Lett.*, 12(8):42604264, 2012.

[216] M. Collins, S. Dasgupta, and R. E. Schapire. A generalization of principal components analysis to the exponential family. In *NIPS*, pages 617–624, 2002.

[217] D. Comaniciu, V. Ramesh, and P. Meer. Kernel-based object tracking. *IEEE Transactions on Pattern Analysis and Machine Intelligence*, 25:564–577, 2003.

[218] P. L. Combettes and J-C. Pesquet. Proximal splitting methods in signal processing. In *Fixed-Point Algorithms for Inverse Problems in Science and Engineering*. Springer, 2011.

[219] Mary L. Comer and Edward J. Delp. The EM/MPM algorithm for segmentation of textured images: analysis and further experimental results. *IEEE Transactions on Image Processing*, 9(10):1731–1744, 2000.

[220] Devis Contarato, Peter Denes, Dionisio Doering, John Joseph, and Brad Krieger. High speed, radiation hard cmos pixel sensors for transmission electron microscopy. *Physics Procedia*, 37:1504–1510, 2012.

[221] Tim F. Cootes and Christopher J. Taylor. Statistical models of appearance for medical image analysis and computer vision. In *Medical Imaging 2001*, pages 236–248. International Society for Optics and Photonics, 2001.

[222] Corinna Cortes and Vladimir Vapnik. Support-vector networks. *Machine Learning*, 20(3):273–297, 1995.

[223] Richard Courant. *Differential and Integral Calculus*, volume 2. John Wiley & Sons, 2011.

[224] Thomas Cover and Joy Thomas. *Elements of Information Theory*. John Wiley and Sons, 1991.

[225] Thomas M. Cover and Joy A. Thomas. *Elements of Information Theory*. Wiley-Interscience, 2006.

[226] D. Cremers, T. Kohlberger, and C. Schnörr. Shape statistics in kernel space for variational image segmentation. *Pattern Recognition*, 36:1929–1943, 2003.

[227] M. W. Crofton. On the theory of local probability, applied to straight lines drawn at random in a plane; the methods used being also extended to the proof of certain new theorems in the integral calculus. *Philosophical Transactions of the Royal Society of London*, 158:181–199, 1868.

[228] James L. Crowley and Alice C. Parker. A representation for shape based on peaks and ridges in the difference of low-pass transform. *Pattern Analysis and Machine Intelligence, IEEE Transactions on*, 6:156–170, 1984.

[229] Gabriella Csurka, Christopher Dance, Lixin Fan, Jutta Willamowski, and Cédric Bray. Visual categorization with bags of keypoints. In *Workshop on Statistical Learning in Computer Vision, ECCV*, volume 1, pages 1–2. Prague, 2004.

[230] Oana G. Cula and Kristin J. Dana. Compact representation of bidirectional texture functions. In *Computer Vision and Pattern Recognition, 2001. CVPR 2001. Proceedings of the 2001 IEEE Computer Society Conference on*, volume 1, pages I–1041. IEEE, 2001.

[231] Alberto G. Curto, Giorgio Volpe, Tim H. Taminiau, Mark P. Kreuzer, Romain Quidant, and Niek F. van Hulst. Unidirectional emission of a quantum dot coupled to a nanoantenna. *Science*, 329(5994):930–933, 2010.

[232] K. Dabov, A. Foi, V. Katkovnik, and K. Egiazarian. Image denoising by sparse 3-d transform-domain collaborative filtering. *Image Processing, IEEE Transactions on*, 16(8):2080–2095, 2007.

[233] W. Dai and O. Milenkovic. Subspace pursuit for compressive sensing signal reconstruction. *IEEE Trans. Information Theory*, 55(5):2230–2249, 2009.

[234] Navneet Dalal and Bill Triggs. Histograms of oriented gradients for human detection. In *Computer Vision and Pattern Recognition, 2005. CVPR 2005. IEEE Computer Society Conference on*, volume 1, pages 886–893. IEEE, 2005.

[235] M. Dallair and D. Furrer. Quantitative metallography of titanium alloys. In *Advanced Materials and Processes*, pages 25–28. ASM International, 2004.

[236] J. G. Dash. Surface melting. *Contemporary Physics*, 30(2):89–100, 1989.

[237] I. Daubechies. *Ten Lectures on Wavelets*. SIAM, New York, 1992.

[238] I. Daubechies, M. Defrise, and C. De Mol. An iterative thresholding algorithm for linear inverse problems with a sparsity constraint. *Comm. Pure Appl. Math.*, 57(11):1413–1457, 2004.

[239] M. A. Davenport and J. Romberg. An overview of low-rank matrix recovery from incomplete observations. *IEEE J. Selected Topics in Sig. Proc.*, 10(4):608–622, 2016.

[240] M. A. Davenport and M. B. Wakin. Analysis of orthogonal matching pursuit using the restricted isometry property. *IEEE Trans. Inform. Theory*, 56(9):4395–4401, September 2010.

[241] G. Davis, S. Mallat, and M. Avellaneda. Adaptive greedy approximations. *Journal of Constructive Approximation*, 13(1):57–98, 1997.

[242] G. Davis, S. Mallat, and M. Avellaneda. Greedy adaptive approximation. *J. Constructive Approx.*, 12:57–98, 1997.

[243] Timothy J. Davis, Mario Hentschel, Na Liu, and Harald Giessen. Analytical model of the three-dimensional plasmonic ruler. *ACS Nano*, 6(2):1291–1298, 2012.

[244] M. De Graef. Lorentz microscopy: Theoretical basis and image simulations. In M. De Graef and Y. Zhu, editors, *Magnetic Microscopy and its Applications to Magnetic Materials*, volume 36 of *Experimental Methods in the Physical Sciences*, chapter 2. Academic Press, 2000.

[245] M. De Graef. *Introduction to Conventional Transmission Electron Microscopy*. Cambridge University Press, 2003.

[246] B. De Man, S. Basu, J.-B. Thibault, Jiang Hsieh, J.A. Fessier, C. Bouman, and K. Sauer. A study of four minimization approaches for iterative reconstruction in X-ray CT. In *2005 IEEE Nuclear Science Symposium Conference Record*, volume 5, pages 2708–2710, Oct. 2005.

[247] P. Debye, H. R. Anderson Jr, and H. Brumberger. Scattering by an inhomogeneous solid. ii. the correlation function and its application. *Journal of Applied Physics*, 28(6):679–683, 1957.

[248] Peter Debye and A. M. Bueche. Scattering by an inhomogeneous solid. *Journal of Applied Physics*, 20(6):518–525, 1949.

[249] Brian L. DeCost, Toby Francis, and Elizabeth A. Holm. Exploring the microstructure manifold: Image texture representations applied to ultrahigh carbon steel microstructures. *Acta Materialia*, 133:30–40, July 2017.

[250] Brian L. DeCost, Matthew D. Hecht, Toby Francis, Bryan A. Webler, Yoosuf N. Picard, and Elizabeth A. Holm. Uhcsdb (ultrahigh carbon steel micrograph database): Tools for exploring large heterogeneous microstructure datasets. *Integrating Materials and Manufacturing Innovation*, 6:197–205, 2017.

[251] Brian L. DeCost and Elizabeth A. Holm. A computer vision approach for automated analysis and classification of microstructural image data. *Computational Materials Science*, 110:126–133, 2015.

[252] Brian L. DeCost and Elizabeth A. Holm. Characterizing powder materials using keypoint-based computer vision methods. *Computational Materials Science*, 126:438–445, January 2017.

[253] Brian L. DeCost, Harshvardhan Jain, Anthony D. Rollett, and Elizabeth A. Holm. Computer vision and machine learning for autonomous characterization of am powder feedstocks. *JOM*, pages 1–10, March 2017.

[254] M. A. Delesse. Procédé mécanique pour détermine la composition des roches. *Comptes rendus hebdomadaires des séances de l'Académie des Sciences*, 25:544–5, 1847.

[255] A. V. Delgado, F. Gonzlez-Caballero, R. J. Hunter, L. K. Koopal, and J. Lyklema. Measurement and interpretation of electrokinetic phenomena. *Journal of Colloid and Interface Science*, 309(2):194–224, 2007.

[256] Hendrix Demers, Nicolas Poirier-Demers, Alexandre Ral Couture, Dany Joly, Marc Guilmain, Niels de Jonge, and Dominique Drouin. Three-dimensional electron microscopy simulation with the CASINO Monte Carlo software. *Scanning*, 33(3):135–146, 2011.

[257] W. Van den Broek, A. Rosenauer, B. Goris, G. T. Martinez, S. Bals, S. Van Aert, and D. Van Dyck. Correction of non-linear thickness effects in HAADF STEM electron tomography. *Ultramicroscopy*, 116:8–12, 2012.

[258] A. J. DeRidder and R. Koch. Forging and processing of high-temperature alloys. In *ASTM STP 672 MiCon 78: Optimization of Processing, Properties, and Service Performance Through Microstructural Control*, pages 547–563. ASM International, 1978.

[259] Satya Dharanipragada and Karthik Visweswariah. Gaussian mixture models with covariances or precisions in shared multiple subspaces. *IEEE Trans. Audio, Speech, and Language Proc.*, 14(4):1255–1266, July 2006.

[260] Caglayan Dicle, Octavia I. Camps, and Mario Sznaier. The way they move: Tracking multiple targets with similar appearance. In *International Conference on Computer Vision*, pages 2304–2311, 2013.

[261] S. J. Dillon, M. Tang, W. C. Carter, and M. P. Harmer. Complexion: A new concept for kinetic engineering in materials science. *Acta Materialia*, 55(18):6208–6218, 2007.

[262] D. Dingley. Progressive steps in the development of electron backscatter diffraction and orientation imaging microscopy. *J. Microsc.*, 213:214–224, 2004.

[263] D. J. Dingley and D. P. Field. Electron backscatter diffraction and orientation imaging microscopy. *Materials Science and Technology*, 13(1):69–78, 1997.

[264] M. N. Do and M. Vetterli. The contourlet transform: An efficient directional multiresolution image representation. *IEEE Trans. Image Process.*, 14(12):2091–2106, 2005.

[265] Peter C. Doerschuk and John E. Johnson. *Ab initio* reconstruction and experimental design for cryo electron microscopy. *IEEE Transactions on Information Theory*, 46(5):1714–1729, August 2000.

[266] Tatiana Domitrovic, Navid Movahed, Brian Bothner, Tsutomu Matsui, Qiu Wang, Peter C. Doerschuk, and John E. Johnson. Virus assembly and maturation: Auto-regulation through allosteric molecular switches. *J. Molecular Biology*, 425(9):1488–1496, 13 May 2013.

[267] Jeff Donahue, Yangqing Jia, Oriol Vinyals, Judy Hoffman, Ning Zhang, Eric Tzeng, and Trevor Darrell. Decaf: A deep convolutional activation feature for generic visual recognition. *arXiv preprint arXiv:1310.1531*, 2013.

[268] C. Dong, C. C. Loy, K. He, and X. Tang. Image super-resolution using deep convolutional networks. *IEEE Trans. Pattern Analysis and Machine Intelligence*, 38(2):295–307, 2016.

[269] Y. Dong, S. Lefebvre, X. Tong, and G. Drettakis. Lazy solid texture synthesis. In *In EGSR 08*, 2008.

[270] D. Donoho. Compressed sensing. *IEEE Trans. Information Theory*, 52(4):1289–1306, 2006.

[271] D. Donoho and I. Johnstone. Ideal spatial adaptation via wavelet shrinkage. *Biometrika*, 81:425–455, 1994.

[272] D. L. Donoho. For most large underdetermined systems of linear equations the minimal l1-norm solution is also the sparsest solution. *Comm. Pure Appl. Math*, 59:797–829, 2004.

[273] D. L. Donoho and M. Elad. Optimally sparse representation in general (non-orthogonal) dictionaries via ℓ_1 minimization. *Proc. Natl. Acad. Sci. USA*, 100:2197–2202, 2003.

[274] D. L. Donoho, M. Elad, and V. N. Temlyakov. Stable recovery of sparse overcomplete representations in the presence of noise. *IEEE Trans. Inform. Theory*, 52(1):6–18, 2006.

[275] D. L. Donoho and X. Huo. Uncertainty principles and ideal atomic decomposition. *IEEE Trans. Inform. Theory*, 47(7):2845–2862, 2001.

[276] D. L. Donoho, A. Maleki, and A. Montanari. Message-passing algorithms for compressed sensing. *Proc. Natl. Acad. Sci. USA*, 106(45):18914–18919, 2009.

[277] D. L. Donoho and P. B. Stark. Uncertainty principles and signal recovery. *SIAM J. Applied Math.*, 49:906–931, 1989.

[278] D. L. Donoho and J. Tanner. Neighborliness of randomly-projected simplices in high dimensions. *Proc. Natl. Acad. Sci. USA*, 102(27):9452–9457, 2005.

[279] D. L. Donoho and J. Tanner. Observed universality of phase transitions in high-dimensional geometry, with implications for modern data analysis and signal processing. *Phil. Trans. Royal Society A*, 367(1906), 2009.

[280] D. L. Donoho, Y. Tsaig, I. Drori, and J. L. Starck. Sparse solution of underdetermined systems of linear equations by stagewise orthogonal matching pursuit. *IEEE Trans. Inform. Theory*, 58(2):1094–1121, 2012.

[281] J. Dooley and M. De Graef. Energy filtered Lorentz microscopy. *Ultramicroscopy*, 67:113–132, 1997.

[282] Jinsong Duan, Dhriti Nepal, Kyoungweon Park, Joy E. Haley, Jarrett H. Vella, Augustine M. Urbas, Richard A. Vaia, and Ruth Pachter. Computational prediction of molecular photoresponse upon proximity to gold nanorods. *J. Phys. Chem. C*, 115(29):13961–13967, 2011.

[283] M. F. Duarte, M. A. Davenport, D. Takhar, J. N. Laska, T. Sun, K. F. Kelly, and R. G. Baraniuk. Single-pixel imaging via compressive sampling. *IEEE Signal Proc. Mag.*, 25(2):83–91, 2008.

[284] Marco F. Duarte and Richard G. Baraniuk. Kronecker compressive sensing. *Image Processing, IEEE Transactions on*, 21(2):494–504, 2012.

[285] M. F. Duarte, M. A. Davenport, D. Takhar, J. N. Laska, Ting Sun, K. F. Kelly, and R. G. Baraniuk. Single-pixel imaging via compressive sampling. *IEEE Signal Processing Magazine*, 25:83–91, 2008.

[286] A. L. Eberle, S. Mikula, R. Schalek, J. Lichtman, M. L. Knothe Tate, and D. Zeidler. High resolution, high throughput imaging with a multibeam scanning electron microscope. *Journal of Microscopy*, 259(2):114–120, 2015.

[287] M. P. Echlin, A. Mottura, C. J. Torbet, and T. M. Pollock. A new tribeam system for three-dimensional multimodal materials analysis. *Review of Scientific Instruments*, 83(2):–, 2012.

[288] Herbert Edelsbrunner and John Harer. Persistent homology—a survey. *Contemporary mathematics*, 453:257–282, 2008.

[289] Editorial. Method of the year 2015. *Nature Methods*, 13, 2016.

[290] B. Efron, T. Hastie, I. Johnstone, and R. Tibshirani. Least angle regression. *Annals of Statistics*, 32:407–499, 2004.

[291] Bradley Efron and David V. Hinkley. Assessing the accuracy of the maximum likelihood estimator: Observed versus expected Fisher information. *Biometrika*, 65(3):457–487, December 1978.

[292] A. Efros and T. Leung. Texture synthesis by non-parametric sampling. In *In International Conference on Computer Vision*, Vol 2, pages 1033–1038, Sep 1999.

[293] Alexei Efros. What makes big data hard? In *IS&T/SPIE Electronic Imaging Conference, Plenary Presentation*, San Francisco, CA, 2014.

[294] Alexei A. Efros and William T. Freeman. Image quilting for texture synthesis and transfer. In *Proceedings of the 28th Annual Conference on Computer Graphics and Interactive Techniques*, pages 341–346. ACM, 2001.

[295] R. F. Egerton. *Electron Energy-Loss Spectroscopy in the Electron Microscope*. Springer, 3rd edition, 2011.

[296] M. T. Eismann, J. Meola, and R. C Hardie. Hyperspectral change detection in the presence of diurnal and seasonal variations. *IEEE Trans. Geoscience and Remote Sensing*, 46:237–249, 2008.

[297] M. T. Eismann, J. Meola, A. D. Stocker, S. G. Beaven, and A. P. Schaum. Airborne hyperspectral detection of small changes. *Applied Optics*, 47:F27–F45, 2008.

[298] M. Elad and M. Aharon. Image denoising via sparse and redundant representations over learned dictionaries. *IEEE Trans. Image Process.*, 15(12):3736–3745, 2006.

[299] M. Elad, P. Milanfar, and R. Rubenstein. Analysis versus synthesis in signal priors. *Inverse Problems*, 23(3):947–968, June 2007.

[300] Michael Elad. Michael Elad personal page. `http://www.cs.technion.ac.il/~elad/Various/KSVD_Matlab_ToolBox.zip`, 2009. [Online; accessed Nov. 2015].

[301] E. Elhamifar and R. Vidal. Sparsity in unions of subspaces for classification and clustering of high-dimensional data. In *49th Annual Allerton Conference on Communication, Control and Computing (Allerton)*, pages 1085–1089, 2011.

[302] Ehsan Elhamifar and René Vidal. Sparse subspace clustering. In *Computer Vision and Pattern Recognition, 2009. CVPR 2009. IEEE Conference on*, pages 2790–2797. IEEE, 2009.

[303] EMBL-EBI. Emdb statistics. `https://www.ebi.ac.uk/pdbe/emdb/statistics\_num\_res.html/`.

[304] K. Engan, S. O. Aase, and J. H. Hakon-Husoy. Method of optimal directions for frame design. In *Proc. IEEE International Conference on Acoustics, Speech and Signal Processing*, pages 2443–2446, 1999.

[305] O. Engler. Deformation and texture of copper–manganese alloys. *Acta Materialia*, 48(20):4827–4840, 2000.

[306] O. Engler, G. Gottstein, J. Pospiech, and J. Jura. Statistics, evaluation, and representation of single grain orientation measurement. *Materials Science Forum*, 157-162:259–274, 1994.

[307] Olaf. Engler and V. Randle. *Introduction to Texture Analysis: Macrotexture, Microtexture, and Orientation Mapping*. CRC Press, 2010.

[308] P. Ercius, T. Caswell, M. W. Tate, A. Ercan, S. M. Gruner, and D. Muller. A pixel array detector for scanning transmission electron microscopy. *Microscopy and Microanalysis*, 14:806–807, 2008.

[309] Peter Ercius, Osama Alaidi, Matthew J. Rames, and Gang Ren. Electron tomography: A three-dimensional analytic tool for hard and soft materials research. *Advanced Materials*, 27(38):5638–5663, 2015.

[310] Peter Ercius, Matthew Weyland, David A. Muller, and Lynne M. Gignac. Three-dimensional imaging of nanovoids in copper interconnects using incoherent bright field tomography. *Applied Physics Letters*, 88:243116, 2006.

[311] Peter Ercius, Matthew Weyland, David A. Muller, and Lynne M. Gignac. Three-dimensional imaging of nanovoids in copper interconnects using incoherent bright field tomography. *Applied Physics Letters*, 88:243116, 2006.

[312] N. Eustathopoulos. Energetics of solid/liquid interfaces of metals and alloys. *International Metals Reviews*, 28(1):189–210, 1983.

[313] Xiaofeng Fan, Weitao Zheng, and David J. Singh. Light scattering and surface plasmons on small spherical particles. *Light Sci. Appl.*, 3:e179, 2014.

[314] A. R. Faruqi, R. Henderson, M. Pryddetch, P. Allport, and A. Evans. Direct single electron detection with a cmos detector for electron microscopy. *Nuclear Instruments and Methods in Physics Research Section A: Accelerators, Spectrometers, Detectors and Associated Equipment*, 546(1-2):170–175, 2005.

[315] P. Feng. *Universal Spectrum Blind Minimum Rate Sampling and Reconstruction of Multiband Signals*. PhD thesis, University of Illinois at Urbana-Champaign, mar 1997. Yoram Bresler, adviser.

[316] P. Feng and Y. Bresler. Spectrum-blind minimum-rate sampling and reconstruction of multiband signals. In *ICASSP*, volume 3, pages 1689–1692, may 1996.

[317] T. G. Ference and R. W. Balluffi. Observation of a reversible grain boundary faceting transition induced by changes of composition. *Scripta Metallurgica*, 22(12):1929–1934, 1988.

[318] J. A. Fessler, E. P. Ficaro, N. H. Clinthorne, and K. Lange. Grouped-coordinate ascent algorithms for penalized-likelihood transmission image reconstruction. *IEEE Transactions on Medical Imaging*, 16(2):166–175, April 1997.

[319] J. A. Fessler. Penalized weighted least-squares image reconstruction for positron emission tomography. *IEEE Trans. on Medical Imaging*, 13(2):290–300, June 1994.

[320] J. A. Fessler and S. D. Booth. Conjugate-gradient preconditioning methods for shift-variant PET image reconstruction. *IEEE Trans. on Image Processing*, 8:688–699, May 1999.

[321] Jeffrey A. Fessler and Donghwan Kim. Axial block coordinate descent (abcd) algorithm for x-ray ct image reconstruction. *Proc. Intl. Mtg. on Fully 3D Image Recon. in Rad. and Nuc. Med*, pages 262–5, 2011.

[322] D. P. Field, R. C. Eames, and T. M. Lillo. The role of shear stress in the formation of annealing twin boundaries in copper. *Scripta Materialia*, 54(6):983–986, 2006.

[323] M. Figueiredo and R. Nowak. An EM algorithm for wavelet-based image restoration. *IEEE Trans. Image Proc.*, 12(8), August 2003.

[324] Andrew J. Fisher and John E. Johnson. Ordered duplex RNA controls capsid architecture in an icosahedral animal virus. *Nature*, 361:176–179, Jan. 14 1993.

[325] M. Fisz. The limiting distribution of a function of two independent random variables and its statistical application. *Colloquium Mathematicum*, 3:138–146, 1955.

[326] American Society for the Testing of Materials. Metals – mechanical testing; elevated and low-temperature tests; metallography. In *Annual Book of ASTM Standards*, volume 3. ASTM, West Conshocken, PA, USA, 2015.

[327] C. Forget. Improved method for E1122 image analysis nonmetallic inclusion ratings. In G.F. Vander Voort, editor, *MiCon 90: Advances in Video Technology for Microstructural Control*, pages 135–150, 1991.

[328] S. Foucart. Stability and robustness of weak orthogonal matching pursuits. In *Recent Adv. in Harm. Analysis and App.*, volume 25, pages 395–405. Springer Proceedings in Math. and Stat., 2011.

[329] F. C. Frank. Orientation mapping. *Metallurgical Transactions A*, 19(3):403–408, 1988.

[330] J. Frank. *Electron Tomography: Methods for Three-Dimensional Visualization of Structures in the Cell*. Springer, 2008.

[331] Joachim Frank. *Three-Dimensional Electron Microscopy of Macromolecular Assemblies*. Academic Press, San Diego, 1996.

[332] M. Frary and C. A. Schuh. Grain boundary networks: Scaling laws, preferred cluster structure, and their implications for grain boundary engineering. *Acta Materialia*, 53(16):4323–4335, 2005.

[333] D. Frączek, W. Olchawa, R. Piasecki, and R. Wiśniowski. Entropic descriptor based reconstruction of three-dimensional porous microstructures using a single cross-section. *ArXiv e-prints*, August 2015.

[334] J. H. Friedman. Regularized discriminant analysis. *J. Am. Statistical Assoc.*, 84:165–175, 1989.

[335] Jerome Friedman, Trevor Hastie, and Robert Tibshirani. *The Elements of Statistical Learning*, volume 1. Springer series in statistics Springer, Berlin, 2001.

[336] P. Fryźlewicz and G. Nason. Poisson intensity estimation using wavelets and the Fisz transformation. Technical report, Department of Mathematics, University of Bristol, United Kingdom, 2001.

[337] J. J. Fuchs. Multipath time-delay detection and estimation. *IEEE Trans. Sig. Proc.*, 47:237–243, 1999.

[338] J. J. Fuchs. On the application of the global matched filter to doa estimation with uniform circular arrays. *IEEE Trans. Signal Proc.*, 49(4):702–709, April 2001.

[339] Keinosuke Fukunaga. *Introduction to Statistical Pattern Recognition*. Academic Press, 2013.

[340] S. D. Fuller, S. J. Butcher, R. H. Cheng, and T. S. Baker. Three-dimensional reconstruction of icosahedral particles–the uncommon line. *J. Struct. Biol.*, 116(1):48–55, January 1996.

[341] David T. Fullwood, Stephen R. Niezgoda, Brent L. Adams, and Surya R. Kalidindi. Microstructure sensitive design for performance optimization. *Progress in Materials Science*, 55(6):477–562, 2010.

[342] D. T. Fullwood, J. A. Basinger, and B. L. Adams. Lattice-based structures for studying percolation in two-dimensional grain networks. *Acta Materialia*, 54(5):1381–1388, 2006.

[343] D. Furrer and H.-J. Fecht. γ' formation in superalloy U720LI. *Scripta Materialia*, 40:1215–1220, 1999.

[344] D. U. Furrer and S. L. Semiatin. Thermomechanical processes for nonferrous alloys. In *ASM Metals Handbook Vol. 14A, Metalworking: Bulk Forming*, pages 375–380. ASM-International, Metals Park, Ohio, 2005.

[345] C. Gammer, C. Rentenberger, H. P. Karnthaler, C. Czarnik, D. Breitlschmidt, S. Pauly, J. Eckert, and A. M. Minor. In situ tem deformation of a bulk metallic glass witha k2-is detector. In *18th International Microscopy Conference*, 2014.

[346] Christoph Gammer, V. Burak Ozdol, Christian H. Liebscher, and Andrew M. Minor. Diffraction contrast imaging using virtual apertures. *Ultramicroscopy*, 155:1–10, 2015.

[347] E. J. Garboczi. Finite element and finite difference programs for computing the linear electric and elastic properties of digital images of random materials. In *NIST Internal Report 6269, http://ciks.cbt.nist.gov/garboczi/, Chapter 2*, 1998.

[348] D. R. Gaskell. *Introduction to the Thermodynamics of Materials*. Taylor & Francis, New York, 5th edition, 2008.

[349] M. Gastpar and Y. Bresler. On the necessary density for spectrum-blind nonuniform sampling subject to quantization. In *ICASSP*, volume 1, pages 348–351, jun 2000.

[350] Leon Gatys, Alexander S Ecker, and Matthias Bethge. Texture synthesis using convolutional neural networks. In *Advances in Neural Information Processing Systems*, pages 262–270, 2015.

[351] Leon A. Gatys, Alexander S. Ecker, and Matthias Bethge. A neural algorithm of artistic style. *arXiv preprint arXiv:1508.06576*, 2015.

[352] J. Gayda and D. Furrer. Dual-microstructure heat treatment. *Advanced Materials and Processes*, 3:36–39, 2003.

[353] M. E. Gehm, R. John, D. J. Brady, R. M. Willett, and T. J. Schulz. Single-shot compressive spectral imaging. *Optics Express*, pages 14013–14027, 2007.

[354] I. M. Gelfand and S. V. Fomin. *Calculus of Variations*. Prentice-Hall, Inc., 1963. Translated from the Russian by Richard A. Silverman.

[355] S. Geman and D. Geman. Stochastic relaxation, gibbs distributions, and the Bayesian restoration of images. *IEEE Transactions on Pattern Analysis and Machine Intelligence*, 6:721–741, 1984.

[356] S. P. Gentry and K. Thornton. Simulating recrystallization in titanium using the phase field method. In *IOP Conference Series: Materials Science and Engineering. 89(1)*, 2015.

[357] Dustin D. Gerrard, David T. Fullwood, Denise M. Halverson, and Stephen R. Niezgoda. Computational homology, connectedness, and structure-property relations. *Computers, Materials, & Continua*, 15(2):129–152, 2010.

[358] V. Y. Gertsman, M. Janecek, and K. Tangri. Grain boundary ensembles in polycrystals. *Acta Materialia*, 44(7):2869–2882, 1996.

[359] Sujit Kumar Ghosh and Tarasankar Pal. Interparticle coupling effect on the surface plasmon resonance of gold nanoparticles: From theory to applications. *Chemical Reviews*, 107(11):4797–4862, 2007.

[360] Robert Ghrist. *Elementary Applied Topology*. Createspace, 2014.

[361] J. W. Gibbs, K. Aditya Mohan, E. B. Gulsoy, A. J. Shahani, X. Xiao, C. A. Bouman, M. De Graef, , and P. W. Voorhees. The three-dimensional morphology of growing dendrites. *Scientific Reports*, 1(2):96 – 111, 2015.

[362] J. W. Gibbs. On the equilibrium of heterogeneous substances. *Transactions of the Connecticut Academy of Arts and Sciences*, III:108–218, 1874-1878. This particular quote taken from page 152.

[363] Anna Gilbert and Piotr Indyk. Sparse recovery using sparse matrices. In *Proceedings of the IEEE 98.6*. Institute of Electrical and Electronics Engineers, 2010.

[364] Ross Girshick, Jeff Donahue, Trevor Darrell, and Jitendra Malik. Rich feature hierarchies for accurate object detection and semantic segmentation. In *Proceedings of the IEEE conference on computer vision and pattern recognition*, pages 580–587, 2014.

[365] N. A. Gjostein and F. N. Rhines. Absolute interfacial energies of [001] tilt and twist grain boundaries in copper. *Acta Metallurgica*, 7(5):319–330, 1959.

[366] S. Gleichman and Yonina C. Eldar. Blind compressed sensing. *IEEE Transactions on Information Theory*, 57(10):6958–6975, 2011.

[367] J. Glover, G. Bradski, and R. B. Rusu. Monte Carlo pose estimation with quaternion kernels and the Bingham distribution. In *Proceedings of Robotics: Science and Systems VII*, Los Angeles, California, USA, June 2011.

[368] G. M. Dilshan Godaliyadda, Dong Hye Ye, Michael D. Uchic, Michael A. Groeber, Gregery T. Buzzard, and Charles A. Bouman. A supervised learning approach for dynamic sampling. In *Computational Imaging XIV*, San Francisco, CA, Feb. 2016.

[369] Joseph Goldstein, Dale E Newbury, D Joy, C Lyman, P Echlin, E Lifshin, L Sawyer, and J Michael. *Scanning Electron Microscopy and X-ray Microanalysis*, volume 306472929 of *ISBN*. Springer, United States, 2003.

[370] D. E. Gómez, K. C. Vernon, P. Mulvaney, and T. J. Davis. Surface plasmon mediated strong exciton–photon coupling in semiconductor nanocrystals. *Nano Lett.*, 10(1):274–278, 2010.

[371] Yunchao Gong, Liwei Wang, Ruiqi Guo, and Svetlana Lazebnik. Multi-scale orderless pooling of deep convolutional activation features. In *Computer Vision–ECCV 2014*, pages 392–407. Springer, 2014.

[372] Yunye Gong, David Veesler, Peter C. Doerschuk, and John E. Johnson. Effect of the viral protease on the dynamics of bacteriophage HK97 maturation intermediates characterized by variance analysis of cryo EM particle ensembles. *J. Struct. Biol.*, 193(3):188–195, 2016. http://doi.org/10.1016/j.jsb.2015.12.012.

[373] R. C. Gonzalez and R. E. Woods. *Digital Image Processing*. Pearson Education, Inc., Third edition, 2008.

[374] S. Gorelova, H. Schaeben, and R. Kawalla. Quantifying texture evolution during hot rolling of magnesium twin roll cast strip. *Materials Science and Engineering: A*, 602:134–142, 2014.

[375] B. Goris, W. Van den Broek, K.J. Batenburg, H. Heidari Mezerji, and S. Bals. Electron tomography based on a total variation minimization reconstruction technique. *Ultramicroscopy*, 113:120 – 130, 2012.

[376] B. Goris, W. Van den Broek, K. J. Batenburg, H. Heidari Mezerji, and S. Bals. Electron tomography based on a total variation minimization reconstruction technique. *Ultramicroscopy*, 113:120–130, 2012.

[377] I. F. Gorodnitsky, J. George, and B. D. Rao. Neuromagnetic source imaging with FOCUSS: A recursive weighted minimum norm algorithm. *Electrocephalography and Clinical Neurophysiology*, 95:231–251, 1995.

[378] Philippe-Henri Gosselin, Naila Murray, Hervé Jégou, and Florent Perronnin. Revisiting the Fisher vector for fine-grained classification. *Pattern Recognition Letters*, 49:92–98, 2014.

[379] S. R. Gottesman and E. E. Fenimore. New family of binary arrays for coded aperture imaging. *Appl. Opt.*, 28:4344–4352, 1989.

[380] G. Gottstein, D. A. Molodov, E. Rabkin, L. S. Shvindlerman, and I. Snapiro. Generation of electrical currents and magnetic fields by grain boundary motion. *Interface Science*, 10(4):279–285, 2002.

[381] Circuit analysis tools (cat) iarpa broad agency announcement iarpa-baa-09-09. Intelligence Advanced Research Projects Agency, Safe and Secure Operations Office, 2009.

[382] Rapid analysis of various emerging nanoelectronics (raven) proposer's day. Intelligence Advanced Research Projects Agency, Safe and Secure Operations Office, 2015.

[383] Rapid analysis of various emerging nanoelectronics (raven) iarpa-baa-15-12. Intelligence Advanced Research Projects Agency, Safe and Secure Operations Office, 2016.

[384] Hayit Greenspan, Bram van Ginneken, and Ronald M Summers. Guest editorial deep learning in medical imaging: Overview and future promise of an exciting new technique. *IEEE Transactions on Medical Imaging*, 35(5):1153–1159, 2016.

[385] K. Gregor and Y. LeCun. Learning fast approximations of sparse coding. In *Proc. Int. Conf. Machine Learning*, 2010.

[386] R. Gribonval and M. Nielsen. Sparse representations in unions of bases. *IEEE Trans. Inform. Theory*, 49:3320–3325, 2003.

[387] R. Gribonval and K. Schnass. Dictionary identification–sparse matrix-factorization via l_1 - minimization. *IEEE Trans. Inform. Theory*, 56(7):3523–3539, 2010.

[388] M. Grigoriu. Nearest neighbor probabilistic model for aluminum polycrystals. *Journal of Engineering Mechanics*, 136(7):821–829, 2010.

[389] G. Grimmett. *Percolation*. Berlin, Springer-Verlag, 1989.

[390] Geoffrey Grimmett. *What Is Percolation?* Springer, 1999.

[391] M. Groeber, S. Ghosh, M. D. Uchic, and M. Dimiduk. A framework for automated analysis and simulation of 3D polycrystalline microstructures. Part2: Synthetic structure generation. *Acta. Mater.*, 56:1274–1287, 2008.

[392] M. A. Groeber, B. K. Haley, M. D. Uchic, D. M. Dimiduk, and S. Ghosh. 3d reconstruction and characterization of polycrystalline microstructures using a fib–sem system. *Materials Characterization*, 57(4–5):259 – 273, 2006.

[393] M. A. Groeber and M. Jackson. DREAM.3D: A digital representation environment for the analysis of microstructure in 3d. *Integrating Materials and Manufacturing Innovation*, 3(1):5, 2014.

[394] G. Grosklos and J. Theiler. Ellipsoids for anomaly detection in remote sensing imagery. *Proc. SPIE*, 9472:94720P, 2015.

[395] J. Gruber, D. C. George, A. P. Kuprat, G. S. Rohrer, and A. D. Rollett. Effect of anisotropic grain boundary properties on grain boundary plane distributions during grain growth. *Scripta Materialia*, 53(3):351–355, 2005.

[396] Ying Gu, Luojia Wang, Pan Ren, Junxiang Zhang, Tiancai Zhang, Olivier J. F. Martin, and Qihuang Gong. Surface-plasmon-induced modification on the spontaneous emission spectrum via subwavelength-confined anisotropic purcell factor. *Nano Lett.*, 12(5):2488–2493, 2012.

[397] Zoher Gueroui and Albert Libchaber. Single-molecule measurements of gold-quenched quantum dots. *Phys. Rev. Lett.*, 93(16):166108, 2004.

[398] Andre Guimer and Gérard Fournet. *Small Angle Scattering of X-rays*. J. Wiley & Sons, New York, 1955.

[399] Yanming Guo, Yu Liu, Ard Oerlemans, Songyang Lao, Song Wu, and Michael S. Lew. Deep learning for visual understanding: A review. *Neurocomputing*, 2015.

[400] Akash Gupta, Ahmet Cecen, Sharad Goyal, Amarendra K. Singh, and Surya R. Kalidindi. Structure–property linkages using a data science approach: Application to a non-metallic inclusion/steel composite system. *Acta Materialia*, 91:239–254, 2015.

[401] V. K. Gupta, D. H. Yoon, H. M. Meyer III, and J. Luo. Thin intergranular films and solid-state activated sintering in nickel-doped tungsten. *Acta Materialia*, 55:3131–3142, 2007.

[402] A. C. Gurbuz, J. H. McClellan, and W. R. Scott. Compressive sensing for subsurface imaging using ground penetrating radar. *Signal Process.*, 89(10):1959–1972, 2009.

[403] K. Güth, T.G. Woodcock, L. Schultz, and O. Gutfleisch. Comparison of local and global texture in HDDR processed Nd-Fe-B magnets. *Acta Materialia*, 59(5):2029 – 2034, 2011.

[404] Saleem Hamady, Abbas Hijazi, and Ali Atwi. Lennard–Jones interactions between nano-rod like particles at an arbitrary orientation and an infinite flat solid surface. *Physica B: Condensed Matter*, 423:26–30, 2013.

[405] J. Hammersley and P. Clifford. Markov fields on finite graphs and lattices. unpublished research, 1971.

[406] Robert M. Haralick, Karthikeyan Shanmugam, and Its'hak Dinstein. Textural features for image classification. *Systems, Man and Cybernetics, IEEE Transactions on*, SMC-3:610–621, 1973.

[407] George Harauz and Marin van Heel. Exact filters for general geometry three dimensional reconstruction. *Optik*, 73(4):146–156, 1986.

[408] G. Harikumar and Y. Bresler. A new algorithm for computing sparse solutions to linear inverse problems. In *ICASSP*, pages 1331–1334, May 1996.

[409] Z. Harmany, R. Marcia, and R. Willett. This is SPIRAL-TAP: Sparse Poisson Intensity Reconstruction ALgorithms – Theory and Practice. *IEEE Trans. Image Process.*, 21(3):1084–1096, 2012.

[410] M. P. Harmer. Interfacial kinetic engineering: How far have we come since Kingery's inaugural Sosman address? *Journal of the American Ceramic Society*, 93(2):301–317, 2010.

[411] Chris Harris and Mike Stephens. A combined corner and edge detector. In *Alvey Vision Conference*, volume 15, page 50. Citeseer, 1988.

[412] G. C. Hasson and C. Goux. Interfacial energies of tilt boundaries in aluminium: Experimental and theoretical determination. *Scripta Metallurgica*, 5(10):889, 1971.

[413] N. Hasson, S. Asulin, S. R. Rotman, and D. Blumberg. Evaluating backgrounds for subpixel target detection: when closer isn't better. *Proc. SPIE*, 9472:94720R, 2015.

[414] T. Hastie, R. Tibshirani, and J. Friedman. *Elements of Statistical Learning: Data Mining, Inference, and Prediction*. Springer-Verlag, New York, 2001. This anomaly detection approach is developed in Chapter 14.2.4, and illustrated in Fig 14.3.

[415] T. Haug, S. Otto, M. Schneider, and J. Zweck. Computer simulation of Lorentz electron micrographs of thin magnetic particles. *Ultramicroscopy*, 96:201–206, 2003.

[416] J. Haupt, W. Bajwa, G. Raz, and R. Nowak. Toeplitz compressed sensing matrices with applications to sparse channel estimation. *IEEE Trans. Inform. Theory*, 56(11):5862–5875, 2010.

[417] S. Hawe, M. Seibert, and M. Kleinsteuber. Separable dictionary learning. In *IEEE Conference on Computer Vision and Pattern Recognition (CVPR)*, pages 438–445, 2013.

[418] A. Hayden, E. Niple, and B. Boyce. Determination of trace-gas amounts in plumes by the use of orthogonal digital filtering of thermal-emission spectra. *Applied Optics*, 35:2802–2809, 1996.

[419] Kaiming He, Xiangyu Zhang, Shaoqing Ren, and Jian Sun. Spatial pyramid pooling in deep convolutional networks for visual recognition. *Pattern Analysis and Machine Intelligence, IEEE Transactions on*, 37(9):1904–1916, 2015.

[420] L. He and L. Carin. Exploiting structure in wavelet-based Bayesian compressive sensing. *IEEE Trans. Sig. Proc.*, 57(9):3488–3497, 2009.

[421] D. J. Heeger and J. R. Bergen. Pyramid-based texture analysis/synthesis. In *In SIGGRAPH 95, pages 229–238*, 1995.

[422] C. M. Hefferan, S. F. Li, J. Lind, U. Lienert, A. D. Rollett, P. Wynblatt, and R. M. Suter. Statistics of high purity nickel microstructure from high energy x-ray diffraction microscopy. *CMC - Computers Materials & Continua*, 14(3):209–219, 2009.

[423] CM Hefferan, SF Li, J. Lind, U. Lienert, AD Rollett, P. Wynblatt, and RM Suter. Statistics of high purity nickel microstructure from high energy x-ray diffraction microscopy. *Computers, Materials, & Continua*, 14(3):209–220, 2010.

[424] A Heinz and P Neumann. Representation of orientation and disorientation data for cubic, hexagonal, tetragonal and orthorhombic crystals. *Acta Crystallographica Section A: Foundations of Crystallography*, 47(6):780–789, 1991.

[425] Mario Hentschel, Vivian E. Ferry, and A. Paul Alivisatos. Optical rotation reversal in the optical response of chiral plasmonic nanosystems: The role of plasmon hybridization. *ACS Photonics*, 2(9):1253–1259, 2015.

[426] C. Herring. Some theorems on the free energies of crystal surfaces. *Physical Review*, 82(1):87–93, 1951.

[427] C. Herring. Surface tension as a motivation for sintering. In K. E. Kingston, editor, *The Physics of Powder Metallurgy*, page 143. McGraw Hill, New York, 1951.

[428] F. Herrmann, Y. A. Erlangga, and T. Y. Lin. Compressive simultaneous full-waveform simulation. *Geophysics*, 74:A35, 2009.

[429] J. C. Heyraud and J. J. Metois. Equilibrium shape and temperature - lead on graphite. *Surface Science*, 128(2-3):334–342, 1983.

[430] R. Hielscher and H. Schaeben. A novel pole figure inversion method: Specification of the *MTEX* algorithm. *Journal of Applied Crystallography*, 41(6):1024–1037, Dec 2008.

[431] D. Hilbert and S. Cohn-Vossen. *Geometry and the Imagination*. AMS Chelsea Publishing Series. AMS Chelsea Pub., 1999.

[432] P. Hill, A. Achim, and D. Bull. The undecimated dual tree complex wavelet transform and its applications to bivariate image denoising using a Cauchy model. In *Proc. IEEE Int. Conf. Image Proc.*, pages 1205–1208, 2012.

[433] John P Hirth and Jens Lothe. *Theory of Dislocations*. John Wiley & Sons, 1982.

[434] Eric M. V. Hoek and Gaurav K. Agarwal. Extended dlvo interactions between spherical particles and rough surfaces. *Journal of Colloid and Interface Science*, 298(1):50–58, 2006.

[435] J. P. Hoffbeck and D. A. Landgrebe. Covariance matrix estimation and classification with limited training data. *IEEE Trans. Pattern Analysis and Machine Intelligence*, 18:763–767, 1996.

[436] D. W. Hoffman and J. W. Cahn. Vector thermodynamics for anisotropic surfaces – i. fundamentals and application to plane surface junctions. *Surface Science*, 31(1):368–388, 1972.

[437] H. Hoffmann. Kernel PCA for novelty detection. *Pattern Recognition*, 40:863–874, 2007.

[438] James M. Hogle. Poliovirus cell entry: Common structural themes in viral cell entry pathways. *Annu. Rev. Microbiol.*, 56:677–702, 15 July 2002.

[439] EA Holm and PM Duxbury. Three-dimensional materials science. *Scripta Materialia*, 54(6):1035–1040, 2006.

[440] EA Holm, GS Rohrer, SM Foiles, AD Rollett, HM Miller, and DL Olmsted. Validating computed grain boundary energies in fcc metals using the grain boundary character distribution. *Acta Materialia*, 59(13):5250–5256, 2011.

[441] Ernest M. Hotze, Tanapon Phenrat, and Gregory V. Lowry. Nanoparticle aggregation: Challenges to understanding transport and reactivity in the environment. *Journal of Environmental Quality*, 39(6):1909–1924, 2010.

[442] Louisa Howard. Dartmouth college electron microscope facility, electron microscopy images. `http://remf.dartmouth.edu/miscellaneous_SEM_P1/`, 2011. [Online; accessed Dec. 2015].

[443] T. E. Hsieh and R. W. Balluffi. Observations of roughening / de-faceting phase transitions in grain boundaries. *Acta Metallurgica*, 37(8):2133–2139, 1989.

[444] Ming-Kuei Hu. Visual pattern recognition by moment invariants. *information Theory, IRE Transactions on*, 8(2):179–187, 1962.

[445] Tao Hu, Juan Nunez-Iglesias, Shiv Vitaladevuni, Lou Scheffer, Shan Xu, Mehdi Bolorizadeh, Harald Hess, Richard Fetter, and Dmitri Chklovskii. Super-resolution using sparse representations over learned dictionaries: Reconstruction of brain structure using electron microscopy. *arXiv preprint arXiv:1210.0564*, 2012.

[446] J. Huang, C. R. Berger, S. Zhou, and J. Huang. Comparison of basis pursuit algorithms for sparse chan- nel estimation in underwater acoustic OFDM. In *Proc. IEEE OCEANS*, pages 1–6, Sydney, Australia, 2010.

[447] Y. Huang, J. Paisley, Q. Lin, X. Ding, X. Fu, and X. P. Zhang. Bayesian nonparametric dictionary learning for compressed sensing MRI. *IEEE Trans. Image Process.*, 23(12):5007–5019, 2014.

[448] D. Hull. *Introduction to Dislocations*. Pergamon, New York, 1965.

[449] E. Humphrey and M. De Graef. On the computation of the magnetic phase shift for magnetic nano- particles of arbitrary shape using a spherical projection model. *Ultramicroscopy*, 129:36–41, 2013.

[450] M. J. Humphry, B. Kraus, A. C. Hurst, A. M. Maiden, and J. M. Rodenburg. Ptychographic elec- tron microscopy using high-angle dark-field scattering for sub-nanometre resolution imaging. *Nature Communications*, 3:730, Mar 2012.

[451] Jyh-Jing Hwang and Tyng-Luh Liu. Contour detection using cost-sensitive convolutional neural net- works. *arXiv preprint arXiv:1412.6857*, 2014.

[452] M. Hykšová, A. Kalousová, and I. Saxl. Early history of geometric probability and stereology. *Image Analysis & Stereology*, 31(1):1–16, 2012.

[453] E. Ising. Beitrag zur Theorie des Ferromagnetismus. *Zeitschrift Physik*, 31:253–258, 1925.

[454] J. D. Jackson. *Classical Electrodynamics, Second Edition*. John Wiley & Sons, 1975.

[455] L. Jacques, P. Vandergheynst, A. Bibet, V. Majidzadeh, A. Schmid, and Y. Leblebici. CMOS com- pressed imaging by random convolution. In *Proc. IEEE Int. Conf. Acoust. Speech Sig. Proc.*, pages 2877–2880, Taipei, Taiwan, 2009.

[456] Laurent Jacques, Jason N. Laska, Petros T. Boufounos, and Richard G. Baraniuk. Robust 1-bit com- pressive sensing via binary stable embeddings of sparse vectors. *Information Theory, IEEE Transac- tions on*, 59(4):2082–2102, 2013.

[457] Sina Jafarpour, Weiyu Xu, Babak Hassibi, and Robert Calderbank. Efficient and robust compressed sensing using optimized expander graphs. *Information Theory, IEEE Transactions on*, 55(9):4299– 4308, 2009.

[458] Anil K. Jain and Farshid Farrokhnia. Unsupervised texture segmentation using gabor filters. In *Systems, Man and Cybernetics, 1990. Conference Proceedings., IEEE International Conference on*, pages 14– 19. IEEE, 1990.

[459] Hervé Jégou, Matthijs Douze, Cordelia Schmid, and Patrick Pérez. Aggregating local descriptors into a compact image representation. In *Computer Vision and Pattern Recognition (CVPR), 2010 IEEE Conference on*, pages 3304–3311. IEEE, 2010.

[460] S. Ji, Y. Xu, and L. Carin. Bayesian compressive sensing. *IEEE Trans. Sig. Proc.*, 56(6):2346–2356, 2008.

[461] Hao Jiang, Sidney Fels, and James J. Little. A linear programming approach for multiple object track- ing. In *IEEE Conference on Computer Vision and Pattern Recognition*, pages 1–8, 2007.

[462] Lin Jiang, Xiaodong Chen, Nan Lu, and Lifeng Chi. Spatially confined assembly of nanoparticles. *Accounts of Chemical Research*, 47(10):3009–3017, 2014.

[463] Lin Jiang, Wenchong Wang, Harald Fuchs, and Lifeng Chi. One-dimensional arrangement of gold nanoparticles with tunable interparticle distance. *Small*, 5(24):2819–2822, 2009.

[464] Z. Jiang, Z. Lin, and L. S. Davis. Label consistent k-svd: Learning a discriminative dictionary for recognition. *IEEE Transactions on Pattern Analysis and Machine Intelligence*, 35(11):2651–2664, 2013.

[465] Yang Jiao, Eric Padilla, and Nikhilesh Chawla. Modeling and predicting microstructure evolution in lead/tin alloy via correlation functions and stochastic material reconstruction. *Acta Materialia*, 61(9):3370–3377, 2013.

[466] K. H. Jin, M. T. McCann, E. Froustey, and M. Unser. Deep convolutional neural network for inverse problems in imaging. *IEEE Trans. Image Proc.*, 26(9):4509–4522, 2017.

[467] Ian Jolliffe. *Principal Component Analysis*. Wiley Online Library, 2002.

[468] J. E. Jones. On the determination of molecular fields. ii. from the equation of state of a gas. *Proceedings of the Royal Society of London A: Mathematical, Physical and Engineering Sciences*, 106(738):463–477, 1924.

[469] Bela Julesz. Visual pattern discrimination. *Information Theory, IRE Transactions on*, 8(2):84–92, 1962.

[470] J. Jura, J. Pospiech, and G. Gottstein. Estimation of the minimum number of single grain orientation measurements for odf determination. *Zeitschrift fur Metallkunde*, 87:476–483, 1996.

[471] Frederic Jurie and Bill Triggs. Creating efficient codebooks for visual recognition. In *Computer Vision, 2005. ICCV 2005. Tenth IEEE International Conference on*, volume 1, pages 604–610. IEEE, 2005.

[472] Tomasz Kaczynski, Konstantin Michael Mischaikow, and Marian Mrozek. *Computational Homology*, volume 157. Springer Science & Business Media, 2004.

[473] A. C. Kak and M. Slaney. *Principles of Computerized Tomographic Imaging*. SIAM, 2001.

[474] S. R. Kalidindi. *Hierarchical Materials Informatics: Novel Analytics for Materials Data*. Butterworth-Heinemann, 2015.

[475] Surya R. Kalidindi, Stephen R. Niezgoda, and Ayman A. Salem. Microstructure informatics using higher-order statistics and efficient data-mining protocols. *Jom*, 63(4):34–41, 2011.

[476] S. V. Kalinin, R. Vasudevan, A. Borisevich, A. Belianinov, R. K. Archibald, C. Symons, E. Lingerfelt, B. Sumpter, L. Vlcek, and S. Jesse. Big, deep, and smart data from atomically resolved images: exploring the origins of materials functionality. *Microscopy and Microanalysis*, 22(S3):1416–1417, 2016.

[477] Rudolph Emil Kalman. A new approach to linear filtering and prediction problems. *Journal of Fluids Engineering*, 82(1):35–45, 1960.

[478] M. E. Kamasak, C. A. Bouman, E. D. Morris, and K. Sauer. Direct reconstruction of kinetic parameter images from dynamic PET data. *IEEE Trans. on Medical Imaging*, 24(5):636 –650, May 2005.

[479] Wayne D. Kaplan, Dominique Chatain, Paul Wynblatt, and W. Craig Carter. A review of wetting versus adsorption, complexions, and related phenomena: The Rosetta Stone of wetting. *Journal of Materials Science*, 48:5681–5717, 2013.

[480] T. Kasetkasem and P. K. Varshney. An image change detection algorithm based on Markov random field models. *IEEE Trans. Geoscience and Remote Sensing*, 40:1815–1823, 2002.

[481] Xiaoxing Ke, Carla Bittencourt, and Gustaaf Van Tendeloo. Possibilities and limitations of advanced transmission electron microscopy for carbon-based nanomaterials. *Beilstein Journal of Nanotechnology*, 6(1):1541–1557, 2015.

[482] Yan Ke and Rahul Sukthankar. Pca-sift: A more distinctive representation for local image descriptors. In *Computer Vision and Pattern Recognition, 2004. CVPR 2004. Proceedings of the 2004 IEEE Computer Society Conference on*, volume 2, pages II–506. IEEE, 2004.

[483] Cynthia L. Kelchner, S. J. Plimpton, and J. C. Hamilton. Dislocation nucleation and defect structure during surface indentation. *Phys. Rev. B*, 58(17):11085, 1998.

[484] Thomas Kemen, Matt Malloy, Brad Thiel, Shawn Mikula, Winfried Denk, Gregor Dellemann, and Dirk Zeidler. Further advancing the throughput of a multibeam sem. In *SPIE Advanced Lithography*, pages 94241U–94241U–6. International Society for Optics and Photonics, 2015.

[485] M. C. Kennedy and A. O'Hagan. Bayesian calibration of computer models. *J. of Royal Statistical Society*, B 63:425–464, 2001.

[486] M. Kerr, M. R. Daymond, R. A. Holt, and J. D. Almer. Mapping of crack tip strains and twinned zone in a hexagonal close packed zirconium alloy. *Acta Materialia*, 58(5):1578 – 1588, 2010.

[487] C. Kervrann and A. Trubuil. An adaptive window approach for Poisson noise reduction and structure preserving in confocal microscopy. In *Proc. IEEE Int. Symp. on Biomed. Imag.*, 2004.

[488] Adarsh Kesireddy and Sara McCaslin. Application of image processing techniques to the identification of phases in steel metallographic specimens. In *New Trends in Networking, Computing, E-learning, Systems Sciences, and Engineering*, pages 425–430. Springer, 2015.

[489] L. G. Khachiyan. Rounding of polytopes in the real number model of computation. *Mathematics of Operations Research*, 21:307–320, 1996.

[490] Bishnu P. Khanal, Anshu Pandey, Liang Li, Qianglu Lin, Wan Ki Bae, Hongmei Luo, Victor I. Klimov, and Jeffrey M. Pietryga. Generalized synthesis of hybrid metalsemiconductor nanostructures tunable from the visible to the infrared. *ACS Nano*, 6(5):3832–3840, 2012.

[491] M. A. Khanesar, M. Teshnehlab, and M. A. Shoorehdeli. A novel binary particle swarm optimization. In *Control & Automation, 2007. MED'07. Mediterranean Conference on*, pages 1–6. IEEE, 2007.

[492] C. S. Kim, A. D. Rollett, and G. S. Rohrer. Grain boundary planes: New dimensions in the grain boundary character distribution. *Scripta Materialia*, 54(6):1005–1009, 2006.

[493] Min Woo Kim, Jiyoung Choi, Liu Yu, Kyung Eun Lee, Sung-Sik Han, and Jong Chul Ye. Cryo-electron microscopy single particle reconstruction of virus particles using compressed sensing theory. In *Proc. SPIE 6498, Computational Imaging V*, volume 6498, pages 64981G–64981G–9, 2007.

[494] Y. Kim, M. S. Nadar, and A. Bilgin. Wavelet-based compressed sensing using gaussian scale mixtures. In *Proc. ISMRM*, page 4856, 2010.

[495] Anika Kinkhabwala, Zongfu Yu, Shanhui Fan, Yuri Avlasevich, Klaus Mullen, and W. E. Moerner. Large single-molecule fluorescence enhancements produced by a bowtie nanoantenna. *Nat. Photon.*, 3(11):654–657, 2009.

[496] Earl J. Kirkland. *Advanced Computing in Electron Microscopy*. Springer, New York, 2010.

[497] E. J. Kirkland. *Advanced Computing in Electron Microscopy*. Plenum Press, New York, 1998.

[498] Christof Klein, Hans Loeschner, and Elmar Platzgummer. 50-kev electron multibeam mask writer for the 11-nm hp node: first results of the proof-of-concept electron multibeam mask exposure tool. *Journal of Micro/Nanolithography, MEMS, and MOEMS*, 11(3):031402–1–031402–7, 2012.

[499] Michael P. Knox. Continuous fiber reinforced thermoplastic composites in the automotive industry. In *Automotive Composites Conference*, 2001.

[500] U. F. Kocks, C. N. Tomé, and H.-R. Wenk, editors. *Texture and Anisotropy: Preferred Orientations in Polycrystals and Their Effect on Materials Properties*. Cambridge University Press, 1998.

[501] Ai Leen Koh, Sang Chul Lee, and Robert Sinclair. A brief history of controlled atmosphere transmission electron microscopy. In *Controlled Atmosphere Transmission Electron Microscopy*, pages 3–43. Springer, 2016.

[502] P. Koiran and A. Zouzias. Hidden cliques and the certification of the restricted isometry property. *IEEE Trans. Inform. Theory*, 60(8):4999–5006, 2014.

[503] E. Kolaczyk. Bayesian multi-scale models for Poisson processes. *Journal of the American Statistical Association*, 94:920–933, 1999.

[504] E. Kolaczyk. Wavelet shrinkage estimation of certain Poisson intensity signals using corrected thresholds. *Statistica Sinica*, 9:119–135, 1999.

[505] E. Kolaczyk and R. Nowak. Multiscale likelihood analysis and complexity penalized estimation. *Annals of Stat.*, 32:500–527, 2004.

[506] S. Kong and D. Wang. A dictionary learning approach for classification: Separating the particularity and the commonality. In *Proceedings of the 12th European Conference on Computer Vision*, pages 186–199, 2012.

[507] Piotr Koniusz, Fei Yan, and Krystian Mikolajczyk. Comparison of mid-level feature coding approaches and pooling strategies in visual concept detection. *Computer Vision and Image Understanding*, 117(5):479–492, 2013.

[508] J. Konrad, S. Zaefferer, and D. Raabe. Investigation of orientation gradients around a hard laves particle in a warm-rolled fe3al-based alloy using a 3d ebsd-fib technique. *Acta Materialia*, 54(5):1369 – 1380, 2006.

[509] J. Kopf, C-W. Fu, D. Cohen-Or, O. Deussen, D. Lischinski, and T-T Wong. Solid texture synthesis from 2D exemplars. In *SIGGRAPH Proceedings, 2, 1–9*, 2007.

[510] A. J. Koster, U. Ziese, A. J. Verkleij, A. H. Janssen, J. de Graaf, J. W. Geus, and K. P. de Jong. Development and application of 3-dimensional transmission electron microscopy (3D-TEM) for the characterization of metal-zeolite catalyst systems. In Sagrario Mendioroz Avelino Corma, Francisco V. Melo and Jos Luis G. Fierro, editors, *12th International Congress on Catalysis Proceedings of the 12th ICC*, volume 130 of *Studies in Surface Science and Catalysis*, pages 329 – 334. Elsevier, 2000.

[511] F. Krahmer, S. Mendelson, and H. Rauhut. Suprema of chaos processes and the restricted isometry property. *Comm. Pure Appl. Math.*, 67(11):1877–1904, 2014.

[512] O. L. Krivanek, A. J. Gubbens, N. Dellby, and C. E. Meyer. Design and applications of a post-column imaging filter. In G. W. Bailey, J. Bentley, and S. A. Small, editors, *50th Ann. Proc. Electron Microsc. Soc. Am.*, pages 1192–1193. San Francisco Press, Inc., 1992.

[513] O. L. Krivanek and P. E. Mooney. Applications of slow-scan ccd cameras in transmission electron microscopy. *Ultramicroscopy*, 49(1):95–108, 1993.

[514] Ondrej L. Krivanek, Tracy C. Lovejoy, Niklas Dellby, Toshihiro Aoki, R. W. Carpenter, Peter Rez, Emmanuel Soignard, Jiangtao Zhu, Philip E. Batson, Maureen J. Lagos, Ray F. Egerton, and Peter A. Crozier. Vibrational spectroscopy in the electron microscope. *Nature*, 514(7521):209–212, 2014.

[515] Alex Krizhevsky, Ilya Sutskever, and Geoffrey E. Hinton. Imagenet classification with deep convolutional neural networks. In *Advances in Neural Information Processing Systems*, pages 1097–1105, 2012.

[516] E. Kröner. Statistical modelling. In *Modelling Small Deformations of Polycrystals*, pages 229–291. Springer, 1986.

[517] Harold W. Kuhn. The Hungarian method for the assignment problem. *Naval Research Logistics Quarterly*, 2(1-2):83–97, 1955.

[518] Maarten Kuijper, Gerald van Hoften, Bart Janssen, Rudolf Geurink, Sacha De Carlo, Matthijn Vos, Gijs van Duinen, Bart van Haeringen, and Marc Storms. Feis direct electron detector developments: Embarking on a revolution in cryo-tem. *Journal of Structural Biology*, 192(2):179–187, 2015.

[519] Olga Kulakovich, Natalya Strekal, Alexandr Yaroshevich, Sergey Maskevich, Sergey Gaponenko, Igor Nabiev, Ulrike Woggon, and Mikhail Artemyev. Enhanced luminescence of cdse quantum dots on gold colloids. *Nano Lett.*, 2(12):1449–1452, 2002.

[520] A. Kumar, V. Sundararaghavan, M. DeGraef, and L. Nguyen. A Markov Random Field Approach for Microstructure Synthesis. *Modelling and Simulation in Materials Science and Engineering*, in press, 2016.

[521] M. Kumar, W. E. King, and A. J. Schwartz. Modifications to the microstructural topology in fcc materials through thermomechanical processing. *Acta Materialia*, 48(9):2081–2091, 2000.

[522] P. Kumar and E. A. Yildirim. Minimum-volume enclosing ellipsoids and core sets. *J. Optimization Theory and Applications*, 126:1–21, 2005.

[523] Animesh Kundu, Kaveh Meshinchi Asl, Jian Luo, and Martin P. Harmer. Identification of a bilayer grain boundary complexion in bi-doped cu. *Scripta Materialia*, 68:146–149, 2013.

[524] Karsten Kunze and Helmut Schaeben. The Bingham distribution of quaternions and its spherical radon transform in texture analysis. *Mathematical Geology*, 36:917–943, 2004.

[525] S. Kusters, M. Seefeldt, and P. Van Houtte. A Fourier image analysis technique to quantify the banding behavior of surface texture components in aa6xxx aluminum sheet. *Materials Science and Engineering: A*, 527(23):6239–6243, 2010.

[526] Anton Kuzyk, Robert Schreiber, Zhiyuan Fan, Gunther Pardatscher, Eva-Maria Roller, Alexander Hogele, Friedrich C. Simmel, Alexander O. Govorov, and Tim Liedl. Dna-based self-assembly of chiral plasmonic nanostructures with tailored optical response. *Nature*, 483(7389):311–314, 2012.

[527] V. Kwatra, I. Essa, A. Bobick, and N. Kwatra. Texture optimization for example-based synthesis. In *ACM Transactions on Graphics (Proc. SIGGRAPH), 24(3), 795-802*, 2005.

[528] H. Kwon, S. Z. Der, and N. M. Nasrabadi. Adaptive anomaly detection using subspace separation for hyperspectral imagery. *Optical Engineering*, 42:3342–3351, 2003.

[529] H. Kwon and N. Nasrabadi. Kernel RX-algorithm: A nonlinear anomaly detector for hyperspectral imagery. *IEEE Trans. Geoscience and Remote Sensing*, 43:388–397, 2005.

[530] Thomas LaGrange, B. Reed, William DeHope, Richard Shuttlesworth, and Glenn Huete. Movie mode dynamic transmission electron microscopy (dtem): Multiple frame movies of transient states in materials with nanosecond time resolution. *Microscopy and Microanalysis*, 17(S2):458–459, 2011.

[531] Clifford Lam and Jianqing Fan. Sparsistency and rates of convergence in large covariance matrix estimation. *Annals of Statistics*, 37(6B):4254–4278, 2009.

[532] Lev Davidovich Landau and Evgenii Mikhailovich Lifshitz. *Course of Theoretical Physics*. Elsevier, 2013.

[533] Gabriel C. Lander, Scott M. Stagg, Neil R. Voss, Anchi Cheng, Denis Fellmann, James Pulokas, Craig Yoshioka, Christopher Irving, Anke Mulder, Pick-Wei Lau, Dmitry Lyumkis, Clinton S. Potter, and Bridget Carragher. Appion: An integrated, database-driven pipeline to facilitate EM image processing. *J. Struct. Biol.*, 166(1):95–102, 2009.

[534] Peter S. Landweber, Emanuel A. Lazar, and Neel Patel. On fiber diameters of continuous maps. *American Mathematical Monthly*, 123(4):392–397, 2016.

[535] J. Lanman, J. Crum, T. J. Deerinck, G. M. Gaietta, A. Schneemann, G. E. Sosinsky, M. H. Ellisman, and J. E. Johnson. Visualizing flock house virus infection in Drosophila cells with correlated fluorescence and electron microscopy. *J. Struct. Biol.*, 161:439–446, 2008.

[536] B. C. Larson, W. Yang, G. E. Ice, J. D. Budai, and J. Z. Tischler. Three-dimensional x-ray structural microscopy with submicrometer resolution. *Nature*, 415(6874):887–890, 2002.

[537] Kurt W. Larson, Hyrum S. Anderson, Matthew G. Blain, Joseph R. Michael, Craig Y. Nakakura, James E. Stevens, Jason W. Wheeler, and David L. Adler. An electron microscope for compressive sensing with flood beam illumination (sand2014-0558). Unpublished Work, 2014.

[538] K. W. Larson, H. S. Anderson, and J. W. Wheeler. Fast electron microscopy via compressive sensing. Technical report, U.S. Patent 8,907,280, 2014.

[539] Jason N. Laska, Petros T. Boufounos, Mark A. Davenport, and Richard G. Baraniuk. Democracy in action: Quantization, saturation, and compressive sensing. *Applied and Computational Harmonic Analysis*, 31(3):429–443, 2011.

[540] E. A. Lazar, J. K. Mason, R. D. MacPherson, and D. J. Srolovitz. Complete topology of cells, grains, and bubbles in three-dimensional microstructures. *Phys. Rev. Lett.*, 109(9):95505, 2012.

[541] Emanuel A. Lazar. Molecular dynamic studies of the fracture of metals. Yeshiva University, 2005. Undergraduate Honors Thesis.

[542] Emanuel A. Lazar. Vorotop: Voronoi cell topology visualization and analysis toolkit. *Modelling and Simulation in Materials Science and Engineering*, 26(1):015011, 2017.

[543] Emanuel A. Lazar, Jian Han, and David J Srolovitz. Topological framework for local structure analysis in condensed matter. *Proceedings of the National Academy of Sciences*, 112(43):E5769–E5776, 2015.

[544] Svetlana Lazebnik and Maxim Raginsky. Supervised learning of quantizer codebooks by information loss minimization. *Pattern Analysis and Machine Intelligence, IEEE Transactions on*, 31(7):1294–1309, 2009.

[545] Svetlana Lazebnik, Cordelia Schmid, and Jean Ponce. Beyond bags of features: Spatial pyramid matching for recognizing natural scene categories. In *Computer Vision and Pattern Recognition, 2006 IEEE Computer Society Conference on*, volume 2, pages 2169–2178. IEEE, 2006.

[546] E. C. Le Ru, P. G. Etchegoin, J. Grand, N. Flidj, J. Aubard, and G. Lvi. Mechanisms of spectral profile modification in surface-enhanced fluorescence. *J. Phys. Chem. C*, 111(44):16076–16079, 2007.

[547] Rowan Leary, Zineb Saghi, Paul A Midgley, and Daniel J Holland. Compressed sensing electron tomography. *Ultramicroscopy*, 131:70–91, 2013.

[548] James M. LeBeau and Susanne Stemmer. Experimental quantification of annular dark-field images in scanning transmission electron microscopy. *Ultramicroscopy*, 108(12):1653 – 1658, 2008.

[549] R. A. Lebensohn, O. Castelnau, R. Brenner, and P. Gilormini. Study of the antiplane deformation of linear 2-D polycrystals with different microstructures. *International Journal of Solids and Structures*, 42(20):5441–5459, 2005.

[550] R. A. Lebensohn and C. N. Tomé. A self-consistent anisotropic approach for the simulation of plastic deformation and texture development of polycrystals: Application to zirconium alloys. *Acta Metallurgica et Materialia*, 41(9):2611 – 2624, 1993.

[551] Y. LeCun, Y. Bengio, and G. Hinton. Deep learning. *Nature*, 521(7553):436–444, 2015.

[552] Yann LeCun, Bernhard Boser, John S. Denker, Donnie Henderson, Richard E. Howard, Wayne Hubbard, and Lawrence D. Jackel. Backpropagation applied to handwritten zip code recognition. *Neural Computation*, 1(4):541–551, 1989.

[553] Yann LeCun, Léon Bottou, Yoshua Bengio, and Patrick Haffner. Gradient-based learning applied to document recognition. *Proceedings of the IEEE*, 86(11):2278–2324, 1998.

[554] J.-S. Lee. Digital image smoothing and the sigma filter. *Computer Vision, Graphics, and Image Processing*, 24(2):255–269, 1983.

[555] Junghoon Lee, Peter C. Doerschuk, and John E. Johnson. Exact reduced-complexity maximum likelihood reconstruction of multiple 3-D objects from unlabeled unoriented 2-D projections and electron microscopy of viruses. *IEEE Transactions on Image Processing*, 16(11):2865–2878, November 2007.

[556] S. Lee, H. R. Piehler, A. D. Rollett, and B. L. Adams. Texture clustering and long-range disorientation representation methods: application to 6022 aluminum sheet. *Metallurgical and Materials Transactions A*, 33(12):3709–3718, 2002.

[557] Tai Sing Lee. Image representation using 2d gabor wavelets. *Pattern Analysis and Machine Intelligence, IEEE Transactions on*, 18(10):959–971, 1996.

[558] S. Lefebvre and H. Hoppe. Appearance-space texture synthesis. In *SIGGRAPH 06, pp. 541–548*, 2006.

[559] E. L. Lehmann and J. P. Romano. *Testing Statistical Hypotheses*. Springer, New York, 2005.

[560] Ralph Leighton. *Surely You're Joking, Mr. Feynman!: Adventures of a Curious Character*. WW Norton, New York, 1985.

[561] Hannes Leipold, Emanuel A. Lazar, Kenneth A. Brakke, and David J. Srolovitz. Statistical topology of perturbed two-dimensional lattices. *J. Stat. Mech.*, 2016:043103, 2016.

[562] Thomas Leung and Jitendra Malik. Representing and recognizing the visual appearance of materials using three-dimensional textons. *International Journal of Computer Vision*, 43(1):29–44, 2001.

[563] E. Levina and P. Bickel. The earth mover's distance is the mallows distance: Some insights from statistics. In *Computer Vision, 2001. ICCV 2001. Proceedings. Eighth IEEE International Conference on*, volume 2, pages 251–256 vol.2, 2001.

[564] Zachary H. Levine. Theory of bright-field scanning transmission electron microscopy for tomography. *Journal of Applied Physics*, 97(3):033101, 2005.

[565] Zachary H. Levine, Anthony J. Kearsley, and John G. Hagedorn. Bayesian tomography for projections with an arbitrary transmission function with an application in electron microscopy. *Journal of Research of the National Institute of Standards and Technology*, 111(6):411 –417, November 2006.

[566] Duo Li, FengChao Wang, ZhenYu Yang, and YaPu Zhao. How to identify dislocations in molecular dynamics simulations? *Science China Physics, Mechanics & Astronomy*, 57(12):2177–2187, 2014.

[567] Hanying Li, Huolin L. Xin, David A. Muller, and Lara A. Estroff. Visualizing the 3D internal structure of calcite single crystals grown in agarose hydrogels. *Science*, 326(5957):1244–1247, 2009.

[568] J. Li, S. J. Dillon, and G. S. Rohrer. Relative grain boundary area and energy distributions in nickel. *Acta Materialia*, 57(14):4304–4311, 2009.

[569] Jing Li and Nigel M. Allinson. A comprehensive review of current local features for computer vision. *Neurocomputing*, 71(10):1771–1787, 2008.

[570] Ju Li. Atomeye: An efficient atomistic configuration viewer. *Modelling and Simulation in Materials Science and Engineering*, 11(2):173, 2003.

[571] X. Li and S. Luo. A compressed sensing-based iterative algorithm for CT reconstruction and its possible application to phase contrast imaging. *BioMedical Engineering OnLine*, 10(1):73, 2011.

[572] Xin Li, Fu-Jen Kao, Chien-Chin Chuang, and Sailing He. Enhancing fluorescence of quantum dots by silica-coated gold nanorods under one- and two-photon excitation. *Optics Express*, 18(11):11335–11346, 2010.

[573] Xueming Li, Shawn Q. Zheng, Kiyoshi Egami, David A. Agard, and Yifan Cheng. Influence of electron dose rate on electron counting images recorded with the k2 camera. *Journal of Structural Biology*, 184(2):251–260, 2013.

[574] H. Y. Liao and G. Sapiro. Sparse representations for limited data tomography. In *Proc. IEEE International Symposium on Biomedical Imaging (ISBI)*, pages 1375–1378, 2008.

[575] Jiunn-Woei Liaw, Hsiao-Yen Tsai, and Chun-Hui Huang. Size-dependent surface enhanced fluorescence of gold nanorod: Enhancement or quenching. *Plasmonics*, 7:543–553, 2012.

[576] Jeff W. Lichtman and Winfried Denk. The big and the small: Challenges of imaging the brain's circuits. *Science*, 334(6056):618–623, 2011.

[577] U. Lienert, J. Almer, B. Jakobsen, W. Pantleon, H. F. Poulsen, D. Hennessy, C. Xiao, and R. M. Suter. 3-dimensional characterization of polycrystalline bulk materials using high-energy synchrotron radiation. *Materials Science Forum*, 539-543:2353–2358, March 2007.

[578] L. C. Lim and T. Watanabe. Grain boundary character distribution controlled toughness of polycrystals — a two-dimensional model. *Scripta Metallurgica*, 23(4):489–494, 1989.

[579] B. Lin, Y. Jin, C. M. Hefferan, S. F. Li, J. Lind, R. M. Suter, M. Bernacki, N. Bozzolo, A. D. Rollett, and G. S. Rohrer. Observation of annealing twin nucleation at triple lines in nickel during grain growth. *Acta Materialia*, 99:63–68, 2015.

[580] T. Lin, E. Lebed, Y. A. Erlangga, and F. J. Herrmann. Interpolating solutions of the helmholtz equation with compressed sensing. In *SEG Technical Program Expanded Abstracts*, volume 27, pages 2122–2126, 2008.

[581] Tsung-Yu Lin and Subhransu Maji. Visualizing and understanding deep texture representations. *arXiv preprint arXiv:1511.05197*, 2015.

[582] Tsung-Yu Lin, Aruni RoyChowdhury, and Subhransu Maji. Bilinear cnn models for fine-grained visual recognition. In *Proceedings of the IEEE International Conference on Computer Vision*, pages 1449–1457, 2015.

[583] S. G. Lingala and M. Jacob. Blind compressive sensing dynamic MRI. *IEEE Transactions on Medical Imaging*, 32(6):1132–1145, 2013.

[584] H. H. Liu, S. Schmidt, H. F. Poulsen, A. Godfrey, Z. Q. Liu, J. A. Sharon, and X. Huang. Three-dimensional orientation mapping in the transmission electron microscope. *Science*, 332(6031):833–834, 2011.

[585] Hongrong Liu, Lei Jin, Sok Boon S. Koh, Ivo Atanasov, Stan Schein, Lily Wu, and Z. Hong Zhou. Atomic structure of human adenovirus by cryo-EM reveals interactions among protein networks. *Science*, 329:1038–1043, 27 August 2010.

[586] Juan Liu and Pierre Moulin. Complexity-regularized denoising of Poisson-corrupted data. In *Proc. Intl. Conf. on Image Processing*, volume 3, pages 254–257, 2000.

[587] Lydia T. Liu, Edgar Dobriban, and Amit Singer. *e* pca: High dimensional exponential family pca. *arXiv preprint arXiv:1611.05550*, 2016.

[588] Xingchen Liu and Vadim Shapiro. Random heterogeneous materials via texture synthesis. *Computational Materials Science*, 99:177–189, 2015.

[589] Stuart Lloyd. Least squares quantization in PCM. *IEEE Transactions on Information Theory*, 28(2):129–137, 1982.

[590] N. Longbotham and G. Camps-Valls. A family of kernel anomaly change detectors. *Proc. 6th IEEE Workshop on Hyperspectral Signal and Image Processing: Evolution in Remote Sensing (WHISPERS)*, 2014.

[591] David G. Lowe. Object recognition from local scale-invariant features. In *Computer Vision, 1999. The Proceedings of the Seventh IEEE International Conference on*, volume 2, pages 1150–1157. IEEE, 1999.

[592] David G. Lowe. Distinctive image features from scale-invariant keypoints. *International Journal of Computer Vision*, 60(2):91–110, 2004.

[593] Binglin Lu and S. Torquato. Lineal-path function for random heterogeneous materials. *Physical Review A*, 45(2):922, 1992.

[594] Guowei Lu, Tianyue Zhang, Wenqiang Li, Lei Hou, Jie Liu, and Qihuang Gong. Single-molecule spontaneous emission in the vicinity of an individual gold nanorod. *J. Phys. Chem. C*, 115(32):15822–15828, 2011.

[595] X. Lu, Y. Yuan, and P. Yan. Alternatively constrained dictionary learning for image superresolution. *IEEE Transactions on Cybernetics*, 44(3):366–377, 2014.

[596] Nicholas Lubbers, Turab Lookman, and Kipton Barros. Inferring low-dimensional microstructure representations using convolutional neural networks. *Phys. Rev. E* 96:052111, 2017.

[597] D. Lunga, S. Prasad, M. M. Crawford, and O. Ersoy. Manifold-learning-based feature extraction for classification of hyperspectral data: A review of advances in manifold learning. *IEEE Signal Processing Magazine*, 31:55–66, Jan 2014.

[598] J. Luo. Stabilization of nanoscale quasi-liquid interfacial films in inorganic materials: A review and critical assessment. *Critical Reviews in Solid State and Material Sciences*, 32(1-2):67–109, 2007.

[599] J. Luo. Liquid-like interface complexion: From activated sintering to grain boundary diagrams. *Current Opinion in Solid State and Materials Science*, 12(5/6):81–88, 2008.

[600] J. Luo and Y.-M. Chiang. Wetting and prewetting on ceramic surfaces. *Annual Review of Materials Research*, 38:227–249, 2008.

[601] J. Luo, V. K. Gupta, D. H. Yoon, and H. M. Meyer. Segregation-induced grain boundary premelting in nickel-doped tungsten. *Applied Physics Letters*, 87:231902, 2005.

[602] Jian Luo. Developing interfacial phase diagrams for applications in activated sintering and beyond: Current status and future directions. *Journal of the American Ceramic Society*, 95(8):2358–2371, 2012.

[603] Jian Luo. Interfacial engineering of solid electrolytes. *Journal of Materiomics*, 1(1):22–32, 2015.

[604] Jian Luo. A short review of high-temperature wetting and complexion transitions with a critical assessment of their influence on liquid metal embrittlement and corrosion. *Corrosion*, 2015.

[605] Jian Luo, Huikai Cheng, Kaveh Meshinchi Asl, Christopher J. Kiely, and Martin P. Harmer. The role of a bilayer interfacial phase on liquid metal embrittlement. *Science*, 333(6050):1730–1733, 2011.

[606] Claude H. P. Lupis. *Chemical Thermodynamics of Materials*. Elsevier Science, 1983, page 581, 1983.

[607] M. Lustig, D. Donoho, and J. Pauly. Sparse MRI: The application of compressed sensing for rapid MR imaging. *Magn. Reson. Med.*, 58(6):1182–1195, 2007.

[608] M. Lustig, J. M. Santos, D. L. Donoho, and J. M. Pauly. k-t SPARSE: High frame rate dynamic MRI exploiting spatio-temporal sparsity. In *Proc. ISMRM*, page 2420, 2006.

[609] Kai Ma, Yunye Gong, Tangi Aubert, Melik Z. Turker, Teresa Kao, Peter C. Doerschuk, and Ulrich Wiesner. Surfactant micelle self-assembly directed highly symmetric ultrasmall inorganic cages. *Nature*, 558: 577–580: 2018. https://doi.org/10.1038/s41586-018-0221-0.

[610] L. Ma, M. M. Crawford, and J. Tian. Anomaly detection for hyperspectral images based on robust locally linear embedding. *J. Infrared Millimeter Terahertz Waves*, 31:753–762, 2010.

[611] S. Ma, K. Meshinchi Asl, C. Tansarawiput, P. R. Cantwell, M. Qi, M. P. Harmer, and J. Luo. A grain-boundary phase transition in si-au. *Scripta Materialia*, 66:203–206, 2012.

[612] Shuailei Ma, Patrick R. Cantwell, Timothy J. Pennycook, Naixie Zhou, Mark P. Oxley, Donovan N. Leonard, Stephen J. Pennycook, Jian Luo, and Martin P. Harmer. Grain boundary complexion transitions in wo3- and cuo-doped tio2 bicrystals. *Acta Materialia*, 61(5):1691–1704, 2013.

[613] I. MacLaren, R. M. Cannon, M. A. Glgn, R. Voytovych, N. P. Pogrion, C. Scheu, U. Tffner, and M. Rhle. Abnormal grain growth in alumina: Synergistic effects of yttria and silica. *Journal of the American Ceramic Society*, 86:650, 2003.

[614] J. P. MacSleyne, J. P. Simmons, and M. De Graef. On the use of 2-d moment invariants for the automated classification of particle shapes. *Acta Materialia*, 56(3):427–437, 2008.

[615] J. P. MacSleyne, J. P. Simmons, and M. De Graef. On the use of moment invariants for the auto-mated analysis of 3d particle shapes. *Modelling and Simulation in Materials Science and Engineering*, 16(4):045008, 2008.

[616] K. Magnusson, J. Jalden, P. Gilbert, and H. Blau. Global linking of cell tracks using the viterbi algo-rithm. *IEEE Transactions on Medical Imaging*, 2014.

[617] P. C. Mahalanobis. On the generalised distance in statistics. *Proc. National Institute of Sciences of India*, 2:49–55, 1936.

[618] J. Mairal, F. Bach, J. Ponce, and G. Sapiro. Online learning for matrix factorization and sparse coding. *J. Mach. Learn. Res.*, 11:19–60, 2010.

[619] J. Mairal, F. Bach, J. Ponce, G. Sapiro, and A. Zisserman. Non-local sparse models for image restora-tion. In *IEEE International Conference on Computer Vision*, pages 2272–2279, Sept 2009.

[620] J. Mairal, M. Elad, and G. Sapiro. Sparse representation for color image restoration. *IEEE Trans. on Image Processing*, 17(1):53–69, 2008.

[621] J. Mairal, G. Sapiro, and M. Elad. Learning multiscale sparse representations for image and video restoration. *SIAM Multiscale Modeling and Simulation*, 7(1):214–241, 2008.

[622] V. Majidzadeh, L. Jacques, A. Schmid, P. Vandergheynst, and Y. Leblebici. A (256x256) pixel 76.7mW CMOS imager/compressor based on real-time in-pixel compressive sensing. In *Proc. IEEE Int. Symp. Circuits and Systems*, pages 2956–2959, 2010.

[623] M. Mäkitalo and A. Foi. Optimal inversion of the Anscombe transformation in low-count Poisson image denoising. *IEEE Trans. Image Process.*, 20(1):99–109, 2011.

[624] K. Malczewski. PET image reconstruction using compressed sensing. In *Signal Processing: Algo-rithms, Architectures, Arrangements and Applications (SPA), 2013*, pages 176–181, Sept 2013.

[625] S. Mallat. *A Wavelet Tour of Signal Processing*. Academic Press, 1999.

[626] S. Mallat. *A Wavelet Tour of Signal Processing: The Sparse Way*. Academic Press, 3rd edition, 2009.

[627] S. G. Mallat and Z. Zhang. Matching pursuits with time-frequency dictionaries. *IEEE Trans. Signal Proc.*, 41(12):3397–3415, December 1993.

[628] S. G. Mallat and Z. Zhang. Matching pursuits with time-frequency dictionaries. *IEEE Transactions on Signal Processing*, 41(12):3397–3415, 1993.

[629] Matt Malloy, Brad Thiel, Benjamin D Bunday, Stefan Wurm, Maseeh Mukhtar, Kathy Quoi, Thomas Kemen, Dirk Zeidler, Anna Lena Eberle, and Tomasz Garbowski. Massively parallel e-beam inspec-tion: enabling next-generation patterned defect inspection for wafer and mask manufacturing. In *SPIE Advanced Lithography*, pages 942319–942319–10. International Society for Optics and Photonics, 2015.

[630] H. S. Malvar and D. H. Staelin. The LOT: Transform coding without blocking effects. *IEEE Trans. Acoustics, Speech, and Signal Proc.*, 37:553–559, April 1989.

[631] A. G. Mamistvalov. n-Dimensional moment invariants and conceptual mathematical theory of recog-nition n-dimensional solids. *IEEE Trans. Pattern Analysis and Machine Intelligence*, 20(8):819–831, August 1998.

[632] L. M. Manevitz and M. Yousef. One-class SVMs for document classification. *J. Machine Learning Res.*, 2:139–154, 2001.

[633] C. Manikopoulos and S. Papavassiliou. Network intrusion and fault detection: a statistical anomaly approach. *IEEE Communications Magazine*, 40(10):76–82, 2002.

[634] D. Manolakis, D. Marden, J. Kerekes, and G. Shaw. On the statistics of hyperspectral imaging data. *Proc. SPIE*, 4381:308–316, 2001.

[635] M. Mansuripur. Computation of electron diffraction patterns in Lorentz electron microscopy of thin magnetic films. *J. Appl. Phys.*, 69:2455–2464, 1991.

[636] C. Manwart, S. Torquato, , and R. Hilfer. Stochastic reconstruction of sandstones. *Physical Review E*, 62:893–899, 2000.

[637] W. Marais and R. Willett. Proximal-gradient methods for Poisson image reconstruction with BM3D-based regularization. In *IEEE International Workshop on Computational Advances in Multi-Sensor Adaptive Processing, CAMSAP'17*, 2017.

[638] M. W. Marcellin, M. J. Gormish, A. Bilgin, and M. P. Boliek. An overview of JPEG-2000. In *Proc. Data Compression Conf.*, pages 523–541, 2000.

[639] R. F. Marcia, Z. T. Harmany, and R. M. Willett. Compressive coded aperture imaging. In *Proc. SPIE Conference on Computational Imaging VII*, volume 7246, pages 72460G–1–13, January 2009.

[640] R. F. Marcia, C. Kim, C. Eldeniz, J. Kim, D. J. Brady, and R. M. Willett. Superimposed video disambiguation for increased field of view. *Opt. Express*, 16(21):16352–16363, 2008.

[641] R. F. Marcia, C. Kim, J. Kim, D. Brady, and R. M. Willett. Fast disambiguation of superimposed images for increased field of view. In *Proc. IEEE Int. Conf. Image Proc.*, pages 2620–2623, 2008.

[642] R. F. Marcia and R. M. Willett. Compressive coded aperture superresolution image reconstruction. In *Proc. IEEE Int. Conf. Acoust. Speech Sig. Proc.*, pages 833–836, April 2008.

[643] Kanti V. Mardia and Peter E. Jupp. *Directional Statistics*, volume 494. John Wiley & Sons, 2009.

[644] G. Mariethoz and S. Lefebvre. Bridges between multiple-point geostatistics and texture synthesis. *Comput. Geosci.*, 66(C):66–80, 2014.

[645] Marcio M. Marim, Michael Atlan, Elsa Angelini, and Jean-Christophe Olivo-Marin. Compressed sensing with off-axis frequency-shifting holography. *Optics Letters*, 35(6):871–873, 2010.

[646] Markos Markou and Sameer Singh. Novelty detection: A review – part 1: statistical approaches. *Signal Processing*, 83:2481–2497, 2003.

[647] Markos Markou and Sameer Singh. Novelty detection: a review – part 2: neural network based approaches. *Signal Processing*, 83:2499–2521, 2003.

[648] Benji Maruyama, Jonathan E. Spowart, Daylond J. Hooper, Herbert M. Mullens, Adriana M. Druma, Calin Druma, and M. Khairul Alam. A new technique for obtaining three-dimensional structures in pitch-based carbon foams. *Scripta Materialia*, 54(9):1709–1713, 2006.

[649] Jeremy K. Mason and Christopher A. Schuh. Representations of texture. In Adam J. Schwartz, Mukul Kumar, Brent L. Adams, and David P. Field, editors, *Electron Backscatter Diffraction in Materials Science*, pages 35–51. Springer US, 2009.

[650] J. K. Mason and C. A. Schuh. Hyperspherical harmonics for the representation of crystallographic texture. *Acta Materialia*, 56(20):6141 – 6155, 2008.

[651] Kremer J. R. and D. N. Mastronarde and J. R. McIntosh. Computer visualization of three-dimensional image data using IMOD. *Journal of Structural Biology*, 116:71–76, August 1996.

[652] Jiri Matas, Ondrej Chum, Martin Urban, and Tomás Pajdla. Robust wide-baseline stereo from maximally stable extremal regions. *Image and Vision Computing*, 22(10):761–767, 2004.

[653] G. F. Mathey. Method of making superalloy turbine disks having graded coarse and fine grains, May 17 1994. US Patent 5,312,497.

[654] Mathworks URL. http://www.mathworks.com/.

[655] Tsutomu Matsui, Gabriel C. Lander, Reza Khayat, and John E. Johnson. Subunits fold at position-dependent rates during maturation of a eukaryotic RNA virus. *Proc. Nat. Acad. Sci. U.S.A.*, 107(32):14111–14115, 10 Aug. 2010.

[656] S. Matteoli, M. Diani, and G. Corsini. Improved estimation of local background covariance matrix for anomaly detection in hyperspectral images. *Optical Engineering*, 49:046201, 2010.

[657] S. Matteoli, M. Diani, and G. Corsini. A tutorial overview of anomaly detection in hyperspectral images. *IEEE A&E Systems Magazine*, 25:5–27, 2010.

[658] S. Matteoli, M. Diani, and G. Corsini. Hyperspectral anomaly detection with kurtosis-driven local covariance matrix corruption mitigation. *IEEE Geoscience and Remote Sensing Lett.*, 8:532–536, 2011.

[659] S. Matteoli, M. Diani, and G. Corsini. Impact of signal contamination on the adaptive detection performance of local hyperspectral anomalies. *IEEE Trans. Geoscience and Remote Sensing*, 52:1948–1968, 2014.

[660] S. Matteoli, M. Diani, and J. Theiler. An overview background modeling for detection of targets and anomalies in hyperspectral remotely sensed imagery. *IEEE J. Sel. Topics in Applied Earth Observations and Remote Sensing*, 7:2317–2336, 2014.

[661] S. Matteoli, T. Veracini, M. Diani, and G. Corsini. Background density nonparametric estimation with data-adaptive bandwidths for the detection of anomalies in multi-hyperspectral imagery. *IEEE Geoscience and Remote Sensing Lett.*, 11:163–167, 2014.

[662] S. Matteoli, T. Veracini, M. Diani, and G. Corsini. A locally adaptive background density estimator: An evolution for RX-based anomaly detectors. *IEEE Geoscience and Remote Sensing Lett.*, 11:323–327, 2014.

[663] S. Matthies and G. W. Vinel. On the reproduction of the orientation distribution function of texturized samples from reduced pole figures using the conception of a conditional ghost correction. *Physica Status Solidi (b)*, 112(2):K111–K114, 1982.

[664] R. Mayer, F. Bucholtz, and D. Scribner. Object detection by using "whitening/dewhitening" to transform target signatures in multitemporal hyperspectral and multispectral imagery. *IEEE Trans. Geoscience and Remote Sensing*, 41:1136–1142, 2003.

[665] Ad-hoc Interagency Group on Advanced Materials. Materials genome initiative for global competitiveness. *National Science and Technology Council OSTP*, pages 1–18, July 2011.

[666] Mark A. McCord, Paul Petric, Upendra Ummethala, Allen Carroll, Shinichi Kojima, Luca Grella, Sameet Shriyan, Charles T Rettner, and Chris F Bevis. Rebl: Design progress toward 16 nm half-pitch maskless projection electron beam lithography. In *SPIE Advanced Lithography*, pages 832311–832311–11. International Society for Optics and Photonics, 2012.

[667] David L. McDowell and Surya R. Kalidindi. The materials innovation ecosystem: A key enabler for the materials genome initiative. *MRS Bulletin*, 41(04):326–337, 2016.

[668] D. L. McDowell, S. Ghosh, and S. R. Kalidindi. Representation and computational structure-property relations of random media. *JOM Journal of the Minerals, Metals and Materials Society*, 63(3):45–51, 2011.

[669] I. M. McKenna, S. O. Poulsen, E. M. Lauridsen, W. Ludwig, and P. W. Voorhees. Grain growth in four dimensions: A comparison between simulation and experiment. *Acta Materialia*, 78:125–134, 2014.

[670] Geoffrey J. McLachlan and Thriyambakam Krishnan. *The EM Algorithm and Extensions*. Wiley-Interscience, 1997.

[671] G. McMullan, A. R. Faruqi, R. Henderson, N. Guerrini, R. Turchetta, A. Jacobs, and G. Van Hoften. Experimental observation of the improvement in mtf from backthinning a cmos direct electron detector. *Ultramicroscopy*, 109(9):1144–1147, 2009.

[672] K. R. Mecke. Morphological characterization of patterns in reaction-diffusion systems. *Physical Review E*, 53(5):4794, 1996.

[673] Nicolai Meinshausen and Peter Bühlmann. High-dimensional graphs and variable selection with the lasso. *Annals of Statistics*, 34(3):1436–1462, 2006.

[674] R. Mendoza, K. Thornton, I. Savin, and P. W. Voorhees. The evolution of interfacial topology during coarsening. *Acta Materialia*, 54(3):743–750, 2006.

[675] J. Meola and M. T. Eismann. Image misregistration effects on hyperspectral change detection. *Proc. SPIE*, 6966:69660Y, 2008.

[676] Domingo Mery and Miguel Angel Berti. Automatic detection of welding defects using texture features. *Insight-Non-Destructive Testing and Condition Monitoring*, 45(10):676–681, 2003.

[677] N. Mevenkamp, P. Binev, W. Dahmen, P. M. Voyles, A. B. Yankovich, and B. Berkels. Poisson noise removal from high-resolution stem images based on periodic block matching. *Advanced Structural and Chemical Imaging*, 1(1):1–19, 2015.

[678] R. R. Meyer, A. I. Kirkland, R. E. Dunin-Borkowski, and J. L. Hutchison. Experimental characterisation of CCD cameras for HREM at 300 kv. *Ultramicroscopy*, 85(1):9–13, 2000.

[679] Joseph R. Michael, Craig Y. Nakakura, Tomasz Garbowski, Anna Lena Eberle, Thomas Kemen, and Dirk Zeidler. High-throughput sem via multi-beam sem: Applications in materials science. *Microscopy and Microanalysis*, 21(S3):697–698, 2015.

[680] P. A. Midgley and M. Weyland. 3D electron microscopy in the physical sciences: The development of z-contrast and eftem tomography. *Ultramicroscopy*, 96(34):413–431, 2003.

[681] Paul A. Midgley and Rafal E. Dunin-Borkowski. Electron tomography and holography in materials science. *Nature Materials*, 8:271–280, 2009.

[682] Krystian Mikolajczyk and Cordelia Schmid. An affine invariant interest point detector. In *Computer VisionECCV 2002*, pages 128–142. Springer, 2002.

[683] Krystian Mikolajczyk and Cordelia Schmid. Scale & affine invariant interest point detectors. *International Journal of Computer Vision*, 60(1):63–86, 2004.

[684] Krystian Mikolajczyk and Cordelia Schmid. A performance evaluation of local descriptors. *IEEE Transactions on Pattern Analysis and Machine Intelligence*, 27(10):1615–1630, 2005.

[685] Anton Milan, Stefan Roth, and Konrad Schindler. Continuous energy minimization for multitarget tracking. *IEEE Transactions on Pattern Analysis and Machine Intelligence*, 36(1):58–72, 2014.

[686] P. Milanfar. A tour of modern image filtering. *IEEE Signal Processing Magazine*, 30(1):106-128, 2013.

[687] Benjamin Kyle Miller. *Development and Application of Operando TEM to a Ruthenium Catalyst for CO Oxidation*. PhD thesis, Arizona State University, 2016.

[688] Michael K. Miller, Thomas F. Kelly, Krishna Rajan, and Simon P. Ringer. The future of atom probe tomography. *Materials Today*, 15(4):158–165, 2012.

[689] Florica Mindru, Tinne Tuytelaars, Luc Van Gool, and Theo Moons. Moment invariants for recognition under changing viewpoint and illumination. *Computer Vision and Image Understanding*, 94(1):3–27, 2004.

[690] Tian Ming, Huanjun Chen, Ruibin Jiang, Qian Li, and Jianfang Wang. Plasmon-controlled fluorescence: Beyond the intensity enhancement. *J. Phys. Chem. Lett.*, 3(2):191–202, 2011.

[691] M. Mishali and Y. Eldar. Xampling: Compressed sensing of analog signals. In Y. Eldar and G. Kutyniok, editors, *Compressed Sensing: Theory and Applications*. Cambridge University Press, 2012.

[692] Y. Mishin, W. J. Boettinger, J. A. Warren, and G. B. McFadden. Thermodynamics of grain boundary premelting in alloys. i. phase-field modeling. *Acta Materialia*, 57(13):3771–3785, 2009.

[693] K. Miyazawa, Y. Iwasaki, K. Ito, and Y. Ishida. Combination rule of Σ values at triple junctions in cubic polycrystals. *Acta Crystallographica Section A: Foundations of Crystallography*, 52(6):787–796, 1996.

[694] Maher Moakher. Means and averaging in the group of rotations. *SIAM Journal on Matrix Analysis and Applications*, 24(1):1–16, 2002.

[695] M. E. Mochel and J. M. Mochel. A ccd imaging and analysis system for the vg hb5 stem. *Proc. of the 44 th Ann. Meet. EMSA*, pages 616–617, 1986.

[696] A. Mohammad-Djafari. Joint estimation of parameters and hyperparameters in a Bayesian approach of solving inverse problems. In *Image Processing, 1996. Proceedings., International Conference on*, volume 1, pages 473–476 vol.2, September 1996.

[697] K. Aditya Mohan, S. V. Venkatakrishnan, J. W. Gibbs, E. B. Gulsoy, X. Xiao, M. De Graef, P. W. Voorhees, and C. A. Bouman. TIMBER: A method for time-space reconstruction from interlaced views. *IEEE Trans. on Computational Imaging*, 1(2):96 – 111, 2015.

[698] D. A. Molodov, G. Gottstein, F. Heringhaus, and L. S. Shvindlerman. Motion of planar grain boundaries in bismuth bicrystals driven by a magnetic field. *Scripta Materialia*, 37(8):1207–1213, 1997.

[699] R. L. Moment and R. B. Gordon. Energy of grain boundaries in halite. *Journal of the American Ceramic Society*, 47(11):570–573, 1964.

[700] H. Monajemi, S. Jafarpour, M. Gavish, and D. L. Donoho. Deterministic matrices matching the compressed sensing phase transitions of Gaussian random matrices. *Proc. Natl. Acad. Sci. USA*, doi: 10.1073/pnas.1219540110, 2013.

[701] T. K. Moon. The expectation-maximization algorithm. *Signal Processing Magazine, IEEE*, 13(6):47–60, 1996.

[702] Paul Mooney. Optimization of image collection for cellular electron microscopy. *Methods in Cell Biology*, 79:661–719, 2007.

[703] Paul E. Mooney, Ming Pan, Gerald Van Hoften, and Felix De Haas. Quantitative evaluation of fiber-optically coupled ccd cameras for use in cryo-microscopy. *Microscopy and Microanalysis*, 10(S02):168–169, 2004.

[704] A. Morawiec. Method to calculate the grain boundary energy distribution over the space of macroscopic boundary parameters from the geometry of triple junctions. *Acta Materialia*, 48(13):3525–3532, 2000.

[705] A. Morawiec. *Orientations and Rotations: Computations in Crystallographic Textures*. Berlin, Heidelberg, New York: Springer-Verlag., 2004.

[706] T. Mori, H. Miura, T. Tokita, J. Haji, and M. Kato. Determination of the energies of [001] twist boundaries in Cu with the shape of boundary SiO_2 particles. *Philosophical Magazine Letters*, 58(1):11–15, 1988.

[707] A. Mousavi and R. G. Baraniuk. Learning to invert: Signal recovery via deep convolutional networks. In *Proc. IEEE Int. Conf. Acoust. Speech Sig. Proc.*, March 2017.

[708] Tim Mueller, Aaron Gilad Kusne, and Rampi Ramprasad. Machine learning in materials science: Recent progress and emerging applications. *Reviews in Computational Chemistry*, 29:186, 2016.

[709] Stefan Mühlig, Alastair Cunningham, José Dintinger, Toralf Scharf, Thomas Bürgi, Falk Lederer, and Carsten Rockstuhl. Self-assembled plasmonic metamaterials. *Nanophotonics: Review*, 2(3):211–240, 2013.

[710] R. Muise. Compressive imaging: An application. *SIAM Journal on Imaging Sciences*, 2:1255–1276, 2009.

[711] W. W. Mullins. Theory of thermal grooving. *Journal of Applied Physics*, 28(3):333–339, 1957.

[712] W. W. Mullins. Grain-boundary grooving by volume diffusion. *Trans. AIME*, 218(4):8, 1960.

[713] W. W. Mullins. Magnetically induced grain-boundary motion in bismuth. *Acta Metallurgica*, 4(4):421–432, 1956.

[714] James R. Munkres. *Elements of Algebraic Topology*, volume 2. Addison-Wesley Reading, 1984.

[715] James R. Munkres. *Topology*. Prentice Hall, 2000.

[716] D. D. Muresan and T. W. Parks. Adaptive principal components and image denoising. In *ICIP*, pages 101–104, 2003.

[717] Y. Naderahmadian, S. Beheshti, and M. A. Tinati. Correlation based online dictionary learning algorithm. *IEEE Transactions on Signal Processing*, 64(3):592–602, 2016.

[718] Vinod Nair and Geoffrey E Hinton. Rectified linear units improve restricted boltzmann machines. In *Proceedings of the 27th International Conference on Machine Learning (ICML-10)*, pages 807–814, 2010.

[719] N. M. Nasrabadi. Regularization for spectral matched filter and RX anomaly detector. *Proc. SPIE*, 6966:696604, 2008.

[720] N. M. Nasrabadi. Kernel subspace-based anomaly detection for hyperspectral imagery. *Proc. 1st IEEE Workshop on Hyperspectral Signal and Image Processing: Evolution in Remote Sensing (WHISPERS)*, 2009.

[721] B. K. Natarajan. Sparse approximate solutions to linear systems. *SIAM J. Comput.*, 24(2):227–234, April 1995.

[722] F. Natterer. *The Mathematics of Computerized Tomography*, volume 32. Society for Industrial and Applied Mathematics, Philadelphia, 2001.

[723] F. Natterer. *The Mathematics of Computerized Tomography*. SIAM, 2001.

[724] D. Needell and J. Tropp. COSAMP: Iterative signal recovery from incomplete and inaccurate measurements. *Appl. and Comp. Harmonic Analysis*, 26(3):301–321, May 2009.

[725] D. Needell and J. A. Tropp. Cosamp: Iterative signal recovery from incomplete and inaccurate samples. *Applied and Computational Harmonic Analysis*, 26(3):301–321, 2009.

[726] D. Needell and R. Vershynin. Signal recovery from incomplete and inaccurate measurements via regularized orthogonal matching pursuit. *IEEE J. Selected Topics in Sig. Proc.*, 4(2):310–316, April 2010.

[727] R. Neelamani, C. E. Krohn, J. R. Krebs, J. Romberg, M. Deffenbaugh, and J. E. Anderson. Efficient seismic forward modeling using simultaneous random sources and sparsity. *Geophysics*, 75(6):WB15–WB27, 2010.

[728] Sahand Negahban and Martin J. Wainwright. Estimation of (near) low-rank matrices with noise and high-dimensional scaling. *Annals of Statistics*, 39(2):1069–1097, 2011.

[729] D. Nepal, M. S. Onses, K. Park, M. Jespersen, C. J. Thode, P. F. Nealey, and R. A. Vaia. Control over position, orientation, and spacing of arrays of gold nanorods using chemically nanopatterned surfaces and tailored particleparticlesurface interactions. *ACS Nano*, 6:5693, 2012.

[730] Dhriti Nepal, Lawrence F. Drummy, Sushmita Biswas, Kyoungweon Park, and Richard A. Vaia. Large scale solution assembly of quantum dotgold nanorod architectures with plasmon enhanced fluorescence. *ACS Nano*, 2013.

[731] Dhriti Nepal, Lawrence F. Drummy, Sushmita Biswas, Kyoungweon Park, and Richard A. Vaia. Large scale solution assembly of quantum dotgold nanorod architectures with plasmon enhanced fluorescence. *ACS Nano*, 7(10):9064–9074, 2013.

[732] Dhriti Nepal, Kyoungweon Park, and Richard A. Vaia. High-yield assembly of soluble and stable gold nanorod pairs for high-temperature plasmonics. *Small*, 8(7):1013–1020, 2012.

[733] Lily Nguyen, Dong Wang, Yunzhi Wang, and Marc De Graef. Quantifying the abnormal strain state in ferroelastic materials: A moment invariant approach. *Acta Materialia*, 94:172–180, 2015.

[734] Carlton W. Niblack, Ron Barber, Will Equitz, Myron D. Flickner, Eduardo H. Glasman, Dragutin Petkovic, Peter Yanker, Christos Faloutsos, and Gabriel Taubin. Qbic project: querying images by content, using color, texture, and shape. In *Proceedings of the SPIE*, volume 1908, pages 173–187, 1993.

[735] A. A. Nielsen, K. Conradsen, and J. J. Simpson. Multivariate alteration detection (MAD) and MAF post-processing in multispectral bi-temporal image data: New approaches to change detection studies. *Remote Sensing of the Environment*, 64:1–19, 1998.

[736] S. R. Niezgoda, D. T. Fullwood, and S. R. Kalidindi. Delineation of the space of 2-point correlations in a composite material system. *Acta Materialia*, 56(18):5285–5292, 2008.

[737] S. R. Niezgoda, Y. C. Yabansu, and S. R. Kalidindi. Understanding and visualizing microstructure and microstructure variance as a stochastic process. *Acta Materialia*, 2011.

[738] Stephen R. Niezgoda and Jared Glover. Unsupervised learning for efficient texture estimation from limited discrete orientation data. *Metallurgical and Materials Transactions A*, 44(11):4891–4905, 2013.

[739] Stephen R. Niezgoda and Jared Glover. Unsupervised learning for efficient texture estimation from limited discrete orientation data. *Metallurgical and Materials Transactions A*, 44(11):4891–4905, 2013.

[740] Stephen R. Niezgoda and Surya R. Kalidindi. Applications of the phase-coded generalized hough transform to feature detection, analysis, and segmentation of digital microstructures. *Computers, Materials, & Continua*, 14(2):79–98, 2010.

[741] Stephen R. Niezgoda, Anand K. Kanjarla, and Surya R. Kalidindi. Novel microstructure quantification framework for databasing, visualization, and analysis of microstructure data. *Integrating Materials and Manufacturing Innovation*, 2(1):1–27, 2013.

[742] Stephen R. Niezgoda, Eric A. Magnuson, and Jared Glover. Symmetrized bingham distribution for representing texture: parameter estimation with respect to crystal and sample symmetries. *Journal of Applied Crystallography*, 49(4), 2016.

[743] Stephen R. Niezgoda, Yuksel C. Yabansu, and Surya R. Kalidindi. Understanding and visualizing microstructure and microstructure variance as a stochastic process. *Acta Materialia*, 59(16):6387–6400, 2011.

[744] B. Nikoobakht and M. A. El-Sayed. Preparation and growth mechanism of gold nanorods using seed-mediated growth method. *Chem. Mater.*, 15(10):1957–1962, 2003.

[745] Manoj Nirmal and Louis Brus. Luminescence photophysics in semiconductor nanocrystals. *Acc. Chem. Res.*, 32(5):407–414, 1998.

[746] Harry Nyquist. Certain topics in telegraph transmission theory. *Transactions of the A.I.E.E.*, 28-29:617–644, 1928.

[747] OIM. *Orientation Imaging Microscopy v6 OIMDC 6.1.3.* TSL EDAX Inc., Mahwah, New Jersey, 2011.

[748] Timo Ojala, Matti Pietikäinen, and Topi Mäenpää. Multiresolution gray-scale and rotation invariant texture classification with local binary patterns. *Pattern Analysis and Machine Intelligence, IEEE Transactions on*, 24(7):971–987, 2002.

[749] Atsuyuki Okabe, Barry Boots, Kokichi Sugihara, and Sung Nok Chiu. *Spatial Tessellations: Concepts and Applications of Voronoi Diagrams*, volume 501. John Wiley & Sons, 2009.

[750] Kenji Okuma, Ali Taleghani, Nando De Freitas, James J Little, and David G Lowe. *A Boosted Particle Filter: Multitarget Detection and Tracking*, pages 28–39. Springer Berlin Heidelberg, 2004.

[751] B. A. Olshausen and D. J. Field. Emergence of simple-cell receptive field properties by learning a sparse code for natural images. *Nature*, 381:607–609, 1996.

[752] B. A. Olshausen and D. J. Field. Emergence of simple-cell receptive field properties by learning a sparse code for natural images. *Nature*, 381(6583):607–609, 1996.

[753] Colm T. O'Mahony, Richard A. Farrell, Tandra Goshal, Justin D. Holmes, and Michael A. Morris. The thermodynamics of defect formation in self-assembled systems, thermodynamics - systems in equilibrium and non-equilibrium. In Juan Carlos Moreno Piraján, editor, *Thermodynamics - Systems in Equilibrium and Non-Equilibrium*, chapter 13. InTech, 2011.

[754] B. Ophir, M. Lustig, and M. Elad. Multi-scale dictionary learning using wavelets. *IEEE Journal of Selected Topics in Signal Processing*, 5(5):1014–1024, 2011.

[755] D. Paganin and K. A. Nugent. Noninterferometric phase imaging with partially coherent light. *Phys. Rev. Lett.*, 80:2586–2589, 1998.

[756] R. Paget and I. Longstaff. Texture synthesis via a noncausal nonparametric multiscale Markov random field. *IEEE Transactions on Image Processing*, 7(6):925–931, June 1998.

[757] M. Pan, C. M. Czarnik, Y. C. Pan, R. Sougrat, K. Li, Y. Han, and J. Ciston. High resolution imaging of zif-8 metal-organic framework structure with a low-dose electron counting direct detection camera. In *18th International Microscopy Conference*, 2014.

[758] Ming Pan and Cory Czarnik. Image detectors for environmental transmission electron microscopy (etem). In *Controlled Atmosphere Transmission Electron Microscopy*, pages 143–164. Springer, 2016.

[759] S. Panaro, F. De Angelis, and A. Toma. Dark and bright mode hybridization: From electric to magnetic fano resonances. *Optics and Lasers in Engineering*, 76:64–69, 2016.

[760] Jitesh H. Panchal, Surya R. Kalidindi, and David L. McDowell. Key computational modeling issues in integrated computational materials engineering. *Computer-Aided Design*, 45(1):4–25, 2013.

[761] João P. Papa, Rodrigo Y. M. Nakamura, Victor Hugo C. De Albuquerque, Alexandre X. Falcão, and João Manuel R. S. Tavares. Computer techniques towards the automatic characterization of graphite particles in metallographic images of industrial materials. *Expert Systems with Applications*, 40(2):590–597, 2013.

[762] F. Papillon, G. S. Rohrer, and P. Wynblatt. Effect of segregating impurities on the grain-boundary character distribution of magnesium oxide. *Journal of the American Ceramic Society*, 92(12):3044–3051, 2009.

[763] Athanasios Papoulis and S. Unnikrishna Pillai. *Probability, Random Variables, and Stochastic Processes*. Tata McGraw-Hill Education, 2002.

[764] N. Parikh and S. Boyd. *Proximal Algorithms.* Foundations and Trends in Optimization. NOW, 2013.

[765] Neal Parikh, Stephen P. Boyd, et al. Proximal algorithms. *Foundations and Trends in Optimization*, 1(3):127–239, 2014.

[766] Kyoungweon Park, Lawrence F. Drummy, Robert C. Wadams, Hilmar Koerner, Dhriti Nepal, Laura Fabris, and Richard A. Vaia. Growth mechanism of gold nanorods. *Chem. Mater.*, 25(4):555–563, 2013.

[767] Kyoungweon Park, Hilmar Koerner, and Richard A. Vaia. Depletion-induced shape and size selection of gold nanoparticles. *Nano Lett.*, 10(4):1433–1439, 2010.

[768] Jason T. Parker, Volkan Cevher, and Philip Schniter. Compressive sensing under matrix uncertainties: An approximate message passing approach. In *Signals, Systems and Computers (ASILOMAR), 2011 Conference Record of the Forty Fifth Asilomar Conference on*, pages 804–808. IEEE, 2011.

[769] Lance Parsons, Ehtesham Haque, and Huan Liu. Subspace clustering for high dimensional data: a review. *ACM SIGKDD Explorations Newsletter*, 6(1):90–105, 2004.

[770] E. Parzen. On estimation of a probability density function and mode. *Ann. Math. Statist.*, 33:1065–1076, 1962.

[771] Y. Pati, R. Rezaiifar, and P. Krishnaprasad. Orthogonal matching pursuit : recursive function approximation with applications to wavelet decomposition. In *Asilomar Conf. on Signals, Systems and Comput.*, pages 40–44 vol.1, 1993.

[772] K. Pawlik. Determination of the orientation distribution function from pole figures in arbitrarily defined cells. *Physica Status Solidi (b)*, 134(2):477–483, 1986.

[773] O. Pele and M. Werman. The quadratic-chi histogram distance family. In Kostas Daniilidis, Petros Maragos, and Nikos Paragios, editors, *Computer Vision – ECCV 2010*, volume 6312 of *Lecture Notes in Computer Science*, pages 749–762. Springer Berlin Heidelberg, 2010.

[774] S. Peleg, M. Werman, and H. Rom. A unified approach to the change of resolution: Space and gray-level. *IEEE Trans. Pattern Anal. Mach. Intell.*, 11(7):739–742, July 1989.

[775] W. B. Pennbaker and J. L. Mitchell. *JPEG Still Image Data Compression Standard*. Springer, 1993.

[776] H Perlwitz, K Lücke, and W Pitsch. Determination of the orientation distribution of the crystallites in rolled copper and brass by electron microscopy. *Acta Metallurgica*, 17(9):1183 – 1195, 1969.

[777] Pietro Perona and Jitendra Malik. Scale-space and edge detection using anisotropic diffusion. *Pattern Analysis and Machine Intelligence, IEEE Transactions on*, 12(7):629–639, 1990.

[778] Florent Perronnin and Christopher Dance. Fisher kernels on visual vocabularies for image categorization. In *Computer Vision and Pattern Recognition, 2007. CVPR'07. IEEE Conference on*, pages 1–8. IEEE, 2007.

[779] Florent Perronnin, Jorge Sánchez, and Thomas Mensink. Improving the fisher kernel for large-scale image classification. In *European Conference on Computer Vision*, pages 143–156. Springer, 2010.

[780] R. Chartrand, personal communication with lead author, 2015.

[781] E. F. Pettersen, T. D. Goddard, C. C. Huang, G. S. Couch, D. M. Greenblatt, E. C. Meng, and T. E. Ferrin. UCSF Chimera—A visualization system for exploratory research and analysis. *J. Comput. Chem.*, 25(13):1605–1612, 2004.

[782] Tanapon Phenrat, Navid Saleh, Kevin Sirk, Robert D. Tilton, and Gregory V. Lowry. Aggregation and sedimentation of aqueous nanoscale zerovalent iron dispersions. *Environmental Science & Technology*, 41(1):284–290, 2007.

[783] James Philbin, Ondřej Chum, Michael Isard, Josef Sivic, and Andrew Zisserman. Lost in quantization: Improving particular object retrieval in large scale image databases. In *Computer Vision and Pattern Recognition, 2008. CVPR 2008. IEEE Conference on*, pages 1–8. IEEE, 2008.

[784] P. J. Phillips, M. Mills, and M. De Graef. Systematic row and zone axis STEM defect image simulations. *Philosophical Magazine A*, 91:2081–2101, 2011.

[785] Ryszard Piasecki. Entropic measure of spatial disorder for systems of finite-sized objects. *Physica A: Statistical Mechanics and Its Applications*, 277(1):157–173, 2000.

[786] Ryszard Piasecki. Microstructure reconstruction using entropic descriptors. In *Proceedings of the Royal Society of London A: Mathematical, Physical and Engineering Sciences*, volume 467, pages 806–820. The Royal Society, 2011.

[787] Y. N. Picard, M. Liu, J. Lammatao, R. Kamaladasa, and M. De Graef. Theory of dynamical electron channeling contrast images of near-surface crystal defects. *Ultramicroscopy*, 146:71–78, 2014.

[788] Mark R. Pickering, John F. Arnold, and Michael R. Frater. An adaptive search length algorithm for block matching motion estimation. *Circuits and Systems for Video Technology, IEEE Transactions on*, 7(6):906–912, 1997.

[789] Ramana M. Pidaparti, Babak Seyed Aghazadeh, Angela Whitfield, A. S. Rao, and Gerald P. Mercier. Classification of corrosion defects in nial bronze through image analysis. *Corrosion Science*, 52(11):3661–3666, 2010.

[790] D. Pimentel-Alarcón, L. Balzano, R. Marcia, R. Nowak, and R. Willett. Group-sparse subspace clustering with missing data. In *Proc. Statistical Signal Processing Workshop*, 2016.

[791] Hamed Pirsiavash, Deva Ramanan, and Charless C. Fowlkes. Globally-optimal greedy algorithms for tracking a variable number of objects. In *IEEE Conference on Computer Vision and Pattern Recognition*, pages 1201–1208, 2011.

[792] Steve Plimpton, Paul Crozier, and Aidan Thompson. Lammps-large-scale atomic/molecular massively parallel simulator. *Sandia National Laboratories*, 18, 2007.

[793] J. P. W. Pluim, J. B. A. Maintz, and M. A. Viergever. Mutual-information-based registration of medical images: A survey. *IEEE Transactions on Medical Imaging*, 22:986–1004, 2003.

[794] P. P. Pompa, L. Martiradonna, A. Della Torre, F. Della Sala, L. Manna, M. De Vittorio, F. Calabi, R. Cingolani, and R. Rinaldi. Metal-enhanced fluorescence of colloidal nanocrystals with nanoscale control. *Nat. Nano.*, 1(2):126–130, 2006.

[795] K. Popat and R. Picard. Novel cluster-based probability model for texture synthesis, classification, and compression. In *Visual Communications and Image Processing*, pages 756–768, 1993.

[796] Alexandra E. Porter, Mhairi Gass, Karin Muller, Jeremy N. Skepper, Paul A. Midgley, and Mark Welland. Direct imaging of single-walled carbon nanotubes in cells. *Nature Nanotechnology*, 2:713 – 717, 2007.

[797] J. Pospiech, K. Sztwiertnia, and F. Haessner. The misorientation function. *Text Microstruct*, 6:201–15, 1986.

[798] James F. Price. Lagrangian and Eulerian representations of fluid flow: Kinematics and the equations of motion. Woods Hole Oceanographic Institution, Woods Hole, MA, 2543, 2006.

[799] M. Protter and Michael Elad. Image sequence denoising via sparse and redundant representations. *IEEE Trans. on Image Processing*, 18(1):27–36, 2009.

[800] M. Prutton, D. K. Wilkinson, Peter G. Kenny, and D. L. Mountain. Data processing for spectrum-images: extracting information from the data mountain. *Applied Surface Science*, 144:1–10, 1999.

[801] C. Przybyla, T. Godar, J. P. Simmons, M. Jackson, L. P. Zawada, and J. Pearce. Statistical characterization of sic/sic ceramic matrix composites at the filament scale with bayesian segmentation hough transform feature extraction, and pair correlation statistics. In *International SAMPE Technical Conference*, pages 859–878, 2013.

[802] Craig Przybyla, Stephen Bricker, Jeffrey P. Simmons, and Russell Hardie. Automated microstructure-properties characterization and simulation in brittle matrix continuous fiber reinforced composites. *American Society for Composites 29th Technical Conference on Composite Materials*, 2014.

[803] Jinyi Qi and Richard M. Leahy. Iterative reconstruction techniques in emission computed tomography. *Physics in Medicine and Biology*, 51:R541, 2006.

[804] H. Qian and J. Luo. Nanoscale surficial films and a surface transition in v2o5-tio2-based ternary oxide systems. *Acta Materialia*, 56:4702–4714, 2008.

[805] H. J. Qian and J. Luo. Vanadia-based equilibrium-thickness amorphous films on anatase (101) surfaces. *Applied Physics Letters*, 91(6):061909, 2007.

[806] H. J. Qian, J. Luo, and Y. M. Chiang. Anisotropic wetting of zno by bi2o3 with and without nanometer-thick surficial amorphous films. *Acta Materialia*, 56(4):862–873, 2008.

[807] C. Qiu, W. Lu, and N. Vaswani. Real-time dynamic MR image reconstruction using Kalman filtered compressed sensing. In *Proc. IEEE International Conference on Acoustics, Speech and Signal Processing*, pages 393–396, 2009.

[808] X. Qu, D. Guo, B. Ning, Y. Hou, Y. Lin, S. Cai, and Z. Chen. Undersampled MRI reconstruction with patch-based directional wavelets. *Magnetic Resonance Imaging*, 30(7):964–977, 2012.

[809] Dierk. Raabe. Continuum scale simulation of engineering materials fundamentals, microstructures, process applications, 2004.

[810] T. Radó. The isoperimetric inequality and the Lebesgue definition of surface area. *Transactions of the American Mathematical Society*, 61:530–555, 1947.

[811] A. Rakotomamonjy. Direct optimization of the dictionary learning problem. *IEEE Transactions on Signal Processing*, 61(22):5495–5506, 2013.

[812] S. Ramani and J. A. Fessler. A splitting-based iterative algorithm for accelerated statistical X-ray CT reconstruction. *IEEE Trans. on Medical Imaging*, 31(3):677–688, March 2012.

[813] I. Ramirez, P. Sprechmann, and G. Sapiro. Classification and clustering via dictionary learning with structured incoherence and shared features. In *Proc. IEEE International Conference on Computer Vision and Pattern Recognition (CVPR) 2010*, pages 3501–3508, 2010.

[814] Trygve Randen and John Hakon Husoy. Filtering for texture classification: A comparative study. *Pattern Analysis and Machine Intelligence, IEEE Transactions on*, 21(4):291–310, 1999.

[815] V. Randle. Mechanism of twinning-induced grain boundary engineering in low stacking-fault energy materials. *Acta Materialia*, 47(15):4187–4196, 1999.

[816] V. Randle. Mechanism of twinning-induced grain boundary engineering in low stacking-fault energy materials. *Acta Materialia*, 47(15-16):4187–4196, 1999.

[817] V. Randle and A. Brown. Development of grain misorientation texture, in terms of coincident site lattice structures, as a function of thermomechanical treatments. *Philosophical Magazine A*, 59(5):1075–1089, 1989.

[818] K. I. Ranney and M. Soumekh. Hyperspectral anomaly detection within the signal subspace. *IEEE Geoscience and Remote Sensing Letters*, 3:312–316, 2006.

[819] N. Raj Rao. Dictionary learning methods for materials science. Manuscript, July 2016.

[820] S. Ratanaphan, D. L. Olmsted, V. V. Bulatov, E. A. Holm, A. D. Rollett, and G. S. Rohrer. Grain boundary energies in body-centered cubic metals. *Acta Materialia*, 88:346–354, 2015.

[821] H. Rauhut. On the impossibility of uniform sparse reconstruction using greedy methods. *Sampl. Tehory Signal Image Process.*, 7(2):197–215, 2008.

[822] H. Rauhut, J. Romberg, and J. Tropp. Restricted isometries for partial random circulant matrices. *Appl. and Comp. Harm. Analysis*, 32(2):242–254, 2012.

[823] Pradeep Ravikumar, Martin J. Wainwright, Garvesh Raskutti, and Bin Yu. High-dimensional covariance estimation by minimizing l_1-penalized log-determinant divergence. *ArXiv:0811.3628v1 [stat.ML]*, pages 1–35, 21 Nov. 2008.

[824] S. Ravishankar and Y. Bresler. MR image reconstruction from highly undersampled k-space data by dictionary learning. *IEEE Trans. Med. Imag.*, 30(5):1028–1041, 2011.

[825] S. Ravishankar and Y. Bresler. Multiscale dictionary learning for MRI. In *Proc. ISMRM*, page 2830, 2011.

[826] S. Ravishankar and Y. Bresler. Learning sparsifying transforms. *IEEE Trans. Signal Process.*, 61(5):1072–1086, 2013.

[827] S. Ravishankar and Y. Bresler. Data-driven adaptation of a union of sparsifying transforms for blind compressed sensing MRI reconstruction. In *Proc. SPIE*, volume 9597, pages 959713–959713–10, 2015.

[828] S. Ravishankar and Y. Bresler. Efficient blind compressed sensing using sparsifying transforms with convergence guarantees and application to magnetic resonance imaging. *SIAM Journal on Imaging Sciences*, 8(4):2519–2557, 2015.

[829] S. Ravishankar, R. R. Nadakuditi, and J. Fessler. Efficient sum of outer products dictionary learning (SOUP-DIL) and its application to inverse problems. *IEEE Transactions on Computational Imaging*, 3(4):694–709, Dec 2017.

[830] S. Ravishankar, R. R. Nadakuditi, and J. A. Fessler. Efficient sum of outer products dictionary learning (SOUP-DIL) - the ℓ_0 method. Preprint: http://arxiv.org/abs/1511.08842, 2015.

[831] L. Raymond. Effect of chemistry and processing on the structure and mechanical properties of Inconel alloy 718. In E. A. Loria, editor, *Superalloy 718-Metallurgy ad Applications*, pages 577–587. TMS, 1989.

[832] Ali Razavian, Hossein Azizpour, Josephine Sullivan, and Stefan Carlsson. Cnn features off-the-shelf: An astounding baseline for recognition. In *Proceedings of the IEEE Conference on Computer Vision and Pattern Recognition Workshops*, pages 806–813, 2014.

[833] B. Recht, M. Fazel, and P. A. Parrilo. Guaranteed minimum-rank solutions of linear matrix equations via nuclear norm minimization. *SIAM Review*, 52(3):471–501, 2010.

[834] D. Reddy, A. Veeraraghavan, and R. Chellappa. P2c2: Programmable pixel compressive camera for high speed imaging. *Computer Vision and Pattern Recognition*, pages 329–336, 2011.

[835] Richard A. Redner and Homer F. Walker. Mixture densities, maximum likelihood and the EM algorithm. *SIAM Review*, 26(2):195–239, April 1984.

[836] I. S. Reed and X. Yu. Adaptive multiple-band CFAR detection of an optical pattern with unknown spectral distribution. *IEEE Trans. Acoustics, Speech, and Signal Processing*, 38:1760–1770, 1990.

[837] Todd R. Reed and J. M. Hans Dubuf. A review of recent texture segmentation and feature extraction techniques. *CVGIP: Image Understanding*, 57(3):359–372, 1993.

[838] D. G. Rees. *Essential Statistics*. Chapman & Hall CRC, 4th edition, 2000.

[839] Donald B. Reid. An algorithm for tracking multiple targets. *IEEE Transactions on Automatic Control*, 24(6):843–854, 1979.

[840] L. Reimer. Scanning electron microscopy: Physics of image formation and microanalysis. In P.W. Hawkes, editor, *Springer Series in Optical Sciences*, volume 45. Springer-Verlag, Berlin, 1985.

[841] Philipp Reineck, Daniel Gmez, Soon Hock Ng, Matthias Karg, Toby Bell, Paul Mulvaney, and Udo Bach. Distance and wavelength dependent quenching of molecular fluorescence by au@sio2 coreshell nanoparticles. *ACS Nano*, 2013.

[842] G. Reinman, T. Ayer, T. Davan, M. Devore, S. Finley, J. Glanovsky, L. Gray, B. Hall, C. Jones, A. Learned, E. Mesaros, R. Morris, S. Pinero, R. Russo, E. Stearns, M. Teicholz, W. Teslik-Welz, and D. Yudichak. Design for variation. *Quality Engineering*, 24:317–345, 2012.

[843] Brian D. Ripley. *Spatial Statistics*, volume 575. John Wiley & Sons, 2005.

[844] Yair Rivenson and Adrian Stern. Compressed imaging with a separable sensing operator. *Signal Processing Letters, IEEE*, 16(6):449–452, 2009.

[845] A. P. Roberts and E. J. Garboczi. Elastic properties of a tungsten-silver composite by reconstruction and computation. *J. Mech. Phys. Solids*, 47:2029–2055, 1999.

[846] A. P. Roberts and S. Torquato. Chord-distribution functions of three-dimensional random media: Approximate first-passage times of Gaussian processes. *Physical Review E*, 59(5):4953–4963, 1999.

[847] Marko Robnik-Šikonja and Igor Kononenko. An adaptation of relief for attribute estimation in regression. In *Machine Learning: Proceedings of the Fourteenth International Conference (ICML97)*, pages 296–304, 1997.

[848] Ryan Robucci, Jordan Gray, Leung Kin Chiu, Justin Romberg, and Paul Hasler. Compressive sensing on a CMOS separable-transform. *Proc. IEEE*, 98(6):1089–1101, 2010.

[849] Angelika Rohde and Alexandre B. Tsybakov. Estimation of high-dimensional low-rank matrices. *Annals of Statistics*, 39(2):887–930, 2011.

[850] G. S. Rohrer. Measuring and interpreting the structure of grain-boundary networks. *Journal of the American Ceramic Society*, 94(3):633–646, 2011.

[851] Gregory S. Rohrer. Grain boundary energy anisotropy: A review. *Journal of Materials Science*, 46(18):5881–5895, 2011.

[852] G. S. Rohrer. Influence of interface anisotropy on grain growth and coarsening. *Annual Review Materials Research*, 35:99–126, 2005.

[853] G. S. Rohrer, J. Gruber, and A. D. Rollett. A model for the origin of anisotropic grain boundary character distributions in polycrystalline materials. In AD Rollett, editor, *Application of Texture Analysis*, pages 343–354. NJ: J. Wiley & Sons, Hoboken, 2009.

[854] G. S. Rohrer, J. Li, S. Lee, A. D. Rollett, M. Groeber, and M. D. Uchic. Deriving grain boundary character distributions and relative grain boundary energies from three-dimensional ebsd data. *Materials Science and Technology*, 26(6):661–669, 2010.

[855] G. S. Rohrer and H. M. Miller. Topological characteristics of plane sections of polycrystals. *Acta Materialia*, 58(10):3805–3814, 2010.

[856] G. S. Rohrer, D. M. Saylor, B. El Dasher, B. L. Adams, A. D. Rollett, and P. Wynblatt. The distribution of internal interfaces in polycrystals. *Zeitschrift für Metallkunde*, 95(4):197–214, 2004.

[857] A. D. Rollett, G. S. Rohrer, and R. M. Suter. Understanding materials microstructure and behavior at the mesoscale. *MRS Bulletin*, 40(11):951–960, 2015.

[858] J. Romberg. Compressive sensing by random convolution. *SIAM J. Imaging Sci.*, 2(4):1098–1128, 2009.

[859] J. Romberg and R. Neelamani. Sparse channel separation using random probes. *Inverse Problems*, 26(11), November 2010.

[860] Andreas Rosenauer, Katharina Gries, Knut Müller, Angelika Pretorius, Marco Schowalter, Adrian Avramescu, Karl Engl, and Stephan Lutgen. Measurement of specimen thickness and composition in $Al_xGa_{1-x}N/GaN$ using high-angle annular dark field images. *Ultramicroscopy*, 109(9):1171–1182, 2009.

[861] A. Rosiwal. über geometrische gesteinsanalysen: ein einfacher weg zur ziffernmässigen feststellung des quantitätsverhältnisses der mineralbestandtheile gemengter gesteine. *Verhandlungen der Kaiserlich-Königlichen Geologischen Reichsanstalt*, 5–6:143–75, 1898.

[862] P. J. Rousseeuw and A. M. Leroy. *Robust Regression and Outlier Detection*. Wiley-Interscience, New York, 1987.

[863] P. J. Rousseeuw and K. Van Driessen. A fast algorithm for the minimum covariance determinant estimator. *Technometrics*, 41:212–223, 1999.

[864] S. T. Roweis and L. K. Saul. Nonlinear dimensionality reduction by locally linear embedding. *Science*, 290:2323–2326, 2000.

[865] Jacob Rubinstein and S. Torquato. Diffusion-controlled reactions: Mathematical formulation, variational principles, and rigorous bounds. *The Journal of Chemical Physics*, 88(10):6372–6380, 1988.

[866] R. Rubinstein, M. Zibulevsky, and M. Elad. Efficient implementation of the k-svd algorithm using batch orthogonal matching pursuit. http://www.cs.technion.ac.il/~ronrubin/Publications/KSVD-OMP-v2.pdf, 2008. Technion, Computer Science Department, Technical Report.

[867] R. Rubinstein, M. Zibulevsky, and M. Elad. Double sparsity: Learning sparse dictionaries for sparse signal approximation. *IEEE Transactions on Signal Processing*, 58(3):1553–1564, 2010.

[868] Y. Rubner, C. Tomasi, and L. J. Guibas. A metric for distributions with applications to image databases. In *Computer Vision, 1998. Sixth International Conference on*, pages 59–66, Jan 1998.

[869] Y. Rubner, C. Tomasi, and L. J. Guibas. The earth mover's distance as a metric for image retrieval. *International Journal of Computer Vision*, 40(2):99–121, 2000.

[870] M. Rudelson and R. Vershynin. On sparse reconstruction from Fourier and Gaussian measurements. *Comm. on Pure and Applied Math.*, 61(8):1025–1045, 2008.

[871] L. Rudin, S. Oscher, and E. Fatemi. Nonlinear total variation based noise removal algorithms. *Phys. D*, 60:259–268, November 1992.

[872] L. I. Rudin, S. Osher, and E. Fatemi. Nonlinear total variation noise removal algorithm. *Physica D*, 60:259–68, 1992.

[873] Leonid I. Rudin, Stanley Osher, and Emad Fatemi. Nonlinear total variation based noise removal algorithms. *Phys. D*, 60(1-4):259–268, November 1992.

[874] D. Rugg, D. Furrer, and N. Brewitt. Textures in titanium alloys: An industrial perspective on deformation, transformation and properties. In A.D. Rollett, editor, *Materials Processing and Texture*, pages 521–533. John Wiley & Sons, Inc., Hoboken, NJ, USA, 2008.

[875] Rachel S. Ruskin, Zhiheng Yu, and Nikolaus Grigorieff. Quantitative characterization of electron detectors for transmission electron microscopy. *Journal of Structural Biology*, 184(3):385–393, 2013.

[876] Chris Rycroft. Voro++: A three-dimensional Voronoi cell library in C++. *Chaos*, 19:041111, 2009.

[877] M. Sadeghi, M. Babaie-Zadeh, and C. Jutten. Dictionary learning for sparse representation: A novel approach. *IEEE Signal Processing Letters*, 20(12):1195–1198, Dec 2013.

[878] M. Sadeghi, M. Babaie-Zadeh, and C. Jutten. Learning overcomplete dictionaries based on atom-by-atom updating. *IEEE Transactions on Signal Processing*, 62(4):883–891, 2014.

[879] A. Saeed-Akbari, J. Imlau, U. Prahl, and W. Bleck. Derivation and variation in composition-dependent stacking fault energy maps based on subregular solution model in high-manganese steels. *Metallurgical and Materials Transactions A*, 40(13):3076–3090, 2009.

[880] Z. Saghi, X. Xu, Y. Peng, B. Inkson, and G. Mobus. Three-dimensional chemical analysis of tungsten probes by energy dispersive X-ray nanotomography. *Applied Physics Letters*, 91:251906, 2007.

[881] Zineb Saghi, Daniel J. Holland, Rowan Leary, Andrea Falqui, Giovanni Bertoni, Andrew J. Sederman, Lynn F. Gladden, and Paul A. Midgley. Three-dimensional morphology of iron oxide nanoparticles with reactive concave surfaces. A compressed sensing-electron tomography (CS-ET) approach. *Nano Letters*, 11(11):4666–4673, 2011.

[882] S. K. Sahoo and A. Makur. Dictionary training for sparse representation as generalization of k-means clustering. *IEEE Signal Processing Letters*, 20(6):587–590, June 2013.

[883] A. A. Salem, M. G. Glavicic, and S. L. Semiatin. A coupled ebsd/eds method to determine the primary- and secondary-alpha textures in titanium alloys with duplex microstructures. *Materials Science and Engineering: A*, 494(1):350 – 359, 2008.

[884] J. Salmon. On two parameters for denoising with non-local means. *IEEE Signal Process. Lett.*, 17:269–272, 2010.

[885] J. Salmon, C-A. Deledalle, R. Willett, and Z. Harmany. Poisson noise reduction with non-local PCA. In *ICASSP*, 2012.

[886] J. Salmon, Z. Harmany, C. Deledalle, and R. Willett. Poisson noise reduction with non-local PCA. *Journal of Mathematical Imaging and Vision*, 48(2):279–294, 2014. doi = 10.1007/s10851-013-0435-6.

[887] S. A. Saltykov. *Stereometrische metallographie*. Leipzig: Deutscher Verlag fur Grundstoffindustrie, 1974.

[888] Jorge Sánchez, Florent Perronnin, and TeóFilo De Campos. Modeling the spatial layout of images beyond spatial pyramids. *Pattern Recognition Letters*, 33(16):2216–2223, 2012.

[889] A. C. Sankaranarayanan, C. Studer, and R. G. Baraniuk. CS-MUVI: Video compressive sensing for spatial-multiplexing cameras. *ICCP*, pages 1–10, 2012.

[890] F. Santosa and W. Symes. Linear inversion of band-limited reflection seismograms. *SIAM J. Sci. Statist. Comput.*, 7:1307–1330, 1986.

[891] S. Sardy, A. Antoniadis, and P. Tseng. Automatic smoothing with wavelets for a wide class of distributions. *Journal of Computational and Graphical Statistics*, 13(2):399–421, 2004.

[892] Prabhat Verma Satoshi Kawata, Yasushi Inouye. Plasmonics for near-field nano-imaging and superlensing. *Nature Photonics*, 3:388–394, 2009.

[893] K. Sauer and C. Bouman. Bayesian estimation of transmission tomograms using segmentation based optimization. *IEEE Trans. on Nuclear Science*, 39:1144–1152, 1992.

[894] K. Sauer and C. A. Bouman. A local update strategy for iterative reconstruction from projections. *IEEE Trans. on Signal Processing*, 41(2):534–548, February 1993.

[895] D. M. Saylor and G. S. Rohrer. Measuring the influence of grain-boundary misorientation on thermal groove geometry in ceramic polycrystals. *Journal of the American Ceramic Society*, 82(6):1529–1536, 1999.

[896] David M. Saylor, Adam Morawiec, and Gregory S. Rohrer. Distribution of grain boundaries in magnesia as a function of five macroscopic parameters. *Acta Materialia*, 51(13):3663–3674, 2003.

[897] David M. Saylor, Adam Morawiec, and Gregory S. Rohrer. The relative free energies of grain boundaries in magnesia as a function of five macroscopic parameters. *Acta Materialia*, 51(13):3675–3686, 2003.

[898] D. M. Saylor. *The Character Dependence of Interfacial Energies in Magnesia*. Carnegie Mellon University, Pittsburgh, USA, 2001.

[899] D. M. Saylor, B. Dasher, Y. Pang, H. M. Miller, P. Wynblatt, A. D. Rollett, and G. S. Rohrer. Habits of grains in dense polycrystalline solids. *Journal of the American Ceramic Society*, 87(4):724–726, 2004.

[900] D. M. Saylor, B. El Dasher, T. Sano, and G. S. Rohrer. Distribution of grain boundaries in SrTiO$_3$ as a function of five macroscopic parameters. *Journal of the American Ceramic Society*, 87(4):670–676, 2004.

[901] D. M. Saylor, B. S. El Dasher, A. D. Rollett, and G. S. Rohrer. Distribution of grain boundaries in aluminum as a function of five macroscopic parameters. *Acta Materialia*, 52(12):3649–3655, 2004.

[902] D. M. Saylor, A. Morawiec, and G. S. Rohrer. Distribution of grain boundaries in magnesia as a function of five macroscopic parameters. *Acta Materialia*, 51(13):3663–3674, 2003.

[903] D. M. Saylor, A. Morawiec, and G. S. Rohrer. The relative free energies of grain boundaries in magnesia as a function of five macroscopic parameters. *Acta Materialia*, 51(13):3675–3686, 2003.

[904] H. Schaeben. Texture approximation or texture modelling with components represented by the von Mises–Fisher matrix distribution on $SO(3)$ and the Bingham distribution on $S^{4+}{}_+$. *Journal of Applied Crystallography*, 29(5):516–525, Oct 1996.

[905] H. Schaeben and H. Siemes. Determination and interpretation of preferred orientation with texture goniometry: An application of indicators to maximum entropy pole- to orientation-density inversion. *Mathematical Geology*, 28:169–201, 1996.

[906] L. L. Scharf. *Statistical Signal Processing: Detection, Estimation, and Time Series Analysis*. Addison-Wesley, 1991.

[907] A. Schaum. Hyperspectral anomaly detection: Beyond RX. *Proc. SPIE*, 6565:656502, 2007.

[908] A. Schaum and E. Allman. Advanced algorithms for autonomous hyperspectral change detection. *IEEE Applied Imagery Pattern Recognition (AIPR) Workshop*, 33:33–38, 2005.

[909] A. Schaum, E. Allman, J. Kershenstein, and D. Alexa. Hyperspectral change detection in high clutter using elliptically contoured distributions. *Proc. SPIE*, 6565:656515, 2007.

[910] A. Schaum and A. Stocker. Long-interval chronochrome target detection. *Proc. ISSSR (Int. Symposium on Spectral Sensing Research)*, 1998.

[911] A. Schaum and A. Stocker. Linear chromodynamics models for hyperspectral target detection. *Proc. IEEE Aerospace Conference*, pages 1879–1885, 2003.

[912] A. Schaum and A. Stocker. Hyperspectral change detection and supervised matched filtering based on covariance equalization. *Proc. SPIE*, 5425:77–90, 2004.

[913] Sjors H. W. Scheres, Haixiao Gao, Mikel Valle, Gabor T. Herman, Paul P. B. Eggermont, Joachim Frank, and Jose-Maria Carazo. Disentangling conformational states of macromolecules in 3D-EM through likelihood optimization. *Nature Methods*, 4(1):27–29, January 2007.

[914] Cordelia Schmid. Constructing models for content-based image retrieval. In *Computer Vision and Pattern Recognition, 2001. CVPR 2001. Proceedings of the 2001 IEEE Computer Society Conference on*, volume 2, pages II–39. IEEE, 2001.

[915] Jürgen Schmidhuber. Deep learning in neural networks: An overview. *Neural Networks*, 61:85–117, 2015.

[916] B. Schölkopf, R. C. Williamson, A. J. Smola, J. Shawe-Taylor, and J. C. Platt. Support vector method for novelty detection. *Advances in Neural Information Processing Systems*, 12:582–588, 1999.

[917] Christian Scholz, Frank Wirner, Jan Götz, Ulrich Rüde, Gerd E Schröder-Turk, Klaus Mecke, and Clemens Bechinger. Permeability of porous materials determined from the Euler characteristic. *Physical Review Letters*, 109(26):264504, 2012.

[918] C. A. Schuh, M. Kumar, and W. E. King. Analysis of grain boundary networks and their evolution during grain boundary engineering. *Acta Materialia*, 51(3):687–700, 2003.

[919] Christopher A. Schuh, Roger W. Minich, and Mukul Kumar. Connectivity and percolation in simulated grain-boundary networks. *Philosophical Magazine*, 83(6):711–726, 2003.

[920] Adam J. Schwartz, Mukul Kumar, Brent L. Adams, and David P. Field. *Electron Backscatter Diffraction in Materials Science*. Springer, New York, 2009.

[921] A. J. Schwartz. The potential engineering of grain boundaries through thermomechanical processing. *JOM*, 50(2):50–55, 1998.

[922] Scientific Instrument Services. Simion, http://simion.com, (accessed april 04, 2016).

[923] C. Scovel, D. Hush, I. Steinwart, and J. Theiler. Radial kernels and their reproducing kernel Hilbert spaces. *J. Complexity*, 26:641–660, 2010.

[924] A.-K. Seghouane and M. Hanif. A sequential dictionary learning algorithm with enforced sparsity. In *IEEE International Conference on Acoustics, Speech and Signal Processing (ICASSP)*, pages 3876–3880, 2015.

[925] I. W. Selesnick. The double-density dual-tree DWT. *IEEE Trans. Sig. Proc.*, 52(5):1304–1314., 2004.

[926] J. F. Shackelford and W. D. Scott. Relative energies of [1100] tilt boundaries in aluminum oxide. *Journal of the American Ceramic Society*, 51(12):688–&, 1968.

[927] C. E. Shannon. A mathematical theory of communication. *Bell Sys. Tech. Journal*, 27, 1948.

[928] Claude E. Shannon. Communication in the presence of noise. *Proceedings of the Institute of Radio Engineers*, 37(1):11, 1949.

[929] X. Shi and J. Luo. Grain boundary wetting and prewetting in ni-doped mo. *Applied Physics Letters*, 94(25):251908, 2009.

[930] X. Shi and J. Luo. Decreasing the grain boundary diffusivity in binary alloys with increasing temperature. *Physical Review Letters*, 105:236102, 2010.

[931] Xiaomeng Shi and Jian Luo. Developing grain boundary diagrams as a materials science tool: A case study of nickel-doped molybdenum. *Physical Review B*, 84(1):014105, 2011.

[932] Heinrich Siemes, Helmut Schaeben, Carlos A. Rosiére, and Horst Quade. Crystallographic and magnetic preferred orientation of hematite in banded iron ores. *Journal of Structural Geology*, 22:1747 – 1759, 2000.

[933] Khe Chai Sim and Mark J. F. Gales. Minimum phone error training of precision matrix models. *IEEE Trans. Audio, Speech, and Language Proc.*, 14(3):882–889, May 2006.

[934] Jeffrey P. Simmons, Craig Przybyla, Stephen Bricker, Dae Woo Kim, and Mary Comer. Physics of mrf regularization for segmentation of materials microstructure images. In *2014 IEEE International Conference on Image Processing (ICIP)*, pages 4882–4886. IEEE, 2014.

[935] J. P. Simmons, P. Chuang, M. Comer, J. E. Spowart, M. D. Uchic, and M. De Graef. Application and further development of advanced image processing algorithms for automated analysis of serial section image data. *Modelling and Simulation in Materials Science and Engineering*, 17(2):025002, 2008.

[936] J. P. Simmons, D. M. Dimiduk, and M. Degraef. Automatic particle coordination recognition using principal component analysis and kohonen neural nets. *Microscopy and Microanalysis*, 11(S02):1634–1635, 2005.

[937] E. P. Simoncelli and J. Portilla. Texture characterization via joint statistics of wavelet coefficient magnitudes. In *In Proc. 5th Intl Conf. on Image Processing Chicago, IL*, 1998.

[938] Karen Simonyan and Andrew Zisserman. Very deep convolutional networks for large-scale image recognition. *arXiv preprint arXiv:1409.1556*, 2014.

[939] H. Singh, A. M. Gokhale, S. I. Lieberman, and S. Tamirisakandala. Image based computations of lineal path probability distributions for microstructure representation. *Materials Science and Engineering: A*, 474(1):104–111, 2008.

[940] Josef Sivic and Andrew Zisserman. Video google: A text retrieval approach to object matching in videos. In *Computer Vision, 2003. Proceedings. Ninth IEEE International Conference on*, pages 1470–1477. IEEE, 2003.

[941] S. D Berger S. J. Pennycook and R. J.Culbertson. Elemental mapping with elastically scatterd electrons. *J. Microsc.*, 144:229–249, 1986.

[942] A. Skodras, C. Christopoulos, and T. Ebrahimi. The JPEG2000 still image compression standard. *IEEE Signal Proc. Mag.*, 18:36–58, Sept 2001.

[943] K. Skretting and K. Engan. Recursive least squares dictionary learning algorithm. *IEEE Transactions on Signal Processing*, 58(4):2121–2130, 2010.

[944] W. Skrotzki, N. Scheerbaum, C.-G. Oertel, R. Arrufiat-Massion, S. Suwas, and L.S. Tóth. Microstructure and texture gradient in copper deformed by equal channel angular pressing. *Acta Materialia*, 55(6):2013–2024, 2007.

[945] Andrew M. Smith and Shuming Nie. Semiconductor nanocrystals: Structure, properties, and band gap engineering. *Acc. Chem. Res.*, 43(2):190–200, 2009.

[946] C. S. Smith, editor. *The Sorby Centennial Symposium on the History of Metallurgy*. Gordon and Beach Science Publishers, 1965.

[947] Cyril Stanley Smith. *A History of Metallography: The Development of Ideas on the Structure of Metals before 1890*. The University of Chicago Press, 1960.

[948] J. R. Smith. *Integrated Spatial and Feature Image Systems: Retrieval, Analysis, and Compression*. PhD thesis, Columbia University, 1997.

[949] L. N. Smith and M. Elad. Improving dictionary learning: Multiple dictionary updates and coefficient reuse. *IEEE Signal Processing Letters*, 20(1):79–82, Jan 2013.

[950] S. M. Smith and J. M. Brady. Susan: A new approach to low level image processing. *Int. J. Comput. Vision*, 23(1):45–78, 1997.

[951] Hichem Snoussi and Ali Mohammad-Djafari. Estimation of structured Gaussian mixtures: The inverse EM algorithm. *IEEE Trans. Sig. Proc.*, 55(7):3185–3191, July 2007.

[952] Munro's electron beam software, http://mebs.co.uk/#p=6 (accessed april 16, 2016).

[953] S. Som and P. Schniter. Compressive imaging using approximate message passing and a Markov-tree prior. *IEEE Trans. Sig. Proc.*, 60(7):3439–3448, 2012.

[954] Bo Song, Ning Xi, Ruiguo Yang, King Wai Chiu Lai, and Chengeng Qu. Video rate atomic force microscopy (afm) imaging using compressive sensing. In *Nanotechnology (IEEE-NANO), 2011 11th IEEE Conference on*, pages 1056–1059. IEEE, 2011.

[955] A. Sousa, M. Hohmann-Marriott, A. Azari, G. Zhang, and R. Leapman. BF STEM tomography for improved 3D imaging of thick biological sections. *Microscopy and Microanalysis*, 15:572–573, 2009.

[956] J. C. H. Spence. *Experimental High-Resolution Electron Microscopy*. Oxford University Press, 1988.

[957] J. C. H. Spence and J. M. Zuo. Large dynamic range, parallel detection system for electron diffraction and imaging. *Review of Scientific Instruments*, 59(9):2102–2105, 1988.

[958] D. A. Spielman, H. Wang, and J. Wright. Exact recovery of sparsely-used dictionaries. In *Proceedings of the 25th Annual Conference on Learning Theory*, pages 37.1–37.18, 2012.

[959] Richard J. Spontak, Michael C. Williams, and David A. Agard. Three-dimensional study of cylindrical morphology in a styrene-butadiene-styrene block copolymer. *Polymer*, 29:387–395, 1988.

[960] J. E. Spowart. Serial sectioning in the micron-plus range and modern techniques for automation. *Microscopy and Microanalysis*, 12:90–91, 8 2006.

[961] Eric A. Stach, Jing Li, Huolin Xin, Dmitri Zakharov, Yo Hon Kwon, and Elsa Reichmanis. Combining post-specimen aberration correction and direct electron detection to image molecular structure in liquid crystal polymers. *Microscopy and Microanalysis*, 22(S3):1924–1925, 2016.

[962] J. L. Starck, M. Elad, and D. L. Donoho. Image decomposition via the combination of sparse representations and a variational approach. *IEEE Trans. Image Proc.*, 14(10):1570–1582, October 2005.

[963] D. W. J. Stein, S. G. Beaven, L. E. Hoff, E. M. Winter, A. P. Schaum, and A. D. Stocker. Anomaly detection from hyperspectral imagery. *IEEE Signal Processing Magazine*, 19:58–69, Jan 2002.

[964] Paul J. Steinhardt, David R. Nelson, and Marco Ronchetti. Bond-orientational order in liquids and glasses. *Phys. Rev. B*, 28(2):784, 1983.

[965] Ernst Steinitz. *Polyeder und Raumeinteilungen*. Teubner, 1916.

[966] I. Steinwart, D. Hush, and C. Scovel. A classification framework for anomaly detection. *J. Machine Learning Research*, 6:211–232, 2005.

[967] I. Steinwart, J. Theiler, and D. Llamocca. Using support vector machines for anomalous change detection. *Proc. IEEE Int. Geoscience and Remote Sensing Symposium (IGARSS)*, pages 3732–3735, 2010.

[968] Andrew Stevens, Hao Yang, Lawrence Carin, Ilke Arslan, and Nigel D. Browning. The potential for bayesian compressive sensing to significantly reduce electron dose in high-resolution STEM images. *Microscopy*, 63(1):41–51, 2014.

[969] James P Stevens. *Applied Multivariate Statistics for the Social Sciences*. Routledge, 2012.

[970] R. Stevenson and E. Delp. Fitting curves with discontinuities. *Proc. of the First International Workshop on Robust Computer Vision*, pages 127–136, October 1990.

[971] D. P. Stewart. Isocon manufacturing of Waspaloy turbine discs. In S. Reichman, D. N. Duhl, G. Maurer, A. Antolovich, and C. Lund, editors, *Superalloys 1988*, pages 545–551. TMS, 1988.

[972] A. D. Stocker, I. S. Reed, and X. Yu. Multi-dimensional signal processing for electro-optical target detection. *Proc. SPIE*, 1305:218–231, 1990.

[973] Ivana Stojanovic, Homer Pien, Synho Do, and W. Clem Karl. Low-dose X-ray CT reconstruction based on joint sinogram smoothing and learned dictionary-based representation. In *9th IEEE International Symposium on Biomedical Imaging (ISBI)*, pages 1012–1015, 2012.

[974] Dietrich Stoyan, Wilfrid S Kendall, Joseph Mecke, and L Ruschendorf. *Stochastic Geometry and Its Applications*. Wiley Chichester, 1995.

[975] G. Strang. *Linear Algebra and Its Applications*. Cengage Learning, 4th edition, 2006.

[976] G. Strang. *Computational Science and Engineering*. Wellesley-Cambridge Press, 2007.

[977] B. Straumal, T. Muschik, W. Gust, and B. Predel. The wetting transition in high and low energy grain boundaries in the cu(in) system. *Acta Metallurgica et Materialia*, 40(5):939–945, 1992.

[978] B. B. Straumal, W. Gust, and D. A. Molodov. Wetting transition on grain boundaries in al contacting with a sn-rich melt. *Interface Science*, 3(2):127–132, 1995.

[979] Boris B. Straumal, Alena S. Gornakova, Olga A. Kogtenkova, Svetlana G. Protasova, Vera G. Sursaeva, and Brigitte Baretzky. Continuous and discontinuous grain-boundary wetting in zn_xal_{1-x}. *Physical Review B*, 78(5):054202, 2008.

[980] J. Stringer. Report of the basic energy sciences advisory committee subpanel review of the electorn beam microcharacterization centers: Past, present, and future. Technical Report DOE/SC-0019, United States Department of Energy, Basic Energy Sciences Advisory Committee, 1999.

[981] Vincent Studer, Jérome Bobin, Makhlad Chahid, Hamed Shams Mousavi, Emmanuel Candes, and Maxime Dahan. Compressive fluorescence microscopy for biological and hyperspectral imaging. *Proceedings of the National Academy of Sciences*, 109(26):E1679–E1687, 2012.

[982] Alexander Stukowski. Visualization and analysis of atomistic simulation data with ovito–the open visualization tool. *Modelling and Simulation in Materials Science and Engineering*, 18(1):015012, 2009.

[983] Alexander Stukowski. Structure identification methods for atomistic simulations of crystalline materials. *Modelling Simul. Mater. Sci. Eng.*, 20(4):045021, 2012.

[984] W. Stumm and J. J. Morgan. *Aquatic Chemistry: Chemical Equilibria and Rates in Natural Waters*. John Wiley & Sons, New York, 1996.

[985] V. Sundararaghavan. Reconstruction of three dimensional anisotropic microstructures from two-dimensional micrographs imaged on orthogonal planes. *Integrating Materials and Manufacturing Innovation*, 3(19):1–11, 2014.

[986] V. Sundararaghavan and N. Zabaras. Classification and reconstruction of three-dimensional microstructures using support vector machines. *Computational Materials Science*, 32(2):223–239, 2005.

[987] V. Sundararaghavan and N. Zabaras. Design of microstructure-sensitive properties in elasto-viscoplastic polycrystals using multi-scale homogenization. *Int. J. Plasticity*, 22:1799–1824, 2006.

[988] Veeraraghavan Sundararaghavan and Nicholas Zabaras. A dynamic material library for the representation of single-phase polyhedral microstructures. *Acta Materialia*, 52(14):4111–4119, 2004.

[989] Subra Suresh. *Fundamentals of Metal-Matrix Composites*. Elsevier, 2013.

[990] A. P. Sutton and R. W. Balluffi. *Interfaces in Crystalline Materials*. Oxford University Press, Oxford, 1995.

[991] M. A. Sutton, N. Li, D. Garcia, N. Cornille, J. J. Orteu, S. R. McNeill, H. W. Schreier, X. Li, and A. P. Reynolds. Scanning electron microscopy for quantitative small and large deformation measurements part ii: Experimental validation for magnifications from 200 to 10,000. *Experimental Mechanics*, 47(6):789–804, 2007.

[992] Michael J. Swain and Dana H. Ballard. Indexing via color histograms. In *Active Perception and Robot Vision*, pages 261–273. Springer, 1992.

[993] M. J. Swain and D. H. Ballard. Color indexing. *International Journal of Computer Vision*, 7(1):11–32, 1991.

[994] Peter R. Swann and Bernd Kraus. Tv system for transmission electron microscopes, April 19 1988. US Patent 4,739,399.

[995] M. Syha, W. Rheinheimer, M. Bäurer, E. M. Lauridsen, W. Ludwig, D. Weygand, and P. Gumbsch. Three-dimensional grain structure of sintered bulk strontium titanate from x-ray diffraction contrast tomography. *Scripta Materialia*, 66(1):1–4, 2012.

[996] Richard Szeliski. *Computer Vision: Algorithms and Applications*. Springer Science & Business Media, 2010.

[997] M. Tanaka and M. Terauchi. *Convergent Beam Electron Diffraction*. JEOL Ltd., 1985.

[998] M. Tanaka, M. Terauchi, and T. Kaneyama. *Convergent Beam Electron Diffraction II*. JEOL Ltd., 1988.

[999] M. Tanaka, M. Terauchi, and K. Tsuda. *Convergent Beam Electron Diffraction III*. JEOL Ltd., 1994.

[1000] Jinghua Tang, Jennifer M. Johnson, Kelly A. Dryden, Mark J. Young, Adam Zlotnick, and John E. Johnson. The role of subunit hinges and molecular "switches" in the control of viral capsid polymorphism. *J. Struct. Biol.*, 154(1):59–67, April 2006.

[1001] Jinghua Tang, Bradley M. Kearney, Qiu Wang, Peter C. Doerschuk, Timothy S. Baker, and John E. Johnson. Dynamic and geometric analyses of *Nudaurelia capensis ω* virus maturation reveal the energy landscape of particle transitions. *J. Molecular Recognition*, 27(4):230–237, 10 February 2014.

[1002] M. Tang, W. C. Carter, and R. M. Cannon. Diffuse interface model for structural transitions of grain boundaries. *Physical Review B*, 73(2):024102, 2006.

[1003] M. Tang, W. C. Carter, and R. M. Cannon. Grain boundary transitions in binary alloys. *Physical Review Letters*, 97(7):075502, 2006.

[1004] D. Tax and R. Duin. Data domain description by support vectors. In M Verleysen, editor, *Proc. ESANN99*, pages 251–256, Brussels, 1999. D. Facto Press.

[1005] David M. J. Tax. *One-Class Classification: Concept-Learning in the Absence of Counter-Examples*. TU Delft, Delft University of Technology, 2001.

[1006] D. M. J. Tax and R. P. W. Duin. Uniform object generation for optimizing one-class classifiers. *J. Machine Learning Res.*, 2:155–173, 2002.

[1007] Maxim N. Tchoul, Scott P. Fillery, Hilmar Koerner, Lawrence F. Drummy, Folusho T. Oyerokun, Peter A. Mirau, Michael F. Durstock, and Richard A. Vaia. Assemblies of titanium dioxide-polystyrene hybrid nanoparticles for dielectric applications. *Chemistry of Materials*, 22(5):1749–1759, 2010.

[1008] W. Telieps and E. Bauer. An analytical reflection and emission uhv surface electron microscope. *Ultramicroscopy*, 17(1):57–65, 1985.

[1009] J. B. Tenenbaum, V. de Silva, and J. C. Langford. A global geometric framework for nonlinear dimensionality reduction. *Science*, 290:2319–2323, 2000.

[1010] A. Tewari, A. M. Gokhale, J. E. Spowart, and D. B. Miracle. Quantitative characterization of spatial clustering in three-dimensional microstructures using two-point correlation functions. *Acta Materialia*, 52(2):307–319, 2004.

[1011] Neil A. Thacker, Adrian F. Clark, John L. Barron, J. Ross Beveridge, Patrick Courtney, William R. Crum, Visvanathan Ramesh, and Christine Clark. Performance characterization in computer vision: A guide to best practices. *Computer Vision and Image Understanding*, 109(3):305–334, 2008.

[1012] Retrieved 2016.

[1013] J. Theiler. Quantitative comparison of quadratic covariance-based anomalous change detectors. *Applied Optics*, 47:F12–F26, 2008.

[1014] J. Theiler. Sensitivity of anomalous change detection to small misregistration errors. *Proc. SPIE*, 6966:69660X, 2008.

[1015] J. Theiler. Subpixel anomalous change detection in remote sensing imagery. *Proc. IEEE Southwest Symposium on Image Analysis and Interpretation*, pages 165–168, 2008.

[1016] J. Theiler. Ellipsoid-simplex hybrid for hyperspectral anomaly detection. *Proc. 3rd IEEE Workshop on Hyperspectral Image and Signal Processing: Evolution in Remote Sensing (WHISPERS)*, 2011.

[1017] J. Theiler. Confusion and clairvoyance: Some remarks on the composite hypothesis testing problem. *Proc. SPIE*, 8390:839003, 2012.

[1018] J. Theiler. The incredible shrinking covariance estimator. *Proc. SPIE*, 8391:83910P, 2012.

[1019] J. Theiler. Matched-pair machine learning. *Technometrics*, 55:536–547, 2013.

[1020] J. Theiler. Spatio-spectral anomalous change detection in hyperspectral imagery. *Proc. 1st IEEE Global Signal and Information Processing Conference*, pages 953–956, 2013.

[1021] J. Theiler. By definition undefined: Adventures in anomaly (and anomalous change) detection. *Proc. 6th IEEE Workshop on Hyperspectral Signal and Image Processing: Evolution in Remote Sensing (WHISPERS)*, 2014.

[1022] J. Theiler. Symmetrized regression for hyperspectral background estimation. *Proc. SPIE*, 9472:94721G, 2015.

[1023] J. Theiler and S. M. Adler-Golden. Detection of ephemeral changes in sequences of images. *IEEE Applied Imagery Pattern Recognition (AIPR) Workshop*, 37, 2009.

[1024] J. Theiler and J. Bloch. Multiple concentric annuli for characterizing spatially nonuniform backgrounds. *The Astrophysical Journal*, 519:372–388, 1999.

[1025] J. Theiler and D. M. Cai. Resampling approach for anomaly detection in multispectral images. *Proc. SPIE*, 5093:230–240, 2003.

[1026] J. Theiler, G. Cao, L. R. Bachega, and C. A. Bouman. Sparse matrix transform for hyperspectral image processing. *IEEE J. Selected Topics in Signal Processing*, 5:424–437, 2011.

[1027] J. Theiler, B. R. Foy, and A. M. Fraser. Characterizing non-Gaussian clutter and detecting weak gaseous plumes in hyperspectral imagery. *Proc. SPIE*, 5806:182–193, 2005.

[1028] J. Theiler and G. Grosklos. Problematic projection to the in-sample subspace for a kernelized anomaly detector. *IEEE Geoscience and Remote Sensing Lett.*, 13:485–489, 2016.

[1029] J. Theiler and D. Hush. Statistics for characterizing data on the periphery. *Proc. IEEE Int. Geoscience and Remote Sensing Symposium (IGARSS)*, pages 4764–4767, 2010.

[1030] J. Theiler and A. Matsekh. Total least squares for anomalous change detection. *Proc. SPIE*, 7695:76951H, 2010.

[1031] J. Theiler and S. Perkins. Proposed framework for anomalous change detection. *ICML Workshop on Machine Learning Algorithms for Surveillance and Event Detection*, pages 7–14, 2006.

[1032] J. Theiler and S. Perkins. Resampling approach for anomalous change detection. *Proc. SPIE*, 6565:65651U, 2007.

[1033] J. Theiler and L. Prasad. Overlapping image segmentation for context-dependent anomaly detection. *Proc. SPIE*, 8048:804807, 2011.

[1034] J. Theiler, C. Scovel, B. Wohlberg, and B. R. Foy. Elliptically-contoured distributions for anomalous change detection in hyperspectral imagery. *IEEE Geoscience and Remote Sensing Lett.*, 7:271–275, 2010.

[1035] J. Theiler and B. Wohlberg. Local co-registration adjustment for anomalous change detection. *IEEE Trans. Geoscience and Remote Sensing*, 50:3107–3116, 2012.

[1036] J. Theiler and B. Wohlberg. Detection of unknown gas-phase chemical plumes in hyperspectral imagery. *Proc. SPIE*, 8743:874315, 2013.

[1037] J. Theiler and B. Wohlberg. Regression framework for background estimation in remote sensing imagery. *Proc. 5th IEEE Workshop on Hyperspectral Image and Signal Processing: Evolution in Remote Sensing (WHISPERS)*, 2013.

[1038] Jean-Baptiste Thibault, K. Sauer, C. Bouman, and J. Hsieh. A three-dimensional statistical approach to improved image quality for multislice helical CT. *Med. Phys.*, 34:4526–4544, 2007.

[1039] John Meurig Thomas, Rowan Leary, Paul A. Midgley, and Daniel J. Holland. A new approach to the investigation of nanoparticles: Electron tomography with compressed sensing. *Journal of Colloid and Interface Science*, 392:7–14, 2013.

[1040] C. V. Thompson. Structure evolution during processing of polycrystalline films. *Annual Review of Materials Science*, 30(1):159–190, 2000.

[1041] M. Thottan and C. Ji. Anomaly detection in IP networks. *IEEE Trans. Signal Processing*, 51:2191–2204, 2003.

[1042] Ye Tian, Tong Wang, Wenyan Liu, Huolin L. Xin, Huilin Li, Yonggang Ke, William M. Shih, and Oleg Gang. Prescribed nanoparticle cluster architectures and low-dimensional arrays built using octahedral dna origami frames. *Nature Nano*, 10(7):637–644, 2015.

[1043] Ye Tian, Jian-Lai Zhou, Hui Lin, and Hui Jiang. Tree-based covariance modeling of hidden Markov models. *IEEE Trans. Audio, Speech, and Language Proc.*, 14(6):2134–2146, November 2006.

[1044] R. Tibshirani. Regression shrinkage and selection via the Lasso. *J. Royal Stat. Soc. Ser. B*, 58(1):267–288, 1996.

[1045] G. A. Tidhar and S. R. Rotman. Target detection in inhomogeneous non-Gaussian hyperspectral data based on nonparametric density estimation. *Proc. SPIE*, 8743:87431A, 2013.

[1046] Mariana Tihova, Kelly A. Dryden, Thucvy L. Le, Stephen C. Harvey, John E. Johnson, Mark Yeager, and Anette Schneemann. Nodavirus coat protein imposes dodecahedral RNA structure independent of nucleotide sequence and length. *J. Virol.*, 78(6):2897–2905, 2004.

[1047] K. Timmermann and R. Nowak. Multiscale modeling and estimation of Poisson processes with application to photon-limited imaging. *IEEE Trans. Inf. Theory*, 45(3):846–862, April, 1999.

[1048] M. J. Todd and E. A. Yildirim. On Khachiyan's algorithm for the computation of minimum-volume enclosing ellipsoids. *Discrete Applied Mathematics*, 155:1731–1744, 2007.

[1049] C. Tomasi and R. Manduchi. Bilateral filtering for gray and color images. In *ICCV*, pages 839–846, 1998.

[1050] X. Tong, J. Zhang, L. Liu, X. Wang, B. Guo, and H.-Y. Shum. Synthesis of bidirectional texture functions on arbitrary surfaces. In *In SIGGRAPH 02, pp. 665–672*, 2002.

[1051] S. Torquato. Nearest-neighbor statistics for packings of hard spheres and disks. *Physical Review E*, 51(4):3170, 1995.

[1052] S. Torquato. *Random Heterogeneous Materials: Microstructure and Macroscopic Properties*. Springer-Verlag, New York, 2002.

[1053] S. Torquato, J. D. Beasley, and Y. C. Chiew. Two-point cluster function for continuum percolation. *The Journal of Chemical Physics*, 88(10):6540–6547, 1988.

[1054] S. Torquato and B. Lu. Chord-length distribution function for two-phase random media. *Physical Review E*, 47(4):2950, 1993.

[1055] Salvatore Torquato. *Random Heterogeneous Materials: Microstructure and Macroscopic Properties*, volume 16. Springer Science & Business Media, 2013.

[1056] J. Tropp and A. Gilbert. Signal recovery from partial information via orthogonal matching pursuit. *IEEE Trans. Inform. Theory*, 53(12):4655–4666, 2007.

[1057] J. A. Tropp. Greed is good: Algorithmic results for sparse approximation. *IEEE Trans. Inform. Theory*, 50(10):2231–2242, October 2004.

[1058] J. A. Tropp, J. N. Laska, M. F. Duarte, J. Romberg, and R. G. Baraniuk. Beyond Nyquist: Efficient sampling of sparse bandlimited signals. *IEEE Trans. Inform. Theory*, 56(1):520–544, January 2010.

[1059] Richard J. Trudeau. *Introduction to Graph Theory*. Courier Corporation, 2013.

[1060] J. Trzasko and A. Manduca. Highly undersampled magnetic resonance image reconstruction via homotopic l_0-minimization. *IEEE Trans. Med. Imaging*, 28(1):106–121, 2009.

[1061] M. A. Tschopp, D. E. Spearot, and D. L. McDowell. Influence of grain boundary structure on dislocation nucleation in fcc metals. In JP Hirth, editor, *Dislocations in Solids*, pages 43–137. Elsevier, 2008.

[1062] Mihran Tuceryan, Anil K Jain, et al. Texture analysis. *Handbook of Pattern Recognition and Computer Vision*, 2:207–248, 1993.

[1063] D. M. Turner and S. R. Kalidindi. Statistical construction of 3-D microstructures from 2-D exemplars collected on oblique sections. *Acta Materialia*, 102:136–148, 2016.

[1064] Mark R Turner. Texture discrimination by gabor functions. *Biological Cybernetics*, 55(2-3):71–82, 1986.

[1065] M. D. Uchic. Serial sectioning methods for generating 3d characterization data of grain- and precipitate-scale microstructures. In Somnath Ghosh and Dennis Dimiduk, editors, *Computational Methods for Microstructure-Property Relationships*, pages 31–52. Springer US, 2011.

[1066] Inc. UES. http://www.ues.com/robo-met-3d.

[1067] A. Ullah. Entropy, divergence and distance measures with econometric applications. *Journal of Statistical Planning and Inference*, 49(1):137 – 162, 1996. Econometric Methodology, Part I.

[1068] S. Umekawa, R. Kotfila, and O. D. Sherby. Elastic properties of a tungsten-silver composite above and below the melting point of silver. *J. Mech. Phys. Solids*, 13(4):229–230, 1965.

[1069] Ervin E. Underwood. *Quantitative Stereology*. Addison-Wesley, 1970.

[1070] M. Unser. Splines: A perfect fit for signal and image processing. *IEEE Signal Proc. Mag.*, pages 22–38, November 1999.

[1071] M. Unser. Sampling — 50 years after Shannon. *Proceedings of the IEEE*, 88(4):569–587, April 2000.

[1072] URL. http://www.python.org/.

[1073] URL. http://www.mpi-forum.org/.

[1074] URL. http://www.mcs.anl.gov/research/projects/mpi/.

[1075] URL. http://www.open-mpi.org/.

[1076] URL. http://ami.scripps.edu/redmine/projects/appion/wiki.

[1077] URL. http://cns.bu.edu/~gsc/ColorHistograms.html.

[1078] URL. Flock House Virus (FHV) web page. http://viperdb.scripps.edu/info_page.php?VDB=2q25.

[1079] VIPERdb URL. http://viperdb.scripps.edu/.

[1080] S. Valiollahzadeh, T. Chang, J. W. Clark, and O. R. Mawlawi. Image recovery in PET scanners with partial detector rings using compressive sensing. In *IEEE Nuclear Science Symposium and Medical Imaging Conference (NSS/MIC)*, pages 3036–3039, Oct 2012.

[1081] Gert Van de Wouwer, Paul Scheunders, and Dirk Van Dyck. Statistical texture characterization from discrete wavelet representations. *Image Processing, IEEE Transactions on*, 8(4):592–598, 1999.

[1082] E. van den Berg and M. P. Friedlander. A solver for large-scale sparse reconstruction, 2009.

[1083] Ewout Van Den Berg and Michael P Friedlander. Probing the pareto frontier for basis pursuit solutions. *SIAM Journal on Scientific Computing*, 31(2):890–912, 2008.

[1084] B. A. Van Der Pluijm, N. C. Ho, and D. R. Peacor. High-resolution x-ray texture goniometry. *Journal of Structural Geology*, 16(7):1029–1032, 1994.

[1085] Jan C. Van Gemert, Jan-Mark Geusebroek, Cor J. Veenman, and Arnold WM Smeulders. Kernel codebooks for scene categorization. In *Computer Vision–ECCV 2008*, pages 696–709. Springer, 2008.

[1086] Jan C. Van Gemert, Cor J. Veenman, Arnold W. M. Smeulders, and Jan-Mark Geusebroek. Visual word ambiguity. *Pattern Analysis and Machine Intelligence, IEEE Transactions on*, 32(7):1271–1283, 2010.

[1087] Luc Van Gool, Theo Moons, and Dorin Ungureanu. Affine/photometric invariants for planar intensity patterns. In *Computer VisionECCV'96*, pages 642–651. Springer, 1996.

[1088] Marin van Heel. Similarity measures between images. *Ultramicroscopy*, 21:95–100, 1987.

[1089] G. Van Tendeloo, S. Turner, and S. Bals. Porous structures in 2 and 3 dimensions. In *18th International Microscopy Conference*, 2014.

[1090] G. Van Tendeloo, D. Van Dyck, and S. Pennycook, editors. *Handbook of Nanoscopy*. Wiley-VCH, 2012.

[1091] G. Vander Voort. *Metallography Principles and Practice*. McGraw-Hill, New York, 1984.

[1092] Manik Varma and Andrew Zisserman. Classifying images of materials: Achieving viewpoint and illumination independence. In *Computer VisionECCV 2002*, pages 255–271. Springer, 2002.

[1093] Andrea Vedaldi and Andrew Zisserman. Efficient additive kernels via explicit feature maps. *Pattern Analysis and Machine Intelligence, IEEE Transactions on*, 34(3):480–492, 2012.

[1094] A. Veeraraghavan, D. Reddy, and R. Raskar. Coded strobing photography: Compressive sensing of high speed periodic videos. *IEEE Trans. Pattern Analysis and Machine Intelligence*, 33:671–686, 2011.

[1095] Ashok Veeraraghavan, Alex V. Genkin, Shiv Vitaladevuni, Lou Scheffer, Shan Xu, Harald Hess, Richard Fetter, Marco Cantoni, Graham Knott, and Dmitri Chklovskii. Increasing depth resolution of electron microscopy of neural circuits using sparse tomographic reconstruction. In *Computer Vision and Pattern Recognition (CVPR), 2010 IEEE Conference on*, pages 1767–1774. IEEE, 2010.

[1096] Singanallur V. Venkatakrishnan, Ming-Siao Hsiao, Nick Garvin, Michael A. Jackson, Marc De Graef, Jeffrey P. Simmons, Charles A. Bouman, and Lawrence F. Drummy. Model-based iterative reconstruction for low-dose electron tomography. *Microscopy and Microanalysis*, 20(SupplementS3):802–803, 2014.

[1097] S. V. Venkatakrishnan, L. F. Drummy, M. Jackson, M. De Graef, J. Simmons, and C. A. Bouman. Model-based iterative reconstruction for bright-field electron tomography. *Computational Imaging, IEEE Transactions on*, 1(1):1–15, March 2015.

[1098] S. V. Venkatakrishnan, L. F. Drummy, M. A. Jackson, C. A. Bouman, J. P. Simmons, and M. De Graef. A phantom-based forward projection approach in support of model-based iterative reconstructions for haadf-stem tomography. *Ultramicroscopy*, 160:7 – 17, 2016.

[1099] S. V. Venkatakrishnan, L. F. Drummy, M. A. Jackson, M. De Graef, J. Simmons, and C.A. Bouman. A model based iterative reconstruction algorithm for high angle annular dark field-scanning transmission electron microscope (HAADF-STEM) tomography. *Image Processing, IEEE Transactions on*, 22(11):4532–4544, Nov 2013.

[1100] R. Venkataramani and Y. Bresler. Further results on spectrum blind sampling of 2D signals. In *Proc. IEEE Int. Conf. Image Proc., ICIP*, volume 2, pages 752–756, October 1998.

[1101] Jaco Vermaak, Arnaud Doucet, and Patrick Pérez. Maintaining multimodality through mixture tracking. In *International Conference on Computer Vision*, pages 1110–1116, 2003.

[1102] R. Vershynin. Introduction to the non-asymptotic theory of random matrices. In Y. Eldar and G. Kutyniok, editors, *Compressed Sensing, Theory and Applications*, pages 210–268. Cambridge University Press, 2012.

[1103] R. Vidal. Subspace clustering. *IEEE Signal Processing Magazine*, 28(2):52–68, 2011.

[1104] Trevor J. Vincent, Jonathan D. Thiessen, Laryssa M. Kurjewicz, Shelley L. Germscheid, Allan J. Turner, Peter Zhilkin, Murray E. Alexander, and Melanie Martin. Longitudinal brain size measurements in app/ps1 transgenic mice. *Magnetic Resonance Insights*, 4:19, 2010.

[1105] Pierre Viste, Jrome Plain, Rodolphe Jaffiol, Alexandre Vial, Pierre Michel Adam, and Pascal Royer. Enhancement and quenching regimes in metal-semiconductor hybrid optical nanosources. *ACS Nano*, 4(2):759–764, 2010.

[1106] Cédric Vonesch, Lanhui Wang, Yoel Shkolnisky, and Amit Singer. Fast wavelet-based single-particle reconstruction in cryo-em. In *Biomedical Imaging: From Nano to Macro, 2011 IEEE International Symposium on*, pages 1950–1953. IEEE, 2011.

[1107] K. Vongsy, M. T. Eismann, and M. J. Mendenhall. Extension of the linear chromodynamics model for spectral change detection in the presence of residual spatial misregistration. *IEEE Trans. Geoscience and Remote Sensing*, 53:3005–3021, 2015.

[1108] G..F. Vander Voort. *Metallography: Principles and Practice*. New York: McGraw-Hill Book Company, 1984.

[1109] G..F. Vander Voort. Examination of some grain size measurement problems. In *Metallography: Past, Present and Future (75th Anniversary Volume) ASTM STP 1165*. Philadelphia: American Society for testing and materials, p. 266, 1993.

[1110] S. Voronin and R. Chartrand. A new generalized thresholding algorithm for inverse problems with sparsity constraints. In *ICASSP*, pages 1636–1640, 2013.

[1111] Georges Voronoï. Nouvelles applications des paramètres continus à la théorie des formes quadratiques. Deuxième mémoire. Recherches sur les parallélloèdres primitifs. *J. Reine Angew. Math.*, 134:198–287, 1908.

[1112] A. Wagadarikar, R. John, R. Willett, and D. Brady. Single disperser design for coded aperture snapshot spectral imaging. *Applied Optics*, 47:B44–B51, 2008.

[1113] Jarrell Waggoner, Youjie Zhou, Jeffrey P. Simmons, Marc De Graef, and Song Wang. 3d materials image segmentation by 2d propagation: a graph-cut approach considering homomorphism. *IEEE Transactions on Image Processing*, 22(12):5282–5293, 2013.

[1114] F. Wagner, H. R. Wenk, C. Esling, and H. J. Bunge. Importance of odd coefficients in texture calculations for trigonal-triclinic symmetries. *Physica Status Solidi (a)*, 67(1):269–285, 1981.

[1115] G. Wahba. *Spline Models for Observational Data*. SIAM, 1990.

[1116] G. Wang and J. Qi. Penalized likelihood PET image reconstruction using patch-based edge-preserving regularization. *IEEE Trans. on Medical Imaging*, PP:1, 2012.

[1117] Lanhui Wang, Yoel Shkolnisky, and Amit Singer. A Fourier-based approach for iterative 3d reconstruction from cryo-em images. *arXiv:1307.5824V1*, 2013.

[1118] Qiu Wang. *Maximum likelihood reconstruction of heterogeneous 3-D objects from 2-D projections of unknown orientation and application to electron microscope images of viruses*. PhD thesis, School of Electrical and Computer Engineering, Cornell University, Ithaca, New York, USA, June 2013.

[1119] Qiu Wang, Tsutomu Matsui, Tatiana Domitrovic, Yili Zheng, Peter C. Doerschuk, and John E. Johnson. Dynamics in cryo EM reconstructions visualized with maximum-likelihood derived variance maps. *J. Struct. Biol.*, 181(3):195–206, March 2013. http://dx.doi.org/10.1016/j.jsb.2012.11.005.

[1120] Tie Wang, Jiaqi Zhuang, Jared Lynch, Ou Chen, Zhongliang Wang, Xirui Wang, Derek LaMontagne, Huimeng Wu, Zhongwu Wang, and Y. Charles Cao. Self-assembled colloidal superparticles from nanorods. *Science*, 338(6105):358–363, 2012.

[1121] Y. Wang, Y. Zhou, and L. Ying. Undersampled dynamic magnetic resonance imaging using patch-based spatiotemporal dictionaries. In *2013 IEEE 10th International Symposium on Biomedical Imaging (ISBI)*, pages 294–297, April 2013.

[1122] T. Wanner, E. R. Fuller, and D. M. Saylor. Homology metrics for microstructure response fields in polycrystals. *Acta Materialia*, 58(1):102–110, 2010.

[1123] O. Watanabe, H. M. Zbib, and E. Takenouchi. Crystal plasticity: micro-shear banding in polycrystals using voronoi tessellation. *International Journal of Plasticity*, 14(8):771–788, 1998.

[1124] T. Watanabe. The impact of grain boundary character distribution on fracture in polycrystals. *Materials Science and Engineering: A*, 176(1):39–49, 1994.

[1125] Ian M. Watt. *The Principles and Practice of Electron Microscopy*. Cambridge University Press, 1997.

[1126] Dartmouth College Electron Microscope Facility, Electron Microscopy Images. https://www.dartmouth.edu/~emlab/gallery/ (accessed 2013, 2015, & 10/31/2018). Dartmouth College, Hanover NH.

[1127] Hui Wei, Zidong Wang, Jiong Zhang, Stephen House, Yi-Gui Gao, Limin Yang, Howard Robinson, Li Huey Tan, Hang Xing, Changjun Hou, Ian M. Robertson, Jian-Min Zuo, and Yi Lu. Time-dependent, protein-directed growth of gold nanoparticles within a single crystal of lysozyme. *Nature Nanotechnology*, 6:93–97, 2011.

[1128] L.-Y. Wei, S. Lefebvre, V. Kwatra, and G. Turk. State of the art in example-based texture synthesis. In *EUROGRAPHICS 2009, State of the Art Report, EG-STAR - 2009*, 2009.

[1129] Li-Yi Wei and Marc Levoy. Fast texture synthesis using tree-structured vector quantization. In *Proceedings of the 27th Annual Conference on Computer Graphics and Interactive Techniques*, pages 479–488. ACM Press/Addison-Wesley Publishing Co., 2000.

[1130] Huanjun Chen, Li Li, Weihai Ni, Zhi Yang and Jianfang Wang. Coupling between molecular and plasmonic resonances in freestanding dyegold nanorod hybrid nanostructures. *Journal of the American Chemical Society*, 130:6692–6693, 2008.

[1131] T. P. Weihs, V. Zinoviev, D. V. Viens, and E. M. Schulson. The strength, hardness and ductility of Ni_3Al with and without boron. *Acta Metallurgica*, 35(5):1109–1118, 1987.

[1132] L. Weinberg. On the maximum order of the automorphism group of a planar triply connected graph. *SIAM J. on Applied Math.*, 14(4):729–738, 1966.

[1133] L. Weinberg. A simple and efficient algorithm for determining isomorphism of planar triply connected graphs. *IEEE Trans. Circuit Theory*, CT13(2):142–148, 1966.

[1134] Shmuel Weinberger. What is... persistent homology? *Notices of the AMS*, 58(1):36–39, 2011.

[1135] Jiang Wen, You Zhisheng, and Li Hui. Segment the metallograph images using gabor filter. In *Speech, Image Processing and Neural Networks, 1994. Proceedings, ISSIPNN'94., 1994 International Symposium on*, pages 25–28. IEEE, 1994.

[1136] H.-R. Wenk and P. Van Houtte. Texture and anisotropy. *Reports on Progress in Physics*, 67(8):1367, 2004.

[1137] Hans-Rudolph Wenk. Texture and anisotropy. *Reviews in Mineralogy and Geochemistry*, 51(1):291–329, 2002.

[1138] H. R. Wenk. *Preferred Orientations in Deformed Metals and Rocks: An Introduction to Modern Texture Analysis*. Academic Press, Orlando, 1985.

[1139] D. Weygand, Y. Breichet, E. Rabkin, B. Straumal, and W. Gust. Solute drag and wetting of a grain boundary. *Philosophical Magazine Letters*, 76(3):133–38, 1997.

[1140] D. Wheeler, D. Brough, T. Fast, S. Kalidindi, and A. Reid. PyMKS: Materials knowledge system in Python (doi: http://dx.doi.org/10.6084/ m9.figshare.1015761), 2014.

[1141] Bruce A. Whitehead and W. A. Hoyt. Function approximation approach to anomaly detection in propulsion system test data. *J. Propulsion and Power*, 11:1074–1076, 1995.

[1142] Hassler Whitney. A set of topological invariants for graphs. *American Journal of Mathematics*, 55(1):231–235, 1933.

[1143] B. A Wilcox and A. H. Clauer. The role of grain size and shape in strengthening of dispersion hardened nickel alloys. *Acta Metallurgica*, 20(5):743–757, 1972.

[1144] R. Willett and R. Nowak. Platelets: a multiscale approach for recovering edges and surfaces in photon-limited medical imaging. *IEEE Transactions on Medical Imaging*, 22(3):332–350, 2003.

[1145] R. Willett and R. Nowak. Multiscale Poisson intensity and density estimation. *IEEE Transactions on Information Theory*, 53(9):3171–3187, 2007.

[1146] David B. Williams and C. Barry Carter. *Transmission Electron Microscopy: A Textbook for Materials Science*. Springer Science+ Business Media, New York, USA, 22:199, 2009.

[1147] D. B. Williams and C. B. Carter. *Transmission Electron Microscopy: A Textbook for Materials Science. Diffraction. II.* Transmission Electron Microscopy: A Textbook for Materials Science. Springer, 1996.

[1148] John Winn, Antonio Criminisi, and Thomas Minka. Object categorization by learned universal visual dictionary. In *Computer Vision, 2005. ICCV 2005. Tenth IEEE International Conference on*, volume 2, pages 1800–1807. IEEE, 2005.

[1149] D. P. Wipf and B. D. Rao. Sparse Bayesian learning for basis selection. *IEEE Trans. Sig. Proc.*, 52(8):2153–2164, 2004.

[1150] N. J. Wittridge and R. D. Knutsen. A microtexture based analysis of the surface roughening behaviour of an aluminium alloy during tensile deformation. *Mater. Sci. Eng. A*, 269(1-2):205–216, 1999.

[1151] D. Wolf and S. Yip. *Materials Interfaces: Atomic-level Structure and Properties*. Chapman & Hall, London, 1992.

[1152] A. P. Woodfield, M. D. Gorman, R. R. Corderman, J. A. Sutliff, and B. Yamrom. Effect of microstructure on dwell fatigue behavior of Ti6242. In P.A. Blenkinsop, W.J. Evans, and H.M. Flower, editors, *Titanium '95: Science and Technology*, pages 1116–1123. The Institute of Materials, The University Press, Cambridge, 1996.

[1153] K. Worden. Structural fault detection using a novelty measure. *J. Sound and Vibration*, 201:85–101, 1997.

[1154] S. Wright, R. Nowak, and M. Figueiredo. Sparse reconstruction by separable approximation. *IEEE Transactions Signal Processing*, 57:2479–2493, 2009.

[1155] S. Wright, R. Nowak, and M. A. T. Figueiredo. Sparse reconstruction by separable approximation. *IEEE Trans. Sig. Proc.*, 57(7):2479–2493, July 2009.

[1156] Stuart I. Wright, Matthew M. Nowell, and John F. Bingert. A comparison of textures measured using x-ray and electron backscatter diffraction. *Metallurgical and Materials Transactions A*, 38:1845–1855, 2007.

[1157] Linxi Wu and Bjorn M. Reinhard. Probing subdiffraction limit separations with plasmon coupling microscopy: concepts and applications. *Chemical Society Reviews*, 43(11):3884–3897, 2014.

[1158] Q. Wu and Y. Yu. Feature matching and deformation for texture synthesis. In *In ACM SIGGRAPH '04, 362–365*, 2004.

[1159] S. Zhu Y. Wu and D. Mumford. Filters, random fields and maximun entropy (FRAME) - towards a unified theory for texture modeling. *International Journal of Computer Vision*, 27(2):107–126, 1998.

[1160] W. Wu, R. F. Giese, and C. J. van Oss. Stability versus flocculation of particle suspensions in water-correlation with the extended dlvo approach for aqueous systems, compared with classical dlvo theory. *Colloids and Surfaces B: Biointerfaces*, 14(14):47–55, 1999.

[1161] P. Wynblatt and D. Chatain. Anisotropy of segregation at grain boundaries and surfaces. *Metallurgical and Materials Transactions a-Physical Metallurgy and Materials Science*, 37A(9):2595–2620, 2006.

[1162] Xianghua Xie. A review of recent advances in surface defect detection using texture analysis techniques. *ELCVIA Electronic Letters on Computer Vision and Image Analysis*, 7(3), 2008.

[1163] Yonghong Xie and Qiang Ji. A new efficient ellipse detection method. In *International Conference on Pattern Recognition*, pages 957–960, 2002.

[1164] Hongyi Xu, Ruoqian Liu, Alok Choudhary, and Wei Chen. A machine learning-based design representation method for designing heterogeneous microstructures. *Journal of Mechanical Design*, 137(5):051403–1–051403–10, 2015.

[1165] L. Xu, J. S. Ren, C. Liu, and J. Jia. Deep convolutional neural network for image deconvolution. In *Advances in Neural Information Processing Systems*, pages 1790–1798, 2014.

[1166] Y. Xu and W. Yin. A fast patch-dictionary method for whole-image recovery. ftp://ftp.math.ucla.edu/pub/camreport/cam13-38.pdf, 2013. UCLA CAM report 13-38.

[1167] M. Yaghoobi, T. Blumensath, and M. Davies. Dictionary learning for sparse approximations with the majorization method. *IEEE Transaction on Signal Processing*, 57(6):2178–2191, 2009.

[1168] J. Yang and Y. Zhang. YALL1 basic models and tests. *SIAM J. Sci. Comput.*, 33:250–278, 2011.

[1169] S. Yang and J. Luo. unpublished results, 2017.

[1170] A. B. Yankovich, B. Berkels, W. Dahmen, P. Binev, S. I. Sanchez, S. Bradley, A. Li, I. Szlufarska, and P. M. Voyles. Picometre-precision analysis of scanning transmission electron microscopy images of platinum nanocatalysts. *Nat. Commun.*, 5(4155), 2014.

[1171] Chenyu Zhang, Albert Oh, Thomas JA Slater, Feridoon Azough, Robert Freer, Sarah J. Haigh, Rebecca Willett, Yankovich, Andrew B. and Paul M. Voyles. Non-rigid registration and non-local principle component analysis to improve electron microscopy spectrum images. *Nanotechnology*, 27(36):364001, 2016. doi = 10.1088/0957-4484/27/36/364001.

[1172] L. P. Yaroslavsky. *Digital Picture Processing*, volume 9 of *Springer Series in Information Sciences*. Springer-Verlag, Berlin, 1985.

[1173] J. C. Ye, Y. Bresler, and P. Moulin. A self-referencing level-set method for image reconstruction from sparse Fourier samples. *Int. J. Computer Vision*, 50(3):253–270, Dec 2002.

[1174] Jong Chul Ye, C. A. Bouman, K. J. Webb, and R. P. Millane. Nonlinear multigrid algorithms for Bayesian optical diffusion tomography. *IEEE Trans. on Image Processing*, 10(6):909 –922, June 2001.

[1175] Xingchen Ye, Linghua Jin, Humeyra Caglayan, Jun Chen, Guozhong Xing, Chen Zheng, Vicky Doan-Nguyen, Yijin Kang, Nader Engheta, Cherie R. Kagan, and Christopher B. Murray. Improved size-tunable synthesis of monodisperse gold nanorods through the use of aromatic additives. *ACS Nano*, 6:28042817, 2012.

[1176] C. L. Y. Yeong and S. Torquato. Reconstructing random media II. Three-dimensional media from two-dimensional cuts. *Physical Review E*, 58(1):224–233, 1998.

[1177] W. Yin, S. Osher, J. Darbon, and D. Goldfarb. Bregman iterative algorithms for compressed sensing and related problems. *SIAM J. Imaging Sciences*, 1(1):143–168, 2008.

[1178] Wotao Yin, Simon Morgan, Junfeng Yang, and Yin Zhang. Practical compressive sensing with toeplitz and circulant matrices. In *Visual Communications and Image Processing 2010*, pages 77440K–77440K–10. International Society for Optics and Photonics, 2010.

[1179] Yadong Yin and A. Paul Alivisatos. Colloidal nanocrystal synthesis and the organic-inorganic interface. *Nature*, 437(7059):664–670, 2005.

[1180] Zhye Yin, Yili Zheng, Peter C. Doerschuk, Padmaja Natarajan, and John E. Johnson. A statistical approach to computer processing of cryo electron microscope images: Virion classification and 3-D reconstruction. *J. Struct. Biol.*, 144(1/2):24–50, 2003.

[1181] J. Yoo, C. Turnes, E. Nakamura, C. Le, S. Becker, E. Sovero, M. Wakin, M. Grant, J. Romberg, A. Emami-Neyestanak, and E. Candès. A compressed sensing parameter extraction platform for radar pulse signal acquisition. *IEEE J. Emerging Topics Cir. and Sys.*, 2(3):626–638, 2012.

[1182] Guoshen Yu, Guillermo Sapiro, and Stéphane Mallat. Image modeling and enhancement via structured sparse model selection. In *Image Processing (ICIP), 2010 17th IEEE International Conference on*, pages 1641–1644. IEEE, 2010.

[1183] Guoshen Yu, Guillermo Sapiro, and Stéphane Mallat. Solving inverse problems with piecewise linear estimators: From Gaussian mixture models to structured sparsity. *IEEE Transactions on Image Processing*, 21(5):2481–2499, 2012.

[1184] Zhou Yu, J. Thibault, C. A. Bouman, K. D. Sauer, and J. Hsieh. Fast model-based X-ray CT reconstruction using spatially nonhomogeneous ICD optimization. *IEEE Trans. on Image Processing*, 20(1):161 –175, January 2011.

[1185] N. Zachariah, J. Flake, B. Mailhe, Q. Wang, X. P. Hu, J. Romberg, and M. S. Nadar. Iterative analysis based non-convex prior (IAN) for enhanced sparse recovery in tight frames. Manuscript, avaiable at jrom.ece.gatech.edu, August 2016.

[1186] Matthew D. Zeiler and Rob Fergus. Visualizing and understanding convolutional networks. In *Computer vision–ECCV 2014*, pages 818–833. Springer, 2014.

[1187] Matthew D. Zeiler, Graham W. Taylor, and Rob Fergus. Adaptive deconvolutional networks for mid and high level feature learning. In *Computer Vision (ICCV), 2011 IEEE International Conference on*, pages 2018–2025. IEEE, 2011.

[1188] S. Zelinka and M. Garland. Jump map based interactive texture synthesis. In *In ACM Trans. Graph.* 23, 930–962, 2004.

[1189] J. Zeman and M. Šejnoha. From random microstructures to representative volume elements. *Modelling and Simulation in Materials Science and Engineering*, 15:S325, 2007.

[1190] B. Zhang, J. Fadili, and J-L. Starck. Wavelets, ridgelets, and curvelets for Poisson noise removal. *IEEE Trans. Image Process.*, 17(7):1093–1108, 2008.

[1191] Hao Zhang, Yi Liu, Dong Yao, and Bai Yang. Hybridization of inorganic nanoparticles and polymers to create regular and reversible self-assembly architectures. *Chemical Society Reviews*, 2012.

[1192] Jianguo Zhang, Marcin Marszałek, Svetlana Lazebnik, and Cordelia Schmid. Local features and kernels for classification of texture and object categories: A comprehensive study. *International journal of computer vision*, 73(2):213–238, 2007.

[1193] Jianguo Zhang and Tieniu Tan. Brief review of invariant texture analysis methods. *Pattern Recognition*, 35(3):735–747, 2002.

[1194] Jie Zhang, Erik Luijten, and Steve Granick. Toward design rules of directional janus colloidal assembly. *Annual Review of Physical Chemistry*, 66(1):581–600, 2015.

[1195] K. S. Zhang, M. S. Wu, and R. Feng. Simulation of microplasticity-induced deformation in uniaxially strained ceramics by 3-D Voronoi polycrystal modeling. *International Journal of Plasticity*, 21(4):801–834, 2005.

[1196] Li Zhang, Yuan Li, and Ramakant Nevatia. Global data association for multi-object tracking using network flows. In *IEEE Conference on Computer Vision and Pattern Recognition*, pages 1–8, 2008.

[1197] R. Zhang, J. Thibault, C. Bouman, K. Sauer, and J. Hsieh. Model-based iterative reconstruction for dual-energy X-ray CT using a joint quadratic likelihood model. *Medical Imaging, IEEE Transactions on*, PP:1–1, 2013.

[1198] Ruoqiao Zhang, Charles A Bouman, Jean-Baptiste Thibault, and Ken D Sauer. Gaussian mixture Markov random field for image denoising and reconstruction. In *Global Conference on Signal and Information Processing (GlobalSIP), 2013 IEEE*, pages 1089–1092. IEEE, 2013.

[1199] T. Zhang. Sparse recovery with orthogonal matching pursuit under RIP. *IEEE Trans. Inform. Theory*, 57(9):6215–6221, September 2011.

[1200] Xing Zhang, Ethan Settembre, Chen Wu, Philip R. Dormitzer, Richard Bellamy, Stephen C. Harrison, and Nikolaus Grigorieff. Near-atomic resolution using electron cryomicroscopy and single-particle reconstruction. *Proc. Nat. Acad. Sci. U.S.A.*, 105(6):1867–1872, 12 February 2008.

[1201] Yuanyao Zhang and Jian Luo. Observation of an unusual case of triple-line instability. *Scripta Materialia*, 88:45–48, 2014.

[1202] Lei Zhao, Tian Ming, Huanjun Chen, Yao Liang, and Jianfang Wang. Plasmon-induced modulation of the emission spectra of the fluorescent molecules near gold nanorods. *Nanoscale*, 3(9):3849–3859, 2011.

[1203] Yibin Zheng and Peter C. Doerschuk. Explicit computation of orthonormal symmetrized harmonics with application to the identity representation of the icosahedral group. *SIAM Journal on Mathematical Analysis*, 32(3):538–554, 2000.

[1204] Yili Zheng. *Novel statistical models and a high-performance computing toolkit for the solution of cryo electron microscopy inverse problems in viral structural biology*. PhD thesis, School of Electrical and Computer Engineering, Purdue University, West Lafayette, Indiana, USA, August 2008.

[1205] Yili Zheng, Qiu Wang, and Peter C. Doerschuk. 3-D reconstruction of the statistics of heterogeneous objects from a collection of one projection image of each object. 29(6):959–970, June 2012.

[1206] Bolei Zhou, Agata Lapedriza, Jianxiong Xiao, Antonio Torralba, and Aude Oliva. Learning deep features for scene recognition using places database. In *Advances in Neural Information Processing Systems*, pages 487–495, 2014.

[1207] Naixie Zhou and Jian Luo. Developing grain boundary diagrams for multicomponent alloys. *Acta Materialia*, 91:202–216, 2015.

[1208] Youjie Zhou, Hongkai Yu, Jeffrey P. Simmons, Craig P Przybyla, and Song Wang. Large-scale fiber tracking through sparsely sampled image sequences of composite materials. *IEEE Transactions on Image Processing*, 25(10):4931–4942, 2016.

[1209] A. K. Ziemann and D. W. Messinger. An adaptive locally linear embedding manifold learning approach for hyperspectral target detection. *Proc. SPIE*, 9472:947200, 2015.

[1210] M. Zontak and M. Irani. Internal statistics of a single natural image. In *CVPR*, pages 977–984, 2011.

[1211] Irina Zubritskaya, Kristof Lodewijks, Nicol Maccaferri, Addis Mekonnen, Randy K. Dumas, Johan kerman, Paolo Vavassori, and Alexandre Dmitriev. Active magnetoplasmonic ruler. *Nano Letters*, 15(5):3204–3211, 2015.

Index